Nanomaterials for Energy and Environmental Applications

Nanomaterials for Energy and Environmental Applications

Topic Editors

Robert A. Varin
Geng Zhong

MDPI

Basel • Beijing • Wuhan • Barcelona • Belgrade • Novi Sad • Cluj • Manchester

Topic Editors

Robert A. Varin
Department of Mechanical and Mechatronics Engineering
University of Waterloo
Waterloo
Canada

Geng Zhong
Tsinghua-Berkeley Shenzhen Institute
Shenzhen International Graduate School
Tsinghua University
Shenzhen
China

Editorial Office

MDPI AG
Grosspeteranlage 5
4052 Basel, Switzerland

This is a reprint of the Topic, published open access by the journals *Applied Sciences* (ISSN 2076-3417), *Energies* (ISSN 1996-1073), *Materials* (ISSN 1996-1944), *Nanoenergy Advances* (ISSN 2673-706X) and *Nanomaterials* (ISSN 2079-4991), freely accessible at: https://www.mdpi.com/topics/O360KW50O1.

For citation purposes, cite each article independently as indicated on the article page online and as indicated below:

Lastname, A.A.; Lastname, B.B. Article Title. *Journal Name* **Year**, *Volume Number*, Page Range.

ISBN 978-3-7258-3885-1 (Hbk)
ISBN 978-3-7258-3886-8 (PDF)
https://doi.org/10.3390/books978-3-7258-3886-8

Contents

materials

MDPI

Article

Mechanical and Electrical Properties of Graphene Oxide Reinforced Copper–Tungsten Composites Produced via Ball Milling of Metal Flakes

Fei Lin [1,2], Ruoyu Xu [3], Mingyu Zhou [3], Robert J. Young [1,2], Ian A. Kinloch [1,2,*] and Yi Ding [4]

1 Department of Materials, National Graphene Institute, University of Manchester, Manchester M13 9PL, UK
2 Department of Materials, Henry Royce Institute, University of Manchester, Manchester M13 9PL, UK
3 Department of High-End Electrical Material, Global Energy Interconnection Research Institute Europe GmbH, 10623 Berlin, Germany
4 State Key Laboratory of Advanced Power Transmission Technology, Global Energy Interconnection Research Institute Co., Ltd., Beijing 102209, China
* Correspondence: ian.kinloch@manchester.ac.uk

Abstract: Copper–tungsten (Cu-W) composites are widely used in high-power and -temperature electrical applications. The combination of these metals, however, leads to compromised physical and electrical properties. Herein, we produce Cu-W-graphene oxide (Cu-W-GO) composites to address this challenge. To ensure uniform density composites, the as-received metal powders were flattened into a flake morphology by ball milling and then mixed with up to 0.5 wt.% GO flakes. The green forms were processed using spark plasma sintering. The GO was found to be well-dispersed amongst the metallic phases in the final composite. The addition of GO reduced the relative density of the composites slightly (4.7% decrease in relative density at 0.5 wt% GO loading for the composites processed at 1000 °C). X-ray diffraction confirmed good phase purity and that no carbide phases were produced. GO was found to improve the mechanical properties of the Cu-W, with an optimal loading of 0.1 wt.% GO found for ultimate compression strength and strain to failure, and 0.3 wt.% optimal loading for the 0.2% offset yield strength. Significantly, the electrical conductivity increased by up to 25% with the addition of 0.1 wt.% GO but decreased with higher GO loadings.

Keywords: nanomaterials; metal matrix composites; mechanical testing; electrical conductivity

Citation: Lin, F.; Xu, R.; Zhou, M.; Young, R.J.; Kinloch, I.A.; Ding, Y. Mechanical and Electrical Properties of Graphene Oxide Reinforced Copper–Tungsten Composites Produced via Ball Milling of Metal Flakes. *Materials* **2022**, *15*, 7736. https://doi.org/10.3390/ma15217736

Academic Editors: Robert A. Varin and Geng Zhong

Received: 8 October 2022
Accepted: 27 October 2022
Published: 3 November 2022

1. Introduction

Copper–tungsten (Cu-W) composites are used in electrical contacts, heat sinks in high voltage power devices and arc-resistant electrodes due to their unique combination of properties. In particular, Cu-W composites possess a high hardness, relatively good electrical and thermal conductivities, a low thermal expansion coefficient, and high oxidation, erosion and arc resistances at extremely high temperatures [1]. Cu-W composites are produced typically by the high temperature infiltration of liquid Cu into a porous W skeleton. This route, however, leads to poor uniformity of the microstructure and thus to relatively low thermal and electrical conductivities due to scattering from defects and voids [2,3]. It is also difficult to fabricate a fully dense Cu-W composite using this method, leading to poor hardness and arc erosion resistance, limiting the operating life of high voltage (e.g., 200 kV) switch gear made from such materials [1]. Hence, innovative methods such as mechanochemical synthesis [4], powder injection moulding [5], mechanical alloying [6], spark plasma sintering [7], and novel additives such as graphene [8] are required to improve the physical properties of the Cu-W composites for all applications.

Graphene, with its good mechanical and electrical properties [9], is potentially an excellent reinforcement for Cu-W composites. Although progress in graphene reinforced metal matrix composites (MMCs) is far behind that of polymer matrices [10–12], interest in graphene-MMCs has grown rapidly over the last decade [13–19], with Al matrix composites [20–22] being the most researched. Powder metallurgy (PM) is the most widely used route for producing graphene-MMCs [21,23–27] with other approaches including molecular level mixing between carbon nanotubes [28] or graphene oxide and the metal ions [13,19], chemical vapour deposition of or graphene on metal foils or powders [14,18,29] and electrodeposition/electroless deposition [27]. Under the right conditions, graphene is found to enhance the properties of the MMCs. For example, Li et al. [30] reported that graphene oxide (GO) could be dispersed in the Al through a direct electrostatic interaction, to give a 18% and 17% increase in elastic modulus and hardness upon the addition of 0.3 wt.% GO. Dong et al. [31–33] improved the arc erosion properties, electrical conductivity and mechanical properties of 70 wt.% W-30 wt.% Cu composites by adding reduced graphene oxide (rGO). More recently, the same group has shown that precoating the graphene oxide with tungsten carbide can control the interface between the Cu and graphene [34].

In this present study we explore a powder production route for Cu-W-GO composites across a range of graphene concentrations. It is well known that there is an optimal loading of graphene to enhance physical properties for both polymer and metal matrix composites. However, previous research generally used a single graphene loading in a given study to establish whether graphene can give benefits in Cu-W composites rather exploring the role of the GO concentration in controlling the structure and properties of the composite. Importantly, ball milling was used in this present study to flatten the Cu and W flakes to ensure a good packing between the flakes and the GO in the green composite. These dispersions were then processed using spark plasma sintering at 1000 and 1100 °C, temperatures chosen to be just below and the above the melting point of the Cu. The microstructure of the resulting materials was characterised using a combination of X-ray diffraction (XRD), Raman spectroscopy and electron microscopy as a function of GO loading. The role of GO loading upon the microstructure, mechanical properties and electrical properties of the composites was then established.

2. Materials and Methods

2.1. Composite Production

Cu and W powders were purchased from Beijing Youyan Powder New Materials Company. Scanning electron microscopy (SEM) demonstrated that the Cu powders comprised dendritic aggregates with a primary particle size of 1–2 μm (Figure 1a), characteristic of materials produced by the electrolysis production route [1]. The W had a polyhedral morphology (Figure 1b), indicating it was produced by the reduction of tungsten oxides [2]. The W powders had a bimodal diameter distribution, with the modes at 2.5 μm and 0.5 μm. Some nanoparticles were seen on the surface of W powders and were assumed to be tungsten oxides.

GO suspensions with a concentration of 1 mg/mL in deionised water were prepared using a modified Hummers method, as described elsewhere [35]. Electron microscopy showed that the GO was primarily monolayer with the diameters of the flakes bimodally distributed with modes at ~20 μm and ~7.5 μm. XPS showed that the C/O ratio in the GO was 2.62.

Preliminary trials found that mixing the as-received materials resulted in a poor dispersion of powder in the green composites (Figure S1). Hence, the as-received metal powders were ball-milled to flatten them so that high packing densities could be achieved during sintering (Figure 1d,c). The Cu and W powders were milled separately in a 500 mL ZrO_2 grinding jar using 2 mm ZrO_2 grinding balls (ball-to-powder weight ratio of 20:1) at a speed of 350 rpm (Retsch PM100) in an ethanol dispersant. A milling cycle of 5 min milling followed by 15 min rest was used to prevent the temperature rising above the boiling point of the ethanol. The powders were processed for a total milling time of an hour, which was found to be the optimal time for obtaining flat, high aspect ratio flakes. The flakes were collected and the mixed in a mass ratio of 20:80 Cu:W in ethanol and stirred for 15 min at 300 rpm. The mixed flakes were finally added to the GO suspension such that the final composites would have nominal loadings of 0, 0.1, 0.3 or 0.5 wt.% GO. This Cu-W-GO mixture was stirred for a further 30 min at 300 rpm and then dried on a hot plate.

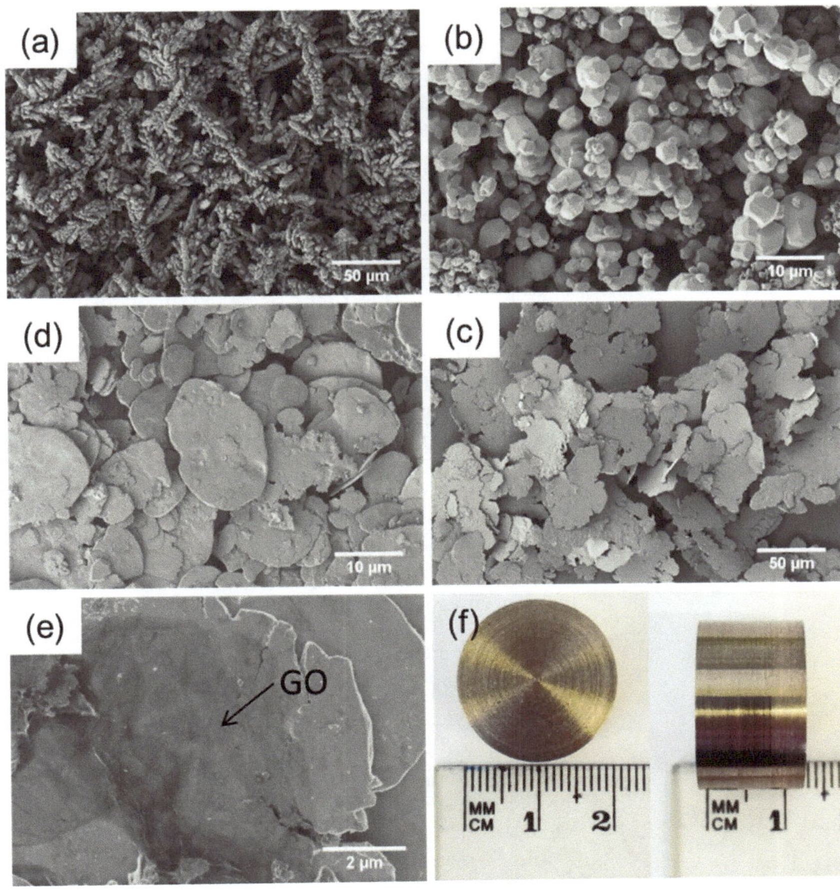

Figure 1. SEM images for raw (**a**) Cu, (**b**) W powders, (**c**) Cu flakes ball milled for 1 h, (**d**) W flakes ball milled for 1 h, (**e**) the flattened Cu-W powders with GO flakes (darker area), (**f**) the final bulk Cu-W-GO composite produced by SPS in the top and side view.

The Cu-W-GO mixtures were consolidated by SPS (FCT HPD 25; FCT Systeme GmbH). Then, 50 g of the mixed Cu-W-GO powder was poured into a graphite mold lined with a graphite foil so that the sample could be easily removed after processing. The green pellets were heated initially from room temperature to the sintering temperature of either 1000 °C or 1100 °C. When the sintering temperature had been reached, the ram applied a pressure of 60 MPa for 5 min, after which the ram pressure was released, and the sample was cooled to room temperature. A ramp rate of 50 °C/min was used for heating and cooling. No loss of Cu was found for samples sintered at 1000 °C, whereas a little Cu was lost for the samples sintered at 1100 °C due to evaporation.

2.2. Characterisation

The morphology and the size of the as-received and milled metal powders were studied using scanning electron microscopy (SEM, Tescan Mira3 FEG-SEM). The relative density of the composite samples was measured by using an Ohaus analytic balance with a density determination kit based on Archimedes' principle. Transmission electron microscopy (TEM, FEI Tecnai TF30 FEG-TEM), SEM (ZEISS Sigma FEG-SEM) and Raman spectroscopy (Renishaw inVia Raman System with 633 nm HeNe laser), X-ray photoelectron spectroscopy (XPS) were used to study the as-produced GO. These samples were prepared by depositing a drop of an aqueous solution of the as-made GO onto the relevant sample holder. SEM and Raman spectroscopy was also used to measure the morphology of polished sections of the sintered composites, including the state of dispersion of GO in the composite. X-ray diffraction (XRD, PANalytical X'Pert Pro X'Celerator X-ray diffractometer with Cu Kα radiation (λ = 0.15418 nm) was used on the polished samples to identify the phase of the sintered materials, including any potential interfacial carbides.

Nanoindentation (MTS nanoindenter XP) tests were conducted on the polished samples to determine the Young's Modulus of the composites. For each sample, 64 points (an array of 8 $\times$ 8) were indented with a spacing of 50 μm and a depth of 1000 nm, at a strain rate of 0.05 s^{-1}. Mechanical compression tests were conducted on the Cu-W-GO composites using an Instron 5569 H1549 50 kN T300 static testing machine at room temperature. Pillar samples with an aspect ratio of 1.5 (3 mm diameter, 4.5 mm height) were cut from SPS bars using electrical discharge machining with their loading axis parallel to the extrusion direction. These pillars were compressed uniaxially at a cross-head displacement rate of 0.3 mm/min until failure. The tests were performed using lubrication in order to minimise friction between the pillar and the anvils. The load–displacement data from the load frame were corrected for machine compliance prior to calculating the stress and strain.

The electrical conductivity of the samples was measured in the length direction using a four-point probe. The spacing between the probes was 1.1 mm. The test specimens were machined to around 15 mm (length) $\times$ 4 mm (width) $\times$ 2 mm (thickness) in dimensions and the accurate size of the specimens measured using a spiral micrometer. The specimens were polished with grinding paper (grit size P1200) before each test. For each GO loading, three specimens were tested.

3. Results and Discussion

3.1. Morphology of the Metal Powders following Ball Milling

A uniform dispersion of the GO within the Cu and W powders is key to obtaining good physical properties. However, it was difficult to produce the composites from the as-received metallic powders due to topological incompatibility between the matrix and the filler. Thus, wet ball milling was introduced to transform the as-received metal powders into flakes. The Cu and W powders were milled for different times (10 min, 20 min, 30 min, 40 min, 50 min, 60 min, 120 min, 180 min and 240 min) and then examined using SEM.

The morphology of the Cu powders varied with the ball milling time and is illustrated in Figures S2 and S3 at low and high magnification, respectively. The dendritic Cu powders went through the processes of deformation, cold welding, squeezing, and finally fracturing during the ball milling. The aim of the milling process in this work was to obtain flakes with the maximum possible aspect ratio before fracture occurred such that the flakes had a high, flat surface area to form an interface with the GO, whilst maintaining the good mechanical properties of the metal. SEM analysis demonstrated that 60 min ball milling was optimal and produced the largest and thinnest flakes, without fracture.

The morphology of the W powders also changed during the ball milling process (Figures S4 and S5). Similar to the Cu, the W powders gradually experienced deformation, cold welding, squeezing and fracturing steps during the ball milling process. However, due to the bimodal powder diameter distribution, the morphological transformation was not uniform within the W sample; the larger W polygonal powders flattened into flakes after only 10 min milling, whereas the smaller W powder particle were flattened much slower and finally became flakes after 40 min milling. The surface of the larger W powders became very smooth after milling for 30 min whereas the surfaces of smaller powder particles was smooth by 60 min. Above 60 min, cracks or defects started to form.

3.2. Microstructure of the GO

TEM imaging and SAED demonstrated that the GO was predominantly monolayer with some folding and wrinkling of the flakes (Figure 2a). SEM showed that the diameters of the flakes were bimodally distributed with modes at ~20 µm and ~7.5 µm. Raman spectroscopy showed the characteristic D and G bands at 1340 cm^{-1} and 1586 cm^{-1} (Figure 2b) upon a fluorescence background. The intensity of the D and G bands are approximately equal which is characteristic of as-made GO [35].

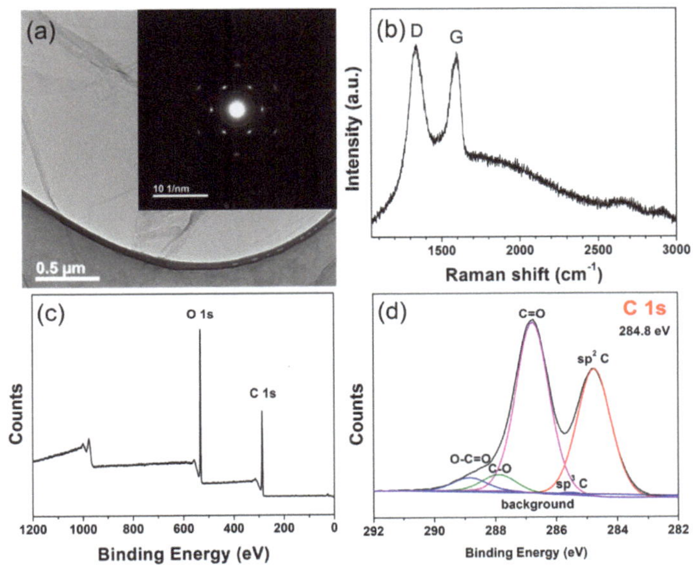

Figure 2. Characterization of the as-produced GO: (**a**) TEM image of GO on lacey carbon film with an inset showing a selected area electron diffraction (SAED) pattern, (**b**) Raman spectrum with the characteristic D and G bands, and (**c**) an XPS survey spectrum and (**d**) an XPS C 1s spectrum.

As is shown in Figure 2c, only the C 1s and O 1s peaks were identified from the survey XPS spectrum, indicating the absence of impurities. The functional groups of GO can be analysed from the C 1s spectrum (Figure 2d). Based on Cao et al. [36], C 1s was assigned to 284.8 eV, and composed of sp^2 C, sp^3 C, C-O, C=O, and O-C=O. Specifically, the C 1s and O 1s accounted for 72.4 at. % and 27.6 at. %, respectively to give a C/O ratio of 2.62. The atomic percentages of sp^2 C, sp^3 C, C-O, C=O, and O-C=O were 39.12%, 0.47%, 51.62%, 4.91%, and 3.87%, respectively.

Maintaining the structural integrity and high aspect ratio of the GO flakes is vital for mechanical reinforcement. Ball milling of Cu, W and GO together was found to result in folding and restacking of the GO. Instead, we added the GO suspensions to the powders after all the ball milling steps had been completed.

3.3. Microstructure of the Cu-W-GO Composites

The Cu-W-GO composite bars produced by SPS were 20 mm in diameter and 14 mm high (Figure 1f). The theoretical density for the unreinforced Cu-W is 15.15 g/cm^3 whereas the measured density of the sample sintered at 1000 °C was 13.9 g/cm^3, giving a relative density of 91.7% (Figure 3). The relative density decreased to 91.1%, 89.7%, and 86.4% upon the addition of 0.1, 0.3 and 0.5 wt.% graphene, respectively. The relative density for unreinforced composite at 1100 °C was significantly higher (97.5%) than that sintered at 1000 °C but the addition to the graphene led to larger decrease in the relative density, with a density of just 76.9% being obtained at 0.5 wt.% GO loading. Similar decreases in relative density have been observed in metal matrix graphene composites produced by powder metallurgy which is due to the graphene frustrating the packing and sintering of the powders, leading to the formation of pores [26]. The larger decrease in relative density that we observed at the higher sintering temperature and with higher GO loadings may be due to the GO keeping the structure more open, allowing Cu vapor to escape.

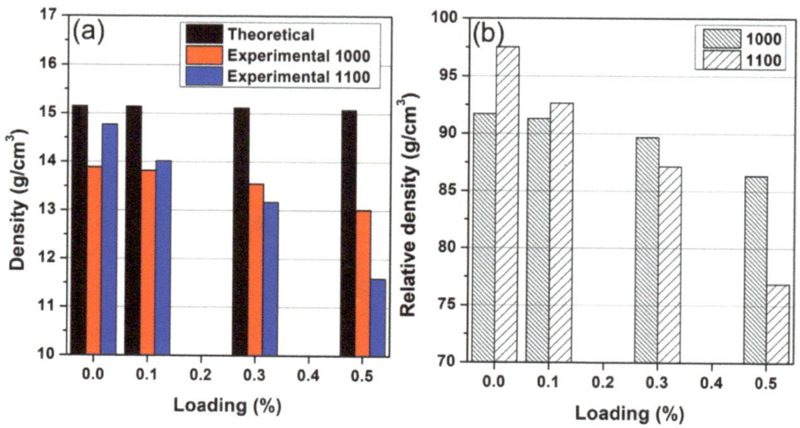

Figure 3. (**a**) Theoretical and experimental densities and (**b**) relative densities for the Cu-W matrix composites with different loadings of GO, processed at 1000 °C and 1100 °C.

SEM images and EDS mappings of the Cu-W and Cu-W-0.1 wt.% GO composites sintered at 1000 °C are shown in Figure 4 and Figures S6–S10, respectively. It can be seen that the initial flake morphology was preserved during sintering, justifying the ball milling processed used to flatten the powders prior to use. The phases are aligned perpendicular to the direction of loading during the sintering processes. Furthermore, it was found that the Cu and W were uniformly dispersed in the bulk composites with the Cu dispersed at the grain boundaries of W and vice versa.

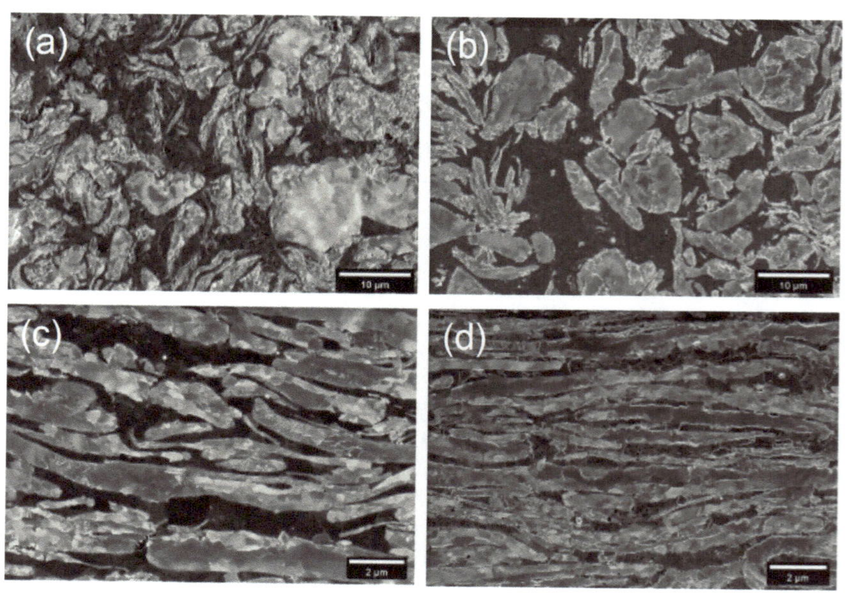

Figure 4. SEM micrographs for polished sections of (**a**,**c**) Cu-W and (**b**,**d**) Cu-W-0.1 wt.% GO sintered at 1000 °C viewed parallel (**a**,**b**) and perpendicular (**c**,**d**) to the extrusion direction.

The EDS results for the Cu-W and Cu-W-0.5 wt.% GO sintered at 1000 °C and 1100 °C are shown in Figures S6–S10, including the SEM, the combined image, and the elemental maps for Cu, W, O and C, and the integrated spectrum for the whole mapping area. Notably, the C element was mapped for comparison with the composites with GO incorporated. In general, the Cu and W phases were uniformly uniformly dispersedin the Cu-W sample, with a good Cu continuous network. This was to the result of the Cu being able to flow around the W during the processing at high temperature, leading to interconnection of the phases. A few oxide particles, which probably formed during the high temperature sintering process, are located at the grain boundaries of the W. The composition of the mapping area was 77.3 wt.% W, 21.9 wt.% Cu, and 0.8 wt.% C, a little lower than the nominal of Cu-W 20/80 ratio.

In comparison, more oxide particles were present in the Cu-W-0.5 wt.% GO composites. GO rather than graphene was used to produce a good slurry and hence a better distribution in the green pieces. However, GO is highly oxidative at high temperature, hence as it was reduced during heating, it would have oxidised the metals. In addition, dispersion of the phases in the 0.5 wt.% sample was less homogeneous than the Cu-W composite; the continuity of the Cu in the composite is interrupted by the pores and the oxides. The elemental composition of the composites was 63.8 wt.% W, 32.3 wt.% Cu, 2.1 wt.% C and 1.8 wt.% O. The percentage ratio of Cu-W is a little higher than the initial 20/80 and relative high oxygen content is believed to be from the tungsten oxides as well as the GO.

The elemental distribution and microstructure for composites processed at 1100 °C is very similar to the composites sintered at 1000 °C. A small difference is that the agglomeration is more severe for composites at 1100 °C than that at 1000 °C, especially for the composite with higher loadings of GO such as 0.5 wt.%.

A good dispersion of the GO (at this stage reduced) in the final sintered Cu-W composite is key in obtaining good physical properties. It is difficult to distinguish GO from the matrix using either optical or electron microscopy. Thus, Raman spectroscopy was used to confirm the existence and the dispersion of GO in the composite [21,37,38]. The Raman spectrum of GO in the bulk composite after consolidation is shown in Figure 5a (c.f. Figure 2b for the spectra from the as-received materials). The spectrum was fitted by 3 Lorentzian type curves, D, G and D′, with positions at 1330 cm^{-1}, 1587 cm^{-1} and 1614 cm^{-1} Raman shift, respectively. The D band results from a disordered structure within the graphene [39], hence Raman spectroscopy is very useful for characterising the disordered structure in sp^2 hybridised carbon materials such as graphene [36]. The G band is from the stretching of the C-C bond in graphene materials and is common to all sp^2 hybridised carbon materials. The G band can, however, split into two bands, G band and D′ band, when defects are randomly distributed in the graphene. This splitting is mainly due to the interaction between the localised vibrational modes of the defects and the extended phonon modes of graphene. The D/G band intensity ratio of the processed GO is about 2.28, much higher than that of the as-produced material (Figure 2b), indicating that composite sintering process introduced defects into the GO [39]. A similar increase in the D/G ratio has been reported upon the ball milling of GO thought also to be due to the introduction of defects [40]. Considering the complicated nature of the G band region, the D band was used as a reference for the Raman mapping. The Raman mapping results are shown in Figure 5b–d, displaying the dispersion of GO in the composites sintered under 1000 °C and 1100 °C, respectively. These maps confirm that the GO uniformly dispersed in the Cu-W matrix composites.

3.4. Phase Purity

XRD patterns for the composites sintered at 1000 °C and 1100 °C are shown in Figure 6, in the 2θ range between 10° and 90°. In this region, some main Bragg reflections from W and Cu are present and labelled in the plots, including W (110), W (200), W (211), W (220), Cu (111), Cu (200), and Cu (220). There are also some reflections from WO_3. The relative peak intensity of Cu is lower at 1100 °C than that at 1000 °C, which is in agreement with the partial loss of Cu when processed at 1100 °C. Finally, the relative peak intensity for WO_3 increases with higher loadings of GO and higher processing temperatures, due to the scavenging of the oxygen from the GO.

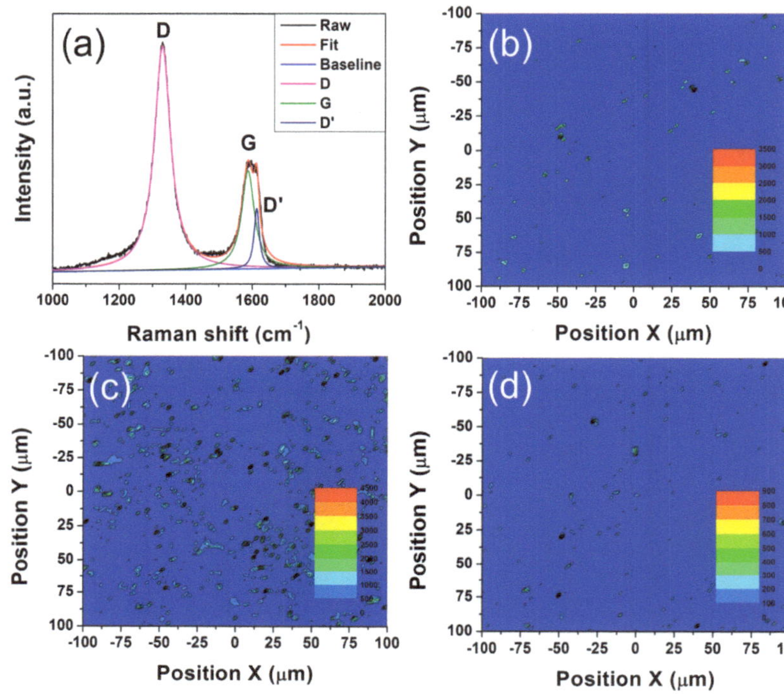

Figure 5. (**a**) Raman spectrum for a Cu-W composite with GO, including the raw spectrum and the curves for the fit spectrum, the baseline, D, G and D′ spectra. Raman mapping of the D band for Cu-W matrix composite with (**b**) 0.3 and (**c**) 0.5 wt.% GO sintered at 1000 °C. Raman mapping (**d**) for Cu-W matrix composite with 0.3 wt.% GO sintered at 1100 °C.

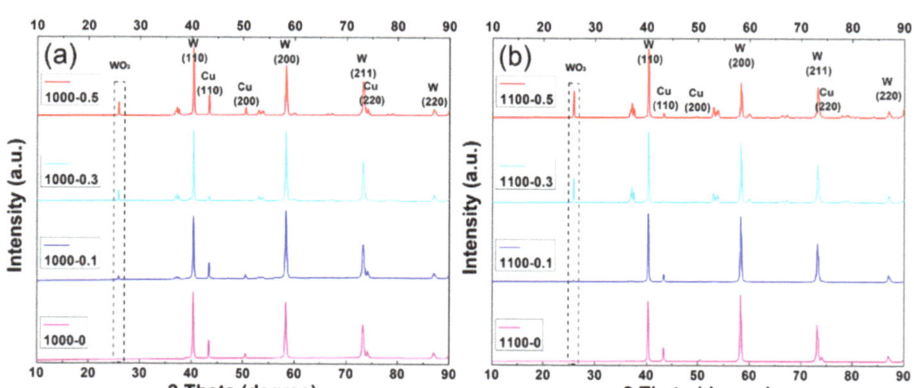

Figure 6. XRD spectra for Cu-W matrix composites with different loadings of GO, sintered at (**a**) 1000 °C and (**b**) 1100 °C. The labelled peaks are from W and Cu, and the other unlabelled peaks are from WO_3, with the main reflection from WO_3 is highlighted to the dotted boxes. Figure S11 gives the reference X-ray reflections for Cu, W and WO_3.

3.5. Mechanical Properties

The Young's modulus the materials was determined using nanoindentation. SEM rather than optical microscopy was used to evaluate the results of the nanoindentation tests due to small indents being produced from such a stiff material. The 8 x 8 arrays from the tests and higher magnification micrographs of typical indents can be seen in Figure 7. The indents that were found to be near defects were excluded from the analysis. To summarise the properties derived from the two samples examined (Cu-W and Cu-W-0.1 wt.% GO), the Young's modulus results obtained from the nanoindentation tests are also shown in Figure 7. The mean value of the Young's modulus of the Cu-W is 218 $\pm$ 89 GPa, which increases to 229 $\pm$ 55 GPa for the Cu-W-0.1 wt% GO. These values compare will those reported in the literature. For example, modulus values of between 160–190 GPa have been reported for a 25:75 Cu:W composite processed in the temperature range 1050–1150 °C [41]. The modulus appears to increase slightly upon the addition of 0.1 wt.% GO but this may not be significant when the errors are considered.

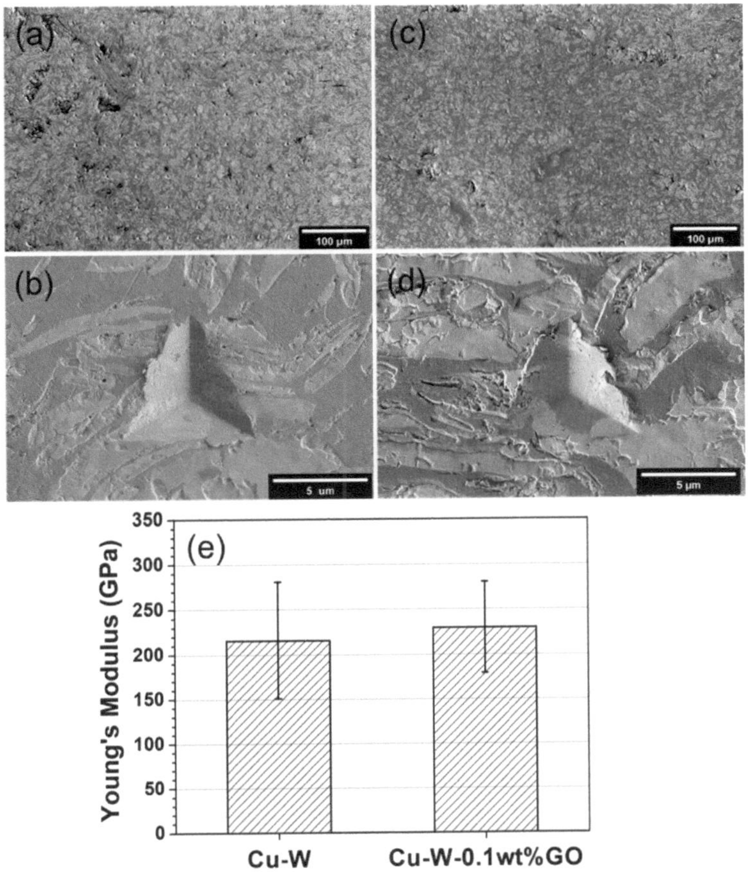

Figure 7. SEM image for the (**a**) Cu-W sample showing the whole indented area, and (**b**) one selected indent; SEM image for the (**c**) Cu-W-0.1 wt.% GO sample showing the whole indented area, and (**d**) one selected indent. (**e**) Values of the Young's moduli of the Cu-W and Cu-W-0.1 wt.% GO materials.

Representative compressive stress–strain curves for all eight different production and sintering conditions are shown in Figure 8. Since the curve in the elastic region is not quite linear a 0.2% offset yield strength (YS) was used. The mean ultimate compressive stress

(UCS) for Cu-W sintered at 1000 °C is 1095 ± 42 MPa, with a corresponding mean strain at failure of 18.3 ± 2.1%. Upon the addition of 0.1 wt.% GO, the compressive strength of the composite increased to 1228 ± 45 MPa and the strain to failure increased to 20.6%. However, at higher GO loadings the UCS and strain to failure dropped, with the GO embrittling the metal matrix composite. The GO did, though, improve the 0.2% offset yield strength compared to that of the control composite at all loadings studied for a 1000 °C processing temperature. However, this improvement in 0.2% offset yield strength was optimal for 0.3 wt.% GO, where it increased by 38% to 560 MPa.

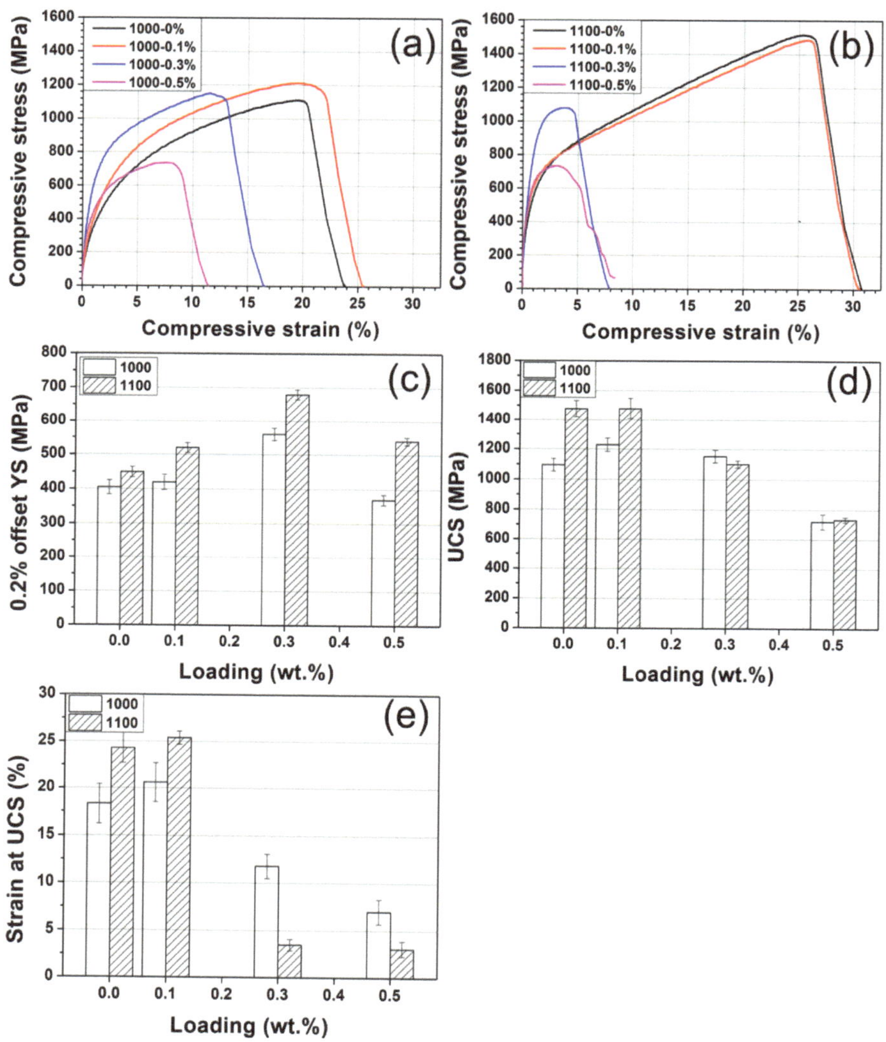

Figure 8. (**a**,**b**) Compressive stress–strain curves for samples with loadings of 0, 0.1, 0.3, and 0.5 wt.% GO at 1000 and 1100 °C, respectively. Each curve in the plot is the most representative one for each sample. The pillars tested had a diameter of 1.5 mm and height of 4.5 mm and 8 specimens were tested for each composition. (**c**) 0.2% offset yield strength, (**d**) Ultimate compressive stress (UCS) and (**e**) strain at UCS for Cu-W matrix composites with different loadings of GO produced by SPS at 1000 and 1100 °C.

The processing temperature had a significant role in the mechanical performance of the composites. Both Cu-W and Cu-W-0.1 wt.% GO processed at 1100 °C showed mechanical properties superior to all samples processed at 1000 °C. For example, Cu-W sample processed at 1000 °C possessed an ultimate compressive stress of 1095 MPa and a strain to failure of 18.3%, whereas the Cu-W sample sintered at 1100 °C had values of 1477 MPa and 24.3%, respectively. The values of UCS compare well with those reported in the literature for Cu-W composites of a similar composition [42]. However, there was significant embrittlement with a smaller plastic region for composites with 0.3 and 0.5 wt.% GO at 1100 °C with the degree of embrittlement increasing with increasing temperature. Interestingly, the Cu-W-0.3 wt.% GO samples had the optimal yield stress, regardless of the sintering temperature used. For the 1100 °C sintered sample, the 0.2% offset yield strength increases to 680 MPa which more than a 50% increase compared with the Cu-W sample.

3.6. Electrical Properties

A four-point probe measurement was used to measure the electrical conductivity of the Cu-W-GO composites (Figure 9). The spacing for each probe was 1.1 mm, which is much larger than the length of the 1000 °C Cu-W phase domains, meaning the results should be typical for the whole composite. The electrical conductivity for composites with 0.1 wt.% GO was higher than the unreinforced sample, with improvements of 25% and 16% at sintering temperatures of 1000 and 1100 °C, respectively. However, the conductivity for composites with 0.3 wt.% and 0.5 wt.% was significantly lower. The effects of GO loading for the electrical conductivity are similar to the ultimate compressive stress and strain. As revealed by EDS and XRD previously, more oxides (WO_3) are formed at higher loadings, which have much lower electrical conductivity than Cu. In addition, the Cu network may be disrupted by these oxides. Finally, more porosity is found (from both relative density and SEM results) in the composites with higher GO loadings, making electron flow less efficient in the composites.

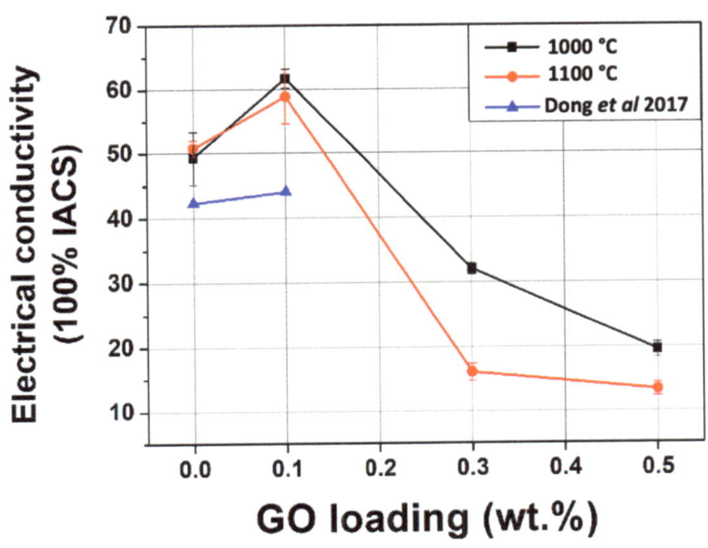

Figure 9. The measured electrical conductivity of samples as function of GO loading and sintering temperature. Additionally, shown are the conductivities reported in [8].

The electrical conductivity results for the Cu-W matrix composites are also compared to the conductivity from Dong et al. [8] in Figure 9. As can be seen, that the conductivity of the composites produced in this work with and without GO is higher than that in their work, possibly due to the ball milling route used allowing a better Cu pathway to be obtained. Further improvements in performance might be obtained through the use of metal-doped reduced GO as demonstrated in a recent study upon arc-erosion resistance in similar materials [43].

4. Conclusions

We have demonstrated that flake powder metallurgy can be successfully employed to produce Cu-W matrix composites with different loadings of GO (0.1, 0.3, 0.5 wt.%) and at different sintering temperature (1000, 1100 °C). The wet ball milling of the as-received flakes was found to flatten them without causing significant fracture, overcoming the morphology incompatibility between the as-received metal powders and GO flakes. This procedure enabled Cu-W composites with densities of 91.7% and 97.5% to be obtained at 1000 and 1100 °C, respectively, with the latter higher due to the Cu being molten during processing. The addition of the GO reduced the density, due to frustrated packing and the introduction of oxides. However, the density of the 1000 °C sintered Cu-W composite dropped by only 5.35% at a 0.5 wt.% GO loading, while the decrease was much higher (20.6%) for the composite sintered at 1100 °C. X-ray diffraction and EDS showed good phase purity, with the content of undesirable oxides increasing with GO loading and sintering temperature. Raman spectroscopy confirmed that the GO was reduced but had survived the sintering process and a good dispersion was achieved even at loadings up to 0.5 wt.% GO. Compression testing found that the composites sintered at 1100 °C has better mechanical properties compared with the composite sintered at 1000 °C due to the higher density at low levels of GO loading. GO was found to improve the mechanical properties, with an optimal loading of 0.1 wt.% loading found for ultimate compressive stress (UCS) and strain to failure and 0.3 wt.% optimal loading for the 0.2% offset yield strength. For higher GO loadings (>0.3 wt%) the strength of the composite was reduced due to an increase in porosity. The electrical conductivity increased with the addition of 0.1 wt.% GO but then decreased at higher loadings. Overall, 0.1 wt.% GO sintered at 1100 °C is considered to be the best composition since the conductivity, yield strength, UCS and strain at the UCS are all improved.

Supplementary Materials: The following supporting information can be downloaded at: https://www.mdpi.com/article/10.3390/ma15217736/s1, Figure S1. SEM of the green composite formed by mixing the as-received materials mixed in a ratio of Cu20W80-0.1 wt.% GO. Figure S2. SEM micrographs for Cu powders after ball milling for 10, 20, 30, 40, 50, 60, 120, 180 and 240 min.; Figure S3. SEM micrographs at high magnification for Cu powders after ball milling for 10, 20, 30, 40, 50, 60, 120, 180 and 240 min.; Figure S4. SEM micrographs for W powders after ball milling for 10, 20, 30, 40, 50, 60, 120, 180 and 240 min.; Figure S5. SEM micrographs at high magnification for W powders after ball milling for 10, 20, 30, 40, 50, 60, 120, 180 and 240 min.; Figure S6. EDS mapping for Cu-W flake powders with 0.1 wt.% GO: SEM image, SEM image combined with elemental mapping, elemental mapping for Cu, W, C and O.; Figure S7. EDS results for Cu-W sintered at 1000 °C, including a secondary electron image, EDS layered image, elemental maps for W, Cu, O, and C, EDS map sum spectrum.; Figure S8. EDS results for Cu-W with 0.5 wt.% GO sintered at 1000 °C, including a secondary electron image, EDS layered image, elemental maps for W, Cu, O, and C, EDS map sum spectrum.; Figure S9. EDS results for Cu-W sintered at 1100 °C, including a secondary electron image, EDS layered image, elemental maps for W, Cu, O, and C, EDS map sum spectrum.; Figure S10. EDS results for Cu-W with 0.5 wt.% GO sintered at 1100 °C, including a secondary electron image, EDS layered image, elemental maps for W, Cu, O, and C, EDS map sum spectrum; Figure S11. X-ray diffraction reference data for the Cu, W and WO_3.

Author Contributions: Conceptualization, R.X., M.Z., R.J.Y., I.A.K. and Y.D.; Funding acquisition, R.X., M.Z. and Y.D.; Investigation, F.L., R.J.Y. and I.A.K.; Supervision, R.J.Y. and I.A.K.; Writing—original draft, F.L.; Writing—review & editing, R.X., M.Z., R.J.Y., I.A.K. and Y.D. All authors have read and agreed to the published version of the manuscript.

Funding: This research was funded by work was supported by the State Grid Corporation of China Science and Technology Foundation (5500-201958505A-0-0-00).

Data Availability Statement: The authors confirm that the data supporting the findings of this study are available within the article and its supplementary material.

Acknowledgments: We thank State Grid Corporation of China Science and Technology Foundation for their funding.

Conflicts of Interest: The authors declare no conflict of interest.

References

1. Dong, L.L.; Ahangarkani, M.; Chen, W.G.; Zhang, Y.S. Recent Progress in Development of Tungsten-Copper Composites: Fabrication, Modification and Applications. *Int. J. Refract. Met. Hard Mater.* **2018**, *75*, 30–42. [CrossRef]
2. Hsu, S.-E.; Chen, C.-I.; Yue, S.; Li, F.K.-W. Mechanical and Thermal Properties of CU-Infiltrated P/M Tungsten Nozzles. *J. Spacecr. Rocket.* **1977**, *14*, 207–211. [CrossRef]
3. Hamidi, A.G.; Arabi, H.; Rastegari, S. A Feasibility Study of W-Cu Composites Production by High Pressure Compression of Tungsten Powder. *Int. J. Refract. Met. Hard Mater.* **2011**, *29*, 123–127. [CrossRef]
4. Dolatmoradi, A.; Raygan, S.; Abdizadeh, H. Mechanochemical Synthesis of W-Cu Nanocomposites via in-Situ Co-Reduction of the Oxides. *Powder Technol.* **2013**, *233*, 208–214. [CrossRef]
5. Cheng, J.; Wan, L.; Cai, Y.; Zhu, J.; Song, P.; Dong, J. Fabrication of W-20 Wt.%Cu Alloys by Powder Injection Molding. *J. Mater. Process. Technol.* **2010**, *210*, 137–142. [CrossRef]
6. Meng, Y.; Zhang, J.; Duan, C.; Chen, C.; Feng, X.; Shen, Y. Microstructures and Properties of W-Cu Functionally Graded Composite Coatings on Copper Substrate via High-Energy Mechanical Alloying Method. *Adv. Powder Technol.* **2015**, *26*, 392–400. [CrossRef]
7. Elsayed, A.; Li, W.; el Kady, O.A.; Daoush, W.M.; Olevsky, E.A.; German, R.M. Experimental Investigations on the Synthesis of W-Cu Nanocomposite through Spark Plasma Sintering. *J. Alloys Compd.* **2015**, *639*, 373–380. [CrossRef]
8. Dong, L.; Chen, W.; Zheng, C.; Deng, N. Microstructure and Properties Characterization of Tungsten–Copper Composite Materials Doped with Graphene. *J. Alloys Compd.* **2017**, *695*, 1637–1646. [CrossRef]
9. Geim, A.K.; Novoselov, K.S. The Rise of Graphene. *Nat. Mater.* **2007**, *6*, 183–191. [CrossRef]
10. Potts, J.R.; Dreyer, D.R.; Bielawski, C.W.; Ruoff, R.S. Graphene-Based Polymer Nanocomposites. *Polymer* **2011**, *52*, 5–25. [CrossRef]
11. Young, R.J.; Liu, M.; Kinloch, I.A.; Li, S.; Zhao, X.; Vallés, C.; Papageorgiou, D.G. The Mechanics of Reinforcement of Polymers by Graphene Nanoplatelets. *Compos. Sci. Technol.* **2018**, *154*, 110–116. [CrossRef]
12. Papageorgiou, D.G.; Kinloch, I.A.; Young, R.J. Graphene/Elastomer Nanocomposites. *Carbon* **2015**, *95*, 460–484. [CrossRef]
13. Nieto, A.; Bisht, A.; Lahiri, D.; Zhang, C.; Agarwal, A. Graphene Reinforced Metal and Ceramic Matrix Composites: A Review. *Int. Mater. Rev.* **2017**, *62*, 241–302. [CrossRef]
14. Cao, M.; Xiong, D.B.; Tan, Z.; Ji, G.; Amin-Ahmadi, B.; Guo, Q.; Fan, G.; Guo, C.; Li, Z.; Zhang, D. Aligning Graphene in Bulk Copper: Nacre-Inspired Nanolaminated Architecture Coupled with in-Situ Processing for Enhanced Mechanical Properties and High Electrical Conductivity. *Carbon* **2017**, *117*, 65–74. [CrossRef]
15. Yang, Z.; Wang, L.; Shi, Z.; Wang, M.; Cui, Y.; Wei, B.; Xu, S.; Zhu, Y.; Fei, W. Preparation Mechanism of Hierarchical Layered Structure of Graphene/Copper Composite with Ultrahigh Tensile Strength. *Carbon* **2018**, *127*, 329–339. [CrossRef]
16. Xiong, D.-B.; Cao, M.; Guo, Q.; Tan, Z.; Fan, G.; Li, Z.; Zhang, D. Graphene-and-Copper Artificial Nacre Fabricated by a Preform Impregnation Process: Bioinspired Strategy for Strengthening-Toughening of Metal Matrix Composite. *ACS Nano* **2015**, *9*, 6934–6943. [CrossRef]
17. Hwang, J.; Yoon, T.; Jin, S.H.; Lee, J.; Kim, T.S.; Hong, S.H.; Jeon, S. Enhanced Mechanical Properties of Graphene/Copper Nanocomposites Using a Molecular-Level Mixing Process. *Adv. Mater.* **2013**, *25*, 6724–6729. [CrossRef]
18. Kim, Y.; Lee, J.; Yeom, M.S.; Shin, J.W.; Kim, H.; Cui, Y.; Kysar, J.W.; Hone, J.; Jung, Y.; Jeon, S.; et al. Strengthening Effect of Single-Atomic-Layer Graphene in Metal–Graphene Nanolayered Composites. *Nat. Commun.* **2013**, *4*, 2114. [CrossRef]
19. Xiong, D.-B.; Cao, M.; Guo, Q.; Tan, Z.; Fan, G.; Li, Z.; Zhang, D. High Content Reduced Graphene Oxide Reinforced Copper with a Bioinspired Nano-Laminated Structure and Large Recoverable Deformation Ability. *Sci. Rep.* **2016**, *6*, 33801. [CrossRef]
20. Li, Z.; Guo, Q.; Li, Z.; Fan, G.; Xiong, D.-B.; Su, Y.; Zhang, J.; Zhang, D. Enhanced Mechanical Properties of Graphene (Reduced Graphene Oxide)/Aluminum Composites with a Bioinspired Nanolaminated Structure. *Nano Lett.* **2015**, *15*, 8077–8083. [CrossRef]
21. Hidalgo-Manrique, P.; Yan, S.; Lin, F.; Hong, Q.; Kinloch, I.A.; Chen, X.; Young, R.J.; Zhang, X.; Dai, S. Microstructure and Mechanical Behaviour of Aluminium Matrix Composites Reinforced with Graphene Oxide and Carbon Nanotubes. *J. Mater. Sci.* **2017**, *52*, 13466–13477. [CrossRef]
22. Shin, S.E.; Choi, H.J.; Shin, J.H.; Bae, D.H. Strengthening Behavior of Few-Layered Graphene/Aluminum Composites. *Carbon* **2015**, *82*, 143–151. [CrossRef]
23. Dutkiewicz, J.; Ozga, P.; Maziarz, W.; Pstruś, J.; Kania, B.; Bobrowski, P.; Stolarska, J. Microstructure and Properties of Bulk Copper Matrix Composites Strengthened with Various Kinds of Graphene Nanoplatelets. *Mater. Sci. Eng. A* **2015**, *628*, 124–134. [CrossRef]
24. Yue, H.; Yao, L.; Gao, X.; Zhang, S.; Guo, E.; Zhang, H.; Lin, X.; Wang, B. Effect of Ball-Milling and Graphene Contents on the Mechanical Properties and Fracture Mechanisms of Graphene Nanosheets Reinforced Copper Matrix Composites. *J. Alloys Compd.* **2017**, *691*, 755–762. [CrossRef]

25. Gao, X.; Yue, H.; Guo, E.; Zhang, H.; Lin, X.; Yao, L.; Wang, B. Preparation and Tensile Properties of Homogeneously Dispersed Graphene Reinforced Aluminum Matrix Composites. *Mater. Des.* **2016**, *94*, 54–60. [CrossRef]
26. Baig, Z.; Mamat, O.; Mustapha, M. Recent Progress on the Dispersion and the Strengthening Effect of Carbon Nanotubes and Graphene-Reinforced Metal Nanocomposites: A Review. *Crit. Rev. Solid State Mater. Sci.* **2016**, *43*, 1–46. [CrossRef]
27. Bakshi, S.R.; Lahiri, D.; Agarwal, A. Carbon Nanotube Reinforced Metal Matrix Composites—A Review. *Int. Mater. Rev.* **2010**, *55*, 41–64. [CrossRef]
28. Cha, S.I.; Kim, K.T.; Arshad, S.N.; Mo, C.B.; Hong, S.H. Extraordinary Strengthening Effect of Carbon Nanotubes in Metal-Matrix Nanocomposites Processed by Molecular-Level Mixing. *Adv. Mater.* **2005**, *17*, 1377–1381. [CrossRef]
29. Chen, Y.; Zhang, X.; Liu, E.; He, C.; Shi, C.; Li, J.; Nash, P.; Zhao, N. Fabrication of In-Situ Grown Graphene Reinforced Cu Matrix Composites. *Sci. Rep.* **2016**, *6*, 19363. [CrossRef]
30. Li, Z.; Fan, G.; Tan, Z.; Guo, Q.; Xiong, D.; Su, Y.; Li, Z.; Zhang, D. Uniform Dispersion of Graphene Oxide in Aluminum Powder by Direct Electrostatic Adsorption for Fabrication of Graphene/Aluminum Composites. *Nanotechnology* **2014**, *25*, 325601. [CrossRef]
31. Dong, L.L.; Huo, W.T.; Ahangarkani, M.; Zhang, B.; Zhao, Y.Q.; Zhang, Y.S. Microstructural Evaluation and Mechanical Properties of In-Situ WC/W-Cu Composites Fabricated by RGO/W-Cu Spark Plasma Sintering Reaction. *Mater. Des.* **2018**, *160*, 1196–1207. [CrossRef]
32. Dong, L.; Chen, W.; Deng, N.; Song, J.; Wang, J. Investigation on Arc Erosion Behaviors and Mechanism of $W_{70}Cu_{30}$ Electrical Contact Materials Adding Graphene. *J. Alloys Compd.* **2017**, *696*, 923–930. [CrossRef]
33. Chen, W.; Dong, L.; Wang, J.; Zuo, Y.; Ren, S.; Fu, Y. Synergistic Enhancing Effect for Mechanical and Electrical Properties of Tungsten Copper Composites Using Spark Plasma Infiltrating Sintering of Copper-Coated Graphene. *Sci. Rep.* **2017**, *7*, 17836. [CrossRef] [PubMed]
34. Dong, L.L.; Fu, Y.Q.; Liu, Y.; Lu, J.W.; Zhang, W.; Huo, W.T.; Jin, L.H.; Zhang, Y.S. Interface Engineering of Graphene/Copper Matrix Composites Decorated with Tungsten Carbide for Enhanced Physico-Mechanical Properties. *Carbon* **2021**, *173*, 41–53. [CrossRef]
35. Rourke, J.P.; Pandey, P.A.; Moore, J.J.; Bates, M.; Kinloch, I.A.; Young, R.J.; Wilson, N.R. The Real Graphene Oxide Revealed: Stripping the Oxidative Debris from the Graphene-like Sheets. *Angew. Chem. Int. Ed.* **2011**, *50*, 3173–3177. [CrossRef]
36. Cao, J.; He, P.; Mohammed, M.A.; Zhao, X.; Young, R.J.; Derby, B.; Kinloch, I.A.; Dryfe, R.A.W. Two-Step Electrochemical Intercalation and Oxidation of Graphite for the Mass Production of Graphene Oxide. *J. Am. Chem. Soc.* **2017**, *139*, 17446–17456. [CrossRef]
37. Li, Z.; Young, R.J.; Kinloch, I.A. Interfacial Stress Transfer in Graphene Oxide Nanocomposites. *ACS Appl. Mater. Interfaces* **2013**, *5*, 456–463. [CrossRef]
38. Papageorgiou, D.G.; Kinloch, I.A.; Young, R.J. Mechanical Properties of Graphene and Graphene-Based Nanocomposites. *Prog. Mater. Sci.* **2017**, *90*, 75–127. [CrossRef]
39. Jorio, A.; Saito, R.; Dresselhaus, G.; Dresselhaus, M.S. *Raman Spectroscopy in Graphene Related Systems*; Wiley-VCH Verlag GmbH & Co. KGaA: Weinheim, Germany, 2011. [CrossRef]
40. Xu, K.X.; Guo, P.Y.; Song, M.S.; Xue, H.; Zhang, L.; Dong, L.L.; Zhang, Y.S. Effects of Ball Milling Times on Microstructure and Properties of Cu matrix Composites Reinforced with Graphene Oxide Nanosheets. *J. Mater. Eng. Perform.* **2022**, *31*, 3437–3444. [CrossRef]
41. Lungu, M.V.; Lucaci, M.; Tsakiris, V.; Brătulescu, A.; Cîrstea, C.D.; Marin, M.; Pătroi, D.; Mitrea, S.; Marinescu, V.; Grigore, F.; et al. Development and Investigation of Tungsten Copper Sintered Parts for Using in Medium and High Voltage Switching Devices. *IOP Conf. Ser. Mater. Sci. Eng.* **2017**, *209*, 012012. [CrossRef]
42. Abu-Oqail, A.; Ghanim, M.; El-Sheikh, M.; El-Nikhaily, A. Effects of Processing Parameters of Tungsten–Copper Composites. *Int. J. Refract. Met. Hard Mater.* **2012**, *35*, 207–212. [CrossRef]
43. Dong, L.L.; Li, L.; Li, X.; Zhang, W.; Fu, Y.Q.; Elmarakbi, A.; Zhang, Y.S. Enhancing Mechanisms of Arc-Erosion Resistance for Copper Tungsten Electrical Contact Using Reduced Graphene Oxides In Situ Modified by Copper Nanoparticles. *Int. J. Refract. Met. Hard Mater.* **2022**, *108*, 105934. [CrossRef]

nanomaterials

MDPI

Review

Advances of Carbon Materials for Dual-Carbon Lithium-Ion Capacitors: A Review

Ying Duan [1,2], Changle Li [1], Zhantong Ye [3], Hongpeng Li [4], Yanliang Yang [1], Dong Sui [1,*] and Yanhong Lu [3,*]

1 Henan Key Laboratory of Function-Oriented Porous Materials, College of Chemistry and Chemical Engineering, Luoyang Normal University, Luoyang 471934, China
2 College of Food and Drug, Luoyang Normal University, Luoyang 471934, China
3 School of Chemistry & Material Science, Langfang Normal University, Langfang 065000, China
4 College of Mechanical Engineering, Yangzhou University, Yangzhou 225127, China
* Correspondence: suidonghy@mail.nankai.edu.cn (D.S.); luyanhong@lfnu.edu.cn (Y.L.)

Abstract: Lithium-ion capacitors (LICs) have drawn increasing attention, due to their appealing potential for bridging the performance gap between lithium-ion batteries and supercapacitors. Especially, dual-carbon lithium-ion capacitors (DC-LICs) are even more attractive because of the low cost, high conductivity, and tunable nanostructure/surface chemistry/composition, as well as excellent chemical/electrochemical stability of carbon materials. Based on the well-matched capacity and rate between the cathode and anode, DC-LICs show superior electrochemical performances over traditional LICs and are considered to be one of the most promising alternatives to the current energy storage devices. In particular, the mismatch between the cathode and anode could be further suppressed by applying carbon nanomaterials. Although great progresses of DC-LICs have been achieved, a comprehensive review about the advances of electrode materials is still absent. Herein, in this review, the progresses of traditional and nanosized carbons as cathode/anode materials for DC-LICs are systematically summarized, with an emphasis on their synthesis, structure, morphology, and electrochemical performances. Furthermore, an outlook is tentatively presented, aiming to develop advanced DC-LICs for commercial applications.

Keywords: lithium-ion capacitor; nanocarbon; carbon nanomaterial; dual-carbon electrode; high energy/power density

Citation: Duan, Y.; Li, C.; Ye, Z.; Li, H.; Yang, Y.; Sui, D.; Lu, Y. Advances of Carbon Materials for Dual-Carbon Lithium-Ion Capacitors: A Review. *Nanomaterials* **2022**, *12*, 3954. https://doi.org/10.3390/nano12223954

Academic Editors: Robert A. Varin and Geng Zhong

Received: 14 October 2022
Accepted: 6 November 2022
Published: 10 November 2022

1. Introduction

Rechargeable energy storage systems are indispensable energy supply components for consumer electronics, electric vehicles, and smart grids [1–5]. More importantly, the reduction of greenhouse gas and the effective utilization of sustainable and clean, but intermittent, solar and wind energy also depend on the development of high-performance energy conversion and storage systems [6–9]. Generally, lithium-ion batteries (LIBs) and supercapacitors (SCs) are the two commonly used energy storage devices and exhibit different electrochemical performances, due to the different energy storage mechanisms. LIBs store/release energy through lithium ions rocking back and forth reversibly between a cathode and anode via a Faradaic reaction, thereby delivering high energy density and high working voltage, as well as low self-discharge [10–12]. By contrast, SCs have the advantages of high power density and long cycling life, based on a fast, non-Faradaic adsorption/desorption mechanism [13]. Nevertheless, both LIBs and SCs have their own shortcomings. For example, LIBs suffer from inferior power performance and limited lifespan, which are ascribed to the severely sluggish Li^+ diffusion in the bulk electrode and structural degradation during repeated charge/discharge processes, respectively [14]. As for the SCs, the physically electrostatic ion adsorption at the electrode/electrolyte interface coupled with small working voltage results in relatively lower energy density, thus hindering their further applications [15]. Therefore, there is an urgent need to develop high-performance energy storage devices that can combine the high energy/working voltage of LIBs with the high power/long lifespan of SCs.

To solve the above-mentioned disadvantages of LIBs and SCs, lithium-ion capacitors (LICs, also named lithium-ion hybrid supercapacitors) offer a promising solution by integrating the advantages of LIBs and SCs into one system. LICs possess not only higher power density than LIBs and other conventional batteries, but also higher energy density than SCs (Figure 1a). As for the typical device configuration, LICs are composed of a capacitor-type electrode from SCs as the cathode to ensure high power output and a battery-type electrode from LIBs as the anode to provide high energy density (Figure 1b) [16]. Lithium salts dissolved in organic solvents are commonly adopted as the electrolytes, delivering a wide working voltage [17]. Compared with LIBs and SCs, LICs follow a hybrid energy storage mechanism. During the charge/discharge process, lithium ions intercalate/de-intercalate into/from the anode through a redox reaction, while anions absorb/desorb onto/from the cathode through formation of an electric double layer at the electrode–electrolyte interface [18]. It is noted that LICs could deliver a much higher output voltage than that of SCs because the anode and cathode work in different potential regions [19]. Therefore, LICs with a hybrid cell configuration can well-bridge the performance gap between LIBs and SCs. Benefiting from the high energy and high power density, LICs have shown great potential in start power, electric vehicles, and frequency modulation in grid and large-scale energy storage systems. So far, LICs have received intensive research interest from both the academic and industrial communities.

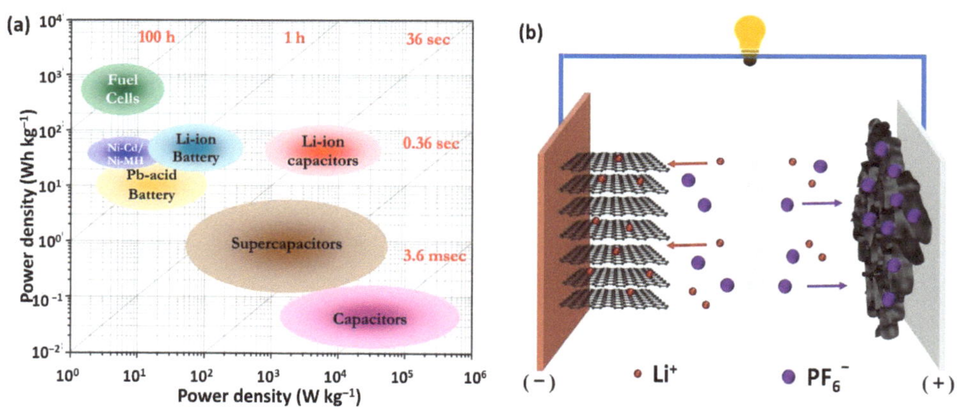

Figure 1. (**a**) Ragone plot of various energy storage devices. Reproduced with permission from Ref. [20]. Copyright 2014, American Chemical Society. (**b**) The configuration and energy storage mechanism of LICs.

Since the electrochemical performances of LICs are significantly influenced by the capacity, rate capability, operation potential, and cycling stability of both the anode and cathode, developing advanced electrode materials is of great importance [17,21,22]. Notably, nanostructured materials have a positive effect on the electrochemical properties of an electrode. For example, a nanosized anode has a better rate, has a longer cycling life, and is a better match with the cathode in kinetics. On the other hand, a cathode with rationally distributed nanoscaled pores has a higher capacity. So far, various cathode and anode materials with stable structure, tailorable microstructure, and rich surface chemistry and matched properties have been investigated in LICs. The capacitor-type cathode is expected to ensure high power and long-term durability, so that porous carbon materials with high electrical conductivity, large specific surface area (SSA), and rich nanosized pores are commonly adopted [22]. To realize high energy and power density, nanosized carbons, such as graphene, carbon nanotubes (CNTs), and their composites with inherently high conductivity and large SSA, have commonly been investigated recently because of the high capacity and excellent rate [23,24]. As compared with the cathode, the battery-type anode materials are the key components to supply high energy. According to the lithium storage mechanisms, anode materials can be roughly divided into three categories: intercalation-type materials (carbonaceous materials, Ti/Nb-based compounds, etc.), conversion-type materials (metal oxide/sulfide/nitride, etc.), and alloying-type materials (Si/Sn-based materials, etc.) [25–29]. LICs using non-carbon materials as anodes usually deliver one superior property at the cost of other performances. For instance, LICs with Ti/Nb-based anodes show high power density and excellent cycling stability, due to their fast Li^+ intercalation/de-intercalation rate and stable structure [30–32]. However, they have a lower specific capacity and higher Li^+ insertion/de-insertion potential than other anodes, leading to unsatisfactory energy density. Although LICs based on conversion/alloying-type anodes could realize high energy density, their intrinsically low conductivity and large volume change during charge/discharge process always result in poor power density and limited cycling life [33–36]. To overcome the above-mentioned drawbacks of non-carbon anodes, developing nanostructured materials, introducing defects, and compositing with nanosized carbons are typically the adopted strategies to enhance the conductivity and/or suppress the volume change [37–39]. Under these circumstances, carbon anodes have drawn inceasing attention because carbon possesses many superior properties over others. The excellent electrical conductivity, relatively high capacity, and stable structure of carbon materials enable LICs with not only high energy/power densities, but also outstanding cycling stability. In addition, the abundant resources, mature material production, and device fabrication technologies of carbon-based materials could promote their commercial application in LICs at a low cost.

Compared with non-carbon anode materials, the carbon anode could well-match with the porous carbon cathode, in terms of capacity, rate, and cycle life. Therefore, tremendous efforts have been devoted to developing dual-carbon lithium-ion capacitors (DC-LICs) by adopting carbon materials as both the cathode and anode (Figure 2). Similar to the typical LICs, cathode materials for DC-LICs are activated carbon (AC), biomass/polymer-derived porous carbons, and nanosized carbons (graphene, CNTs, and composites) [23,40–42]. The anode materials of DC-LICs include traditional carbon materials (graphite, hard carbon, and soft carbon), and nanosized carbons (graphene, CNTs, graphdiyne, and their composites) [22,37,43]. Benefitting from the low cost, abundance in raw materials, tunable surface chemistry and composition, adjustable structure and morphology, and excellent physical/chemical/electrochemical stability of carbon materials, DC-LICs have demonstrated superior electrochemical performances. It should be noted that nanocarbon-based DC-LICs show superior electrochemical properties over the traditional carbons because of the enhanced capacity and rate capability [13,19,44]. During the past two decades, great progresses of electrode materials, electrolytes, cell configuration, and pre-lithiation technology for LICs have been achieved and extensively reviewed [22,24,45–47]. However, a systematical summary of the advances in electrode materials for DC-LICs is highly needed [48,49]. Therefore, in this review, the latest progresses of electrode materials ranging from traditional carbons to novel nanosized carbons for DC-LICs are comprehensively overviewed. The material synthesis, structure, morphology, electrochemical performance, and merits and demerits are particularly discussed, aiming to develop high-performance carbon-based electrode materials for industrial-scale applications. Furthermore, the outlooks for DC-LICs are tentatively presented. This review will provide some constructive guidelines for fabricating advanced DC-LICs for both the scientific community and industry.

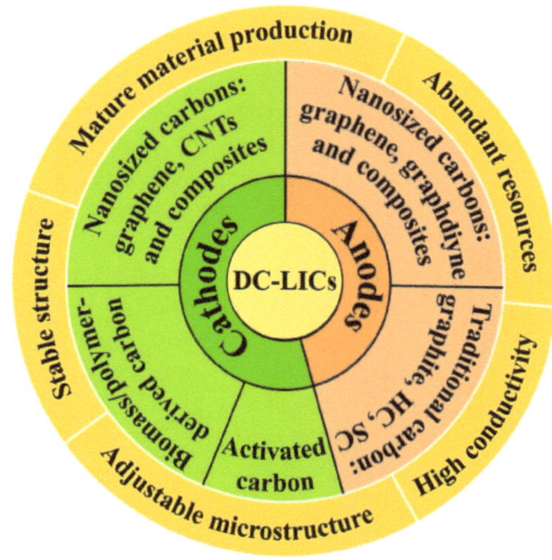

Figure 2. Typical carbon cathode and anode materials for DC-LICs.

2. Progresses of Carbon Materials as Cathode for DC-LICs

Based on the energy storage mechanism of fast ion adsorption/desorption at the electrode/electrolyte interface, the capacitor-type cathodes are expected to provide high power density and ensure long-term durability in LICs. Bearing this in mind, porous carbon materials with high SSA, rich interconnected pores, excellent electrical conductivity, and stable structure are investigated as cathodes. Thus, traditional porous carbons (AC, biomass/polymer-derived porous carbon) and novel nanosized carbons (graphene, CNTs and their composites) are commonly adopted. The typical examples are summarized in Table 1.

Table 1. Typical carbon cathodes and the electrochemical performances of DC-LICs.

Carbon Cathodes		Performances of DC-LICs					
Materials	Specific Capacity (mAh g^{-1})	Rate Capability (mAh g^{-1})	Voltage (V)	Maximum Energy Density (Wh kg^{-1})	Maximum Power Density (kW kg^{-1})	Cycling Stability	Ref.
Commercial AC (YP-50F)	48	35 @ 1 A g^{-1}	2.0–4.0	73.0	/	85% after 1000 cycles	[50]
KHPC−K	100.5	75 @ 10 A g^{-1}	0.01–4.0	169.0	97.0	77.7% after 5000 cycles	[51]
NHCN-2	125	98 @ 15 A g^{-1}	2.0–4.5	146.0	52.0	91% after 40,000 cycles	[52]
LPCs-3	80	26 @ 10 A g^{-1}	2.0–4.0	97.0	11.4	92.3% after 5000 cycles	[53]
DPC-MK	149	69 @ 50 A g^{-1}	1.0–4.2	160.6	39.7	95.6% after 8000 cycles	[54]
0.1-BNC	113	63 @ 10 A g^{-1}	0.02–4.5	220.0	22.5	81% after 5000 cycles	[55]
S-NPC-40	95.9	43.2 @ 10 A g^{-1}	0–4.0	176.1	20.0	82% after 20,000 cycles	[56]
CHPC	132	100 @ 8 A g^{-1}	0.01–4.2	220	66.9	70% after 3000 cycles	[57]
N/S-CNF0.25	133	102 @ 10 A g^{-1}	1.0–4.3	154.0	18.6	92% after 6000 cycles	[58]
PHNCNB	72	51 @ 10 A g^{-1}	1.0–4.0	148.5	25.0	90% after 8000 cycles	[59]
NHPCS	74	64 @ 5 A g^{-1}	0–4.2	151.0	10.7	96.3% after 3000 cycles	[60]
NCNs-2	115	62 @ 10 A g^{-1}	0–4.5	218.4	22.5	84.5% after 10,000 cycles	[61]
BNC	75.2	51.4 @ 5 A g^{-1}	1.0–4.0	115.5	10.0	71.6% after 2000 cycles	[62]
ANCS	113	67 @ 10 A g^{-1}	0–4.5	206.7	22.5	86.6% after 10,000 cycles	[63]
NPCS-1	97.4	51.3 @ 10 A g^{-1}	0–4.0	135.6	10.0	82% after 10,000 cycles	[64]
NPCNF	122	53 @ 100 A g^{-1}	1.0–4.3	143.0	45.0	83.1% after 10,000 cycles	[65]
URGO	35	29 @ 1.1 A g^{-1}	2.0–4.0	106.0	4.2	~100% after 1000 cycles	[66]
PRGO	171	92.3 @ 8.71 A g^{-1}	0.01–4.0	262.0	9.0	91% after 4000 cycles	[67]
NGF-0	82	61 @ 8 A g^{-1}	1.0–4.0	147	48.9	87% after 10,000 cycles	[68]
GPC	95	40 @ 5 A g^{-1}	0–4.2	142.9	12.1	88% after 5000 cycles	[18]
SGCs	257.1	147.7 @ 6 A g^{-1}	0–4.0	249.9	19.62	95.4% after 10,000 cycles	[69]
A-N-GS	104	57 @ 5 A g^{-1}	0–4.5	187.9	11.25	93.5% after 3000 cycles	[70]
MRPG/CNT	108	83.3 @ 5 A g^{-1}	0–4.5	232.6	45.2	86% after 5000 cycles	[71]
$Zn_{90}Co_{10}$-APC	118.8	50 @ 5 A g^{-1}	2.0–4.0	108	15.0	86% after 10,000 cycles	[72]

2.1. Traditional Porous Carbon Cathode

2.1.1. Activated Carbon

As a conventional porous carbon material used in SCs, AC has the advantages of acceptable SSA, porous structure, low cost, and mature fabrication technology. Typically, ACs are prepared from carbon-rich materials through activation with KOH, H_3PO_4, $ZnCl_2$, H_2O, and/or CO_2 at high temperature [73]. In the early studies of LICs, AC was usually used as the cathode. For example, the first protype of LIC was fabricated with AC as the cathode and nanostructured $Li_4Ti_5O_{12}$ as the anode [74]. The obtained asymmetric hybrid system shows high energy density, an extended cycle life, and fast charge/discharge capability. Later, several groups developed DC-LICs by pairing AC with graphite [75], hard carbon [40,76], or graphene [77]. However, conventional ACs still suffer from low capacity and poor rate capability, due to small SSA and low conductivity, resulting in inferior energy and power densities. Hence, it is highly urgent to develop advanced ACs with high conductivity and desired microstructure.

To further improve the capacity and rate of AC, compositing with nanosized carbons is a feasible method. For example, Ma et al. designed graphene/activated carbon (G/AC) composites through a fast, self-propagating, high-temperature synthesis (SHS) process, which combined the advantages of the two components [78]. The conductivity of the prepared G/AC was largely enhanced from 389 to 2941 S m^{-1}, and a remarkable rate performance with 84% capacity retention at 10 A g^{-1} was achieved, compared with 65% for pure AC. Based on this cathode, the assembled LICs with graphene/soft carbon anode demonstrated a greatly improved energy density of 152 Wh kg^{-1} and power density of 18.9 kW kg^{-1}.

2.1.2. Biomass-Derived Porous Carbon

Recently, biomass-derived porous carbons (BDPCs) are emerging as cost-effective electrode materials for DC-LICs, due to their abundance, renewability, and sustainability [53,79]. Biomass is an economical and environmentally friendly raw material for preparing porous carbon and has the following merits: (1) the diversity of biomass with different nanostructure and morphology provides many choices for preparing porous carbon with desired electrochemical performances; (2) the inherent ordered porous structure of biomass facilitates the activation process to form porous carbon with high SSA and large pore volume; (3) the heteroatom doped porous carbon could be facilely realized via a self-doping strategy [80–82]. Particularly, heteroatom doping has multiple advantages. The study by Lee et al. revealed that heteroatom doping is beneficial to the betterment of electrical conductivity and pore generation, which eventually enhances the electrochemical properties of carbonaceous materials [83].

Similar to conventional ACs, BDPCs are commonly prepared via a typical carbonization and activation processes [51,84]. Additionally, BDPCs have porous morphology, stable structure, heteroatom dopants, and high conductivity, endowing SCs or LICs based on them with high energy/power density and good cycling stability [82,84,85]. Several works have explored agricultural waste-derived porous carbons in LICs by pairing with non-carbon anodes, which demonstrate that BDPCs could serve as excellent cathodes [86–90]. Then, the applications of BDPCs in DC-LICs were investigated. Wang et al. prepared a sponge-like carbon (SLC) from gulfweed through KOH activation and used as electrode material for DC-LICs (Figure 3a) [91]. As shown in Figure 3b, the obtained sample exhibited a sponge-like structure with rich porosity. Benefiting from the astonishing structure, rational pore size distribution (PSD), and heteroatom doping, the optimized device with SLC as both cathode and anode delivered a high energy density of 127 Wh kg^{-1} and a peak power of 33.57 kW kg^{-1} (Figure 3c). More importantly, an ultra-stable cycling performance with 99% capacity retention was achieved after 100,000 cycles (Figure 3d). Ma et al. reported a cathode of N-doped hierarchical carbon nanolayer (NHCNs) through a facile, one-step template carbonization/activation from naturally abundant and renewable chitosan biomass [52]. NHCNs has a high capacity of 125 mAh g^{-1}, which could be ascribed to its hierarchically porous structure with large SSA (2350 m^2 g^{-1}) and high nitrogen doping level. Moreover, DC-LICs based on NHCNs show a maximum power density of 52 kW kg^{-1} and an ultra-long cycling life of 40,000 cycles. Anyway, graphene is commonly introduced to further enhance the conductivity of BDPCs by forming composites, aiming to achieve higher specific capacity and better rate capability [86,92,93]. It should be noted that biomass-derived carbons have the disadvantages of uncontrollable impurity and element composition, due to the diversity of raw materials. Tedious post treatment is usually needed to obtain products with high purity and good batch stability.

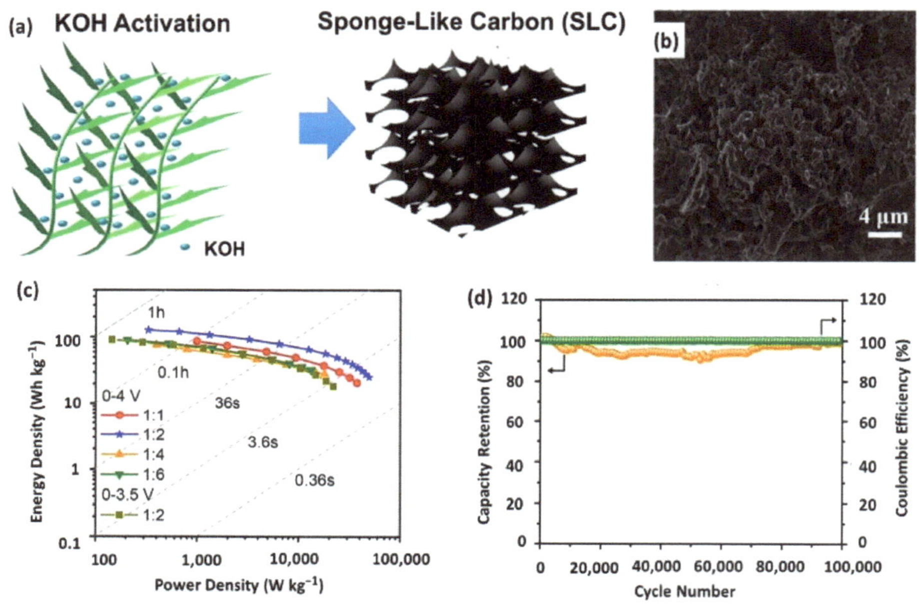

Figure 3. (**a**) Schematic illustration of the preparation process of SLC; (**b**) SEM image of SLC; (**c**) Ragone plots and (**d**) cycling stability of DC-LICs based on SLC electrodes. Reproduced with permission from Ref. [91]. Copyright 2019, American Chemical Society.

2.1.3. Polymer-Derived Porous Carbon

Porous carbon materials could also be prepared through pyrolysis and the activation of polymers. Compared with the uncontrollable impurity and element composition of biomass-derived carbons, polymers could be designed and adjusted from the precursors, so that the structure and composition could be well-controlled. Thus, polymer-derived porous carbons receive increasing interests in energy storage devices [41,94]. For instance, Hung et al. designed a hierarchical porous activated carbon (H-HPAC) material through pyrolysis of polyvinylpyrrolidone (PVP)-derived hydrogel, using K_2CO_3 as both the initiator for hydrogel formation and the activator [95]. In the preparing process, the numerous water molecules captured within the PVP function as a green template during the formation of hydrogel, making this method a facile and eco-friendly strategy to design highly porous carbon. As a cathode, H-HPAC demonstrates a capacitance of 128.7 F g^{-1}, due to its high SSA (2012 m^2 g^{-1}) and large pore volume (1.16 cm^3 g^{-1}). Wang et al. proposed a nitrogen-doped activated porous carbon by carbonization and activation of polypyrrole [96]. The product has a hierarchically porous structure with plenty of mesopores created by surfactant and micropores generated by activation. The all-carbon LIC delivers a high energy density of 167 Wh kg^{-1}, which still remains 88.9 Wh kg^{-1} at an ultrahigh power density of 13.2 kW kg^{-1}. Similarly, Ajuria et al. reported an activated carbon from furfuryl alcohol-based polymer by a process of polymerization, carbonization, and activation [97].

The obtained cathode material showed a sponge-like surface with high SSA and broad pore distribution. Combined with a hard carbon anode made from the same furfuryl alcohol-derived polymer, the obtained DC-LICs offer a medium energy density of 110 Wh kg^{-1} at 7 kW kg^{-1} and keep 50 Wh kg^{-1}, even at an ultrahigh power of 50 kW kg^{-1} (discharge in less than 10 s). Although polymer-derived porous carbons have demonstrated advantages over biomass, in terms of purity and controllable composition, the high cost and tedious synthesis process should not be ignored.

As a class of organic porous polymers, conjugated microporous polymers (CMPs), covalent organic frameworks (COFs), and their derivatives have been applied in various energy storage devices [98–103]. CMPs and COFs have the merits of well-defined crystal structure, plenty of nanopores, large SSA, and rich heteroatoms, making them the ideal platform to prepare high-performance carbon materials [104,105]. Therefore, CMPs and COFs could serve as excellent precursors to prepare self-doping porous carbons via a simple pyrolysis process. CMPs/COFs-derived porous carbons have been successfully used as electrode materials in LIBs and SCs [106–109]. However, there are very few reports covering the application of CMPs/COFs-derived porous carbons in LICs or DC-LICs. Hence, more efforts could focus on this field in future research.

2.2. *Nanosized Carbon-Based Cathode*

Traditional AC and biomass/polymer-derived porous carbons usually suffer from low capacity, due to too many inaccessible micropores, resulting in inferior energy density. Moreover, their amorphous structure with numerous pores and defects leads to low electrical conductivity and, thus, poor power output. Therefore, nanosized carbons, such as graphene, CNTs, and their composites, are emerging as excellent cathodes for DC-LICs because of their inherently high conductivity and large theoretical SSA.

2.2.1. Graphene-Based Cathode

Graphene, as a novel two-dimensional (2D) nanosized carbon material, has drawn intensive attention, ever since its discovery and has found numerous applications in energy conversion and storage [110–112]. Graphene possesses the following advantages. First of all, the layered structure with long-range conjugation endows graphene with large SSA, high electrical conductivity, and outstanding mechanical strength [113]. Anyway, the rich surface chemistry coupled with the 2D structure also makes graphene an excellent substrate for forming composites with other materials [13,114,115]. As a result, graphene-based materials have been adopted as an excellent cathode in LICs [23,24].

Graphene oxide (GO) is an appealing precursor of graphene-based cathode. GO can be easily dispersed in polar solvent, due to the abundant oxygen-containing groups on the surface or at the edge [116]. Moreover, the oxygen-containing functional groups could provide extra capacity via redox reaction [111]. Generally, GO should be reduced by high temperature, hydrothermal reaction, or reductants to recover the conjugation structure before being used as cathode [66,67,117]. After the treatment, the reduced GO demonstrates enhanced electrical conductivity. Besides the typical double-layer adsorption/desorption charge storage, the remaining functional groups exhibit reversible Li^+ binding and facilitate electrolyte infiltration, which largely improves the specific capacity and rate capability [66]. Several reports have explored the application of reduced GO-based cathodes in LICs [118–121]. As a typical example, Dubal et al. prepared a partially reduced GO (PRGO) by thermal annealing [67]. The obtained sample presented three-dimensional (3D) interconnected networks with open-porous nanosheets, which is expected to contribute a fast charge/discharge rate (Figure 4a). Moreover, the partial reduction strategy maintains a substantial amount of C=O redox groups, which could undergo redox reaction with consequent Li^+ uptake [122]. Accordingly, PRGO shows an astonishing capacity of 171 mAh g^{-1}, with an excellent rate of 92.3 mAh g^{-1} at 8.71 A g^{-1} (Figure 4b). The DC-LICs using PRGO as cathode deliver an ultrahigh energy density of 262 Wh kg^{-1}, which keeps 78 Wh kg^{-1} at a high power density of 9 kW kg^{-1} (Figure 4c).

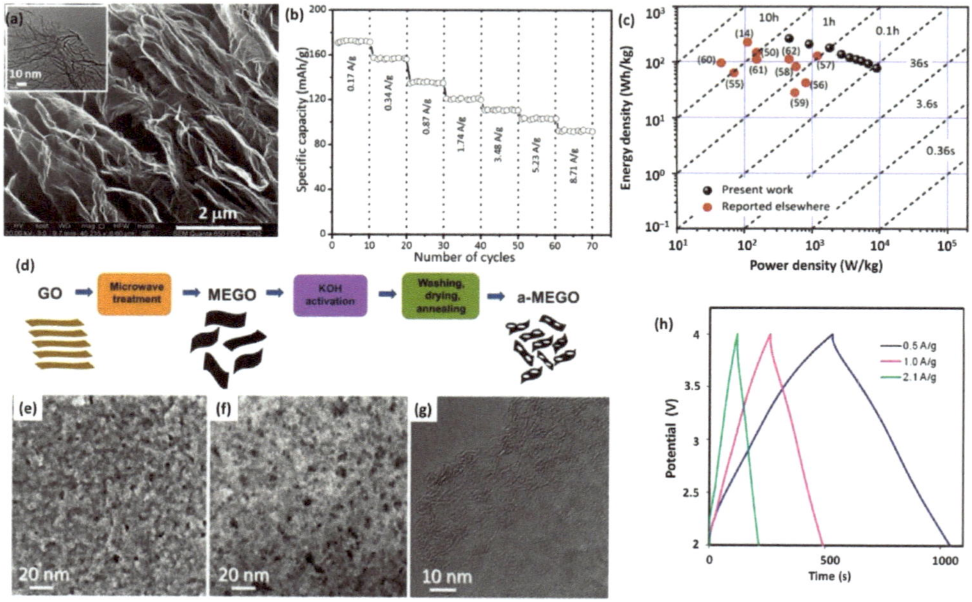

Figure 4. (**a**) FESEM of PRGO (inset shows TEM image); (**b**) Rate capability of PRGO; (**c**) Ragone plot for PRGO-based LICs. Reproduced with permission from Ref. [67]. Copyright 2018, Elsevier. (**d**) Schematic showing the microwave exfoliation/reduction of GO and the following chemical activation of MEGO with KOH; Images of (**e**) high-resolution SEM, (**f**) ADF-STEM and (**g**) high-resolution phase contrast electron micrograph of a-MEGO. Reproduced with permission from Ref. [110]. Copyright 2011, American Association for the Advancement of Science. (**h**) Charge/discharge curves for a-MEGO/graphite LICs at various current densities. Reproduced with permission from Ref. [123]. Copyright 2012, Royal Society of Chemistry.

However, the reduced GO suffers from small SSA and poor conductivity because of the unavoidable sheet restacking and incomplete recovery of the conjugation structure, resulting in inferior electrochemical performance. Therefore, highly porous graphene cathodes with higher conductivity and larger SSA are developed by activation or the template-assisted chemical vapor deposition (CVD) method. As a pioneering work, Ruoff et al. developed a graphene cathode (a-MEGO) for LICs through the chemical activation of microwave-expanded graphite oxide to obtain a dense network of nanometer-scale pores surrounded by highly curved carbon layers (Figure 4d) [110]. As shown in Figure 4e–g, the a-MEGO demonstrated the porous morphology with a pore size distribution of sizes between ~1 and ~10 nm. More importantly, a-MEGO had a very high SSA of 3100 m^2 g^{-1}, while retaining the high conductivity of graphene. As expected, the a-MEGO cathode exhibited a nearly symmetric charge/discharge curve and delivered a specific capacitance of 266 and 213 F g^{-1} at 1.0 and 2.5 A g^{-1}, respectively (Figure 4h). By pairing with graphite, the dual carbon-based devices presented a high energy density of 147.8 Wh kg^{-1} [123]. Similarly, Zhang et al. fabricated porous graphene by activation of reduced GO [124]. The obtained product had a highly crumpled morphology and porous structure with abundant mesopores that contributed to a high SSA (2103 m^2 g^{-1}) and large pore volume (1.8 cm^3 g^{-1}). This structure provides highly exposed active sites for ion adsorption/desorption and fast transport path for ions and electrons, leading to enhanced capacity and rate.

Besides chemical activation, porous graphene with high conductivity and outstanding structural stability could also be fabricated by chemical vapor deposition with the assistance of hard template [125–127]. For example, Xiao et al. synthesized S-doped graphene nano-capsules (SGCs) by CVD with the presence of a MgO template [69]. As displayed in Figure 5a,b, the SGCs exhibited an integrated nano-capsule structure with a uniform size of 50 nm, and no obvious cracked products were observed, indicating that SGCs have a good structural rigidity and stability. SGCs show an extremely high capacity of 257.1 mA h g^{-1} at 1 A g^{-1} and an appealing rate capability of 147.7 mA h g^{-1} at 6 A g^{-1}, both of which are far better than those of the control samples (Figure 5c). With SGCs as both cathode and anode, the assembled symmetric LIC delivers an ultrahigh energy density of 249.9 Wh kg^{-1} at a high power density of 2.12 kW kg^{-1}, which still retains 149.8 Wh kg^{-1}, even at 14.99 kW kg^{-1} (Figure 5d). As can be seen from Figure 5e, the SGCs//SGCs also presents excellent long-term cycling stability with a retention of 95.4% after 10,000 cycles. It should be noted that the electrode still keeps an intactness and undamaged nano-capsules morphology, even after cycling for 10,000 times, verifying the superior structural stability of SGCs (inset of Figure 5e). The above excellent properties could be ascribed to the hollowed and stable structure with an abundant mesopore-dominant porosity, good electronical conductivity, enlarged interlayer spacing, and S-doping of SGCs.

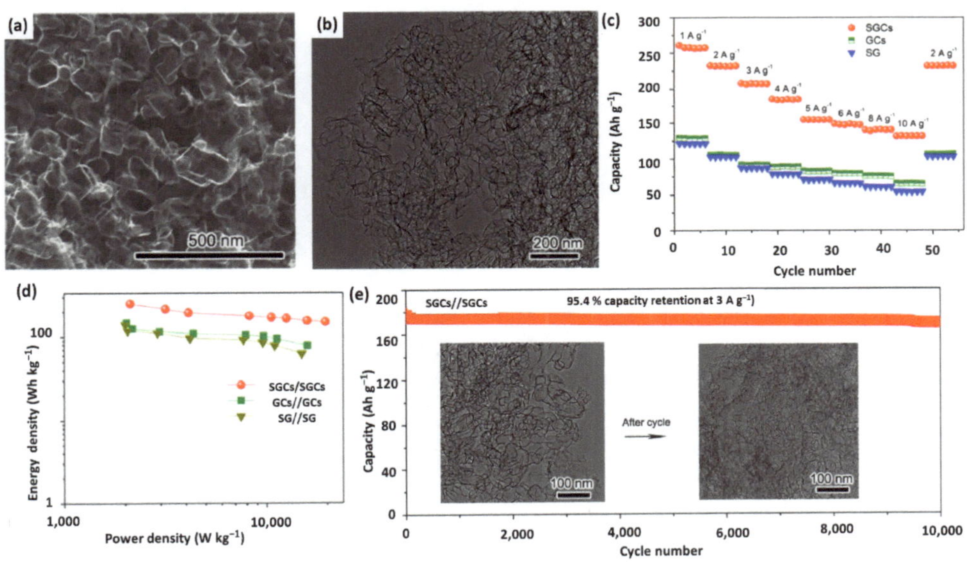

Figure 5. (**a**) SEM and (**b**) TEM images of SGCs; (**c**) Rate performance of SGCs cathode, compared with GCs and SG; (**d**) Ragone plots of SGCs//SGCs, compared with GCs//GCs and SG//SG; (**e**) Cyclic performance of SGCs//SGCs (the insets are the before and cycled TEM images of SGCs after 10,000 cycles). Reproduced with permission from Ref. [69]. Copyright 2022, Elsevier.

Despite the superb electrochemical performance for pure graphene-based cathodes, the high cost and tedious preparation procedure promotes researchers to explore graphene-based composites. This strategy overcomes the high cost of graphene, but keeps its inherently excellent properties. Moreover, forming composites with other materials could prevent the restacking and agglomerating of graphene. Chen's group presented a simple and green, but very efficient, approach to prepare 3D graphene-based porous materials through in-situ hydrothermal polymerization/carbonization of the mixture of cheap biomass or industry carbon sources with GO, followed by chemical activation (Figure 6a) [13,128]. The optimal product presented a sponge-like morphology and porous microstructure (Figure 6b,c). More importantly, it had an ultrahigh SSA (3523 m^2 g^{-1}), mesopore-dominated porosity, and excellent bulk conductivity (up to 303 S m^{-1}), thus contributing an outstanding electrode for SCs and LICs [24,33]. As shown in Figure 6d, the 3D porous graphene-based cathode demonstrated a high capacity and outstanding rate [129].

Benefiting from the dual graphene-based electrodes, the obtained all-graphene LIC with optimized cathode/anode ratio delivered a maximum energy density of 142.9 Wh kg^{-1} and a peak power energy of 12.1 kW kg^{-1} (Figure 6e). Anyway, heteroatom doping was applied to enhance the electrochemical performances of graphene-based materials. For instance, Wang et al. designed 3D porous activated nitrogen-doped graphene sheet (A-N-GS) by aniline polymerization with GO and then KOH activation [70]. A-N-GS demonstrated a sheet-like structure with a rough surface and 3D interconnected porous network. Coupled with the 3D highly conductive pathway and high-level nitrogen doping, A-N-GS showed a much improved capacity and rate performance, compared with the non-doped or non-activated samples. With a graphene-based anode, the all graphene LICs could reach an ultrahigh energy density of 187.9 Wh kg^{-1} at a power density of 2.25 kW kg^{-1}, which still remained at 111.4 Wh kg^{-1}, even at 11.25 kW kg^{-1}.

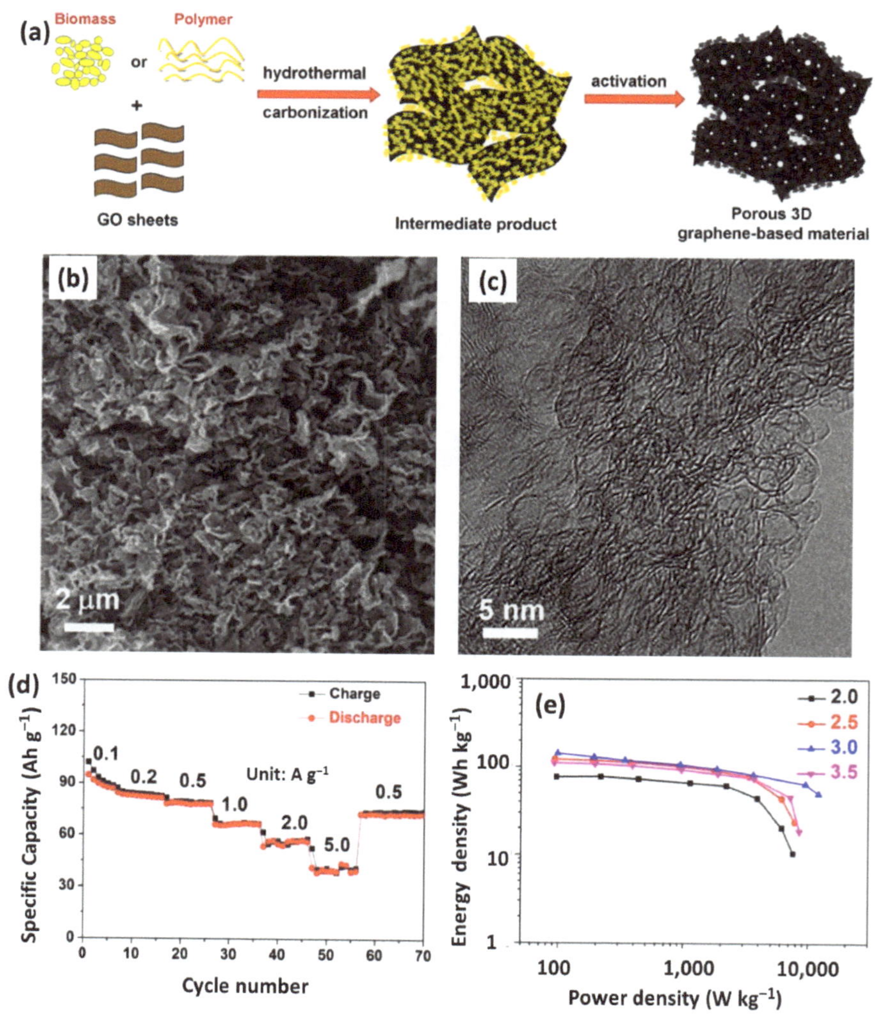

Figure 6. (**a**) A preparation schematic of porous 3D graphene-based materials; (**b**) SEM and (**c**) TEM images of porous 3D graphene-based products. Reproduced with permission from Ref. [13]. Copyright 2013, Springer Nature. (**d**) Rate performance of 3D graphene-based cathode; (**e**) Ragone plots of all-graphene-based LICs with different mass ratios. Reproduced with permission from Ref. [129]. Copyright 2021, IOP Publishing.

2.2.2. Carbon Nanotube-Based Cathode

As a typical one-dimensional nanosized carbon material, the carbon nanotube has exceptional conductivity, large aspect ratio, flexibility, and excellent mechanical strength [130]. All these merits make CNTs superior electrode material in various energy conversion and storage systems [131]. Particularly, CNTs are excellent substrates for fabricating flexible devices [132,133]. Nonetheless, similar graphene, CNTs usually demonstrate mediocre electrochemical performance, due to the severe agglomeration issue. As a consequence, forming composites with other materials is a feasible method, in which CNTs serve as either a spacer to increase the SSA or a conductive additive to enhance the electrical conductivity [134].

Several reports have verified that CNTs/graphene composites can solve the restacking issue of both the two components and form a 3D conductive network [135–137]. Additionally, the increased SSA and the well-distributed nanopores promote electrolyte infiltration and diffusion, endowing the obtained product-enhanced electrolyte accessibility. Hence, the high capacity and excellent rate could be expected for CNTs-based composites. Bai et al. proposed a CNT supported porous graphene (MRPG/CNT) by a facile microwave irradiation method [71]. From the SEM and TEM images in Figure 7a,b, CNT was uniformly distributed between graphene sheets to support the layer structure. The obtained MRPG/CNT cathode showed much improved capacity and rate capability, compared with MRPG (Figure 7c). In addition, the DC-LIC with symmetric electrodes achieved a maximum energy density of 232.6 Wh kg^{-1} and an extremely high power density of 45.2 kW kg^{-1}. The authors deemed that the excellent electrochemical properties could be explained by the 3D ion/electron channel model, as illustrated in Figure 7d. CNT intercalation into graphene sheets inhibited the restacking of graphene, expanded the layer space, and improved the electrode conductivity, forming a well in-plane and cross-plane channels for both the ion and electron migration. This model is supported by the Nyquist plot and Bode plot (Figure 7e,f). MRPG/CNT//MRPG/CNT LIC has a small equivalent series resistance of 18.0 Ω, which is beneficial for high rate of output and long-term cycling. Furthermore, the characteristic relaxation time constant τ was calculated to be 1.21 s, indicating the fast reaction kinetics and high-power capability of the full cell.

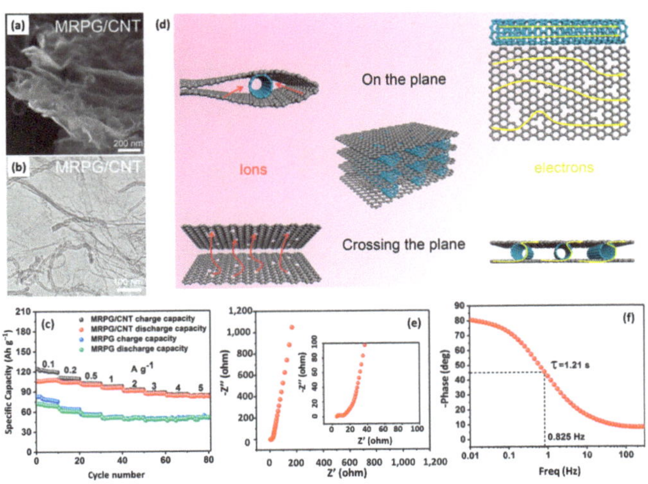

Figure 7. (**a**) SEM and (**b**) TEM images of MRPG/CNT; (**c**) Rate performanceof MRPG and MRPG/CNT; (**d**) Schematic illustration of the ion channels and electron channels in MRPG/CNT; (**e**) Nyquist impedance and (**f**) Bode plots of MRPG/CNT//MRPG/CNT LIC. Reproduced with permission from Ref. [71]. Copyright 2021, American Chemical Society.

Overall, graphene/CNT-based cathodes have advantages in achieving high capacity and outstanding rate performance. However, nanosized carbons usually have relatively lower density (~0.3 g cm^{-3}) than commercial AC (~0.5 g cm^{-3}) because of their low tapping density, resulting in actually the same or even lower volumetric energy density than commercial ACs [138]. To overcome this dilemma, practical graphene technologies, such as the capillary drying process and rapid drying process, provide a promising solution to fabricate high tap density graphene-based composites [139–141].

3. Progresses of Carbon Anode Materials for DC-LICs

As the battery-type electrode for DC-LICs, the anode plays a crucially important role in increasing energy density. On the other hand, both the power output and cycling stability are highly influenced by the anode, which are still the bottlenecks for advanced DC-LICs. Hence, developing anode materials with high capacities, high rates, and stable structures is of great significance to realize practical DC-LICs. As summarized in Table 2, carbon materials ranging from traditional carbon (graphite, hard carbon, soft carbon, etc.) to nanosized carbons (graphene, carbon nanotubes, graphdiyne, and their composites) have been applied as anodes.

Table 2. Typical carbon anodes and the electrochemical performances of DC-LICs.

Carbon Anodes		Performances of DC-LICs					
Materials	Specific Capacity (mAh g^{-1})	Rate Capability (mAh g^{-1})	Voltage (V)	Maximum Energy Density (Wh kg^{-1})	Maximum Power Density (kW kg^{-1})	Cycling Stability	Ref.
HC	423	100 @ 50 C	1.5−4.2	110.0	25.0	81% after 10,000 cycles	[97]
HC-rGO	450	162 @ 10 C	1.5−4.2	130.0	5.5	88% after 10,000 cycles	[92]
PHC-4	1040.2	231.7 @ 6.4 A g^{-1}	2.0−4.0	104.0	11.9	84.7% after 5000 cycles	[142]
KHPC-600	1064	280 @ 10 A g^{-1}	0.01−4.0	169.0	97.0	77.7% after 5000 cycles	[51]
HNBC	1392	620 @ 1 A g^{-1}	0−4.5	186.31	11.25	81.9% after 10,000 cycles	[143]
G/SC	360	200 @ 4 A g^{-1}	2.0−4.0	151.0	18.9	93.8% after 10,000 cycles	[78]
SLC	829	148 @ 10 A g^{-1}	0−4.0	127.0	33.57	99% after 100,000 cycles	[91]
NOPCNS	810	249 @ 50 A g^{-1}	2.0−4.2	184.0	78.1	70% after 10,000 cycles	[144]
GOCAF	398	195 @ 10 C	1.5−4.2	100.0	9.0	80% after 15,000 cycles	[93]
NOPCNS	810	249 @ 50 A g^{-1}	0−4.0	184.0	78.1	70% after 10,000 cycles	[144]
HNBC	1392	300 @ 5 A g^{-1}	0−4.5	186.31	11.25	81.9% after 10,000 cycles	[143]
NPC	1740	369 @ 10 A g^{-1}	0−4.5	203	90.0	80% after 20,000 cycles	[145]
HNCNBs	850	321 @ 20 A g^{-1}	1.0−4.0	148.5	25.0	90% after 8000 cycles	[59]
BDC	1018	564 @ 5 A g^{-1}	2.0−4.5	207.0	17.06	88% after 15,000 cycles	[146]
NDPC-0.5	1000	295 @ 5 A g^{-1}	0−4.0	116.9	10.0	81% after 8000 cycles	[147]
FRGO	660	220 @ 3.72 A g^{-1}	0−4.2	148.3	7.8	80% after 3000 cycles	[18]
PRGO	982	166 @ 20 A g^{-1}	0.01−4.0	262.0	9.0	91% after 4000 cycles	[67]
PDA-GN	1150	371 @ 5 A g^{-1}	0−4.2	135.6	21.0	65% after 3000 cycles	[124]
SHSG	854	333 @ 10 C	2.0−4.5	146.0	52.0	~91% after 40,000 cycles	[52]
NPG	859	758 @ 2 A g^{-1}	1.0−4.0	195.0	14.98	~100% after 5000 cycles	[148]
G-COOH	450	145 @ 10 A g^{-1}	1.0–4.2	120.8	53.55	98.9% after 50,000 cycles	[127]
rGO800-P	461	185 @ 10 C	1.5–4.5	91.0	26.0	76% after 10,000 cycles	[117]
F-GDY	1825.9	979.2 @ 5 A g^{-1}	2.0–4.0	200.2	13.117	80% after 6000 cycles	[149]
N-GDY	1096.1	440 @ 4 A g^{-1}	2.0−4.0	174.0	11.25	89.7% after 2000 cycles	[150]
GC1100	354	222 @ 2 A g^{-1}	2.0−4.0	104.0	6.628	96.5 % after 3000 cycles	[151]
GNS-13	356	66.7 @ 5 A g^{-1}	2.0−4.0	112.0	19.6	96.5% after 5000 cycles	[152]
GMC	119	378 @ 1 A g^{-1}	0−4.5	190.63	11.25	81.8% after 10,000 cycles	[153]

3.1. *Traditional Carbon Anode*

3.1.1. Graphite

Graphite is the mainstream anode for the current LIBs industry, owing to its low cost, excellent electrical conductivity, and relatively high capacity. In addition, graphite with a long-range ordered structure has a nearly flat charge/discharge profile at low potential vs. Li/Li^+, which thus enables DC-LICs with a high working voltage by combining the graphite with a capacitor-type cathode [62,75,77,154]. Considering the relatively high taping density, graphite has the advantage of achieving high volumetric capacity. Recently, recycling graphite from spent LIBs has become an important research topic, owing to sustainability and low cost [155–157]. For instance, Aravindan et al. proposed an efficient route to reutilize the recovered graphite (RG) as the anode to fabricate DC-LIC [158]. After a simple treatment, the RG demonstrates a comparable capacity and rate with pristine graphite. The obtained full cell delivers a maximum energy density of 185.5 Wh kg^{-1}, with excellent low/high-temperature performance.

Although graphite demonstrates appealing properties as an anode for DC-LICs, pristine graphite is not suitable for high-rate and long-cycling DC-LICs, due to the sluggish kinetics of the intercalation reaction and ion diffusion [159]. To achieve high-rate graphite, reducing the particle size and defect engineering are two typical strategies. For example, Pandolfo et al. evaluated the rate capability of various commercial graphite materials and confirmed that reducing the graphite particle size showed enhanced rate capability, but increased the irreversible capacity loss [160]. Microcrystalline graphite (MG) consists of many closely packed graphitic nanocrystals with small inter-particle voids [161]. This structure enables MG with fast ion diffusion of nanostructured materials and high mass density of bulk materials simultaneously. MG is regarded to break the limitations of the slow Faradaic reaction and Li^+ diffusion of bulk material, which well-matches the cathode by providing an ultrafast capacitor-like response. Hence, Kang et al. adopted high-temperature-purified MG as anode in LICs [162]. As expected, the MG-based LICs showed an extremely high power density of 352 kW kg^{-1}. Chen's work demonstrated that intrinsic lattice defect engineering could effectively enhance the fast-charging capability of the graphite [163]. The graphite thermally treated in CO_2 had a superior rate capability of 209 mAh g^{-1} at 10 C (only 15 mAh g^{-1} for the pristine graphite), which is attributed to the defect-induced enhancement of the kinetics and pseudocapacitive storage.

3.1.2. Hard/Soft Carbon

Besides graphite, hard carbon (HC) and soft carbon (SC) are two important anodes in high-power LIBs, which also show great potential in DC-LICs [60,164,165]. HC and SC are commonly prepared by the thermal pyrolysis of carbon-rich precursors [166]. Hard carbon, known as nongraphitizable carbon material, consists of a large fraction of curved and randomly distributed graphitic sheets and cannot be reshaped into graphite, even at a temperature of 3000 °C, because of the interlayer cross-linking of the precursors [167,168]. Notably, the disordered structure of HC allows for Li^+ insertion on both sides of the graphene sheets within the microscopic graphite-like regions, leading to a potentially higher capacity [97]. Soft carbons, on the other hand, contain a semi-graphitic structure and have relatively fewer defects and higher crystallinity than HC [169]. Particularly, the typical structure of SC consists of a high strain region and a low strain region. The former is a disordered region with unorganized carbon, while the latter is a graphitic region with high crystallinity carbon. This novel structure enables SC with an excellent ion diffusion rate and high conductivity. Normally, HC and SC exhibit better rate performance and long-term cycling stability, due to their enlarged interlayer distance and increased electrode/electrolyte interface. Therefore, they have been recognized as promising anodes for high-power and long-life Li/Na/K-ion batteries and LICs [48,49,167,168,170].

Particle size has a profound effect on the rate capability of HC. For example, HC was prepared through simple polymerization of furfuryl alcohol at room temperature, followed by a calcination step [97]. To realize the desired high power, ball milling was performed on the HC to achieve a sub-micron particle size. In addition, the morphology of HCs can largely affect the electrochemical performances of LICs. By comparing the physical and electrochemical behaviors of spherical HC and irregular HC, Shi et al. found that irregular HC presented a distinct Li^+ intercalation plateau at a low potential [76]. This helps the sufficient utilization of AC cathode and the cycling stability of LICs. Balducci et al. firstly proposed to apply petroleum coke-based SC in LICs [171]. The prepared sample had a moderate capacity of 250 mAh g^{-1} at 0.1 C, but a better rate capability than that of many graphite electrodes at high current densities. Later, the same group examined the cycling stability of petroleum coke-based LICs [172]. When operating at a working voltage of 4.0 V, the device could maintain an extremely stable cycling life of 50,000, with energy and power of 48 Wh kg^{-1} and 9 kW kg^{-1}, respectively.

Compared with the graphite, however, HC and SC have relatively lower electrical conductivity, owing to their disordered structure and abundant defects, which is detrimental to rate performance. Forming composites with conductive polymers or low-dimensional nanocarbons is an effective strategy to overcome this drawback. For example, Park et al. demonstrated that the incorporation of electrically conductive poly(3,4-ethylenedioxythiophene)-poly-styrene sulfonate (PEDOT-PSS) into soft carbon had a positive effect [173]. Thanks to the lowered electron and charge transfer resistance, the SC anode exhibited an enhanced charge capacity retention of 64% at 5 C with only 1.0 wt% PEDOT-PSS added (Figure 8a). In addition, better conductive network leads to an enhanced utilization of the electrode material, which can be observed from the higher capacity retention after prolonged cycling (Figure 8b). Numerous reports have verified that graphene could significantly improve the electrochemical performance of the composites by increasing conductivity and reducing contact resistance [93,122]. Ajuria et al. prepared a composite of recycled olive pit-derived hard carbon embedded in a reduced GO matrix (HC-rGO) [92]. The graphene sheets well-wrapped the HC particles and formed a 3D interconnected carbon network. This structure facilitated the electrolyte diffusion and enhanced the conductivity. The electrochemical impedance spectroscopy (EIS) analysis revealed that charge-transfer resistance at the electrode–electrolyte interphase and the charge transport resistance within the electrode were considerably decreased when adding reduced GO. Consequently, HC-rGO anodes displayed a much-improved rate capability, compared with bare HC, and LICs-based on them had higher energy density at high power density. Ma et al. designed graphene/soft carbon (G/SC) composites through a large-scale SHS strategy, which combined the advantages of high conductivity of graphene and the low cost of SC (Figure 8c) [78]. As shown in Figure 8d, SC was evenly coated by graphene, leading to enhanced reaction interface and electrical conductivity. The G/SC showed a reversible capacity of 360 mAh g^{-1} at 0.1 A g^{-1}, which remained at 200 mAh g^{-1} at 4 A g^{-1} (38% higher than that of pure SC). The EIS result also verified that the reaction resistance was decreased by the additional interface area provided by the SHS-prepared graphene network (Figure 8e). Based on the above discussions, introducing highly conductive materials is a facile method to further enhance the performance of traditional carbon anodes.

In summary, conventional carbon anodes have received intensive research interest in the early study of DC-LICs because of their low cost, mature manufacturing process, and successful application in commercial energy storage devices. However, all of them have limitations. The sluggish reaction kinetics of graphite, unsatisfactory capacity of SC, voltage hysteresis, and large irreversible capacity of HC are still the main obstacles before large-scale commercialization. Possible research directions might be particle size reduction, interlay distance enlargement, and/or compositing with nanostructured carbons.

3.2. Nanosized Carbon-Based Anode

Nanosized carbons, such as graphene, CNTs, and graphdiyne, have been investigated as anodes in LIBs, owing to their unique structural, mechanical, and electrical properties [174,175]. The long-range π–π conjugation structure of nanocarbons ensures highly intrinsic conductivity, while the adjustable layer distance, tailorable porosity, and rich surface chemistry provide easy electrolyte penetration and fast ion diffusion rates. These superior properties enable them to have a high capacity and excellent rate capability. Moreover, nanocarbons can serve as effective buffering components or backbones in anode materials, benefiting the structural integrity and long-term cycling stability. Considering the sluggish kinetics of graphite, high irreversible loss of HC, and low specific capacity of SC, novel nanocarbon-based anodes provide an appealing alternative for DC-LICs.

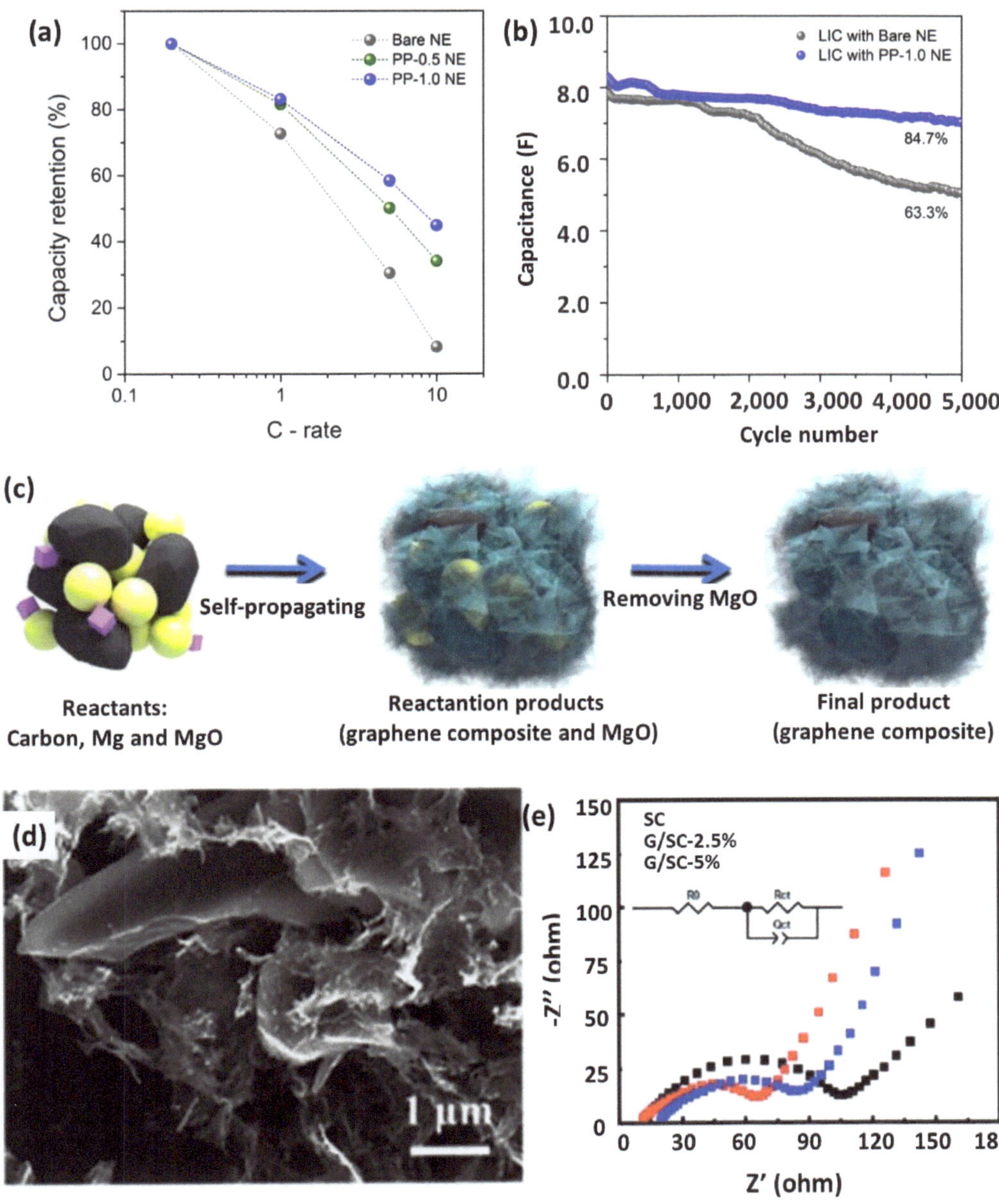

Figure 8. (**a**) Capacity retentions at various current densities during charge; (**b**) Capacitance retention of LICs employing bare NE and PP-1.0 NE. Reproduced with permission from Ref. [173]. Copyright 2015, Elsevier. (**c**) Schematic diagram of preparing G/SC; (**d**) SEM image of the G/SC composite; (**e**) EIS curves of pure SC and G/SC. Reproduced with permission from Ref. [78]. Copyright 2021, Royal Society of Chemistry.

3.2.1. Graphene-Based Anode

Benefiting from the 2D structure and adjustable interlay distance, both sides of graphene sheets can host lithium ions, while the abundant in-plane defects and exposed edges of graphene further offer more active sites [176,177]. Coupled with high conductivity, graphene has been regarded as an advanced anode with high specific capacity and fast charge/discharge rates for LICs. As a pioneering work, Zhou et al. adopted the prelithiated graphene nanosheets as anode materials and compared them with conventional graphite [77]. The graphene anode has a wide pore size distribution, with the co-existence of micropores and mesopores, which are responsible for charge storage and ion transport from electrolyte to electrode interface, respectively. Benefiting from the reduced charge transfer impedance and enhanced Li^+ diffusion rate, the full cells based on graphene anode have higher energy and power densities than those of graphite-based devices.

Binder-free, self-supporting graphene films are regarded as promising anodes for DC-LICs, due to the adjustable interlayer distance and their potential to prepare flexible devices. By multiple exposure of a free-standing GO film in a focused camera photoflash, Huang et al. prepared a flash-reduced GO (FRGO) [18]. As shown in Figure 9a of the side-view optical image, the thickness of the flash-reduced part expanded distinctly, compared with the unreduced GO film. This could be further verified by the cross-sectional SEM images (Figure 9b,c). The GO sheets of the pristine film were tightly stacked, while the FRGO showed a quite swelled structure with large voids and pathways formed between the highly warped graphene. The increased pressure between the GO sheets during thermal heating-induced de-oxygenation accounts for the loose and open-pore structure, facilitating a rapid Li^+ diffusion and, thus, enabling FRGO with enhanced ion intercalation kinetics at high charge/discharge rates. FRGO has a reversible capacity of more than 660 mAh g^{-1} at 1 C and excellent rate of 220 mAh g^{-1} at 10 C (Figure 9d). Zheng et al. synthesized an anode of edge-carbonylated graphene nanosheets (G-COOH) with enlarged interlayer distance via a ball milling method [127]. As shown in Figure 9e, the charge storage abides by the following mechanisms: (i) the rich porosity enabling rapid lithium-ion diffusion and, thus, fast reaction kinetics; (ii) the high degree edge-carboxylation structure storing lithium ions via reaction between C=O and Li^+; (iii) the carboxyl modification enlarging the graphite layer space for fast lithium-ion insertion and desertion. All the features endow G-COOH with fast pseudocapacitive and lithium-inserted capacity, as well as long-term durability, which are far better than HC (Figure 9f,g). Integrated with a porous graphene cathode, the assembled all-graphene LIC delivers an ultrahigh power density of 53.55 kW kg^{-1} and an unprecedented cycling stability of 98.9% retention after 50,000 cycles (Figure 9h,i). Based on the above discussions, the merits of tunable interlayer space, rich surface/edge chemistry, and high electrical conductivity endow graphene-based anodes with a high capacity and excellent rate capability. On the other hand, the high cost and tedious preparation process promote researchers to develop graphene-based composites.

As discussed in Section 2.2.1, pure graphene has the disadvantage of unavoidable restacking and a high cost. Under these circumstances, CNTs are usually used as a spacer to prohibit the agglomeration of graphene [134,178]. Hydrothermal treatments or chemical reduction reactions are adopted to form graphene/CNTs composites [135,136,179]. However, the composites prepared by these two methods suffer from the insufficient reduction of GO and destruction of the CNT structure, leading to increased ohmic resistance. Consequently, Tour et al. seamlessly designed well-connected graphene CNT carpets (GCNT) through CNT growth from graphene substrate (Figure 10a) [137]. Thanks to the continuous electrical path from the active material to the current collector, the obtained binder-free GCNT anode has a high capacity and excellent rate capability (Figure 10b). By applying GCNT as both the anode and cathode, the obtained binder-free DC-LICs exhibited high energy density (~120 Wh kg^{-1}) and outstanding power density (~20.5 kW kg^{-1}). Another group reported a graphene-based electrode material via microwave irradiation of GO/CNT mixture (Figure 10c) [71]. The CNT intercalation into graphene prevented the restacking of graphene and enhanced the conductivity of the composite. Benefiting from a highly porous structure and better conductivity, MRPG/CNT showed a higher capacity and better rate performance than pure MRPG (Figure 10d). Specifically, MRPG/CNT showed high capacities of 1249 and 370 mAh g^{-1} at 0.1 and 2 A g^{-1}, respectively.

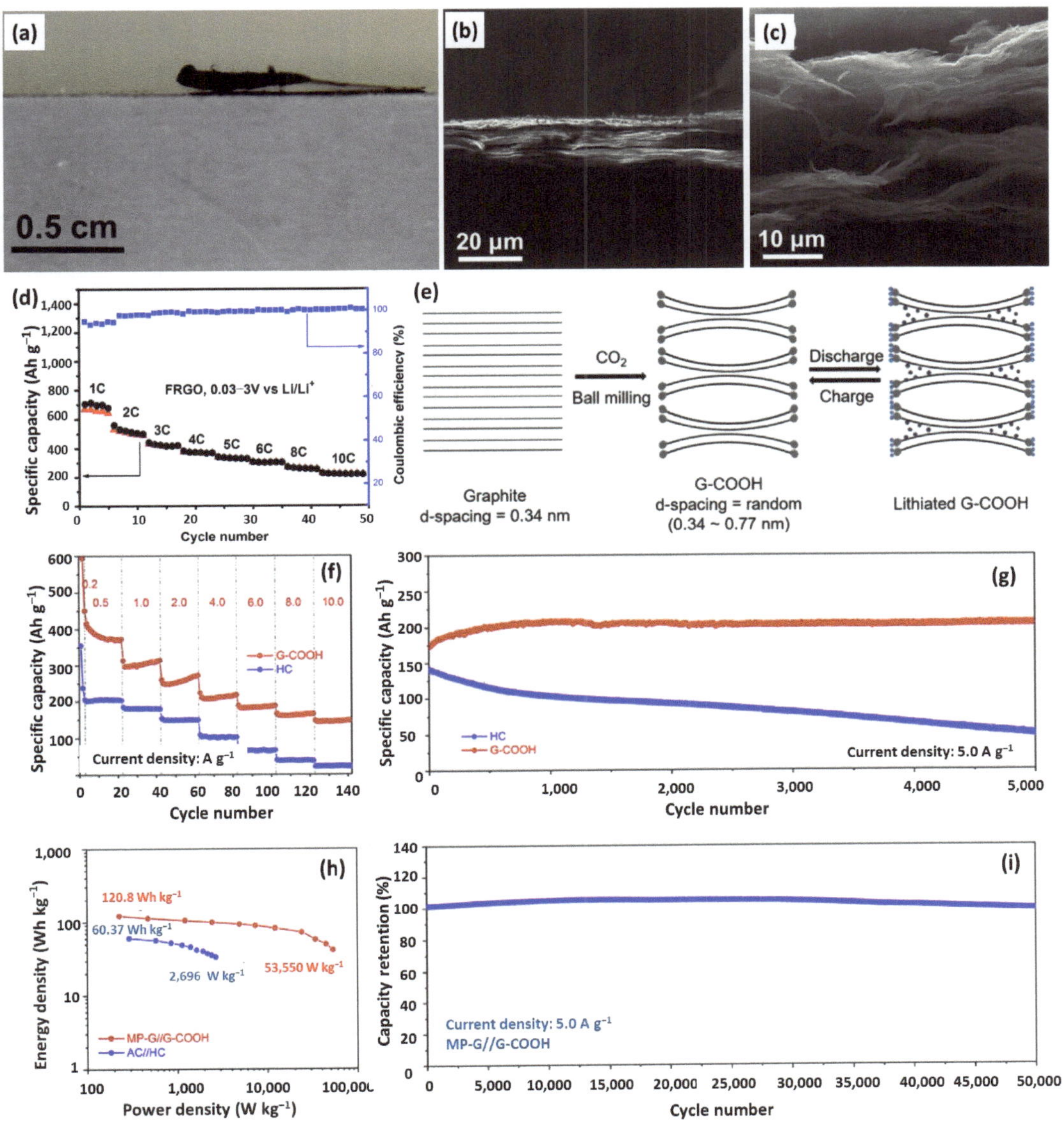

Figure 9. (**a**) Optical image of side view of partially reduced GO film disc; Side view SEM images of GO film (**b**) and FRGO film (**c**); (**d**) Rate performance of FRGO. Reproduced with permission from Ref. [18]. Copyright 2015, Elsevier. (**e**) Lithium-ion storage mechanism scheme of G-COOH; (**f**) Rate and (**g**) cycling performances of G-COOH and HC; (**h**) Ragone Plot and (**i**) long cycling performance of MP-G//G-COOH. Reproduced with permission from Ref. [127]. Copyright 2019, Elsevier.

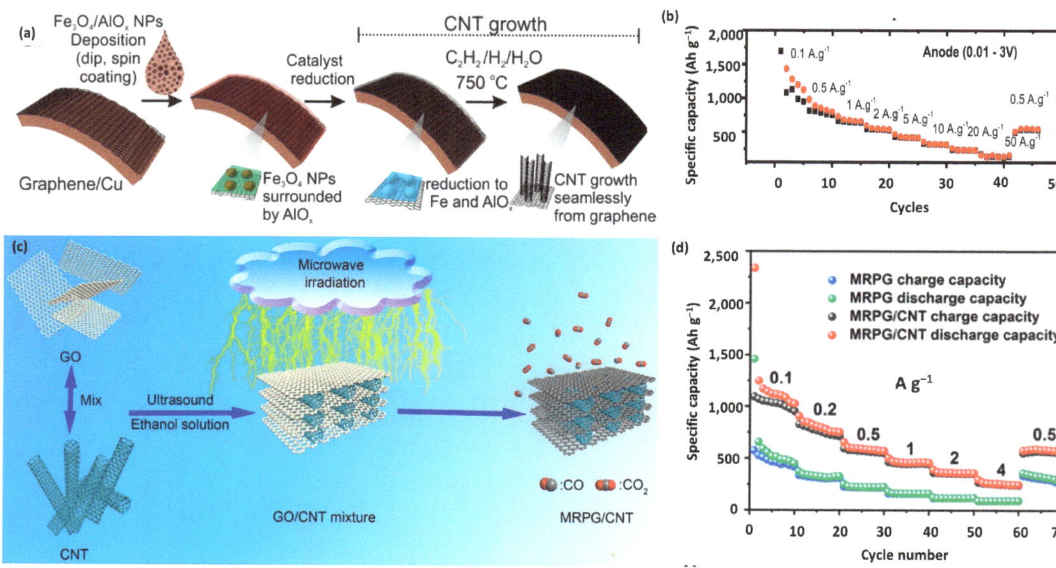

Figure 10. (**a**) Scheme for the growth of GCNT hybrid materials; (**b**) Rate testing of GCNT. Reproduced with permission from Ref. [137]. Copyright 2017, American Chemical Society. (**c**) Schematic illustration of the synthesis process of MRPG/CNT; (**d**) Rate performance of MRPG and MRPG/CNT. Reproduced with permission from Ref. [71]. Copyright 2019, American Chemical Society.

3.2.2. Graphdiyne-Based Anode

As a new 2D nanocarbon allotrope, consisting of a diethynyl group and a benzene ring, graphdiyne (GDY) was synthesized for the first time in 2010 by an in-situ chemical reaction of hexaethynylbenzene on a copper surface [180]. Stemming from its unique molecular structure with a high degree of π conjugation, GDY and its composites possess the characteristics of both a 2D material and a 3D porous material, such as high SSA, uniform porous channels, fast electron transport, wide electrochemical potential windows, and large double-layer capacitance [181,182]. GDY can store energy via two mechanisms, i.e., the Faradaic process in the bulk material and non-Faradaic process via reversible ion adsorption/desorption on the surface of pores [183]. All these features make GDY-based materials exhibit promising electrochemical properties in LICs.

Huang's group firstly proposed using GDY as an anode in LICs by pairing with AC [183]. The 2D atomic layer structure coupled with abundant meso-/micro-pores enables GDY Li^+ diffusion (both in-plane and out-of-plane), high electrolyte ion transport rate, and an abundance of electroactive sites (Figure 11a). As an anode, GDY delivers a moderate specific capacity of 572.5 (mAh g^{-1}), and the LICs have acceptable energy density (112.2 Wh kg^{-1}). To further improve the capacity, the authors prepared N-doped graphdiyne (N-GDY) by nitriding under NH_3 (Figure 11b) and fluorine-enriched graphdiyne (F-GDY) by solvothermal reaction (Figure 11c) [149,150]. For example, F-GDY possessed an ultrahigh capacity of 1825.9 mAh g^{-1} at 0.1 A g^{-1} and outstanding rate capability of 979.2 mAh g^{-1} at 2 A g^{-1} (Figure 11d). F-GDY-based LIC released a maximum energy and power density of 200.2 Wh kg^{-1} and 13.12 kW kg^{-1}, respectively. The astonishing electrochemical performances could be ascribed to the 3D porous structure, high conductivity, enlarged interlayer spacing, and enhanced wettability due to the incorporation of heteroatom dopants.

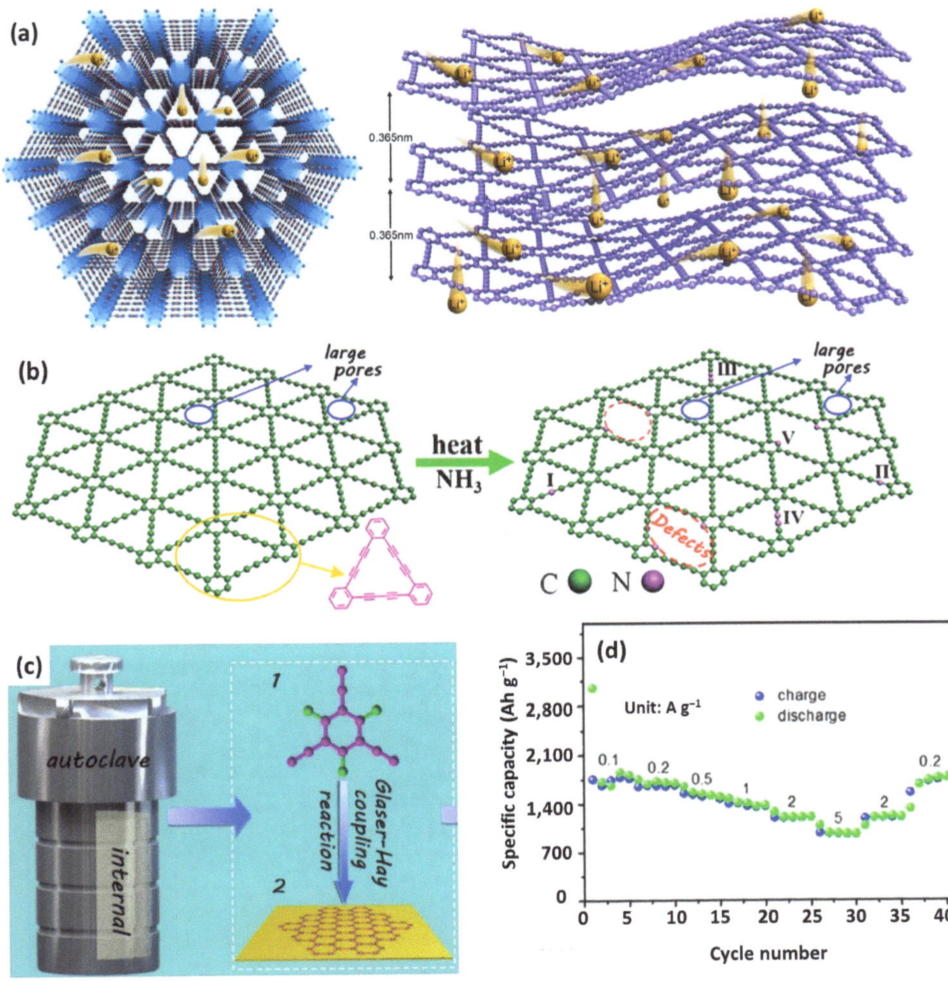

Figure 11. (**a**) Scheme of lithium diffusion in GDY layers. Reproduced with permission from Ref. [183]. Copyright 2016, Elsevier. (**b**) Schematic illustration of preparation of N-GDY. Reproduced with permission from Ref. [149]. Copyright 2019, Wiley-VCH. (**c**) The schematic of synthesis of F-GDY; (**d**) Rate capability of F-GDY. Reproduced with permission from Ref. [150]. Copyright 2018, Wiley-VCH.

In general, nanosized carbons with novel structures have the merits of high capacity and well-matched rates with cathodes, making them very promising anodes for high-energy and high-power DC-LICs. Nevertheless, several problems should be carefully investigated and solved before their commercial utilization, including the large irreversible capacity and low Coulomb efficiency during the initial charge/discharge, low volumetric capacity/energy density, sophisticated synthesis process, and high cost. These nettlesome issues are even worse for graphene-based anode. Currently, compositing nanosized carbon with conventional carbons is emerging as a promising solution by combining their advantages and overcoming the shortcomings of both. Anyway, elaborately regulating the microstructure (porosity and SSA) is another strategy for balancing the electrochemical properties, aiming to suppress the side reactions, but keep their merits.

4. Summary and Outlook

As discussed in the previous sections, carbon materials demonstrate superior properties as electrodes in DC-LICs, due to their inimitable advantages of high electrical conductivity, tunable microstructure, and physical/chemical/electrochemical stability. Overall, traditional carbons have advantages in material preparation, cost, and abundant resources, while nanosized carbons, such as graphene-based materials, show superiority in capacity and rate. Benefiting from the novel cell configuration and hybrid energy storage mechanism, DC-LICs demonstrate superior electrochemical performances over LIBs and SCs and could bridge the performance gap between them. Hence, DC-LICs show great potential in the application scenarios where high energy density and high power output are both required.

Although DC-LICs have achieved great progress in material design and preparation strategies over the past two decades, it is still a difficult and sophisticated project to achieve high energy/power densities and long lifespans simultaneously. Especially, the mismatches in capacity, kinetics, and cycling life between the cathode and anode still remain. Anyway, there is a long way to go for scalable and efficient pre-lithiation technology. Additionally, studies in energy/thermal management systems are rare, and they are indispensable components for practical applications. Last, but not least, the practical application of DC-LICs needs to be investigated in depth, and more application scenario should be explored. Therefore, the following challenges should be thoroughly investigated and solved prior to the commercialization of DC-LICs.

(1) Developing a high-capacity carbon cathode. LICs are supposed to be able to deliver high energy and power densities. However, high energy density is always achieved at the cost of power density and vice versa, which is owed to the mismatched capacity and kinetics between cathodes and anodes. Hence, developing a high-capacity cathode is the top priority. To optimize the capacity of the carbon cathode, porous carbons with the desired microstructure, such as hierarchical porous carbon with large SSA and rational pore size distribution, should be given more attention. Furthermore, designing novel carbon and carbon-based composites with controlled morphologies and structures for high-capacity cathodes is an urgent need. For example, forming composites with traditional carbons and nanosized carbons is an effective method for further enhancing the SSA and conductivity of porous carbons

(2) Designing high-rate and long-lifespan carbon anodes. Poor rates and limited cycling lives are the two main drawbacks for the carbon anode, which are ascribed to the well-known sluggish Faradaic reaction and structure degradation during the repeated charge/discharge processes, respectively. Numerous reports have verified that elaborately designed nanostructure help to enhance ion diffusion and alleviate volume changes, while heteroatom doping is beneficial for enlarging the interlayer distance and improving wettability, which enables carbon anodes with improved rates and prolonged cycling lives. However, a large irreversible capacity and low Coulombic efficiency, owing to the high SSA and rich defects for nanostructured materials, should not be ignored. Developing carbon-based anode materials with optimal surface chemical properties, inner microstructures, and structural stabilities should be received more attention.

(3) Exploring feasible and efficient pre-lithiation technologies. With the large amount electrolyte consumption forming SEI at the anode surface, pre-lithiation is of great significance for maintaining high performance by compensating for the loss of active charge carriers and extending the working voltage. However, most of the pre-lithiation technologies are unsafe, time-consuming, or not cost-efficient. Therefore, developing safe and commercial-scale pre-lithiation technology is of critical importance for practical applications.

(4) Investigating advanced electrolytes. The stable electrochemical window of an electrolyte determines the working voltage and, thus, influences the energy and power densities of the obtained device. Moreover, the electrolyte stability also affects cycling performance. Generally, the continuous reduction of electrolytes at the anode can be effectively prohibited by forming SEI. However, oxidation by losing electrons at high

potential can hardly be avoided, and the generated gas and byproducts lead to increased resistance. With this, it is imperative to investigate the anti-oxidized electrolyte additives. Anyway, other types of electrolytes, such as "water-in salts" and gel electrolytes, should be developed and applied in LICs, due to their high safety.

(5) Developing suitable energy/thermal management systems for LICs. Energy storage devices need the aid of energy/thermal management systems to provide stable and high-quality power. The current energy/thermal management systems for LIBs and SCs are not suitable for LICs because of their different electrochemical properties. Unfortunately, very few reports about LICs have covered this topic. This is a very important research direction for practical applications to which both academia and industry should pay more attention.

(6) Other dual-carbon-based metal-ion capacitors. Lithium faces big obstacles, due to its limited reserves and uneven distribution, promoting researchers to investigate other systems. Sodium/potassium-based hybrid capacitors are deemed to be the most competitive candidates because they share similar electrochemical performances and working mechanism with LICs, but have a huge advantage in natural abundance. Anyway, novel multivalent metal-ion capacitors (such as Zn, Mg, Ca, and Al) have become the research hotspots, in that they have the potential to provide twice or triple the amount of electrons per unit of active materials and are less sensitive to air and water, not to mention their abundance. These systems offer a promising alternative to LICs and, hence, more efforts should be devoted to investigating electrode materials, electrolytes, device configurations, and energy storage mechanisms.

In summary, this review systematically overviews the recent developments of carbon cathodes and anodes, in which their physical and chemical properties, electrochemical performances, and advantages and disadvantages are discussed. DC-LICs have achieved remarkable progress, thanks to the extensive investigation of carbon electrode materials. Although many challenges still remain, we believe the drawbacks of current DC-LICs will be overcome with the cooperation of academic and industrial communities, and we hope this article could provide some guidelines for future researchers.

Author Contributions: Conceptualization, D.S. and Y.L.; validation, C.L. and Y.D.; data curation, Z.Y. and H.L.; writing—original draft preparation, Y.D. and D.S.; resources, Y.Y. and C.L.; project administration, D.S. and Y.L.; funding acquisition, Y.Y. and Y.L. All authors have read and agreed to the published version of the manuscript.

Funding: This research was funded by National Natural Science Foundation of China [NSFC, 21902071], Henan Provincial Science and Technology Research Project [192102310202], and the Natural Science Foundation of Hebei Province of China [E2020408004, B2022408005].

Conflicts of Interest: The authors declare no conflict of interest.

References

1. Dunn, B.; Kamath, H.; Tarascon, J.M. Electrical Energy Storage for the Grid: A Battery of Choices. *Science* **2011**, *334*, 928–935. [CrossRef]
2. Yadlapalli, R.T.; Alla, R.R.; Kandipati, R.; Kotapati, A. Super capacitors for energy storage: Progress, applications and challenges. *J. Energy Storage* **2022**, *49*, 104194. [CrossRef]
3. Liu, G.; Li, M.; Wu, N.; Cui, L.; Huang, X.; Liu, X.; Zhao, Y.; Chen, H.; Yuan, W.; Bai, Y. Single-Crystalline Particles: An Effective Way to Ameliorate the Intragranular Cracking, Thermal Stability, and Capacity Fading of the $LiNi_{0.6}Co_{0.2}Mn_{0.2}O_2$ Electrodes. *J. Electrochem. Soc.* **2018**, *165*, A3040–A3047. [CrossRef]
4. Li, H.; Guo, C.; Zhang, T.; Xue, P.; Zhao, R.; Zhou, W.; Li, W.; Elzatahry, A.; Zhao, D.; Chao, D. Hierarchical Confinement Effect with Zincophilic and Spatial Traps Stabilized Zn-Based Aqueous Battery. *Nano Lett.* **2022**, *22*, 4223–4231. [CrossRef]
5. Guo, D.; Wang, F.; Yang, M.; Hu, G.; Liu, G.; Wu, N.; Qin, A.; Liu, X. Constructing abundant oxygen vacancies in NaVPO4F@C for boosting sodium storage kinetics. *Electrochim. Acta* **2022**, *424*, 140695. [CrossRef]
6. Sui, D.; Chang, M.; Wang, H.; Qian, H.; Yang, Y.; Li, S.; Zhang, Y.; Song, Y. A Brief Review of Catalytic Cathode Materials for Na-CO_2 Batteries. *Catalysts* **2021**, *11*, 603. [CrossRef]
7. Gopalakrishnan, R.; Goutam, S.; Miguel Oliveira, L.; Timmermans, J.-M.; Omar, N.; Messagie, M.; Van den Bossche, P.; van Mierlo, J. A Comprehensive Study on Rechargeable Energy Storage Technologies. *J. Electrochem. Energy Convers. Storage* **2017**, *13*, 040801. [CrossRef]

8. Poizot, P.; Dolhem, F. Clean energy new deal for a sustainable world: From non-CO_2 generating energy sources to greener electrochemical storage devices. *Energy Environ. Sci.* **2011**, *4*, 2003–2019. [CrossRef]
9. Yan, W.B.; Sui, D.; Yang, Y.L.; Chang, M.J. Bi- or multi-functional charge transporting materials for perovskite solar cells. *Dye. Pigment.* **2022**, *200*, 110102. [CrossRef]
10. Goodenough, J.B.; Park, K.-S. The Li-Ion Rechargeable Battery: A Perspective. *J. Am. Chem. Soc.* **2013**, *135*, 1167–1176. [CrossRef]
11. Liu, G.; Xiao, F.; Zhang, T.; Gu, Y.; Li, J.; Guo, D.; Xu, M.; Wu, N.; Cao, A.; Liu, X. In-situ growth of MoO_2@N doped carbon on Mo_2C-MXene for superior lithium storage. *Appl. Surf. Sci.* **2022**, *597*, 153688. [CrossRef]
12. Wu, N.; Wu, H.; Kim, J.-K.; Liu, X.; Zhang, Y. Restoration of Degraded Nickel-Rich Cathode Materials for Long-Life Lithium-Ion Batteries. *ChemElectroChem* **2018**, *5*, 78–83. [CrossRef]
13. Zhang, L.; Zhang, F.; Yang, X.; Long, G.; Wu, Y.; Zhang, T.; Leng, K.; Huang, Y.; Ma, Y.; Yu, A.; et al. Porous 3D graphene-based bulk materials with exceptional high surface area and excellent conductivity for supercapacitors. *Sci. Rep.* **2013**, *3*, 1408. [CrossRef]
14. Tarascon, J.M.; Armand, M. Issues and challenges facing rechargeable lithium batteries. *Nature* **2001**, *414*, 359–367. [CrossRef]
15. Huang, S.; Zhu, X.; Sarkar, S.; Zhao, Y. Challenges and opportunities for supercapacitors. *APL Mater.* **2019**, *7*, 100901. [CrossRef]
16. Liang, J.; Wang, D.-W. Design Rationale and Device Configuration of Lithium-Ion Capacitors. *Adv. Energy Mater.* **2022**, *12*, 2200920. [CrossRef]
17. Li, B.; Zheng, J.; Zhang, H.; Jin, L.; Yang, D.; Lv, H.; Shen, C.; Shellikeri, A.; Zheng, Y.; Gong, R.; et al. Electrode Materials, Electrolytes, and Challenges in Nonaqueous Lithium-Ion Capacitors. *Adv. Mater.* **2018**, *30*, 1705670. [CrossRef]
18. Zhang, T.; Zhang, F.; Zhang, L.; Lu, Y.; Zhang, Y.; Yang, X.; Ma, Y.; Huang, Y. High energy density Li-ion capacitor assembled with all graphene-based electrodes. *Carbon* **2015**, *92*, 106–118. [CrossRef]
19. Lang, J.; Zhang, X.; Liu, B.; Wang, R.; Chen, J.; Yan, X. The roles of graphene in advanced Li-ion hybrid supercapacitors. *J. Energy Chem.* **2018**, *27*, 43–56. [CrossRef]
20. Aravindan, V.; Gnanaraj, J.; Lee, Y.-S.; Madhavi, S. Insertion-Type Electrodes for Nonaqueous Li-Ion Capacitors. *Chem. Rev.* **2014**, *114*, 11619–11635. [CrossRef]
21. Jagadale, A.; Zhou, X.; Xiong, R.; Dubal, D.P.; Xu, J.; Yang, S. Lithium ion capacitors (LICs): Development of the materials. *Energy Storage Mater.* **2019**, *19*, 314–329. [CrossRef]
22. Wang, X.; Liu, L.; Niu, Z. Carbon-based materials for lithium-ion capacitors. *Mater. Chem. Front.* **2019**, *3*, 1265–1279. [CrossRef]
23. Sui, D.; Chang, M.; Peng, Z.; Li, C.; He, X.; Yang, Y.; Liu, Y.; Lu, Y. Graphene-Based Cathode Materials for Lithium-Ion Capacitors: A Review. *Nanomaterials* **2021**, *11*, 2771. [CrossRef] [PubMed]
24. Ma, Y.; Chang, H.; Zhang, M.; Chen, Y. Graphene-Based Materials for Lithium-Ion Hybrid Supercapacitors. *Adv. Mater.* **2015**, *27*, 5296–5308. [CrossRef]
25. Li, C.; Zhang, X.; Sun, C.; Wang, K.; Sun, X.; Ma, Y. Recent progress of graphene-based materials in lithium-ion capacitors. *J. Phys. D Appl. Phys.* **2019**, *52*, 143001. [CrossRef]
26. Si, L.; Yan, K.; Li, C.; Huang, Y.; Pang, X.; Yang, X.; Sui, D.; Zhang, Y.; Wang, J.; Charles Xu, C. Binder-free SiO_2 nanotubes/carbon nanofibers mat as superior anode for lithium-ion batteries. *Electrochim. Acta* **2022**, *404*, 139747. [CrossRef]
27. Wu, N.; Qiao, X.; Shen, J.; Liu, G.; Sun, T.; Wu, H.; Hou, H.; Liu, X.; Zhang, Y.; Ji, X. Anatase inverse opal TiO_{2-x}@N-doped C induced the dominant pseudocapacitive effect for durable and fast lithium/sodium storage. *Electrochim. Acta* **2019**, *299*, 540–548. [CrossRef]
28. Guo, D.; Yang, M.; Zhang, L.; Li, Y.; Wang, J.; Liu, G.; Wu, N.; Kim, J.-K.; Liu, X. Cr_2O_3 nanosheet/carbon cloth anode with strong interaction and fast charge transfer for pseudocapacitive energy storage in lithium-ion batteries. *RSC Adv.* **2019**, *9*, 33446–33453. [CrossRef]
29. Sun, X.; Geng, L.; Yi, S.; Li, C.; An, Y.; Zhang, X.; Zhang, X.; Wang, K.; Ma, Y. Effects of carbon black on the electrochemical performances of SiO_x anode for lithium-ion capacitors. *J. Power Sources* **2021**, *499*, 229936. [CrossRef]
30. Leng, K.; Zhang, F.; Zhang, L.; Zhang, T.; Wu, Y.; Lu, Y.; Huang, Y.; Chen, Y. Graphene-based Li-ion hybrid supercapacitors with ultrahigh performance. *Nano Res.* **2013**, *6*, 581–592. [CrossRef]
31. Li, Y.; Wang, R.; Zheng, W.; Zhao, Q.; Sun, S.; Ji, G.; Li, S.; Fan, X.; Xu, C. Design of Nb_2O_5/graphene hybrid aerogel as polymer binder-free electrodes for lithium-ion capacitors. *Mater. Technol.* **2020**, *35*, 625–634. [CrossRef]
32. Yang, C.; Lan, J.-L.; Ding, C.; Wang, F.; Siyal, S.H.; Yu, Y.; Yang, X. Three-dimensional hierarchical ternary aerogels of ultrafine TiO_2 nanoparticles@porous carbon nanofibers-reduced graphene oxide for high-performance lithium-ion capacitors. *Electrochim. Acta* **2019**, *296*, 790–798. [CrossRef]
33. Zhang, F.; Zhang, T.; Yang, X.; Zhang, L.; Leng, K.; Huang, Y.; Chen, Y. A high-performance supercapacitor-battery hybrid energy storage device based on graphene-enhanced electrode materials with ultrahigh energy density. *Energy Environ. Sci.* **2013**, *6*, 1623–1632. [CrossRef]
34. Wang, R.; Wang, S.; Jin, D.; Zhang, Y.; Cai, Y.; Ma, J.; Zhang, L. Engineering layer structure of MoS_2-graphene composites with robust and fast lithium storage for high-performance Li-ion capacitors. *Energy Storage Mater.* **2017**, *9*, 195–205. [CrossRef]
35. Shao, R.; Niu, J.; Zhu, F.; Dou, M.; Zhang, Z.; Wang, F. A facile and versatile strategy towards high-performance Si anodes for Li-ion capacitors: Concomitant conductive network construction and dual-interfacial engineering. *Nano Energy* **2019**, *63*, 103824. [CrossRef]

36. Liu, C.; Ren, Q.; Zhang, S.; Yin, B.-S.; Que, L.; Zhao, L.; Sui, X.; Yu, F.; Li, X.; Gu, D.; et al. High energy and power lithium-ion capacitors based on Mn_3O_4/3D-graphene as anode and activated polyaniline-derived carbon nanorods as cathode. *Chem. Eng. J.* **2019**, *370*, 1485–1492. [CrossRef]
37. Sui, D.; Si, L.; Li, C.; Yang, Y.; Zhang, Y.; Yan, W. A Comprehensive Review of Graphene-Based Anode Materials for Lithium-ion Capacitors. *Chemistry* **2021**, *3*, 1215–1246. [CrossRef]
38. Liu, W.; Zhang, X.; Xu, Y.; Wang, L.; Li, Z.; Li, C.; Wang, K.; Sun, X.; An, Y.; Wu, Z.-S.; et al. 2D Graphene/MnO Heterostructure with Strongly Stable Interface Enabling High-Performance Flexible Solid-State Lithium-Ion Capacitors. *Adv. Funct. Mater.* **2022**, *32*, 2202342. [CrossRef]
39. Guo, D.; Yang, M.; Wang, F.; Cheng, Y.; Zhang, A.; Liu, G.; Wu, N.; Cao, A.; Mi, H.; Liu, X. Regulating the electronic structure of $MoO_2/Mo_2C/C$ by heterostructure and oxygen vacancies for boosting lithium storage kinetics. *Dalton Trans.* **2022**, *51*, 12620–12629. [CrossRef]
40. Cao, W.J.; Zheng, J.P. Li-ion capacitors with carbon cathode and hard carbon/stabilized lithium metal powder anode electrodes. *J. Power Sources* **2012**, *213*, 180–185. [CrossRef]
41. Wang, H.; Shao, Y.; Mei, S.; Lu, Y.; Zhang, M.; Sun, J.-K.; Matyjaszewski, K.; Antonietti, M.; Yuan, J. Polymer-Derived Heteroatom-Doped Porous Carbon Materials. *Chem. Rev.* **2020**, *120*, 9363–9419. [CrossRef] [PubMed]
42. Sennu, P.; Aravindan, V.; Ganesan, M.; Lee, Y.-G.; Lee, Y.-S. Biomass-Derived Electrode for Next Generation Lithium-Ion Capacitors. *ChemSusChem* **2016**, *9*, 849–854. [CrossRef] [PubMed]
43. Han, C.; Li, H.; Shi, R.; Xu, L.; Li, J.; Kang, F.; Li, B. Nanostructured Anode Materials for Non-aqueous Lithium Ion Hybrid Capacitors. *Energy Environ. Mater.* **2018**, *1*, 75–87. [CrossRef]
44. Liu, W.; Zhang, X.; Xu, Y.; Li, C.; Wang, K.; Sun, X.; Su, F.; Chen, C.-M.; Liu, F.; Wu, Z.-S.; et al. Recent Advances on Carbon-Based Materials for High Performance Lithium-Ion Capacitors. *Batter. Supercaps* **2021**, *4*, 407–428. [CrossRef]
45. Su, F.; Hou, X.; Qin, J.; Wu, Z.-S. Recent Advances and Challenges of Two-Dimensional Materials for High-Energy and High-Power Lithium-Ion Capacitors. *Batter. Supercaps* **2020**, *3*, 10–29. [CrossRef]
46. Hou, R.; Liu, B.; Sun, Y.; Liu, L.; Meng, J.; Levi, M.D.; Ji, H.; Yan, X. Recent advances in dual-carbon based electrochemical energy storage devices. *Nano Energy* **2020**, *72*, 104728. [CrossRef]
47. Li, F.; Cao, Y.; Wu, W.; Wang, G.; Qu, D. Prelithiation Bridges the Gap for Developing Next-Generation Lithium-Ion Batteries/Capacitors. *Small Methods* **2022**, *6*, 2200411. [CrossRef]
48. Zhang, S.S. Dual-Carbon Lithium-Ion Capacitors: Principle, Materials, and Technologies. *Batter. Supercaps* **2020**, *3*, 1137–1146. [CrossRef]
49. Li, G.; Yang, Z.; Yin, Z.; Guo, H.; Wang, Z.; Yan, G.; Liu, Y.; Li, L.; Wang, J. Non-aqueous dual-carbon lithium-ion capacitors: A review. *J. Mater. Chem. A* **2019**, *7*, 15541–15563. [CrossRef]
50. Yang, Z.; Guo, H.; Li, X.; Wang, Z.; Yan, Z.; Wang, Y. Natural sisal fibers derived hierarchical porous activated carbon as capacitive material in lithium ion capacitor. *J. Power Sources* **2016**, *329*, 339–346. [CrossRef]
51. Qiu, D.; Kang, C.; Li, M.; Wei, J.; Hou, Z.; Wang, F.; Yang, R. Biomass-derived mesopore-dominant hierarchical porous carbon enabling ultra-efficient lithium ion storage. *Carbon* **2020**, *162*, 595–603. [CrossRef]
52. Li, C.; Zhang, X.; Wang, K.; Sun, X.; Ma, Y. High-power and long-life lithium-ion capacitors constructed from N-doped hierarchical carbon nanolayer cathode and mesoporous graphene anode. *Carbon* **2018**, *140*, 237–248. [CrossRef]
53. Yang, Z.; Guo, H.; Yan, G.; Li, X.; Wang, Z.; Guo, Y.; Wang, X.; Wu, Y.; Wang, J. High-Value Utilization of Lignin To Prepare Functional Carbons toward Advanced Lithium-Ion Capacitors. *ACS Sustain. Chem. Eng.* **2020**, *8*, 11522–11531. [CrossRef]
54. Huang, X.; Le, T.; Chen, M.; Yang, C.; Yang, Y. Regulating the Pore Structure and Heteroatom Doping of Hollow Carbon Fiber Based on a Trifunctional Template Method for High-Performance Lithium-Ion Capacitors. *ACS Appl. Energy Mater.* **2022**, *5*, 3034–3042. [CrossRef]
55. Xia, Q.; Yang, H.; Wang, M.; Yang, M.; Guo, Q.; Wan, L.; Xia, H.; Yu, Y. High Energy and High Power Lithium-Ion Capacitors Based on Boron and Nitrogen Dual-Doped 3D Carbon Nanofibers as Both Cathode and Anode. *Adv. Energy Mater.* **2017**, *7*, 1701336. [CrossRef]
56. Hao, J.; Bai, J.; Wang, X.; Wang, Y.; Guo, Q.; Yang, Y.; Zhao, J.; Chi, C.; Li, Y. S, O dual-doped porous carbon derived from activation of waste papers as electrodes for high performance lithium ion capacitors. *Nanoscale Adv.* **2021**, *3*, 738–746. [CrossRef]
57. Han, L.; Kang, S.; Zhu, X.; Li, J.; Wang, Q.; Jia, X. High-Performance Lithium-Ion Capacitors Produced by Atom-Thick Carbon Cathode and Nitrogen-Doped Porous Carbon Anode. *Energy Fuels* **2021**, *35*, 16894–16902. [CrossRef]
58. Chen, M.; Le, T.; Zhou, Y.; Kang, F.; Yang, Y. Thiourea-Induced N/S Dual-Doped Hierarchical Porous Carbon Nanofibers for High-Performance Lithium-Ion Capacitors. *ACS Appl. Energy Mater.* **2020**, *3*, 1653–1664. [CrossRef]
59. Liang, T.; Wang, H.; Fei, R.; Wang, R.; He, B.; Gong, Y.; Yan, C. A high-power lithium-ion hybrid capacitor based on a hollow N-doped carbon nanobox anode and its porous analogue cathode. *Nanoscale* **2019**, *11*, 20715–20724. [CrossRef]
60. Xu, Y.; Jiang, J.; Li, Z.; Yang, Z.; Zhang, Y.; An, Y.; Zhu, Q.; Dou, H.; Zhang, X. Aerosol-assisted preparation of N-doped hierarchical porous carbon spheres cathodes toward high-stable lithium-ion capacitors. *J. Mater. Sci.* **2020**, *55*, 13127–13140. [CrossRef]
61. Li, Z.; Cao, L.; Chen, W.; Huang, Z.; Liu, H. Mesh-Like Carbon Nanosheets with High-Level Nitrogen Doping for High-Energy Dual-Carbon Lithium-Ion Capacitors. *Small* **2019**, *15*, 1805173. [CrossRef] [PubMed]
62. Jiang, J.; Nie, P.; Fang, S.; Zhang, Y.; An, Y.; Fu, R.; Dou, H.; Zhang, X. Boron and nitrogen dual-doped carbon as a novel cathode for high performance hybrid ion capacitors. *Chin. Chem. Lett.* **2018**, *29*, 624–628. [CrossRef]

63. Sun, F.; Liu, X.; Wu, H.B.; Wang, L.; Gao, J.; Li, H.; Lu, Y. In Situ High-Level Nitrogen Doping into Carbon Nanospheres and Boosting of Capacitive Charge Storage in Both Anode and Cathode for a High-Energy 4.5 V Full-Carbon Lithium-Ion Capacitor. *Nano Lett.* **2018**, *18*, 3368–3376. [CrossRef] [PubMed]
64. Zou, K.; Deng, Y.; Wu, W.; Zhang, S.; Chen, G. A novel eutectic solvent precursor for efficiently preparing N-doped hierarchically porous carbon nanosheets with unique surface functional groups and micropores towards dual-carbon lithium-ion capacitors. *J. Mater. Chem. A* **2021**, *9*, 13631–13641. [CrossRef]
65. Chen, M.; Le, T.; Zhou, Y.; Kang, F.; Yang, Y. Enhanced Electrode Matching Assisted by In Situ Etching and Co-Doping toward High-Rate Dual-Carbon Lithium-Ion Capacitors. *ACS Sustain. Chem. Eng.* **2021**, *9*, 10054–10061. [CrossRef]
66. Lee, J.H.; Shin, W.H.; Ryou, M.-H.; Jin, J.K.; Kim, J.; Choi, J.W. Functionalized Graphene for High Performance Lithium Ion Capacitors. *ChemSusChem* **2012**, *5*, 2328–2333. [CrossRef]
67. Dubal, D.P.; Gomez-Romero, P. All nanocarbon Li-Ion capacitor with high energy and high power density. *Mater. Today Energy* **2018**, *8*, 109–117. [CrossRef]
68. Li, C.; Zhang, X.; Wang, K.; Sun, X.; Xu, Y.; Su, F.; Chen, C.-M.; Liu, F.; Wu, Z.-S.; Ma, Y. Nitrogen-enriched graphene framework from a large-scale magnesiothermic conversion of CO_2 with synergistic kinetics for high-power lithium-ion capacitors. *NPG Asia Mater.* **2021**, *13*, 59. [CrossRef]
69. Xiao, Z.; Yu, Z.; Gao, Z.; Li, B.; Zhang, M.; Xu, C. S-doped graphene nano-capsules toward excellent low-temperature performance in Li-ion capacitors. *J. Power Sources* **2022**, *535*, 231404. [CrossRef]
70. Wang, R.; Zhao, Q.; Zheng, W.; Ren, Z.; Hu, X.; Li, J.; Lu, L.; Hu, N.; Molenda, J.; Liu, X.; et al. Achieving high energy density in a 4.5 V all nitrogen-doped graphene based lithium-ion capacitor. *J. Mater. Chem. A* **2019**, *7*, 19909–19921. [CrossRef]
71. Xiao, Y.; Liu, J.; He, D.; Chen, S.; Peng, W.; Hu, X.; Liu, T.; Zhu, Z.; Bai, Y. Facile Synthesis of Graphene with Fast Ion/Electron Channels for High-Performance Symmetric Lithium-Ion Capacitors. *ACS Appl. Mater. Interfaces* **2021**, *13*, 38266–38277. [CrossRef] [PubMed]
72. Zou, K.; Cai, P.; Wang, B.; Liu, C.; Li, J.; Qiu, T.; Zou, G.; Hou, H.; Ji, X. Insights into Enhanced Capacitive Behavior of Carbon Cathode for Lithium Ion Capacitors: The Coupling of Pore Size and Graphitization Engineering. *Nano-Micro Lett.* **2020**, *12*, 121. [CrossRef] [PubMed]
73. Wang, J.; Kaskel, S. KOH activation of carbon-based materials for energy storage. *J. Mater. Chem.* **2012**, *22*, 23710–23725. [CrossRef]
74. Amatucci, G.G.; Badway, F.; Du Pasquier, A.; Zheng, T. An Asymmetric Hybrid Nonaqueous Energy Storage Cell. *J. Electrochem. Soc.* **2001**, *148*, A930. [CrossRef]
75. Sivakkumar, S.R.; Pandolfo, A.G. Evaluation of lithium-ion capacitors assembled with pre-lithiated graphite anode and activated carbon cathode. *Electrochim. Acta* **2012**, *65*, 280–287. [CrossRef]
76. Zhang, J.; Liu, X.; Wang, J.; Shi, J.; Shi, Z. Different types of pre-lithiated hard carbon as negative electrode material for lithium-ion capacitors. *Electrochim. Acta* **2016**, *187*, 134–142. [CrossRef]
77. Ren, J.J.; Su, L.W.; Qin, X.; Yang, M.; Wei, J.P.; Zhou, Z.; Shen, P.W. Pre-lithiated graphene nanosheets as negative electrode materials for Li-ion capacitors with high power and energy density. *J. Power Sources* **2014**, *264*, 108–113. [CrossRef]
78. An, Y.; Liu, T.; Li, C.; Zhang, X.; Hu, T.; Sun, X.; Wang, K.; Wang, C.; Ma, Y. A general route for the mass production of graphene-enhanced carbon composites toward practical pouch lithium-ion capacitors. *J. Mater. Chem. A* **2021**, *9*, 15654–15664. [CrossRef]
79. Li, B.; Xiao, Z.; Chen, M.; Huang, Z.; Tie, X.; Zai, J.; Qian, X. Rice husk-derived hybrid lithium-ion capacitors with ultra-high energy. *J. Mater. Chem. A* **2017**, *5*, 24502–24507. [CrossRef]
80. Dubey, P.; Shrivastav, V.; Maheshwari, P.H.; Sundriyal, S. Recent advances in biomass derived activated carbon electrodes for hybrid electrochemical capacitor applications: Challenges and opportunities. *Carbon* **2020**, *170*, 1–29. [CrossRef]
81. Zhang, G.; Liu, X.; Wang, L.; Fu, H. Recent advances of biomass derived carbon-based materials for efficient electrochemical energy devices. *J. Mater. Chem. A* **2022**, *10*, 9277–9307. [CrossRef]
82. Luo, X.; Chen, S.; Hu, T.; Chen, Y.; Li, F. Renewable biomass-derived carbons for electrochemical capacitor applications. *SusMat* **2021**, *1*, 211–240. [CrossRef]
83. Sennu, P.; Arun, N.; Madhavi, S.; Aravindan, V.; Lee, Y.-S. All carbon based high energy lithium-ion capacitors from biomass: The role of crystallinity. *J. Power Sources* **2019**, *414*, 96–102. [CrossRef]
84. Zhang, L.; Zhang, F.; Yang, X.; Leng, K.; Huang, Y.; Chen, Y. High-Performance Supercapacitor Electrode Materials Prepared from Various Pollens. *Small* **2013**, *9*, 1342–1347. [CrossRef] [PubMed]
85. Lin, Y.-T.; Chang-Jian, C.-W.; Hsieh, T.-H.; Huang, J.-H.; Chu Weng, H.; Hsiao, Y.-S.; Syu, W.-L.; Chen, C.-P. High-performance Li-Ion capacitor constructed from biomass-derived porous carbon and high-rate Li4Ti5O12. *Appl. Surf. Sci.* **2021**, *543*, 148717. [CrossRef]
86. Li, B.; Zhang, H.; Zhang, C. Agricultural waste-derived activated carbon/graphene composites for high performance lithium-ion capacitors. *RSC Adv.* **2019**, *9*, 29190–29194. [CrossRef] [PubMed]
87. Yang, S.; Zhang, L.; Sun, J.; Li, K.; Zhao, S.; Zhao, D.; Wang, J.; Yang, C.; Wang, X.; Cao, B. Corncob-Derived Hierarchical Porous Activated Carbon for High-Performance Lithium-Ion Capacitors. *Energy Fuels* **2020**, *34*, 16885–16892. [CrossRef]
88. Sun, J.; Yang, S.; Ai, J.; Yang, C.; Jia, Q.; Cao, B. Hierarchical Porous Activated Carbon Obtained by a Novel Heating-Rate-Induced Method for Lithium-Ion Capacitor. *ChemistrySelect* **2019**, *4*, 5300–5307. [CrossRef]

89. Babu, B.; Lashmi, P.G.; Shaijumon, M.M. Li-ion capacitor based on activated rice husk derived porous carbon with improved electrochemical performance. *Electrochim. Acta* **2016**, *211*, 289–296. [CrossRef]
90. Lu, Q.; Lu, B.; Chen, M.; Wang, X.; Xing, T.; Liu, M.; Wang, X. Porous activated carbon derived from Chinese-chive for high energy hybrid lithium-ion capacitor. *J. Power Sources* **2018**, *398*, 128–136. [CrossRef]
91. Zhang, H.; Xu, X.; Wang, H.; Lyu, Y.; Liu, X.; Zhao, Y.; Shi, J.; Liu, W.; Paek, E.; Mitlin, D. Lithium Ion Capacitor with Identical Carbon Electrodes Yields 6 s Charging and 100 000 Cycles Stability with 1% Capacity Fade. *ACS Sustain. Chem. Eng.* **2019**, *7*, 2867–2877. [CrossRef]
92. Ajuria, J.; Zarrabeitia, M.; Arnaiz, M.; Urra, O.; Rojo, T.; Goikolea, E. Graphene as Vehicle for Ultrafast Lithium Ion Capacitor Development Based on Recycled Olive Pit Derived Carbons. *J. Electrochem. Soc.* **2019**, *166*, A2840–A2848. [CrossRef]
93. Gómez-Urbano, J.L.; Moreno-Fernández, G.; Arnaiz, M.; Ajuria, J.; Rojo, T.; Carriazo, D. Graphene-coffee waste derived carbon composites as electrodes for optimized lithium ion capacitors. *Carbon* **2020**, *162*, 273–282. [CrossRef]
94. Zheng, B.; Lin, X.; Zhang, X.; Wu, D.; Matyjaszewski, K. Emerging Functional Porous Polymeric and Carbonaceous Materials for Environmental Treatment and Energy Storage. *Adv. Funct. Mater.* **2020**, *30*, 1907006. [CrossRef]
95. Abdelaal, M.M.; Hung, T.-C.; Mohamed, S.G.; Yang, C.-C.; Hung, T.-F. Two Birds with One Stone: Hydrogel-Derived Hierarchical Porous Activated Carbon toward the Capacitive Performance for Symmetric Supercapacitors and Lithium-Ion Capacitors. *ACS Sustain. Chem. Eng.* **2022**, *10*, 4717–4727. [CrossRef]
96. Zheng, W.; Li, Z.; Han, G.; Zhao, Q.; Lu, G.; Hu, X.; Sun, J.; Wang, R.; Xu, C. Nitrogen-doped activated porous carbon for 4.5 V lithium-ion capacitor with high energy and power density. *J. Energy Storage* **2022**, *47*, 103675. [CrossRef]
97. Arnaiz, M.; Nair, V.; Mitra, S.; Ajuria, J. Furfuryl alcohol derived high-end carbons for ultrafast dual carbon lithium ion capacitors. *Electrochim. Acta* **2019**, *304*, 437–446. [CrossRef]
98. Xu, Y.; Jin, S.; Xu, H.; Nagai, A.; Jiang, D. Conjugated microporous polymers: Design, synthesis and application. *Chem. Soc. Rev.* **2013**, *42*, 8012–8031. [CrossRef]
99. Geng, K.; He, T.; Liu, R.; Dalapati, S.; Tan, K.T.; Li, Z.; Tao, S.; Gong, Y.; Jiang, Q.; Jiang, D. Covalent Organic Frameworks: Design, Synthesis, and Functions. *Chem. Rev.* **2020**, *120*, 8814–8933. [CrossRef]
100. Weng, T.-H.; Mohamed, M.G.; Sharma, S.U.; Chaganti, S.V.; Samy, M.M.; Lee, J.-T.; Kuo, S.-W. Ultrastable Three-Dimensional Triptycene- and Tetraphenylethene-Conjugated Microporous Polymers for Energy Storage. *ACS Appl. Energy Mater.* **2022**. [CrossRef]
101. Mohamed, M.G.; Sharma, S.U.; Yang, C.-H.; Samy, M.M.; Mohammed, A.A.K.; Chaganti, S.V.; Lee, J.-T.; Wei-Kuo, S. Anthraquinone-Enriched Conjugated Microporous Polymers as Organic Cathode Materials for High-Performance Lithium-Ion Batteries. *ACS Appl. Energy Mater.* **2021**, *4*, 14628–14639. [CrossRef]
102. Zhang, C.; Qiao, Y.; Xiong, P.; Ma, W.; Bai, P.; Wang, X.; Li, Q.; Zhao, J.; Xu, Y.; Chen, Y.; et al. Conjugated Microporous Polymers with Tunable Electronic Structure for High-Performance Potassium-Ion Batteries. *ACS Nano* **2019**, *13*, 745–754. [CrossRef] [PubMed]
103. Hu, Y.; Wayment, L.J.; Haslam, C.; Yang, X.; Lee, S.-h.; Jin, Y.; Zhang, W. Covalent organic framework based lithium-ion battery: Fundamental, design and characterization. *EnergyChem* **2021**, *3*, 100048. [CrossRef]
104. Huang, N.; Wang, P.; Jiang, D. Covalent organic frameworks: A materials platform for structural and functional designs. *Nat. Rev. Mater.* **2016**, *1*, 16068. [CrossRef]
105. Amin, K.; Ashraf, N.; Mao, L.; Faul, C.F.J.; Wei, Z. Conjugated microporous polymers for energy storage: Recent progress and challenges. *Nano Energy* **2021**, *85*, 105958. [CrossRef]
106. Zhang, X.; Zhu, G.; Wang, M.; Li, J.; Lu, T.; Pan, L. Covalent-organic-frameworks derived N-doped porous carbon materials as anode for superior long-life cycling lithium and sodium ion batteries. *Carbon* **2017**, *116*, 686–694. [CrossRef]
107. Zhang, Q.; Sun, H.; Wang, X.; Zhu, Z.; Liang, W.; Li, A.; Wen, S.; Deng, W. Conjugated Microporous Polymer-Derived Porous Hard Carbon as High-Rate Long-Life Anode Materials for Lithium Ion Batteries. *Energy Technol.* **2013**, *1*, 721–725. [CrossRef]
108. Ibrahim, M.; Abdelhamid, H.N.; Abuelftooh, A.M.; Mohamed, S.G.; Wen, Z.; Sun, X. Covalent organic frameworks (COFs)-derived nitrogen-doped carbon/reduced graphene oxide nanocomposite as electrodes materials for supercapacitors. *J. Energy Storage* **2022**, *55*, 105375. [CrossRef]
109. Yuan, K.; Hu, T.; Xu, Y.; Graf, R.; Shi, L.; Forster, M.; Pichler, T.; Riedl, T.; Chen, Y.; Scherf, U. Nitrogen-doped porous carbon/graphene nanosheets derived from two-dimensional conjugated microporous polymer sandwiches with promising capacitive performance. *Mater. Chem. Front.* **2017**, *1*, 278–285. [CrossRef]
110. Zhu, Y.; Murali, S.; Stoller, M.D.; Ganesh, K.J.; Cai, W.; Ferreira, P.J.; Pirkle, A.; Wallace, R.M.; Cychosz, K.A.; Thommes, M.; et al. Carbon-Based Supercapacitors Produced by Activation of Graphene. *Science* **2011**, *332*, 1537–1541. [CrossRef]
111. Jang, B.Z.; Liu, C.; Neff, D.; Yu, Z.; Wang, M.C.; Xiong, W.; Zhamu, A. Graphene Surface-Enabled Lithium Ion-Exchanging Cells: Next-Generation High-Power Energy Storage Devices. *Nano Lett.* **2011**, *11*, 3785–3791. [CrossRef] [PubMed]
112. Sui, D.; Wu, M.; Shi, K.; Li, C.; Lang, J.; Yang, Y.; Zhang, X.; Yan, X.; Chen, Y. Recent progress of cathode materials for aqueous zinc-ion capacitors: Carbon-based materials and beyond. *Carbon* **2021**, *185*, 126–151. [CrossRef]
113. Geim, A.K.; Novoselov, K.S. The rise of graphene. *Nat. Mater.* **2007**, *6*, 183–191. [CrossRef] [PubMed]
114. Ma, Y.; Chen, Y. Three-dimensional graphene networks: Synthesis, properties and applications. *Natl. Sci. Rev.* **2014**, *2*, 40–53. [CrossRef]

115. Mao, S.; Lu, G.; Chen, J. Three-dimensional graphene-based composites for energy applications. *Nanoscale* **2015**, *7*, 6924–6943. [CrossRef]
116. Becerril, H.A.; Mao, J.; Liu, Z.; Stoltenberg, R.M.; Bao, Z.; Chen, Y. Evaluation of Solution-Processed Reduced Graphene Oxide Films as Transparent Conductors. *ACS Nano* **2008**, *2*, 463–470. [CrossRef]
117. Moreno-Fernández, G.; Granados-Moreno, M.; Gómez-Urbano, J.L.; Carriazo, D. Phosphorus-Functionalized Graphene for Lithium-Ion Capacitors with Improved Power and Cyclability. *Batter. Supercaps* **2021**, *4*, 469–478. [CrossRef]
118. Aravindan, V.; Mhamane, D.; Ling, W.C.; Ogale, S.; Madhavi, S. Nonaqueous Lithium-Ion Capacitors with High Energy Densities using Trigol-Reduced Graphene Oxide Nanosheets as Cathode-Active Material. *ChemSusChem* **2013**, *6*, 2240–2244. [CrossRef]
119. Wang, H.; Guan, C.; Wang, X.; Fan, H.J. A High Energy and Power Li-Ion Capacitor Based on a TiO_2 Nanobelt Array Anode and a Graphene Hydrogel Cathode. *Small* **2015**, *11*, 1470–1477. [CrossRef]
120. Li, H.; Shen, L.; Wang, J.; Fang, S.; Zhang, Y.; Dou, H.; Zhang, X. Three-dimensionally ordered porous $TiNb_2O_7$ nanotubes: A superior anode material for next generation hybrid supercapacitors. *J. Mater. Chem. A* **2015**, *3*, 16785–16790. [CrossRef]
121. Ye, L.; Liang, Q.; Lei, Y.; Yu, X.; Han, C.; Shen, W.; Huang, Z.-H.; Kang, F.; Yang, Q.-H. A high performance Li-ion capacitor constructed with $Li_4Ti_5O_{12}$/C hybrid and porous graphene macroform. *J. Power Sources* **2015**, *282*, 174–178. [CrossRef]
122. Sui, D.; Xu, L.; Zhang, H.; Sun, Z.; Kan, B.; Ma, Y.; Chen, Y. A 3D cross-linked graphene-based honeycomb carbon composite with excellent confinement effect of organic cathode material for lithium-ion batteries. *Carbon* **2020**, *157*, 656–662. [CrossRef]
123. Stoller, M.D.; Murali, S.; Quarles, N.; Zhu, Y.; Potts, J.R.; Zhu, X.; Ha, H.-W.; Ruoff, R.S. Activated graphene as a cathode material for Li-ion hybrid supercapacitors. *Phys. Chem. Chem. Phys.* **2012**, *14*, 3388–3391. [CrossRef] [PubMed]
124. Yang, Y.; Lin, Q.; Ding, B.; Wang, J.; Malgras, V.; Jiang, J.; Li, Z.; Chen, S.; Dou, H.; Alshehri, S.M.; et al. Lithium-ion capacitor based on nanoarchitectured polydopamine/graphene composite anode and porous graphene cathode. *Carbon* **2020**, *167*, 627–633. [CrossRef]
125. Ma, X.; Gao, D. High Capacitive Storage Performance of Sulfur and Nitrogen Codoped Mesoporous Graphene. *ChemSusChem* **2018**, *11*, 1048–1055. [CrossRef]
126. Ma, X.; Zhao, L.; Song, X.; Yu, Z.; Zhao, L.; Yu, Y.; Xiao, Z.; Ning, G.; Gao, J. Superior capacitive behaviors of the micron-sized porous graphene belts with high ratio of length to diameter. *Carbon* **2018**, *140*, 314–323. [CrossRef]
127. Jin, L.; Guo, X.; Gong, R.; Zheng, J.; Xiang, Z.; Zhang, C.; Zheng, J.P. Target-oriented electrode constructions toward ultra-fast and ultra-stable all-graphene lithium ion capacitors. *Energy Storage Mater.* **2019**, *23*, 409–417. [CrossRef]
128. Zhang, M.; Sun, Z.; Zhang, T.; Qin, B.; Sui, D.; Xie, Y.; Ma, Y.; Chen, Y. Porous asphalt/graphene composite for supercapacitors with high energy density at superior power density without added conducting materials. *J. Mater. Chem. A* **2017**, *5*, 21757–21764. [CrossRef]
129. Sui, D.; Wu, M.; Liu, Y.; Yang, Y.; Zhang, H.; Ma, Y.; Zhang, L.; Chen, Y. High performance Li-ion capacitor fabricated with dual graphene-based materials. *Nanotechnology* **2020**, *32*, 015403. [CrossRef]
130. Yu, Y.; Cui, C.; Qian, W.; Xie, Q.; Zheng, C.; Kong, C.; Wei, F. Carbon nanotube production and application in energy storage. *Asia-Pac. J. Chem. Eng.* **2013**, *8*, 234–245. [CrossRef]
131. Frackowiak, E.; Béguin, F. Electrochemical storage of energy in carbon nanotubes and nanostructured carbons. *Carbon* **2002**, *40*, 1775–1787. [CrossRef]
132. Zhang, Y.; Ye, T.; Li, L.; Peng, H. Carbon Nanotubes for Flexible Fiber Batteries. In *Nanoporous Carbons for Soft and Flexible Energy Devices*; Borghi, F., Soavi, F., Milani, P., Eds.; Springer International Publishing: Cham, Switzerland, 2022; pp. 1–22. [CrossRef]
133. Wen, L.; Li, F.; Cheng, H.-M. Carbon Nanotubes and Graphene for Flexible Electrochemical Energy Storage: From Materials to Devices. *Adv. Mater.* **2016**, *28*, 4306–4337. [CrossRef] [PubMed]
134. Wang, Y.; Wu, Y.; Huang, Y.; Zhang, F.; Yang, X.; Ma, Y.; Chen, Y. Preventing Graphene Sheets from Restacking for High-Capacitance Performance. *J. Phys. Chem. C* **2011**, *115*, 23192–23197. [CrossRef]
135. Wang, J.-A.; Li, S.-M.; Wang, Y.-S.; Lan, P.-Y.; Liao, W.-H.; Hsiao, S.-T.; Lin, S.-C.; Lin, C.-W.; Ma, C.-C.M.; Hu, C.-C. Preparation and Properties of NrGO-CNT Composite for Lithium-Ion Capacitors. *J. Electrochem. Soc.* **2017**, *164*, A3657–A3665. [CrossRef]
136. Aphirakaramwong, C.; Phattharasupakun, N.; Suktha, P.; Krittayavathananon, A.; Sawangphruk, M. Lightweight Multi-Walled Carbon Nanotube/N-Doped Graphene Aerogel Composite for High-Performance Lithium-Ion Capacitors. *J. Electrochem. Soc.* **2019**, *166*, A532–A538. [CrossRef]
137. Salvatierra, R.V.; Zakhidov, D.; Sha, J.; Kim, N.D.; Lee, S.-K.; Raji, A.-R.O.; Zhao, N.; Tour, J.M. Graphene Carbon Nanotube Carpets Grown Using Binary Catalysts for High-Performance Lithium-Ion Capacitors. *ACS Nano* **2017**, *11*, 2724–2733. [CrossRef] [PubMed]
138. Simon, P.; Gogotsi, Y. Capacitive Energy Storage in Nanostructured Carbon–Electrolyte Systems. *Acc. Chem. Res.* **2013**, *46*, 1094–1103. [CrossRef] [PubMed]
139. Han, D.; Weng, Z.; Li, P.; Tao, Y.; Cui, C.; Zhang, L.; Lin, W.; Gao, Y.; Kong, D.; Yang, Q.-H. Electrode thickness matching for achieving high-volumetric-performance lithium-ion capacitors. *Energy Storage Mater.* **2019**, *18*, 133–138. [CrossRef]
140. Zong, J.; Ni, W.; Xu, H.; Ding, F.; Wang, T.; Feng, W.; Liu, X. High tap-density graphene cathode material for lithium-ion capacitors via a mass-scalable synthesis method. *Chem. Eng. J.* **2019**, *360*, 1233–1240. [CrossRef]
141. Jia, Y.; Zhang, J.; Kong, D.; Zhang, C.; Han, D.; Han, J.; Tao, Y.; Lv, W.; Yang, Q.-H. Practical Graphene Technologies for Electrochemical Energy Storage. *Adv. Funct. Mater.* **2022**, *32*, 2204272. [CrossRef]

142. Yang, Z.; Gao, Y.; Zhao, Z.; Wang, Y.; Wu, Y.; Wang, X. Phytic acid assisted formation of P-doped hard carbon anode with enhanced capacity and rate capability for lithium ion capacitors. *J. Power Sources* **2020**, *474*, 228500. [CrossRef]
143. Zeng, D.; Xiong, H.; Qi, K.; Guo, X.; Qiu, Y. Constructing N-doping biomass-derived carbon with hierarchically porous architecture to boost fast reaction kinetics for higfh-performance lithium storage. *J. Colloid Interface Sci.* **2022**, *605*, 741–751. [CrossRef] [PubMed]
144. Liu, M.; Zhang, Z.; Dou, M.; Li, Z.; Wang, F. Nitrogen and oxygen co-doped porous carbon nanosheets as high-rate and long-lifetime anode materials for high-performance Li-ion capacitors. *Carbon* **2019**, *151*, 28–35. [CrossRef]
145. Cui, Y.; Liu, W.; Lyu, Y.; Zhang, Y.; Wang, H.; Liu, Y.; Li, D. All-carbon lithium capacitor based on salt crystal-templated, N-doped porous carbon electrodes with superior energy storage. *J. Mater. Chem. A* **2018**, *6*, 18276–18285. [CrossRef]
146. Sun, F.; Wu, H.B.; Liu, X.; Liu, F.; Han, R.; Qu, Z.; Pi, X.; Wang, L.; Gao, J.; Lu, Y. A high-rate and ultrastable anode enabled by boron-doped nanoporous carbon spheres for high-power and long life lithium ion capacitors. *Mater. Today Energy* **2018**, *9*, 428–439. [CrossRef]
147. Zou, K.; Guan, Z.; Deng, Y.; Chen, G. Nitrogen-rich porous carbon in ultra-high yield derived from activation of biomass waste by a novel eutectic salt for high performance Li-ion capacitors. *Carbon* **2020**, *161*, 25–35. [CrossRef]
148. Luan, Y.; Hu, R.; Fang, Y.; Zhu, K.; Cheng, K.; Yan, J.; Ye, K.; Wang, G.; Cao, D. Nitrogen and Phosphorus Dual-Doped Multilayer Graphene as Universal Anode for Full Carbon-Based Lithium and Potassium Ion Capacitors. *Nano-Micro Lett.* **2019**, *11*, 30. [CrossRef] [PubMed]
149. Shen, X.; He, J.; Wang, K.; Li, X.; Wang, X.; Yang, Z.; Wang, N.; Zhang, Y.; Huang, C. Fluorine-Enriched Graphdiyne as an Efficient Anode in Lithium-Ion Capacitors. *ChemSusChem* **2019**, *12*, 1342–1348. [CrossRef]
150. Shen, X.; Yang, Z.; Wang, K.; Wang, N.; He, J.; Du, H.; Huang, C. Nitrogen-Doped Graphdiyne as High-capacity Electrode Materials for Both Lithium-ion and Sodium-ion Capacitors. *ChemElectroChem* **2018**, *5*, 1435–1443. [CrossRef]
151. Yang, Z.; Guo, H.; Li, X.; Wang, Z.; Wang, J.; Wang, Y.; Yan, Z.; Zhang, D. Graphitic carbon balanced between high plateau capacity and high rate capability for lithium ion capacitors. *J. Mater. Chem. A* **2017**, *5*, 15302–15309. [CrossRef]
152. Li, G.; Huang, Y.; Yin, Z.; Guo, H.; Liu, Y.; Cheng, H.; Wu, Y.; Ji, X.; Wang, J. Defective synergy of 2D graphitic carbon nanosheets promotes lithium-ion capacitors performance. *Energy Storage Mater.* **2020**, *24*, 304–311. [CrossRef]
153. Zeng, D.; Xiong, H.; Wu, L.; Zhang, Y.; Qi, K.; Guo, X.; Qiu, Y. High-energy graphite microcrystalline carbon for high-performance lithium-ion capacitor: Diffusion kinetics and lithium-storage mechanism. *J. Colloid Interface Sci.* **2022**, *623*, 1190–1199. [CrossRef]
154. Kumagai, S.; Ishikawa, T.; Sawa, N. Cycle performance of lithium-ion capacitors using graphite negative electrodes at different pre-lithiation levels. *J. Energy Storage* **2015**, *2*, 1–7. [CrossRef]
155. Gao, Y.; Wang, C.; Zhang, J.; Jing, Q.; Ma, B.; Chen, Y.; Zhang, W. Graphite Recycling from the Spent Lithium-Ion Batteries by Sulfuric Acid Curing–Leaching Combined with High-Temperature Calcination. *ACS Sustain. Chem. Eng.* **2020**, *8*, 9447–9455. [CrossRef]
156. Ma, X.; Chen, M.; Chen, B.; Meng, Z.; Wang, Y. High-Performance Graphite Recovered from Spent Lithium-Ion Batteries. *ACS Sustain. Chem. Eng.* **2019**, *7*, 19732–19738. [CrossRef]
157. Kayakool, F.A.; Gangaja, B.; Nair, S.; Santhanagopalan, D. Li-based all-carbon dual-ion batteries using graphite recycled from spent Li-ion batteries. *Sustain. Mater. Technol.* **2021**, *28*, e00262. [CrossRef]
158. Divya, M.L.; Natarajan, S.; Lee, Y.-S.; Aravindan, V. Achieving high-energy dual carbon Li-ion capacitors with unique low- and high-temperature performance from spent Li-ion batteries. *J. Mater. Chem. A* **2020**, *8*, 4950–4959. [CrossRef]
159. Wu, Y.; Huang, X.; Huang, L.; Chen, J. Strategies for Rational Design of High-Power Lithium-ion Batteries. *Energy Environ. Mater.* **2021**, *4*, 19–45. [CrossRef]
160. Sivakkumar, S.R.; Nerkar, J.Y.; Pandolfo, A.G. Rate capability of graphite materials as negative electrodes in lithium-ion capacitors. *Electrochim. Acta* **2010**, *55*, 3330–3335. [CrossRef]
161. Huang, L.; Li, H.; Wang, X.; Ding, Y.; Wang, J.; Jiang, L. High-efficiency, self-grinding exfoliation of small graphene nanosheets from microcrystalline graphite driven by microbead milling as conductive additives. *Sci. China Mater.* **2022**, *65*, 2463–2471. [CrossRef]
162. Yu, X.; Zhan, C.; Lv, R.; Bai, Y.; Lin, Y.; Huang, Z.-H.; Shen, W.; Qiu, X.; Kang, F. Ultrahigh-rate and high-density lithium-ion capacitors through hybriding nitrogen-enriched hierarchical porous carbon cathode with prelithiated microcrystalline graphite anode. *Nano Energy* **2015**, *15*, 43–53. [CrossRef]
163. Wang, M.; Wang, J.; Xiao, J.; Ren, N.; Pan, B.; Chen, C.-s.; Chen, C.-h. Introducing a Pseudocapacitive Lithium Storage Mechanism into Graphite by Defect Engineering for Fast-Charging Lithium-Ion Batteries. *ACS Appl. Mater. Interfaces* **2022**, *14*, 16279–16288. [CrossRef] [PubMed]
164. Pendashteh, A.; Orayech, B.; Suhard, H.; Jauregui, M.; Ajuria, J.; Silván, B.; Clarke, S.; Bonilla, F.; Saurel, D. Boosting the performance of soft carbon negative electrode for high power Na-ion batteries and Li-ion capacitors through a rational strategy of structural and morphological manipulation. *Energy Storage Mater.* **2022**, *46*, 417–430. [CrossRef]
165. Sun, X.; Zhang, X.; Liu, W.; Wang, K.; Li, C.; Li, Z.; Ma, Y. Electrochemical performances and capacity fading behaviors of activated carbon/hard carbon lithium ion capacitor. *Electrochim. Acta* **2017**, *235*, 158–166. [CrossRef]
166. Xie, F.; Xu, Z.; Jensen, A.C.S.; Au, H.; Lu, Y.; Araullo-Peters, V.; Drew, A.J.; Hu, Y.-S.; Titirici, M.-M. Hard–Soft Carbon Composite Anodes with Synergistic Sodium Storage Performance. *Adv. Funct. Mater.* **2019**, *29*, 1901072. [CrossRef]

167. Dou, X.; Hasa, I.; Saurel, D.; Vaalma, C.; Wu, L.; Buchholz, D.; Bresser, D.; Komaba, S.; Passerini, S. Hard carbons for sodium-ion batteries: Structure, analysis, sustainability, and electrochemistry. *Mater. Today* **2019**, *23*, 87–104. [CrossRef]
168. Xie, L.; Tang, C.; Bi, Z.; Song, M.; Fan, Y.; Yan, C.; Li, X.; Su, F.; Zhang, Q.; Chen, C. Hard Carbon Anodes for Next-Generation Li-Ion Batteries: Review and Perspective. *Adv. Energy Mater.* **2021**, *11*, 2101650. [CrossRef]
169. Cao, B.; Liu, H.; Xu, B.; Lei, Y.; Chen, X.; Song, H. Mesoporous soft carbon as an anode material for sodium ion batteries with superior rate and cycling performance. *J. Mater. Chem. A* **2016**, *4*, 6472–6478. [CrossRef]
170. Zhao, L.-F.; Hu, Z.; Lai, W.-H.; Tao, Y.; Peng, J.; Miao, Z.-C.; Wang, Y.-X.; Chou, S.-L.; Liu, H.-K.; Dou, S.-X. Hard Carbon Anodes: Fundamental Understanding and Commercial Perspectives for Na-Ion Batteries beyond Li-Ion and K-Ion Counterparts. *Adv. Energy Mater.* **2021**, *11*, 2002704. [CrossRef]
171. Schroeder, M.; Winter, M.; Passerini, S.; Balducci, A. On the Use of Soft Carbon and Propylene Carbonate-Based Electrolytes in Lithium-Ion Capacitors. *J. Electrochem. Soc.* **2012**, *159*, A1240–A1245. [CrossRef]
172. Schroeder, M.; Winter, M.; Passerini, S.; Balducci, A. On the cycling stability of lithium-ion capacitors containing soft carbon as anodic material. *J. Power Sources* **2013**, *238*, 388–394. [CrossRef]
173. Lim, Y.-G.; Park, M.-S.; Kim, K.J.; Jung, K.-S.; Kim, J.H.; Shahabuddin, M.; Byun, D.; Yu, J.-S. Incorporation of conductive polymer into soft carbon electrodes for lithium ion capacitors. *J. Power Sources* **2015**, *299*, 49–56. [CrossRef]
174. de las Casas, C.; Li, W. A review of application of carbon nanotubes for lithium ion battery anode material. *J. Power Sources* **2012**, *208*, 74–85. [CrossRef]
175. Al Hassan, M.R.; Sen, A.; Zaman, T.; Mostari, M.S. Emergence of graphene as a promising anode material for rechargeable batteries: A review. *Mater. Today Chem.* **2019**, *11*, 225–243. [CrossRef]
176. Yoo, E.; Kim, J.; Hosono, E.; Zhou, H.-s.; Kudo, T.; Honma, I. Large Reversible Li Storage of Graphene Nanosheet Families for Use in Rechargeable Lithium Ion Batteries. *Nano Lett.* **2008**, *8*, 2277–2282. [CrossRef] [PubMed]
177. Han, S.; Wu, D.; Li, S.; Zhang, F.; Feng, X. Graphene: A Two-Dimensional Platform for Lithium Storage. *Small* **2013**, *9*, 1173–1187. [CrossRef] [PubMed]
178. Wu, Y.; Zhang, T.; Zhang, F.; Wang, Y.; Ma, Y.; Huang, Y.; Liu, Y.; Chen, Y. In situ synthesis of graphene/single-walled carbon nanotube hybrid material by arc-discharge and its application in supercapacitors. *Nano Energy* **2012**, *1*, 820–827. [CrossRef]
179. Sun, Y.; Tang, J.; Qin, F.; Yuan, J.; Zhang, K.; Li, J.; Zhu, D.-M.; Qin, L.-C. Hybrid lithium-ion capacitors with asymmetric graphene electrodes. *J. Mater. Chem. A* **2017**, *5*, 13601–13609. [CrossRef]
180. Li, G.; Li, Y.; Liu, H.; Guo, Y.; Li, Y.; Zhu, D. Architecture of graphdiyne nanoscale films. *Chem. Commun.* **2010**, *46*, 3256–3258. [CrossRef]
181. Gao, X.; Liu, H.; Wang, D.; Zhang, J. Graphdiyne: Synthesis, properties, and applications. *Chem. Soc. Rev.* **2019**, *48*, 908–936. [CrossRef]
182. Chen, X.; Jiang, X.; Yang, N. Graphdiyne Electrochemistry: Progress and Perspectives. *Small* **2022**, *18*, 2201135. [CrossRef] [PubMed]
183. Du, H.; Yang, H.; Huang, C.; He, J.; Liu, H.; Li, Y. Graphdiyne applied for lithium-ion capacitors displaying high power and energy densities. *Nano Energy* **2016**, *22*, 615–622. [CrossRef]

materials MDPI

Review

Application of Nanotechnology in Extinguishing Agents

Anna Rabajczyk *, Maria Zielecka and Justyna Gniazdowska

Scientific and Research Center for Fire Protection—National Research Institute, Nadwiślańska 213, 05-420 Józefów, Poland

* Correspondence: arabajczyk@cnbop.pl; Tel.: +48-22-769-32-69

Abstract: Extinguishing agents are a very important tool in the field of security, both in terms of private and social aspects. Depending on the type of burning substance and place of fire, appropriately prepared and developed solutions should be used. We can distinguish, among others, materials, powders or foaming agents. Modifications introduced into them, including ones based on the achievements in the field of nanotechnology, can improve their safety of use and extend their service life. Such amendments also reduce the costs of production and neutralization of the area after a fire, and increase the fire extinguishing effectiveness. The introduction of nanoparticles allows, e.g., shortening of the fire extinguishing time, reduction of the risk of smoke emission and the toxic substances contained in it, and an increase in the specific surface of particles and thus increasing the sorption of pollutants. The elaborations use metal nanoparticles, e.g., NP-Ag, metal oxides such as NP-SiO_2, as well as particles of substances already present in extinguishing agents but treated and reduced to nanosize. It should be noted, however, that all changes must lead to obtaining a tool that meets the relevant legal requirements and has appropriate approvals.

Keywords: extinguishing agents; nanotechnology; application

Citation: Rabajczyk, A.; Zielecka, M.; Gniazdowska, J. Application of Nanotechnology in Extinguishing Agents. *Materials* **2022**, *15*, 8876. https://doi.org/10.3390/ma15248876

Academic Editors: Robert A. Varin and Geng Zhong

Received: 10 November 2022
Accepted: 8 December 2022
Published: 12 December 2022

1. Introduction—Extinguishing Agents

A fire is defined as a chemical combustion process at which high-temperature oxidation of a combustible material (or fuel) takes place [1]. Oxygen, temperature, fuel and chemical reactions are required for the fire to catch and spread [2]. A fire can occur both inside a building and outdoors, and it can be at one point or may occupy a large area. It can include natural, artificial, modified, simple and complex substances. The problem of fires, especially in relation to large areas, is expanding from year to year as a consequence of climate change and the increasingly frequent high temperatures and hydrological droughts. Tidey [3] showed that there was a fourfold increase in the number of forest fires, as there were over 1900 fires between January and July 2022, while the average for 2006–2021 for this period is equal to 520. He also indicated that 2021 was the worst season in the EU since 2000 in terms of the number of forest fires and the area covered by the fire. It should be noted, however, that fires also occur in construction facilities, industrial plants, waste depots, material heaps or other facilities, which determines the number of fires in general. Over the last five years (2017–2021), just in Poland the number of fires amounted to an average of 132,813 [4], causing a number of socio-economic losses, i.e., losses in infrastructure, property, environment and above all losses in people.

The selection of an extinguishing agent is, therefore, a necessary element not only in respect to the prevention system, but also in respect to the subsequent fire extinguishing actions. The extinguishing agent must be adapted to the potential fire hazards, as it allows for adequate protection and minimization of the fire consequences. The choice of an extinguishing agent is also subject to the relevant regulations and constitutes a legal requirement, the purpose of which is to protect the place of the incident against the spread of fire and to maximally reduce the time of fire extinguishing. This obligation applies not only to firefighters during the firefighting action, but also to the owners and administrators of real property. It should be remembered that the combustion reaction releases energy, various substances and heat. Therefore, when extinguishing fires, it is important to select an appropriate extinguishing agent as well as extinguishing method and technique. This is defined by a number of parameters characteristic to each specific case, in particular by the fire class, and by the properties of substances and materials involved in the fire (especially by their physical state). A special position among fires is occupied by those involving vegetable and animal fats (class F of fire) [5,6]. Class A fires include fires of solids of organic origin, during which the phenomenon of incandescence occurs. Class B fires are fires of flammable liquids and substances that melt due to the heat generated by a fire, while class C fires include fires of gases, e.g., methane, acetylene, etc. Class D fire is a fire of metals, including magnesium, uranium and sodium [5,6]. Fires caused by electrical equipment are marked with the symbol of an electric spark, not the letter E [5,6]. It should be noted, however, that in some countries, such as the UK, there are fire classes such as:

- Class A—the presence of solids, including materials such as paper, wood and plastic.
- Class B—the presence of liquids, such as paraffin, petrol and oil.
- Class C—flammable gases, such as propane, butane and methane.
- Class D—the presence of metal products, such as aluminum, magnesium and titanium.
- Class E—caused by electricity or ones that involve electrical equipment and apparatus.
- Class F—most commonly occur in kitchens and food preparation facilities, and involve cooking oil or fat [7].

Due to the variety of substances, elements that are affected by the fire, and the scale and possibilities of the fire spreading, the fire suppression mechanism of extinguishing agents also has to be diverse. Its effect can be obtained by removing heat from the surface of the burning material, cooling, cutting off the access of oxygen as well as by inhibiting radical reactions taking place in the flame. For this reason, there are numerous extinguishing agents (Figure 1) based on water, loose materials and gases.

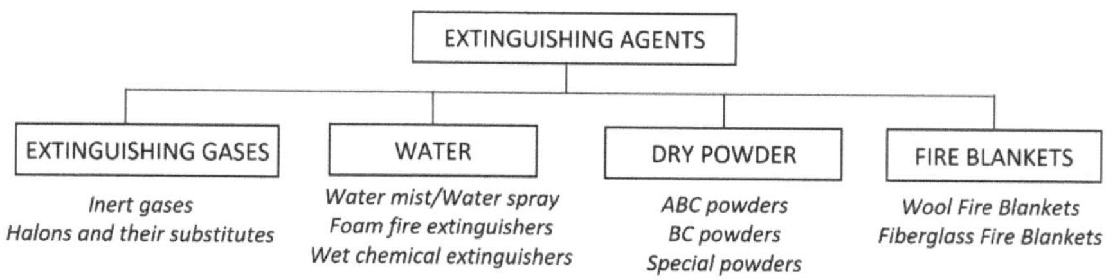

Figure 1. Division of extinguishing agents.

Taking into account the above, extinguishing agents are subject to modifications and continuous improvement in order to better protect people, property and infrastructure in the event of a fire. Modifications introduced in recent years are increasingly based on nanotechnology (Figure 2). On the one hand, new substances are used, the sizes of which are in the nanoscale. On the other hand, scientists take advantage of the already known compounds, the particle sizes of which are getting smaller, which allows obtaining new properties of the same compounds.

The introduced innovative solutions allow for the faster and more precise action of extinguishing agents. Nanoparticles are characterized by a larger adsorption surface, which results in faster reactions with compounds emitted during a fire. Nanoparticles often have the ability to produce reactive oxygen particles, which allows the degradation of organic, toxic compounds into simpler, less toxic or neutral compounds. The use of nanoparticles can therefore reduce the risk of smoke spreading along with the toxic substances contained in it, minimizing the threat to human life and health.

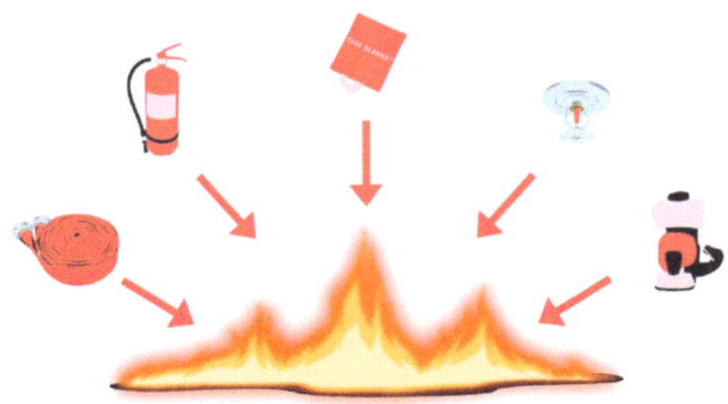

Figure 2. Extinguishing media covered by the work on modification based on nanotechnology.

Therefore, continuous research in this area is essential. It allows for the development of tools adequate to the changing materials, structures and substances that surround us. Tracking the effects of the introduced innovations also allows for determining further directions of the introduced solutions and constant improvement of tools in the area of fire protection.

The aim of this review is to present the most important information on nanotechnology-based innovations in extinguishing agents, based on a literature review using the following keywords: extinguishing agents, fire extinguisher, foam, fire blanket, nanoparticles, fire protection, fire, smoke. The literature review was carried out using the following databases: Web of Knowledge, Scopus and Google Scholar. This study also covered Espacenet, Patentscope and Google Patents, which resulted in the presentation of selected patents related to the subject. In this paper, the analysis covers both literature reports and patent databases, reports of companies whose activities are related to fire protection. The information obtained will allow determination of the directions of introduced changes as well as their legitimacy in the context of effective fire protection. It should be noted that products used in fire protection must meet the relevant legal requirements in a given country and be adequate to the fire situation. In addition, such products must have a certificate of approval issued by the appropriate approval body.

2. Powder Extinguishing Agents

The EN 615 standard [8] defines dry powder as an extinguishing medium in the form of finely divided solid chemicals, consisting of one or more main components and additives improving its properties [9]. The main advantage of powders is the combination of physical and chemical action on the fire, which makes the extinguishing process relatively quick. It should be noted, however, that powders indicate a negative impact on the environment, cause losses after a fire due to the elements being covered by a cloud of powder, and pose a threat to humans through the possibility of getting into the eyes or having an allergic effect on the skin. Moreover, they are characterized by a high price.

Depending on the criterion, several divisions can be found. One of them takes into account the division into classes of fires for which they were intended. It includes "BC" and "ABC" powders as well as "D" special powder intended for extinguishing class D fires, i.e., metal fires. Since powder extinguishing agents are compositions containing powdered chemical compounds characterized by flame inhibition, the extinguishing capabilities of powders depend on their chemical composition and determine the use of a given powder for extinguishing certain types of fires (Table 1) [10].

Table 1. Chemical composition of the base of the powder fire extinguishing agent vs. application to extinguish different groups of fires.

Extinguishing Powder Base	Chemical Formula	Class of Fires Being Extinguished
Potassium hydrogen carbonate	$KHCO_3$	BC
Urea + Potassium hydrogen carbonate	$NH_2CONH_2 + KHCO_3$	BC
Sodium hydrogen carbonate	$NaHCO_3$	BC
Sodium tetraborate	$Na_2B_4O_7 \cdot 10H_2O$	D
Potassium chloride	KCl	D
Sodium chloride	NaCl	D

The effectiveness of extinguishing powders is related to their properties, in particular the ability to deactivate free radicals, which then affects the interruption of the chain combustion reaction. Additionally, the low melting point is important. It allows the powder to melt quickly, and form a protective layer on the surface of the burning material, reducing oxygen access to the burning surface. What is more, the access of oxygen is limited by the clouds of extinguishing powder formed above the burning surface. At the same time, with a sufficiently high pressure of the powder stream, the flames are mechanically blown out. It should be noted that alkali metal salts are characterized by a very good ability to inhibit free radical reactions; however, the extinguishing efficiency of salts containing the same anion depends on the presence of the appropriate alkali metal cation and decreases with its atomic mass [11]. Still, potassium and sodium are the most frequently used for economic reasons. The use of a given alkali metal salt determines the extinguishing efficiency, which varies with the anion used, according to the series: $PO_4^{3-} < SO_4^{2-} < Cl^- < Br^- < J^- < CO_3^{2-} < CN^- < C_2O_4^{2-}$. Salts containing oxalates and cyanides are not used due to their strong toxic properties [11].

Carbonate powders are used to extinguish class B and C fires, while phosphate powders are used to extinguish fires of solids, flammable liquids and gases (ABC). Chlorine powders, on the other hand, based on metal chlorides (most often NaCl or KCl), are used to extinguish metal fires. The greatest efficiency in extinguishing fires is shown by the BC type extinguishing powders, i.e., powders based on carbonates and urea powders. The salts of these powders, such as sodium hydrogen carbonate $NaHCO_3$, have an inhibitory effect on the flame, thus slowing down the course of free radical reactions during a fire. ABC type extinguishing powders, based on phosphate salts, also create a glassy layer on the surface of the combusted solid material, isolating the fuel from the oxidant. The separating layer is formed by the thermal decomposition of phosphate salts and the formation of polyphosphates [12]. A feature of ABC powders containing ammonium dihydrogen phosphate $(NH_4)H_2PO_4$ is hindering the process of reignition of cellulose materials. Metaphosphoric acid $(HPO_3)_3$, which has a low melting point, appears when the dust comes into contact with a flame, forms a glassy film on solid surfaces and cuts off contact with oxygen in class (A) fires that burn with embers [13]. This process is responsible for the fire suppression effect. The dry powder also has a cooling effect. However, the thermal energy required to decompose dry chemical powders is strongly related to the extinguishing capacity of the substance. As a result, all dry chemicals need to be heat sensitive in order for the substance to become chemically active and absorb ("swallow") the heat. Additionally, spraying of dry chemical powder creates a dust cloud between the flame and the fuel [14]. This cloud, to some extent, isolates the fuel from the heat emitted by the flame.

The inhibitory effect of dry powders occurs according to two processes, i.e., homophasic inhibition or heterophasic inhibition. In the first case of homophase inhibition, the extinguishing powder acting as the inhibitor, as well as the fuel and the oxidant, are in the same gas phase [15]. In the second case of heterophase inhibition, the inhibitor, i.e., the extinguishing powder and its decomposition products, as well as the reactants, i.e., the fuel and the oxidant, are in the gas phase. The inhibitor may be gaseous metal hydroxide or gaseous metal atoms. The gaseous metal hydroxide is produced by the pyrolysis of oxygen salts in a two-step reaction to form a liquid metal oxide which then reacts with water vapor [15–17]. On the other hand, anaerobic salts decompose in the presence of hydrogen ions to form metal cations which, when reacted with water vapor, form metal hydroxide. The latter reaction is reversible, which explains the lower inhibitory effect of anaerobic salts in extinguishing flame fires. It is, in fact, a chain reaction that limits the combustion process. This is possible due to the strong reducing properties of alkali ions that arise through the recombination of these ions until they react with carbon dioxide. In this case, the inhibition process takes place with the formation of many transition complexes [15,16].

In the process of heterophasic inhibition, the extinguishing powder is in the solid phase and the reactants (fuel and oxidant) in the gas phase. Radicals present in the gas phase collide with solid powder particles; this in turn prevents their further participation in the combustion process due to energy loss as a result of the collisions. Such a course of the process causes the advantage of recombination reactions over propagation reactions, and the combustion process is inhibited [15,16].

Another mechanism influencing the effective operation of extinguishing powders is the isolation of the fuel from the oxidant. This mechanism is characteristic of fire-extinguishing powders containing salts of phosphoric acid [18,19]. Such powders are particularly useful in extinguishing fires of solids forming glowing coals, classified as group A, which takes place during the combustion of materials of organic origin. Extinguishing powder containing orthophosphates when applied to a burning surface undergoes a cycle of changes leading to the formation of poly metaphosphoric acid. Under fire conditions, it is a liquid substance of high viscosity, which, after the fire stops, forms a glassy layer that insulates the fuel from the oxidant. This layer significantly reduces the risk of glowing coals reignition (e.g., in the case of cellulosic materials).

An important influence on the effectiveness of the extinguishing powder, apart from the composition, which is a mixture of various metal salts and water-repellent agents, is also the size of the powder grains. Smaller powder particles increase the surface area and break down particles of various substances faster with the release of decomposition products neutralizing free radicals. Consequently, they provide better interaction with the burning material and positively influence the fire extinguishing capacity. It was found that the powders with various particle sizes were the most effective, i.e., containing:

- from 60 to 80% by weight of the total powder in the range of 20 ÷ 60 μm;
- from 10 to 15% by weight of the total powder in the range of 100 ÷ 200 μm [20].

It was also found that the content of the fraction with a particle size of 100 ÷ 200 μm positively influences the range of the powder stream and the effectiveness of its penetration into the fire [20]. On the other hand, a too small particle size of the powder may cause entrainment of these particles by volatile substances released during a fire, which may determine a reduction in extinguishing efficiency [15,16].

The characteristics and structure of the materials used are changing faster and faster. The requirements regarding the need to ensure greater security and better and more effective security are also changing. All this requires work aimed at increasing the effectiveness of extinguishing powders, including the modification of extinguishing powders with nanoparticles immobilized in the structure of extinguishing powder particles. Ni et al. [21] developed a new type of extinguishing powder by immobilizing $NaHCO_3$ nanoparticles on porous zeolite [21]. The new powder based on potassium salts with organic and inorganic additives was prepared using a ball mill [22]. The results showed that the modified powders are superior in fire suppression compared to commercial products. Research work focused on the use of magnesium hydroxide, including those containing nanoparticles [23]. It has been shown that the developed powder is characterized by higher extinguishing efficiency compared to commercial BC and ABC powders. Very good results of the extinguishing efficiency were obtained for the extinguishing powder consisting of nanoparticles of magnesium hydroxide with the addition of melamine cyanurate and phosphorus-based ODOPB (phosphorus ODOPB—10-dihydro-9-oxa-10-phosphaphenanthrene-10-oxide) [24]. The components of the extinguishing powder were mixed wet, dry and ultrasonically in order to increase the homogeneity of the nanocomposite. The fire extinguishing time with the nanocomposite was 45.2% shorter than that of the commercial ABC–MAP powder [24]. Moreover, the amount of nanocomposite used was 63.2% lower than in the case of commercial powder. The possible mechanism of action of this nanocomposite fire extinguishing powder was also discussed. On the basis of the results obtained, it was found that the extinguishing mechanism in this case is complex and consists of simultaneous chemical and physical processes as well as a cooling and flame suppression effect [24].

It should be noted that dry chemical powders have good stability at low and normal temperatures. At higher temperatures, some additives melt and cause stickiness; therefore, it is recommended that the storage temperature not exceed 50 °C. At fire temperatures, the active substances contained in the dry chemical powder decompose and perform their extinguishing functions. Apart from basic substances, the composition of extinguishing powders also includes various additives that are responsible for improving the performance of fire extinguishers, including resistance to wetting with water, fluidity, resistance to caking and facilitating encapsulation. Literature reports indicate that work is still underway to improve the extinguishing properties of extinguishing powders.Nie et al. [25] introduced nanometer-sized, very-fine sodium bicarbonate powder into the pore structure of zeolite to solve the problem of powder agglomeration. This allowed the powders to penetrate the flame efficiently and achieve better extinguishing performance [25,26]. Whereas Kuang et al. [22] found that the extinguishing performance of the ultra-fine powder was related not only to the size of the powder surface structure, but also to the main constituents of the extinguishing agent. The experiment showed that the nanometric magnesium hydroxide powder had a higher extinguishing efficiency than the commercial dry ammonium hydrogen carbonate powder [26,27]. Koshiba et al. [28] conducted an experimental study of combustion inhibition and found that metallocene powder has greater fire-fighting benefits than the commonly used ammonium dihydrogen phosphate. The minimum extinguishing concentration of ferrocene was 11 times lower than that of ammonium dihydrogen phosphate, indicating that the metallocene, represented by ferrocene, was more effective. The effect of compounds containing manganese and zinc on the combustion rate of a methane–air flame was investigated experimentally by Linteris et al. [26,29] and compared with iron pentacarbonyl $Fe(CO)_5$ and CF_3Br bromotrifluoromethane. It was found that powders containing nanoparticles of manganese or zinc were more effective in extinguishing the flame, and the effectiveness of stopping fires of zinc compounds was twice as high as CF_3Br [26].

A typical extinguishing powder available on the market (e.g., ABC extinguishing powder: specific surface 0.34–0.46 m^2/g) may prove insufficient due to the sedimentation time and small contact surface, e.g., when extinguishing a gas flame (Figure 3). The solution for extinguishing this type of fire are nanopowders (grain diameter approx. 100 nm, specific surface area 25–100 m^2/g), which fall in the air at a speed of 7.3 cm/day, creating an aerosol cloud, thanks to which the extinguishing efficiency is approx. 30 times greater than common powder [30,31]. However, it should be remembered that not all nanopowders have quenching properties, an example of which is a powder containing NiO nanoparticles. For extinguishing purposes, ZrO_2 works well, as it inhibits methane combustion. Extinguishing aerosols are also an interesting solution, i.e., fire extinguishing agents produced as a result of the reaction of solid combustion in special aerosol generators. This method does not reduce the oxygen level in the air. The task of extinguishing aerosols is to bind free radicals resulting from combustion processes with a highly efficient and effective extinguishing aerosol. Its micro- or nanoparticles present in active surfaces break the chain of physicochemical reactions. The use of these measures has an impact on the environment, as they leave a trace of contaminants at the site of the fire. In the products generated in the production of aerosols, there are substances such as ammonia, nitrogen oxides, carbon monoxide and hydrogen cyanide which have a harmful effect on the human body and the environment [32].

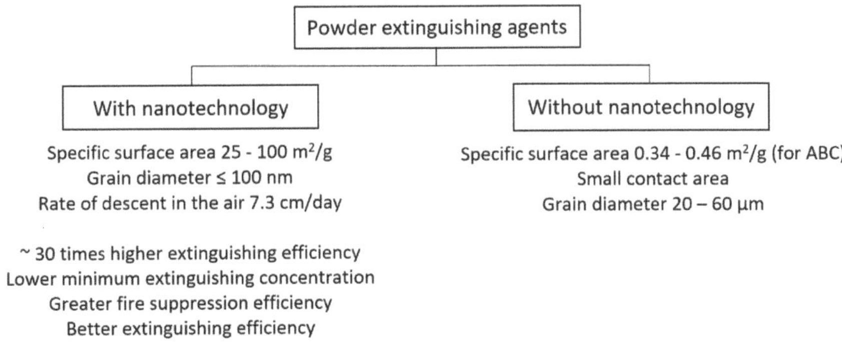

Figure 3. Comparison of standard and modified extinguishing powders with the use of nanotechnology.

Although the extinguishing aerosols do not contain any corrosive substances, in an aqueous environment they can reach a pH of 8–10, which can damage materials and equipment sensitive to high pH. Due to the alkaline reaction of the depositing aerosols after discharge, they are not recommended for use in "clean rooms". It should also be ensured that at high temperatures in the stream, the firefighter does not get burned during the discharge. An additional hazard may be the release of metal nanoparticles and metal oxides during the event. Many of these compounds are highly reactive [33].

Fire extinguishing sprays are based on potassium carbonates and nitrates. They have a grain size in the range of 200 nm and at aerosol release temperatures, i.e., > 1000 °C, they can be active centers and react immediately with radicals supporting the flame combustion reaction [32,34].

The aerosols are applicable to fires of class A, B, C and F and electrical devices with voltage up to 36 kV. They are also used during fires in archives and monuments (especially with difficult access), the armaments industry and in the event of maritime incidents. They are produced by extinguishing aerosol generators, which can be part of a permanent fire extinguishing system or constitute a stand-alone extinguishing device, and therefore must meet certain requirements [35]. Depending on how the fire generators are triggered, the aerosol may act locally or volumetrically. Its effectiveness may also vary depending on environmental conditions. The developed surface of the solid and the type of chemical reaction are also important. In the case of nanoparticles, which have a significantly increased surface area, a higher fire suppression efficiency is observed than in the case of particles of relatively larger sizes. However, the reduction of the particle size affects the possibility of agglomeration of the particles, which may result in the difficult contact of the particle surface with the flame radicals. Biel et al. [32] also found that one of the parameters determining the atomization of the nanosize of an aerosol is environmental humidity, which has an effect, inter alia, on the grain distribution and thus on the surface activity of the FP-40C type extinguishing aerosol [32]. The results of the research showed that in dry (30% humidity) and very humid (90% humidity) environments the number of aerosol nanoparticles decreases with time due to both aggregation and lumping, and the most favorable conditions for extinguishing aerosol atomization occur in an environment with a humidity of approx. 70%. This may be related to the more effective capture of water radicals from chain reactions taking place in the flames [32].

Common additives to extinguishing compositions are also silicas [36] (flame or precipitated), metal stearates, talc [34]. The search for the optimal content of functional additives in powder compositions is quite a difficult task. Too low a proportion of the additive causes deterioration of the utility properties of the powder and shortens the guaranteed shelf life. If the amount of the functional additive is excessive in relation to the surface area of the powder particles, the flow of the composition deteriorates due to the increase in the number of contacts and the additional frictional forces acting between the additive particles [37,38]. The data obtained under dynamic conditions make it possible to predict the behavior of powders at high flow rates [39]. The effectiveness of hydrophobized nanosilica with a particle size of about 65 nm as an additive to ammonium phosphate fire extinguishing powders was a subject of a study by Shamsutdinov et al. [40]. The surface of the extinguishing component was evenly covered with spherical hydrophobized nanosilica nanoparticles. The addition of nanosilica made it possible to obtain a hydrophobic coating on the particles of the extinguishing powder. The apparent contact angle of the coated particles was found to be greater than 160°. Dynamic flow resistance of commercial and tested extinguishing powders was compared. The specific energy and flow energy of the aerated test powder were relatively low, which indicates a poor aerodynamic interaction between particles in dynamic processes. It was found that the spray resistance of the tested powder was the lowest among the tested extinguishing powders (including commercial ones), and therefore it was characterized as having the best flow. Similar results were obtained with the use of mesoporous nanosilicas of various structures (MCM-41, MCM-48 and SBA-15) with a large specific surface of over 1400 m^2/g [41]. Such nanosilicas have been found to improve the fluidity of ammonium phosphate-based extinguishing powders.

The aforementioned nanosilica positively influenced the improvement in fluidity of extinguishing powders based on ammonium phosphate. It has been shown that with the reduction of the particle size of the powder (or agglomerates), the internal friction between the powder particles decreases [42,43].

The particle size distribution is also important. It has been found that the narrow particle size distribution makes it possible to obtain a powder with better flow properties [44]. Veregin and Bartha [45] showed that the particle size of the functional nanoadditive affects the flow properties of powder materials and depends on the forces of interparticle interactions and the contact radius of the powders' surface.

The reduction in particle size is accompanied by a reduction in their contact surface, which leads to a decrease in the influence of interparticle forces. It was found that increasing the mixing time of the powders improves the properties of the powders [46]. This effect results from the possibility of obtaining an increase in the homogeneity of the coating of extinguishing powder particles with silica nanoparticles and reducing the possibility of agglomeration of silica nanoparticles.

It has been found that the use of hydrophobic additives significantly improves the fluidity of the materials by eliminating capillary bridges between the particles, thus ensuring easy breaking of agglomerates of particles in powder compositions [47,48]. It should be emphasized that if the amount of silica nanoparticles exceeds the surface area of the powdered extinguishing material, the flow properties of the composition decrease due to the increasing contact and friction forces between the nanoparticle and particles [49,50].

Modification of the properties of extinguishing powders by introducing nanoparticles enables a significant increase in their extinguishing efficiency (Table 2) and makes it possible to improve the safety of people and property during an incident.

The introduction of nanoparticles to the extinguishing powder changes the properties of the extinguishing powder by:

- changing the composition of the extinguishing powder particles after incorporation of nanoparticles, which modifies the powder's mechanism of action and enables the extinguishing efficiency of such powder to be increased;
- immobilization of nanoparticles on the surface of extinguishing powder particles, which changes their physical properties, such as fluidity or hydrophobicity, significantly improving the extinguishing efficiency of the powder.

It should be noted that nanosubstances are released into the environment, characterized by diverse activity, which is conditioned by the size of the surface, and the shape and size of the grain or structure. Therefore, they can have various impacts on human health and environmental safety. It is therefore necessary, when introducing new solutions, to carry out an assessment in the field of toxicology and biodegradation or interaction with compounds present in the environment.

Table 2. Possibilities of using nanotechnology to modify powder extinguishing agents.

Nanoparticles	Characteristic	Effect	Ref.
$NaHCO_3$	NP-$NaHCO_3$ on porous zeolite	improving fire suppression efficiency	[21,25]
Ferrocen	sized nanoparticles < 100 nm	greater efficiency of extinguishing flames and stopping fires	[28]
Mn, Zn	sized nanoparticles < 100 nm	greater efficiency of extinguishing flames and stopping fires	[29]
$Mg(OH)_2$	NP-$Mg(OH)_2$ with the addition of melamine cyanurate and phosphorus ODOPB—10-dihydro-9-oxa-10-phosphaphenanthrene-10-oxide	higher extinguishing efficiency, shorter fire extinguishing time by about 45%, lower consumption of extinguishing powder	[24]
SiO_2	hydrophobized nanoparticles with a size of about 65 nm as an additive to extinguishing powders based on ammonium phosphates	improving the fluidity of powders	[40]
SiO_2	mesoporous nanoparticles of various structures (MCM-41, MCM-48 and SBA-15), characterized by a large specific surface (>1400 m^2/g)	improving the fluidity of powders	[41]

3. Foam Concentrates

Foam concentrates used in rescue and firefighting operations are used to generate foam, i.e., bubbles made of liquid. The EN 1568 series of standards specifies the requirements for the physicochemical properties and minimum effectiveness of extinguishing foams intended for the production of medium-, light- and heavy-expansion foams suitable for surface application to liquids immiscible and/or miscible with water. The EN 1568 standard [51] distinguishes types of foam extinguishing agents such as: synthetic (S), protein (P), fluoroprotein (FP), alcohol-resistant (AR), aqueous film-forming concentrates (AFFF), fluoroprotein water film-forming agents (FFFP) and fluorine free foam concentrates (F3).

The composition of synthetic foaming agents (S) is based on surfactants that do not contain organofluorine compounds. Protein agents (P) contain hydrolysed protein substances of animal origin, and fluoroprotein (FP) and film-forming fluoroprotein concentrates (FFFP) additionally fluorinated surfactants. The composition AFFF is based on mixtures of hydrocarbons and fluorinated foaming agents, thanks to which they have the ability to form a water film on the surfaces of some fuels. Alcohol-resistant foaming agents (AR) in their composition contain a mixture of the above agents, additionally containing substances that make the foams resistant to polar liquids having the ability to form a polymer film on the alcohol surface. Fluorine free foam concentrates (F3) are intended for applications analogous to AFFF and/or AR foams. They are based on mixtures of hydrocarbon surfactants and non-fluorine stabilizers [52]. The mixtures also contain corrosion inhibitors and substances that lower the freezing point [53,54].

Pursuant to the Regulation of the European Union Commission [55], from July 4th 2020, the following may not be manufactured or placed on the market: foaming agents containing perfluorooctanoic acid (PFOA) or its salts in a concentration equal to or higher than 25 ppb PFOA, including its salts or 1000 ppb of one derivative or a combination thereof. The environmental problem of using hazardous substances is solved by adding other surfactants that do not contain fluorine or by adding a small amount of short chain fluorocarbon surfactant to replace the long chain fluorocarbon surfactant.

Foam is used both to extinguish fires, mainly class B, but also class A. In the event of a fire of polar liquids, it is necessary to use foam produced by foam-producing agents specially developed for this purpose. Fire extinguishing foams are completely useless for extinguishing gas fires. Additional hazards may arise when extinguishing fires of substances that react with water, such as organoaluminum compounds, metals or carbide [56].

Therefore, fire foams are primarily used to extinguish flammable liquids in chemical, food or coke industry plants, as well as refineries and petrochemical plants.

The properties of extinguishing foams are determined by the ratio of the foam volume to the volume of the solution from which the foam was formed, expressed by the so-called foaming number. The use of appropriate foaming agents makes it possible to obtain light-, medium- and heavy-expansion foam (Figure 4).

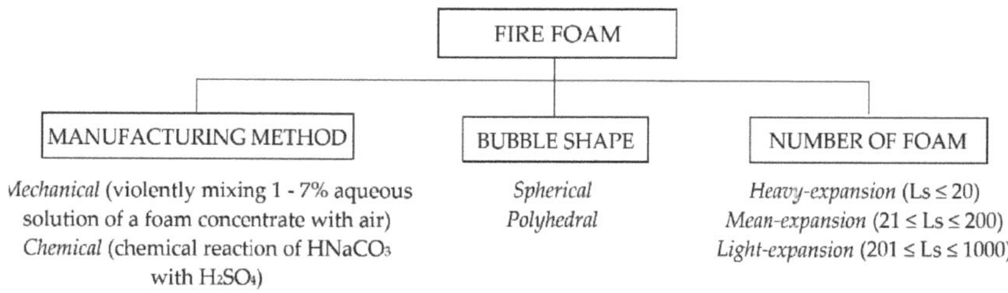

Figure 4. Division of fire foams.

The properties of foams are also determined, in addition to the number of foaming, by parameters such as dispersion, foam durability or fluidity. The degree of disintegration of the foam bubbles determines the dispersibility. The test results showed that the larger the bubble diameter, the smaller the dispersibility [57]. The ability of the foam to maintain its properties, which it obtained at the time of production, is determined on the basis of the draining time of the aqueous solution of the foaming agent therefrom. Durability depends not only on the number of foaming or dispersibility, but also on such elements as the concentration and properties of the foaming agent, and the quality of the water which is the basis for the production of the foam. The study results showed that the slower the bubble destruction process takes place, the more durable the foam, and the faster it spreads over the surface of the burning material, the less foam is destroyed and the faster fire is extinguished [57,58]. Therefore, when developing the agent, one should also take into account shaping the behavior of the extinguishing foam depending on the temperature of the fire and air, and the type of environment with which the foam comes into contact [57,58].

Fire foams, thanks to the creation of a low surface tension of the water solution and the foam concentrate, allow the water film, although heavier than the flaming liquid, to freely spread over its surface. This prevents flammable gases and vapors from entering the combustion zone, allowing the fire zone to cool down. A properly applied layer of foam allows for the elimination of the combustion zone and the cooling of the fire zone. Therefore, the effectiveness of a fire extinguishing action with the use of foam extinguishing agent depends not only on its proper application, but also on the quality of the foam concentrate [58].

Therefore, it should be stated that the extinguishing foam is a thermodynamically unstable dispersion system and the foam stability in difficult conditions is still an important element determining the effectiveness and safety of its use. Therefore, work is underway on the stability of foam extinguishing agents by introducing new agents, including nanoparticles [59]. For example, Li et al. [60] used a titanium dioxide (H-TiO_2)/gel system with a three-dimensional lattice structure and a flame retardant effect. H-TiO_2 foam gel was obtained by mixing acrylic acid (AA) and acrylamide (AM) with sodium borate and H-TiO_2, the optimal gelation time was 18 min at 40 °C, and the concentrations of the cross-linking agent and initiator were 0.3 wt.% and 0.4 wt.%. The results obtained show that for the coal sample the ignition time (TTI) of the H-TiO_2 foam gel was up to 12 times longer than for the raw coal samples, while the maximum heat release rate (HHR) was reduced by 56.59%, and the minimum total emission heat (THR) was reduced by 32.04% [60,61].

Aluminum hydroxide nanoparticles (Nano-ATH) were used as an additive to the composite foaming agent solution to prepare nano-ATH foam with high stability [62]. The results of the work indicated that after the addition of nano-ATH, the viscoelastic modulus of the foam liquid film increased from 0.91 mN/m to 3.08 mN/m, and the foam volume increased from 380 mL to 700 mL. There was also an increase in the mechanical strength of the liquid film at a mass concentration of 2%, as the half-life of the foam increased from 78 sec to 453 s. In the samples of carbon with nano-ATH foam, the ignition time increased from 26 sec to 176 sec, the combustion sustained time was reduced from 520 sec to 211 sec, the peak heat release rate decreased from 117 to 58 kW/m^2 and the peak smoke production rate decreased from 0.084 to 0.0049 m^2/s during combustion, indicating good flame retardant properties and smoke suppression [62].

Additionally, nano-foam extinguishing agents with a non-toxic LC50/LD50 certificate have been developed. They are characterized by high resistance to oil and fire, as well as effective insulation when mixed with water to extinguish an electric fire [63]. The size of the nanoparticles used in the devices ranges from 80 to 150 nm on average. Due to the addition of a fluorosurfactant, the foam has a good fluidity and the nanofoam extinguishing agents are sprayed with high pressure gas to form a thin film on the surface of the oil in case of oil contamination. Thanks to the material refinement technology used, the foam gets better parameters and covers a larger area, and thus effectively blocks gas combustion, suppresses fuel vapors after extinguishing a fire and prevents the backflow of the air.

The idea of using nanoparticles was used in the research on the improvement of aqueous film-forming concentrates (AFFF). A structure has been developed that consists of 3–12 parts by weight of a complex surfactant, 10–150 parts of nanoparticles, 0.3–1 parts of stabilizer, 15–30 parts of dispersant, 3–9 parts of antifreeze, 1–10 parts suspending agent and water [64]. For the preparation of the suspension, nanoparticles of silicon dioxide, aluminum hydroxide, aluminum oxide, magnesium hydroxide, magnesium oxide, antimony oxide, calcium carbonate, iron oxide and antimony trioxide were used. To prepare the foam, the nanoparticles in the form of a suspension were evenly mixed with the complex surfactant solution. The extinguishing agent was added with a short chain fluorocarbon surfactant or a complex surfactant combined with an organosilicon surfactant and a hydrocarbon surfactant to form a compact foam layer and a closed water film layer, and obtain adequate stability and thermal insulation of the foam. The addition of nanoparticles has improved the thermal insulation properties of the foam layer and has avoided rapid cracking of the foam caused by too high an external temperature. Appropriate dispersion and stability of the foam containing nanoparticles dispersed in other components has a positive effect on the extinguishing efficiency and control of solid and liquid fires [64].

It should be added that silica is widely used in foams, which is characterized by a large surface and the ability to absorb water. The foam covers the surfaces, protecting them against the spread of the flame front, and separating the fuel from the oxygen. Additionally, low-expansion foams provide significant cooling [65–67]. Large-size ultrafast gelled foams have been developed, which are characterized by exceptional thermal stability, mechanical strength and biocompatibility. They allow the production of controlled gelling hybrid silica foams in the viscosity range from 2 to 30 Pa·s to 100 Pa·s. The foam consists of ordered silica nanoparticles with a narrow particle size distribution of ~10–20 nm. The obtained results indicate that the extinguishing efficiency of silica-based sol-gel foams is almost 50 times higher than that of ordinary water. Importantly, the biodegradation index is only 3 days, while even for conventional foaming agents this index is several times higher [68].

Strong foam stabilizing effect with the use of SiO_2 nanoparticle-cationic surfactant tetradecyltrimethylammonium bromide (TTAB) mixtures was obtained at concentrations of SiO_2 nanoparticles as low as 2 wt% [69]. Binks et al. [70] found that silica nanoparticles, modified with chloro- and methyl- silanes, can stabilize the foam even for several days without disproportionation. Solid particles increase the flow resistance of the fluid in the liquid film, so they can stabilize the foam [71,72].

Work is underway on foams used in the event of fires of hydrocarbon fuels. Current fire protection formulations, based on perfluorinated surfactants, are highly toxic and have extremely poor biodegradability, spread through the food chain and affect many species over an extended period of time. The work covers composites of metal carbonate nanoparticles and liquid ionic surfactants, surfactants in the form of nanocapsulated metal carbonate ions (NEIL), so that it is possible to create a stable foam and release CO_2, a compound depriving fire of oxygen [73,74].

The analysis of the results of work on the application of nanotechnology for the modernization of foaming agents showed that silica nanoparticles are most often used in the works (Table 3).

Table 3. Possibilities of using metal nanoparticles and their compounds in extinguishing foams.

Nanoparticles	Characteristic	Effect	Ref.
TiO_2	H-TiO_2/gel system with a three-dimensional network structure	flame retardant effect, longer ignition time	[60]
$Al(OH)_3$	addition to the composite foaming agent solution, mass concentration 2%	high foam stability, good flame retardant and smoke suppression properties	[62]
SiO_2, $Al(OH)_3$, Al_2O_3, $Mg(OH)_2$, MgO, FeO, Sb_2O_3	evenly mixed with the complex surfactant solution	improving the thermal insulation properties of the foam layer, affects the efficiency of extinguishing and controlling solid and liquid fires	[64]
SiO_2	ordered nanoparticles with a narrow particle size distribution of ~10−20 nm	50-times higher extinguishing efficiency, biodegradation index—3 days	[68]
SiO_2	SiO2-TTAB mixture, with an NP-SiO2 concentration of 2 wt%.	strong foam stabilizing effect	[69]
SiO_2	NP-SiO_2 modified with chloro- and methyl- silanes	stabilization of the foam even for several days without disproportionation	[70]
$Me_x(CO_3)_y$	surfactants in the form of nanoencapsulated metal carbonate ions (NEIL)	formation of a stable foam and the release of CO_2	[73,74]

The addition of nanoparticles improves the properties of fire-extinguishing foams in terms of stability, extinguishing efficiency, and smoke suppression. and thus improves the safety of both people, the environment and property. However, fire extinguishing foams may, to a greater or lesser extent, affect the environment and human health. The presence of synthetic surfactants and other chemical substances constituting the foam composition may have an irritating and toxic effect on humans and constitute a significant environmental pollution. Nanoparticles can also have a significant impact on humans and environmental processes. As indicated with surfactants, non-polar substances, such as hydrocarbons, which are persistent carcinogenic compounds, may migrate to soil and groundwater, posing a threat to human health and the quality of the environment. Therefore, the modifications should consider not only the effectiveness in terms of effective and quick fire extinguishing, but also minimizing the risk to environmental safety.

4. Other Fire Extinguishing Agents

The most frequently used extinguishing agent, mainly due to its availability, low price and relatively good extinguishing properties, is water. The extinguishing property of water is related to its cooling effect. It lowers the temperature of the burning material and the combustion zone. The water vapor generated during the extinguishing inhibits the reactions of free radicals with flammable gases, thanks to which it dilutes the combustion zone and has an insulating effect [75].

A contraindication to the use of water to extinguish fires are those fires of flammable materials which react violently in high temperatures in contact with water, causing additional danger. These are, i.a., fires of metals which react with water at room temperature or at combustion temperature, releasing large amounts of heat [15,76]. The increase in the extinguishing efficiency of water is possible by addition of chemical compounds that change its physical properties. The purpose of these compounds is to lower the surface tension of water, which results in better absorption of water into the surface of solid combustible materials. The compounds lowering the surface tension of water include, among others, wetting agents, i.e., compounds containing in their molecule a hydrophilic chain saturated with fluorine atoms, attached, for example, to the potassium salt of an aliphatic acid sulphamide, which are added to water in an amount of 0.5 to 2%, thickeners (15 to 25%) or thickeners—wetting [77].

For example, Dali et al. [78] showed that the quenching time of liquid hydrocarbons by suspensions containing carbon nanostructures (CNS), such as functionalized multi-walled carbon nanotubes (MWCNTs), is on average 3.5–5.0 times shorter than the quenching time of liquids with finely divided water. Wetting agents and additives increase the intensity of the heat sink, creating a film on the surface of the burning oil product. Carbon nanotubes increase the thermal conductivity and change the rheological properties of liquids at low concentrations [79]. It has also been found that the CNS water-based slurries are extinguishing agents with a predominantly cooling and diluting effect. Suspension droplets enter the combustion zone, which causes an intense heating to the boiling point. Such a process causes evaporation and cooling of the combustion zone, and the flame is extinguished with a sufficient amount of water vapor in the combustion zone [78,80]. However, it should be remembered that too high a concentration of nanoparticles may lead to aggregation of nanoparticles, which reduces the effective thermal conductivity of the suspensions and the specific heat of vaporization.

It should be noted that due to the good fluidity of the water, most of it is lost when sprayed onto the fire site. Large amounts of water are required for large-scale fires, rapid fire spread, and when extinguishing poses significant difficulties. Often, in the case of strong winds and the need to drop the agent from a height, it is hard to control the effectiveness of operations and some water resources are wasted. Therefore, research is being carried out on water retention in appropriate structures. Such a solution is, among others, superabsorbent resin, which is a three-dimensional network polymer. It has hydrophilic groups in its structure and is slightly cross-linked, which means that it can absorb a large amount of water, swell and keep water from flowing out. High-molecular polyacrylic acid superabsorbent resin hydrogel is widely used in the field of firefighting. This substance has a significant heat capacity under high-temperature conditions. However, some superabsorbent resins have a low self-adhesive capacity, and have difficulty in adhering to vertical wall surfaces, wooden furniture surfaces or surfaces containing stainless steel such as oil tanker surfaces. These objects cannot, therefore, be effectively covered, which prevents successful extinguishing and fire protection. To improve these properties, a polymeric hydrogel extinguishing agent has been developed. It contains 0.1–0.5 wt. % of a polyacrylate superabsorbent resin and/or an acrylate–acrylamide copolymer, and 0.01–0.5 wt. % inorganic nanoparticles and water. Inorganic nanoparticles, mainly diatomite, kaolin, bentonite, attapulgite, colloidal silica and precipitated silica, added to a special superabsorbent resin improve its viscosity, adhesion of hydrogel extinguishing agent and extinguishing effect [81].

Water can also be used in the form of water mist. Pressurized water is transported through pipes to special nozzles, which then spray water in the form of mist over the area. The water mist extinguishing system allows the temperature to be reduced quickly, the humidity to increase and prevents the oxygen fire from finding a new source. The developed Water Mist technology allows the production of water nanoparticles that facilitate the fogging process and reduce the amount of water needed to extinguish the fire [82]. Very-fine water mist can suppress the flame in case of fires covered with obstructions in tunnels, thanks to better fluidity and obstacle avoidance [83,84]. Kudo et al. [85] investigated the effect of frequency on the size distribution of ultrasonic fog and proposed a mechanism for generating nanometric ultrasonic fog based on the amount of water vapor around the liquid column. It has been found that there is a relationship between the fog size distribution and the ultrasound frequency. Increasing the power intensity and density by changing the surface diameter of the ultrasonic oscillator influenced the numerical concentration and size distribution of the nano-sized fog. Using this technique, the diameter of the fog can be controlled by changing the frequency of the ultrasonic transducer [85].

Similar results were obtained by Jiang et al. [84], who conducted experimental and numerical studies of the effect of nanoparticle water mist with phosphorus-containing compounds (PCC) on the flame in the event of a CH_4/coal dust explosion. The additives used were dimethyl methylphosphonate (DMMP) and phytic acid (PA). The results obtained showed that the water mist in the form of nanoparticles can act as a thermal barrier and effectively reduce the flame temperature. A system containing DMMP, compared to PA, can effectively lower the concentration of toxic (CO) and flammable gases (CH_4 and H_2) in CH_4 hybrid and coal dust explosions. Additionally, water mist containing DMMP can significantly slow down the speed of the flame and cut off the flame. The results of the research also showed that PCC additives can also increase the heat absorption capacity of the water mist [84].

In order to improve the damping capacity of water mist in the form of nanoparticles, various chemical additives in different concentrations are added to the water. The conducted research allowed for the selection of various inorganic salts and organic compounds, including compounds containing Na, K, P, Ca, Fe and surfactants. According to experiments and computational studies, it has been proved that the addition of salt additives can improve the effectiveness of chemical suppression [84,86–88].

Another solution in the field of firefighting is a fire blanket, handheld firefighting equipment used to mechanically cut off the air supply to burning materials. Its use involves tightly covering a small, burning object, e.g., a container with a burning substance or burning clothes. When using a blanket, however, one should remember to cover the source of the fire from his or her side to avoid burning. At the same time, it should be noted that the blanket can be used effectively only to extinguish small sources of fire located close to the person putting out the fire. However, it is possible to use it multiple times and not to destroy the extinguished objects. High-temperature fire blankets are made of non-flammable or fire-resistant fabric materials (e.g., aramids, glass fiber, amorphous silica, pre-oxidized carbon and mineral fibers) [89]. Fire blankets made of polyester or wool are useless in the event of a fat fire, as they are at risk of spontaneous combustion or ignition. Blankets can also be made of fiberglass in the form of nano fiberglass. The Nano Fiberglass fabric [90] helps to extinguish the most difficult flames, regardless of the cause, in a very short time. It can be used to extinguish all types of fires, including fires of lubricants, liquids and gases.

An interesting solution is the invention [91], which allows the use of nanocrystalline particles with a relatively large surface area to reduce the amount of various substances generated during fires and to suppress the fire itself. The results of the research showed that nanocrystalline particles can come from the group consisting of metal oxides, metal hydroxides, carbonates, bicarbonates, phosphorus, inorganic phosphorus compounds, boron compounds, antimony compounds, molybdenum compounds, titanium compounds, zirconium compounds, zinc compounds, amidosulfonates, sulfates, bromine compounds, chlorine compounds and mixtures thereof. Metal oxides and metal hydroxides Mg, Sr, Ba, Ca, Ti, Zr, Fe, V, Mn, Ni, Cu, Al, Si, Zn, Ag, Mo, Sb and mixtures thereof are the most preferred nanocrystalline materials. However, sodium, aluminum, magnesium and calcium hydroxides, carbonates and bicarbonates are most preferred. It has been shown that the size of the nanocrystalline particles should be less than 25 nm; nonetheless, the most optimal particles are about 1–20 nm, especially between about 2 and 10 nm. In contrast, the values characterizing the multipoint Brunauer–Emmett–Teller (BET) surface area should be at least about 15 m^2/g, with the most optimal being at least about 70 m^2/g and most preferably in the range of about 200–850 m^2/g [91].

The effectiveness of smoke removal is determined by the amount of nanocrystalline particles dispensed in a given area (i.e., mass concentration of nanoparticles), aerodynamic geometric mean diameter (GMD) of the particles and the settling velocity of the particles. The most optimal amount of nanocrystalline particles to be dispensed in a region is between about 1–5 g/m^3. The nanocrystalline particles at first must be scattered in the area affected by the smoke to absorb some of the smoke, especially the carbon particles, which tend to obscure visibility. The developed solution also allows a reduction of the amount of various toxic compounds, such as acrolein, toluene diisocyanate, formaldehyde, isocyanates, HCN, CO, NO, HF and HCl generated during a fire [91].

Nanotechnology creates opportunities to improve the effectiveness of firefighting (Table 4) and thus the safety of users and people involved in the event.

Table 4. Possibilities of using nanotechnology to improve the efficiency of firefighting.

Nanoparticles	Characteristic	Effect	Ref.
CNS	suspension of carbon nanostructures (CNS) in the form of functionalized multi-walled carbon nanotubes (MWCNT)	shorter fire extinguishing time, mainly cooling and diluting effect	[78]
diatomite, kaolin, bentonite, attapulgite, SiO_2	inorganic nanoparticles added to a special superabsorbent resin	improving the adhesion of hydrogel extinguishing agent, better extinguishing effect	[81]
mist of H_2O	Water Mist technology	reducing the amount of water needed to extinguish the fire, the ability to avoid obstacles, quickly lowering the temperature	[82,83]
	ultrasonic fragmentation of H_2O particles		[85]
	system containing dimethyl methylphosphonate (DMMP)	reducing the concentration of toxic and combustible gases and coal dust, the thermal barrier effectively reduces the temperature of the flame	[84]
water glass	nano fiberglass	extinguishing all types of fires (including fires of lubricants, liquids and gases)	[90]
hydroxides, carbonates and bicarbonates of Na, Al, Mg and Ca	nanocrystalline particles with a size of about 2-10 nm, with a surface area of 200–850 m^2/g	reduction in part of the smoke (especially carbon particles) and the amount of various toxic compounds formed during a fire	[91]

The use of nanoparticles shortens the fire extinguishing time, reduces the concentration of toxic and combustible gases and reduces the amount of dust, which significantly reduces pollution and migration of pollutants along with smoke over longer distances. It should also be noted that the fragmentation of water particles to sizes < 100 nm allows a reduction of the amount of water needed to extinguish the fire and the amount of fire extinguishing sewage.

5. Conclusions

The fire protection system must be constantly improved as a consequence of changes taking place in the economic, social and legal space. In addition to preventive measures, research is necessary in the field of tools used in rescue operations. Solutions based on nanotechnology are being introduced more and more often. They are aimed at improving the key properties of individual extinguishing agents and thus their effectiveness in a firefighting situation. In this context, the requirements for efficiency improvement play a major role, which are becoming more and more demanding.

The analysis of the type of nanoparticles used in the processes of modification of extinguishing agents shows that inorganic nanoparticles, mainly silica and oxides, hydroxides and bicarbonates/carbonates of alkali metals and aluminum, are the most widely used. The introduction of these nanoparticles to standardly used powders or foaming agents allows for faster fire extinguishing, greater efficiency, reduction in the emission of toxins into the environment and reduction in heat emission to the environment. An interesting solution is also the reduction of water particles to the size of nanoparticles, using, among others, ultrasound, which creates a thermal barrier and reduces the time of extinguishing the fire, as well as reduces the consumption of the extinguishing agent. Thus, it improves the safety of people involved in the event and allows you to reduce the costs of occurring fires.

However, selection of nanoparticles must be appropriate, because at higher concentrations they may be subject to aggregation, which in turn reduces the effectiveness of the quenching process. At the same time, it should be noted that foams and extinguishing powders may have a negative impact on the environment due to the use of various types of chemical compounds with greater or lesser chemical activity in their production. As a result of the application of such measures, some areas can become contaminated and require further treatment, consisting in the neutralization of substances created during the fire and extinguishing, at least. Therefore, in order to further improve the extinguishing efficiency and the resistance of the extinguishing agent, as well as to minimize environmental pollution, it is necessary to develop extinguishing agents that can extinguish various complex fires and, at the same time, remain stable and environmentally friendly.

Solutions implemented on the basis of nanotechnology contribute to faster cooling of the environment, improved visibility in case of high smoke and more effective fire extinguishing action. However, work carried out in this scope and the modifications introduced need to consider the toxicity of nanoparticles, the possibility of their migration and the negative impact both on the environment and on humans, including those involved in the incident.

Author Contributions: Conceptualization, A.R.; writing—original draft preparation, A.R., M.Z. and J.G.; writing—review and editing, A.R., M.Z. and J.G.; project administration, A.R.; funding acquisition, J.G. All authors have read and agreed to the published version of the manuscript.

Funding: This work was supported by a grant from the Ministry of Science and Technology for maintaining research potential.

Institutional Review Board Statement: Not applicable.

Informed Consent Statement: Not applicable.

Data Availability Statement: Not applicable.

Conflicts of Interest: The authors declare no conflict of interest.

References

1. Tępiński, J.; Klapsa, W.; Cygańczuk, K.; Lesiak, P.; Lewak, M.W. Testing of Large Scale Pool Fire of Technical Ethanol. *SFT* **2022**, *59*, 96–109. [CrossRef]
2. Yilmaz-Atay, H.; Wilk-Jakubowski, J.Ł. A Review of Environmentally Friendly Approaches in Fire Extinguishing: From Chemical Sciences to Innovations in Electrical Engineering. *Polymers* **2022**, *14*, 1224. [CrossRef] [PubMed]
3. Tidey, A. Fighting Europe's Fires: Inside the EU's Emergency Response Centre. Available online: https://www.euronews.com/my-europe/2022/07/25/fighting-europes-fires-inside-the-eus-emergency-response-centre (accessed on 6 September 2022).
4. Statistics Poland. Available online: https://stat.gov.pl/en/search/search.html?query=fire (accessed on 23 August 2022).
5. Antonov, A.; Skorobagatko, T.; Yakovchuk, R.; Sviatkevych, O. Interaction of Fire-Extinguishing Agents with Flame of Diesel Bio Fuel and Its Mixtures. *Sci. Pap. Main Sch. Fire Serv.* **2020**, *2020*, 73. [CrossRef]
6. Park, C. Fire Extinguishers. 5 Types of Fire Extinguishers: A Guide to Using the Right Class. IFSEC GLOBAL. Available online: https://www.ifsecglobal.com/global/choose-right-type-fire-extinguisher/ (accessed on 6 September 2022).
7. City Fire Protection in London. Available online: https://www.cityfire.co.uk/news/fire-extinguishers-types/ (accessed on 6 September 2022).
8. *EN 615:2009*; Fire Protection—Fire Extinguishing Media—Specifications for Powders (Other Than Class D Powders). ISS Z021; European Norm: Newark, DE, USA, 2009.
9. *ISO 7202:2018*; Fire Protection—Fire Extinguishing Media—Powder. Technical Committee: ISO/TC 21/SC 6 Foam and Powder Media and Firefighting Systems Using Foam and Powder. ICS 13.220.10 Fire-Fighting; ISO: Geneva, Switzerland, 2018.

10. Izak, P.; Biel, M.; Mastalska-Popławska, J.; Janik, P.; Mortka, P.; Lesiak, P. The Effect of Magnesium Hydroxide Addition on the Extinguishing Efficiency of Sodium Bicarbonate Powders. *Materials* **2022**, *15*, 3449. [CrossRef]
11. Zhang, X.; Ismail, M.H.S.; Ahmadun, F.-R.; Abdullah, N.; Hee, C. Hot Aerosol Fire Extinguishing Agents and The Associated Technologies: A Review. *Braz. J. Chem. Eng.* **2015**, *32*. [CrossRef]
12. Li, H.; Fenf, L.; Du, D.; Guo, X.; Hua, M.; Pan, X. Fire suppression performance of a new type of composite superfine dry powder. *Fire Mater.* **2019**, *43*, 905–916. [CrossRef]
13. Li, Y.; Qi, L.; Liu, Y.; Qiao, J.; Wang, M.; Liu, X.; Li, S. Recent Advances in Halogen-Free Flame Retardants for Polyolefin Cable Sheath Materials. *Polymers* **2022**, *14*, 2876. [CrossRef] [PubMed]
14. Fatsa Fire, Most Effective Fire Extinguishing Methods. 2022. Available online: https://www.fatsafire.com/most-effective-fire-extinguishing-methods/ (accessed on 26 September 2022).
15. Izak, P.; Kidoń, A.; Mastalska-Popławska, J. Mechanism of Fire-extinguishing Aerosol's Action. *BiTP* **2017**, *46*, 56–71. [CrossRef]
16. Wilczkowski, S. Inhibition effect of chemical compounds in selected extinguishing agents. *Saf. Fire Technol.* **2010**, *19*, 99–105.
17. Velencoso, M.M.; Battig, A.; Markwart, J.C.; Schartel, B.; Wurm, F.R. Molecular Firefighting-How Modern Phosphorus Chemistry Can Help Solve the Challenge of Flame Retardancy. *Angew. Chem. Int. Ed.* **2018**, *57*, 10450–10467. [CrossRef]
18. Li, H.; Hua, M.; Pan, X.; Li, S.; Guo, X.; Zhang, H.; Jiang, J. The reaction pathway analysis of phosphoric acid with the active radicals: A new insight of the fire-extinguishing mechanism of ABC dry powder. *J. Mol. Model.* **2019**, *25*, 255. [CrossRef] [PubMed]
19. Wang, Q.; Wang, F.; Li, C.; Li, Z.; Li, R. Fire extinguishing performance and mechanism for several typical dry water extinguishing agents. *RSC Adv.* **2021**, *11*, 9827–9836. [CrossRef] [PubMed]
20. Rakowska, J.; Radwan, K.; Ślosorz, Z. Comparative Study of the Results of the Extinguishing Powder Grain Size Analysis Carried out by Different Methods. *BiTP* **2014**, *34*, 57–64. [CrossRef]
21. Ni, X.; Kuang, K.; Yang, D.; Jin, X.; Liao, G. A new type of fire suppressant powder of $NaHCO_3$/zeolite nanocomposites with core–shell structure. *Fire Saf. J.* **2009**, *44*, 968–975. [CrossRef]
22. Kuang, K.; Chow, W.K.; Ni, X.; Yang, D.; Zeng, W.; Liao, G. Fire suppressing performance of superfine potassium bicarbonate powder. *Fire Mater.* **2011**, *35*, 353–366. [CrossRef]
23. Haiqiang, L.; Ruowen, Z.; Jiaxin, G.; Siuming, L.; Yuan, H. Good dry powder to suppress high building fires. *APCBEE Procedia* **2014**, *9*, 291–295. [CrossRef]
24. Ibrahim, H.; Patruni, J.R. Experimental investigation on extinguishing performance of a novel nanocomposite for gaseous fires. *J. Loss Prev. Process Ind.* **2020**, *65*, 104143. [CrossRef]
25. Ni, X.; Kuang, K.; Wang, X.; Liao, G. A New Type of BTP/Zeolites Nanocomposites as Mixed-phase Fire Suppressant: Preparation, Characterization, and Extinguishing Mechanism Discussion. *J. Fire Sci.* **2009**, *28*, 5–25. [CrossRef]
26. Du, D.; Pan, X.; Hua, M. Experimental Study on Fire Extinguishing Properties of Compound Superfine Powder. *Procedia Eng.* **2018**, *211*, 142–148. [CrossRef]
27. Kaiqian, K.; Xin, H.; Guangxuan, L. A comparison between superfine magnesium hydroxide powders and commercial dry powders on fire suppression effectiveness. *Process Saf. Environ. Prot.* **2008**, *86*, 182–188. [CrossRef]
28. Koshiba, Y.; Takahashi, Y.; Ohtani, H. Flame Suppression Ability of Metallocene (nickelocene, cobaltocene, ferrocene, manganocene, and chromocene). *Fire Saf. J.* **2012**, *51*, 10–17. [CrossRef]
29. Linteris, G.T.; Knyazev, V.D.; Babushok, V.I. Inhibition of Premixed Methane Flames by Manganese and Tin Compounds. *Combust. Flame* **2002**, *129*, 221–238. [CrossRef]
30. Łukaszczuk, P. The Application of Nanotechnology in Fire Protection. *BiTP* **2016**, *42*, 95–102. [CrossRef]
31. Rabajczyk, A.; Zielecka, M.; Popielarczyk, T.; Sowa, T. Nanotechnology in Fire Protection—Application and Requirements. *Materials* **2021**, *12*, 7849. [CrossRef] [PubMed]
32. Biel, M.; Izak, P.; Skubacz, K.; Stempkowska, A.; Mastalska-Popławska, J. The Effect of Humidity on the Atomization Process and Structure of Nanopowder Designed for Extinguishment. *Materials* **2021**, *14*, 3329. [CrossRef]
33. Rabajczyk, A.; Zielecka, M.; Porowski, R.; Hopke, P.K. Metal nanoparticles in the air: State of the art and future perspectives. *Environ. Sci. Nano* **2020**, *7*, 3233. [CrossRef]
34. Kunin, A.V.; Smirnov, S.A.; Lapshin, D.N.; Semenov, A.D.; Il'in, A.P. Technology development for the production of ABCE fire extinguishing dry powders. *Russ. J. Gen. Chem.* **2016**, *86*, 450–459. [CrossRef]
35. *CEN/TR 15276-1*; Fixed firefighting systems—Condensed aerosol extinguishing systems—part. 1: Requirements and test methods for components. European Committee For Standardization, Management Centre: Brussels, Belgium, 2009.
36. Xie, J.; Zhang, J.; Ding, C.; Wang, X. Hydrophobic nano SiO_2 as flow-enhancing additives and flame retardant synergizes with $CaCO_3$ to suppress gas explosion. *RSC Adv.* **2021**, *11*, 4672–4681. [CrossRef]
37. Qu, L.; Morton, D.A.V.; Zhou, Q. Particle engineering via mechanical dry coating in the design of pharmaceutical solid dosage forms. *Curr. Pharm. Des.* **2015**, *21*, 5802–5814. [CrossRef]
38. Zimmerman, I.; Eber, M.; Meyer, K. Nanomaterials as flow regulators in dry powders. *Z. Phys. Chem.* **2004**, *218*, 51–102. [CrossRef]
39. Hare, C.; Zafar, U.; Ghadiri, M.; Freeman, T.; Clayton, J.; Murtagh, M.J. Analysis of the dynamics of the FT4 Powder Rheometer. *Powder Technol.* **2015**, *285*, 123–127. [CrossRef]
40. Shamsutdinov, A.S.; Kondrashova, N.B.; Valtsifer, I.V.; Bormashenko, E.; Huo, Y.; Saenko, E.V.; Pyankova, A.V.; Valtsifer, V.A. Manufacturing, Properties, and Application of Nanosized Superhydrophobic Spherical Silicon Dioxide Particles as a Functional Additive to Fire Extinguishing Powders. *Ind. Eng. Chem. Res.* **2021**, *60*, 11905–11914. [CrossRef]

41. Saenko, E.V.; Huo, Y.; Shamsutdinov, A.S.; Kondrashova, N.B.; Valtsifer, I.V.; Valtsifer, V.A. Mesoporous Hydrophobic Silica Nanoparticles as Flow-Enhancing Additives for Fire and Explosion Suppression Formulations. *ACS Appl. Nano Mater.* **2020**, *3*, 2221–2233. [CrossRef]
42. Foster, J.; Doll, J. Particle size effect on talc lubricant activity. In *American Association of Pharmaceutical Scientists Annual Meeting Poster Session*; American Association of Pharmaceutical Scientists; Specialty Minerals Inc.: Easton PA, USA, 2004; p. 18.
43. Yang, J.; Sliva, A.; Banerjee, A.; Dave, R.N.; Pfeffer, R. Dry particle coating or improving the flowability of cohesive powders. *Powder Technol.* **2005**, *158*, 21–33. [CrossRef]
44. Zegzulka, J.; Jezerska, L.; Zadrapa, F.; Zidek, M.; Gelnar, D. Flow characterization methods of glidants. In Proceedings of the 8th International Conference on Nanomaterials-Research & Application, Brno, Czech Republic, 19–21 October 2016.
45. Veregin, R.P.N.; Bartha, R. A quantitative cohesive force mechanism for powder flow: The role of metal oxide surface additives in xerographict oner. *J. Imaging Sci. Technol.* **2001**, *45*, 571–578.
46. Reding, N.S.; Shiflett, M.B. Characterization of Thermal Stability and Heat Absorption for Suppressant Agent/Combustible Dust Mixtures via Thermogravimetric Analysis/Differential Scanning Calorimetry. *Ind. Eng. Chem. Res.* **2019**, *58*, 4674–4687. [CrossRef]
47. Park, E.J.; Cho, Y.K.; Kim, D.H.; Jeong, M.-G.; Kim, Y.H.; Kim, Y.D. Hydrophobic Polydimethylsiloxane (PDMS) Coating of Mesoporous Silica and Its Use as a Preconcentrating Agent of Gas Analytes. *Langmuir* **2014**, *30*, 10256–10262. [CrossRef]
48. Jonat, S. Mechanism of Hydrophilic and Hydrophobic Colloidal Silicon Dioxide Types as Glidants. PhD Dissertation, Universität Tübingen, Tübingen, Germany, 2005.
49. Kojima, T.; Elliott, J.A. Incipient flow properties of two component fine powder systems and their relationships with bulk density and particle contacts. *Powder Technol.* **2012**, *228*, 359–370. [CrossRef]
50. Kojima, T.; Elliott, J.A. Effect of silica nanoparticles on the bulk flow properties of fine cohesive powders. *Chem. Eng. Sci.* **2013**, *101*, 315–328. [CrossRef]
51. *EN 1568-1:2018*; Fire Extinguishing Media—Foam Concentrates—Part 1: Specification for Medium Expansion Foam Concentrates for Surface Application to Water-Immiscible Liquids. European Committee for Standardization, Management Centre: Brussels, Belgium.
52. Roth, J.; Abusallout, I.; Hill, T.; Holton, C.; Thapa, U.; Hanigan, D. Release of Volatile Per- and Polyfluoroalkyl Substances from Aqueous Film-Forming Foam. *Environ. Sci. Technol. Lett.* **2020**, *7*, 164–170. [CrossRef]
53. Jiang, N.; Sheng, Y.; Li, C.; Lu, S. Surface activity, foam properties and aggregation behavior of mixtures of short-chain fluorocarbon and hydrocarbon surfactants. *J. Mol. Liq.* **2018**, *268*, 249–255. [CrossRef]
54. Wendong Kang, W.; Yan, L.; Ding, F.; Guo, X.; Xu, Z. Experimental study on fire-extinguishing efficiency of protein foam in diesel pool fire. *Case Stud. Therm. Eng.* **2019**, *16*, 100557. [CrossRef]
55. Commission Regulation (EU) 2017/1000 of 13 June 2017 Amending Annex XVII to Regulation (EC) No 1907/2006 of the European Parliament and of the Council on Registration, Evaluation, Authorization and Use of Chemical Restrictions (REACH) for Perfluorooctanoic Acid (PFOA), Its Salts and Derivatives. Available online: https://www.legislation.gov.uk/eur/2017/1000/annex?view=plain (accessed on 5 October 2022).
56. Kuti, R. Advantages of Water Fog Use as a Fire Extinguisher. *AARMS* **2015**, *14*, 259–264. [CrossRef]
57. Denkov, N.; Tcholakova, S.; Politova-Brinkova, N. Physicochemical control of foam properties. *Curr. Opin. Colloid Interface Sci.* **2020**, *50*, 101376. [CrossRef]
58. Dlugogorski, B.Z.; Kennedy, E.M.; Schaefer, T.H.; Vitali, J.A. What Properties Matter in Fire-Fighting Foams? Whitepaper. In Proceedings of the National Research Institute of Fire and Disaster 2nd NRIFD Symposium—Conference Proceedings, Tokyo, Japan, 17–19 July 2002; Available online: https://www.kappetijn.eu/wp-content/uploads/2019/05/Solberg-what-properties-matther-in-foam.pdf (accessed on 5 October 2022).
59. Fan, X.; Guan, X.; Zhang, M.; Liu, Y.; Li, Y. Aqueous foam synergistically stabilized by the composite of lignin nanoparticles and surfactant. *Colloids Surf. A Physicochem. Eng. Asp.* **2022**, *643*, 128727. [CrossRef]
60. Li, Y.; Xiao, G.; Li, F.; Guo, Y.; Chen, C.; Chen, C.; Li, R.; Yang, Z. A novel H-TiO_2/gel co-stabilized three-dimensional network synergistic fire-retardant foam gel for coal-pile. *Colloids Surf. A Physicochem. Eng. Asp.* **2022**, *650*, 129642. [CrossRef]
61. Sheng, Y.; Peng, Y.; Zhang, S.; Guo, Y.; Ma, L.; Wang, Q.; Zhang, H. Study on Thermal Stability of Gel Foam Co-Stabilized by Hydrophilic Silica Nanoparticles and Surfactants. *Gels* **2022**, *8*, 123. [CrossRef] [PubMed]
62. Zhang, L.; Yunpeng Bian, Y.; Kuai, D. Preparation and flame retardant property of nano-aluminum hydroxide foam for preventing spontaneous coal combustion. *Fuel* **2021**, *304*, 121494. [CrossRef]
63. DAZZEON. Available online: http://www.dazzeon.com/en/firesavior/ (accessed on 5 October 2022).
64. Youije, S.; Hanling, Z.; Yunchuan, P.; Li, H.; Zhenping, W.; Yang, L.; Canbin, Y. Efficient Environment-Friendly NP-Foam Extinguishing Agent and Preparation Method Thereof. China Patent CN114748828A, 15 July 2022. Available online: https://patents.google.com/patent/CN114748828A/en (accessed on 5 October 2022).
65. Mosina, K.S.; Nazarova, E.A.; Vinogradov, A.V.; Vinogradov, V.V.; Krivoshapkina, E.F.; Krivoshapkin, P.V. Alumina Nanoparticles for Firefighting and Fire Prevention. *ACS Appl. Nano Mater.* **2020**, 3, 4386–4393. [CrossRef]
66. Martin, T.J. Fire-Fighting Foam Technology. In *Foam Engineering: Fundamentals and Applications*; John Wiley & Sons: Chichester, UK, 2012.

67. Zhou, R.; Dou, X.; Lang, X.; He, L.; Liu, J.; Mu, S. Foaming Ability and Stability of Silica Nanoparticle-Based Triple-Phase Foam for Oil Fire Extinguishing: Experimental. *Soft Mater.* **2018**, *16*, 327–338. [CrossRef]
68. Vinogradov, A.V.; Kuprin, D.S.; Abduragimov, I.M.; Kuprin, G.N.; Serebriyakov, E.; Vinogradov, V.V. Silica Foams for Fire Prevention and Firefighting. *ACS Appl. Mater. Interfaces* **2016**, *8*, 294–301. [CrossRef] [PubMed]
69. Carn, F.; Colin, A.; Pitois, O.; Vignes-Adler, M.; Backov, R. Foam Drainage in the Presence of Nanoparticle−Surfactant Mixtures. *Langmuir* **2009**, *25*, 7847–7856. [CrossRef] [PubMed]
70. Binks, B.P.; Rodrigues, J.A. Enhanced stabilization of emulsions due to surfactant-induced nanoparticle flocculation. *Langmuir* **2007**, *23*, 7436–7439. [CrossRef]
71. Li, X.; Guo, R.; Qian, X. Research on the influence of wollastonite fibers on the stability of foam extinguishment agent and its effect on the extinguishing efficiency of pool fire. *Fire Mater.* **2020**, *44*, 1053–1063. [CrossRef]
72. Hill, C.; Eastoe, J. Foams: From nature to industry. *Adv. Colloid Interface Sci.* **2017**, *247*, 496–513. [CrossRef]
73. SERDP-ESRCP. Innovative Nano-Encapsulated Ionic Liquid-Based Surfactants for Fluorine-Free Fire Extinguishing Foams, WP20-1539. Available online: https://serdp-estcp.org/projects/details/44074bfa-601f-48c2-8250-3305ce801082 (accessed on 5 October 2022).
74. SERDP-ESRCP. Innovative Nano-Encapsulated Ionic Liquid Based Surfactants for PFAS-Free Fire Extinguishing Foams, WP18-1597. Available online: https://serdp-estcp.org/projects/details/0e41dbfb-f716-408e-bf3c-5e443c43ba0e (accessed on 5 October 2022).
75. Shen, J.; Liang, J.; Lin, X.; Lin, H.; Yu, J.; Wang, S. The Flame-Retardant Mechanisms and Preparation of Polymer Composites and Their Potential Application in Construction Engineering. *Polymers* **2022**, *14*, 82. [CrossRef] [PubMed]
76. Maguire, J.F.; Woodcock, L.V. Thermodynamics of Tower-Block Infernos: Effects of Water on Aluminum Fires. *Entropy* **2020**, *22*, 14. [CrossRef]
77. Fire Safety Devices Pvt. Ltd. Wetting Agent. Available online: http://fcfsd.com/wetting-agents.html (accessed on 5 October 2022).
78. Dali, F.A.; Shidlovsky, G.L.; Khasikhanov, M.S.; Zalaev, R.U.; Tagirova, P.R.; Saidulaev, S.S.; Masaeva, L.M.; Erzhapova, R.S. The use of carbon nanotubes in the fire extinguishing of oil and oil products. *IOP Conf. Ser. Mater. Sci. Eng.* **2020**, *905*, 012011. [CrossRef]
79. Yu, W.; Xie, H. A review on nanofluids: Preparation, stability mechanisms, and applications. *J. Nanomater.* **2012**, 435873. [CrossRef]
80. Toropov, D.; Ivanov, A.; Dali, F.; Perlin, A.; Lebedev, A.; Shidlovsky, G. Extinguishing characteristics of water suspensions with carbon nanostructures at extinguishing liquid hydrocarbons fires (oil and gas industry). *Fire Prot. Saf. Sci. J.* **2019**, *13*, 22–31. [CrossRef]
81. Xu, D.H.; Ma, R.; Li, S.S.; Shi, T.F.; Zhou, H.W.; Pan, G.; Fu, Z.L.; Wang, X.W.; Li, S.J.; Tang, H.Q. High-Molecular Hydrogel Fire Extinguishing Agent and Preparation Method Thereof. CN109966688B, 17 August 2019.
82. SMH SISTEMAS. Fire Suppression System by Water Mist. Available online: https://www.smh.com.br/en/fire-suppression-system-by-water-mist/ (accessed on 5 October 2022).
83. Liu, Y.P.; Wang, X.S.; Liu, T.; Ma, J.; Li, G.C.; Zhao, Z.H. Preliminary study on extinguishing shielded fire with water mist. *Process Saf. Environ. Protect.* **2020**, *141*, 344–354. [CrossRef]
84. Jiang, H.; Bi, M.; Huang, L.; Zhou, Y.; Gao, W. Suppression mechanism of ultrafine water mist containing phosphorus compounds in methane/coal dust explosions. *Energy* **2022**, *239*, 121987. [CrossRef]
85. Kudo, T.; Sekiguchi, K.; Sankoda, K.; Namiki, N.; Nii, S. Effect of ultrasonic frequency on size distributions of nanosized mist generated by ultrasonic atomization. *Ultrason. Sonochem.* **2017**, *37*, 16–22. [CrossRef] [PubMed]
86. Linteris, G.T.; Rumminger, M.D.; Babushok, V.I. Catalytic inhibition of laminar flames by transition metal compounds. *Prog. Energy Combust. Sci.* **2008**, *34*, 288–329. [CrossRef]
87. Tianwei, Z.; Hao, L.; Zhiyue, H.; Zhiming, D.; Yong, W. Active substances study in fire extinguishing by water mist with potassium salt additives based on thermoanalysis and thermodynamics. *Appl. Therm. Eng.* **2017**, *122*, 429–43838. [CrossRef]
88. Dombrovsky, L.A.; Dembele, S.; Wen, J.X. A simplified model for the shielding of fire thermal radiation by water mists. *Int. J. Heat Mass Transf.* **2016**, *96*, 199–209. [CrossRef]
89. Takahashi, F. Fire blanket and intumescent coating materials for failure resistance. *MRS Bull.* **2021**, *46*, 429–434. [CrossRef]
90. WowZone Buyer's Protection. FlameCape™ Fire Blanket. Available online: https://wowzonegadgets.com/products/flamecape-fire-blanket (accessed on 5 October 2022).
91. Mulukutla, R.S.; Malchesky, P.S.; Maghirang, R.; Klabunde, J.S.; Klabunde, K.J.; Koper, O. Metal Oxide Nanoparticles for Smoke Clearing and Fire Suppression. US20080210444A1, 4 September 2010.

nanomaterials

Article

Zinc Oxide Nanoparticles—Solution-Based Synthesis and Characterizations

Khagendra P. Bhandari [1,*], Dhurba R. Sapkota [2], Manoj K. Jamarkattel [2], Quenton Stillion [1] and Robert W. Collins [2]

1 Department of Physics and Astronomy, Ohio Northern University, Ada, OH 45810, USA; q-stillion@onu.edu
2 Wright Center for Photovoltaics Innovation & Commercialization, University of Toledo, Toledo, OH 43606, USA; dhurbaraj.sapkota@rockets.utoledo.edu (D.R.S.); manoj.jamarkattel@rockets.utoledo.edu (M.K.J.); robert.collins@utoledo.edu (R.W.C.)
* Correspondence: k-bhandari@onu.edu; Tel.: +1-419-772-2739

Abstract: Zinc oxide (ZnO) nanoparticles have shown great potential because of their versatile and promising applications in different fields, including solar cells. Various methods of synthesizing ZnO materials have been reported. In this work, controlled synthesis of ZnO nanoparticles was achieved via a simple, cost-effective, and facile synthetic method. Using transmittance spectra and film thickness of ZnO, the optical band gap energies were calculated. For as-synthesized and annealed ZnO films, the bandgap energies were found to be 3.40 eV and 3.30 eV, respectively. The nature of the optical transition indicates that the material is a direct bandgap semiconductor. Spectroscopic ellipsometry (SE) analysis was used to extract dielectric functions where the onset of optical absorption of ZnO was observed at lower photon energy due to annealing of the nanoparticle film. Similarly, X-ray diffraction (XRD) and scanning electron microscopy (SEM) data revealed that the material is pure and crystalline in nature, with the average crystallite size of ~9 nm.

Keywords: zinc oxide; nanoparticle; absorbance; absorption coefficient; bandgap; ellipsometry; crystal structure; surface morphology

Citation: Bhandari, K.P.; Sapkota, D.R.; Jamarkattel, M.K.; Stillion, Q.; Collins, R.W. Zinc Oxide Nanoparticles—Solution-Based Synthesis and Characterizations. *Nanomaterials* **2023**, *13*, 1795. https://doi.org/10.3390/nano13111795

Academic Editors: Robert A. Varin and Geng Zhong

Received: 5 May 2023
Revised: 26 May 2023
Accepted: 30 May 2023
Published: 2 June 2023

1. Introduction

In materials science, the ZnO is a wide bandgap II–VI semiconductor compound, the chemical bonding of which is largely covalent but with a substantial contribution from ionic bonding [1]. The ZnO can crystallize in the wurtzite, zinc blende, and rock salt crystal structures. The compound semiconductor has a tetrahedral bonding configuration, where each anion is surrounded by four cations at the corners of a tetrahedron, and each cation is surrounded by four anions, corresponding to the sp^3 covalent bonding [2]. The compound zinc oxide (ZnO) is one of the most functional materials with remarkable and unique physical and chemical properties, such as strong chemical stability [3–5], high electrochemical coupling coefficients [6,7], and a broad spectrum of radiation absorption [8], along with high photostability [9].

Among the diversity of materials and needs, the oxide semiconductor ZnO is of great interest and is one of the versatile materials in materials science. The research interest in ZnO is growing every year, as can be observed from the rapid growth in the number of publications about the materials [10]. The material has a non-toxic nature and is cost-effective in producing both small and large scales. ZnO is considered as a leading candidate material for the next generation of the electronics industry, and as a practical material for medical devices [11,12]. Zinc oxide has also been considered as a promising anode material for Li-ion batteries because of its high theoretical capacity, environmental friendliness, abundance, and economical nature [13,14]. ZnO is recognized as an n-type, multi-functional semiconductor material, with a direct bandgap energy of ~3.37 eV and exciton binding energy up to 60 eV, even at room temperature, which is even higher than that of GaN [2,15–17]. This exceptional characteristic of ZnO makes it a lucrative material for room-temperature ultraviolet lasing devices [18,19]. Other applications of ZnO include optoelectronic and spintronic devices [20], UV light emitters [21], varistors [22,23], ceramic positive temperature coefficient thermistors [24,25], transparent high-power electronics [26,27], surface acoustic wave guides [28], piezoelectric transducers [29], chemical and gas sensor [30], solar cells [31–33], and piezoelectric nanogenerators [34]. As a result of its high emission efficiency, it is an important material for solid-state lighting technology [35]. It also provides opportunities for the formation of effective new optical and magnetic devices, such as spin-polarized solar cells, spin light-emitting diodes, and magneto-optical switches [36].

Apart from bulk ZnO and thin films/epilayers [37], several nanostructured polymorphic shapes of ZnO are available, such as nanorods [38,39], nanowires [33,40], nanoflowers [41,42], and nanoparticles [43–45]. These nanostructures have proven to be important for optoelectronic devices because of their large surface-to-volume ratios [46,47]. These various structures achieved by the nanostructures of ZnO make these materials unique, with a diverse set of characteristics and prospective uses in a variety of nanotechnology disciplines. Several methods for the synthesis of ZnO nanostructures have been reported, such as microemulsion synthesis [48,49], sol-gel techniques [50–52], mechanochemical processing [53,54], spray pyrolysis and drying [55,56], thermal decomposition of organic precursor [57,58], RF plasma synthesis [59,60], supercritical water processing [61,62], self-assembling [63], hydrothermal processing [64–66], vapor transport process [67,68], microwave assisted synthesis [69,70], direct precipitation, and homogeneous precipitation [71,72].

Out of all these deposition methods, hydrothermal, precipitation, colloidal, and sol-gel belong to the main category of the liquid-phase method of synthesis of nanomaterials. These methods are important because of their convenient operation procedures, simple synthetic route, and controllable particle size distribution. Hydrothermal synthesis can be used to fabricate nanomaterials at all temperature ranges and is considered one of the best methods for high-temperature synthesis. Similarly, the precipitation method is used to generate a pure and homogeneous nanomaterial. Product separation is necessary after precipitation and constant product quality throughout the precipitation process is sometimes challenging.

In the chemical vapor deposition (CVD) method, solid material is deposited using vapors obtained from the chemical reaction of the materials occurring on or in the locality of a normally heated substrate's surface [73]. The generated thin film has uniform thickness with low porosity, even on a substrate with a complicated shape. This method is important to fabricate selective deposition, even on patterned substrates, and provides a pure material with economical production. On the other hand, CVD requires an expensive high-vacuum deposition system, making this deposition method not appropriate for all users.

In this work, we investigate the structural, morphological, and optical properties of ZnO nanoparticles obtained from the solvothermal synthesis route. The solvothermal method helps to speed up the reaction among the reactants and to enhance the crystal growth, resulting in self-assembly of nanoparticles in the solution. The synthesis is completed using zinc acetate and potassium hydroxide in the presence of methanol (Section 2). ZnO nanoparticles obtained from this method have already been used in quantum dot solar cells [74,75]. However, the authors realize that a detailed characterization, including a spectroscopic ellipsometry study, is necessary to further investigate its applications. We found this method to be a cost-effective and facile method for the synthesis of ZnO nanoparticles. The structure, morphology, and optical properties of the nanoparticles were investigated using XRD, SEM, SE, and a UV/Vis spectrophotometer.

2. Materials and Methods

2.1. Materials

Zinc acetate ($ZnAc_2$, 99.99%, Sigma Aldrich, St. Louis, MO, USA), potassium hydroxide (KOH, certified ACS pellets, Fisher Scientific, Waltham, MA, USA), chloroform (anhydrous, 99+%, Sigma Aldrich), and methanol (MeOH, anhydrous, 99.8%, Sigma Aldrich) were purchased and used as received.

2.2. Nanoparticle Synthesis

Synthesis of ZnO nanoparticles was conducted using a modified form of the solvothermal method described elsewhere [74–76]. Two solutions were prepared in separate containers: solution 1 was prepared by dissolving 0.55 g of zinc acetate ($Zn(CH_3COO)_2$) in 25 mL of methanol (MeOH). The solution was stirred for an hour at a hot plate temperature of 60 °C. Solution 2 was similarly prepared by dissolving 0.28 g of potassium hydroxide (KOH) in 12.5 mL of methanol by stirring at the same hotplate temperature as for solution 1. Solution 2 was added to solution 1 using a titration process with the help of a burette and the reaction was left to run for 2 h at 60 °C for the nucleation and growth of the nanoparticles. After 2 h, the solution mixture was centrifuged to extract nanoparticles and re-suspended in methanol. This process was repeated three times, and finally, dry ZnO nanoparticles were suspended in chloroform for characterization in solution and thin films. The synthesis of ZnO nanoparticles was summarized using the following balanced equation:

$$\mathrm{Zn(CH_3COO)_2 + 2KOH \rightarrow Zn(OH)_2 + 2CH_3COOK}$$
$$\mathrm{Zn(OH)_2 \rightarrow ZnO + H_2O}$$

2.3. Thin Film Fabrication

ZnO nanoparticles thin films were fabricated using the spin coating process. A ZnO nanoparticle solution of the desired concentration was prepared in chloroform. The obtained solution was spin-coated on the Fisher brand borosilicate glass and soda lime glass substrates at 1000 rpm for 10 s and 4000 rpm for 25 s at room temperature (23 °C). The prepared films were annealed at different hotplate temperatures for different times.

2.4. Characterization Methods

A field emission scanning electron microscope (Hitachi S-4800, Japan) was used to acquire SEM images of ZnO surface. The XRD patterns of the ZnO films were taken using a Rigaku Ultima III X-ray diffractometer (Rigaku, USA) fitted with a small-angle X-ray scattering at 40 kV accelerating voltage and 44 mA current. Dielectric functions were obtained using a spectroscopic ellipsometry study. To obtain SE data, a rotating-compensator ellipsometer J. A. Woollam Co., (St. Lincoln, NE, USA) M-2000 DI was used. Unpolarized absorbance and transmittance spectra of the films were taken using a Shimadzu UV 2401PC Spectrophotometer. Spectroscopic ellipsometry data were taken to identify their dielectric functions. It is important to note that all these measurements were conducted at room temperature.

3. Results and Discussions

3.1. Phase Analysis and Surface Morphology

To understand the purity and crystallinity of the ZnO thin films, XRD data were collected from as-synthesized film and an annealed film using Cu Kα radiation (λ = 1.54059 Å) in focused beam geometry. The XRD patterns obtained from the as-synthesized film are shown in Figure S1. Figure 1 shows XRD spectra of ZnO film annealed at 300 °C for 30 min. By comparing Figure 1 and Figure S1, it can be seen that the annealed film is more crystalline as compared to the as-synthesized film. All XRD peaks are indexed by the hexagonal wurtzite structure (#PDF 98-000-0483) of ZnO matching exactly with the standard peaks for the wurtzite phase represented by vertical lines, as shown in Figure 1. These standard reference peaks were obtained from MDI JADE (Jade^{TM} computer software from Materials Data Inc., St. Livermore, CA, USA). There is no evidence of traces of any other materials and sharp diffraction peaks are indicative of the good crystallinity and phase purity of the materials. The intense and sharp peaks of the XRD pattern indicate phase purity of ZnO nanoparticles. Further analysis of the material was done using Bragg's law to identify the lattice constants, lattice spacing, and crystallite sizes.

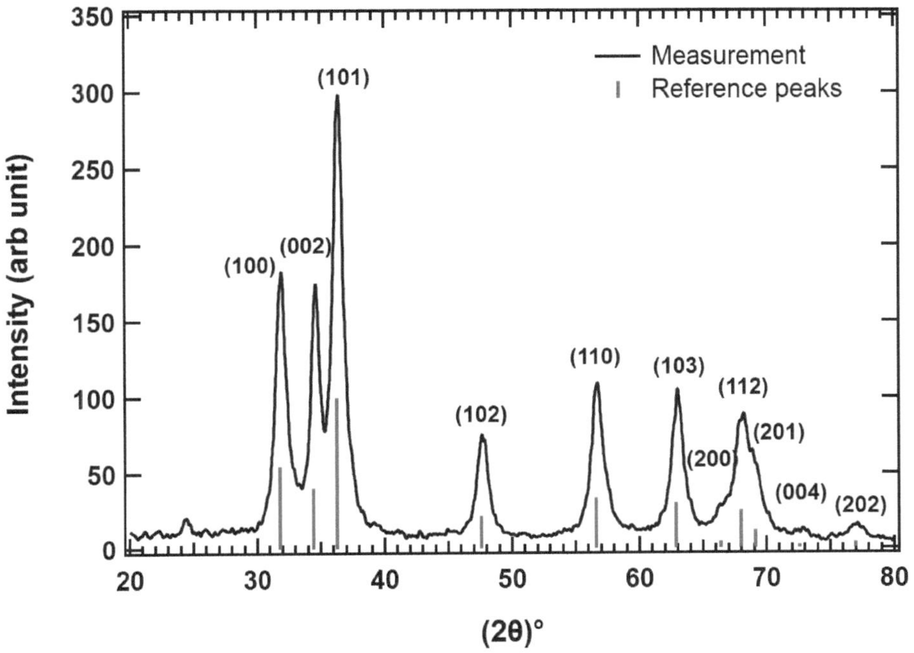

Figure 1. XRD patterns of annealed film of ZnO nanoparticles deposited on borosilicate glass substrate (Fisher brand microscope slide). The film was heated at 300 °C for 30 min.

The equation representing Bragg's law of diffraction [77] is

$$2d_{hkl}sin\theta = n\lambda \tag{1}$$

where n is the order of diffraction, λ is X-ray wavelength, and d_{hkl} is the lattice interplanar spacing. For hexagonal structure, the plane spacing d_{hkl} is related to the lattice constants $a = b$, c and the Miller indices (h, k, l) by the relation [77]:

$$\frac{1}{d^2_{(hkl)}} = \frac{4}{3}\left(\frac{h^2 + hk + k^2}{a^2}\right) + \frac{l^2}{c^2} \tag{2}$$

Using Bragg's law for $n = 1$, Equation (3) reduces to

$$sin^2\theta = \frac{\lambda^2}{4a^2}\left[\frac{4}{3}\left(h^2 + hk + k^2\right) + \left(\frac{a}{c}\right)^2 l^2\right] \tag{3}$$

The above equation for $\langle 100 \rangle$ and $\langle 002 \rangle$ planes, for example, becomes $a = \frac{\lambda}{\sqrt{3}sin\theta}$ and $c = \frac{\lambda}{sin\theta}$. Using λ = 1.54059 Å and 2θ for two planes 31.767° and 34.419°, lattice constants were calculated as $a = b = 3.2499$ Å and $c = 5.2070$ Å, respectively. These values match very well with the previous work calculated by Bindu et al. [78]. The interplanar spacing d_{hkl} were calculated using Bragg's law from XRD patterns for some major XRD peaks corresponding to $\langle hkl \rangle$ planes and were compared with reference data, as shown in Table 1.

The crystallite size (D) was also calculated in accordance with the Debye–Scherrer formula [44,79]:

$$D = \frac{k\lambda}{\beta cos\theta} \tag{4}$$

Table 1. Interplanar spacing from XRD, reference for corresponding $\langle hkl \rangle$ planes, FWHM, and crystallite size. The average crystallite size is ~9 nm.

$\langle hkl \rangle$	d_{hkl} Å	d_{Ref} Å	FWHM, β (°)	D (nm)
⟨100⟩	2.8028	2.8146	1.003	8
⟨002⟩	2.5894	2.6035	0.714	12
⟨101⟩	2.4686	2.4760	0.793	11
⟨102⟩	1.9058	1.9112	0.935	9
⟨110⟩	1.6221	1.6250	1.041	9
⟨103⟩	1.4745	1.4774	0.981	10
⟨112⟩	1.3734	1.3785	1.632	6
⟨201⟩	1.2369	1.3598	1.290	8

In the above formula, k is a dimensionless shape factor (~0.90), λ is the wavelength of the incident X-ray, D is the crystallite size, θ is the Bragg's angle, and β is the full width at half maxima (FWHM) of the corresponding peak. The β values were calculated using IGOR-Pro 9 Scientific data analysis software by Wave Metrics. The crystallite size was rounded to closest whole number in nanometer, as shown in Table 1. The average crystallite size for synthesized ZnO nanoparticle is 9 ± 2 nm. The average crystallite size calculated in this work is one of the smallest sizes obtained for ZnO nanoparticles [80–82]. The crystallite size is assumed to be the size of the smallest crystal, whereas the particle size indicates size of nanoparticle.

Figure 2 shows SEM results of as-synthesized ZnO nanoparticles deposited on glass substrate at two different magnifications. As shown in Figure 2, these nanoparticles have very small sizes with no clear shapes but in terms of sizes they are larger than their crystallite sizes estimated by the Debye–Scherrer equation, as nanoparticles are an agglomerated form of crystallite. All these images were taken at 5.0 kV accelerating voltage using a lower SE detector to minimize charging resulting from the low carrier concentration of undoped nanoparticles.

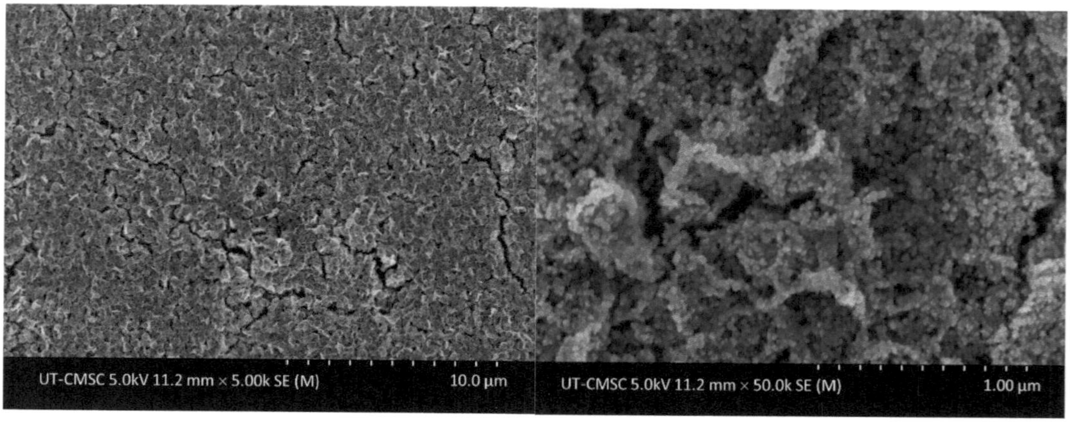

Figure 2. Scanning electron microscopy (SEM) images of ZnO nanoparticles. The nanoparticle thin film was spin-coated and SEM images were taken without the prior heat treatment.

ZnO nanoparticles can have different shapes and sizes based on the adoption of precursors and experimental procedures. The formation of nanoparticles usually takes place through the nucleation and growth processes. According to the Gibbs–Curies–Wulff theorem, the shape of a crystal depends on the surface free energy of individual crystallographic faces and the final shape of the nanoparticles is determined in such a way that the total free energy of the system is minimized [83]. In general, spherical nanoparticles are expected as the spherical shape, has the minimum surface energy for a given volume. By adjusting the reaction time, amount of precursors, temperature and pH, and the type of material used for the synthesis, nanostructures with a controlled size and shape, such as nanorods, nanowires, nanobelts, and nanostars, can be synthesized [84,85].

3.2. Optical Properties Using Spectroscopic Ellipsometry

Through careful analysis of XRD and SEM results, we see that ZnO nanoparticles can be considered as a suitable material for photovoltaic (PV) applications. Another way of finding the suitability of these ZnO nanoparticles in PV applications is to study their optical properties using dielectric functions. Optical responses of nanocrystalline ZnO thin films, in the form of energy-dependent complex dielectric functions, $\epsilon(E) = \epsilon_1(E) + i\epsilon_2(E)$, enables findings of optical and morphological properties applicable to PV and other electronic devices. To find dielectric constants at room temperature, external spectroscopic ellipsometry (ex-situ SE) data were collected from a ZnO thin film deposited on soda lime glass using the spin coating method. The ZnO thin film was annealed at a hot plate temperature of 300 °C for 30 min before taking the SE data. The SE spectra were collected ex situ at a 70° angle of incidence using a M2000 SE system with a back-side tape collecting 635 data points over a spectral range from 0.734 to 4.00 eV.

The complex dielectric functions for the ZnO thin film were parameterized by using two critical point parabolic bands (CPPB) and one Tauc-Lorentz [86] oscillator, with the detailed procedure described elsewhere [87]. Similarly, the film surface roughness layer was modeled using a Bruggeman effective medium approximation (EMA) [88,89]. From the SE analysis, the surface roughness and bulk layer thickness of the nanoparticle thin film sample are d_s = 57 ± 1 nm and d_b = 339 ± 1 nm, respectively. Similarly, total effective thickness of the film and bandgap energy of the material are d_{eff} = 380 ± 1 nm and 3.22 ± 0.01 eV, respectively.

After finding thicknesses from the parametric models, numerical inversion [90] was used to extract dielectric constants at each spectral point for the sample over the measured spectral range. The complex dielectric response function $\epsilon(E)$ in the spectral range of 0.734 eV to 4.00 eV for the thin film of ZnO nanoparticles, as obtained from fitting the model, is shown in Figure 3. The dielectric function shown is comparable to the previously published work [91–93]. The imaginary part of dielectric function ϵ_2 shows a characteristic peak at ~3.22 eV, with an onset of energy at ~3.0 eV for the nanoparticle film. The onset of absorption corresponds to the absorption edge of ZnO, a direct bandgap semiconductor, which matches well with the values calculated in Figure 4. The onset of energy found in this work is ~0.36 eV smaller than Girish Lakhwani et al. [91] for a similar ZnO nanoparticle film, but the value is similar to the single crystalline ZnO [92]. Even though the onset of absorption was expected to be higher for nanoparticle thin film due to the quantum confinement effect, it is smaller and similar to the bulk value due the annealing effect. The $\epsilon_1(E)$ and $\epsilon_2(E)$ spectra of the ZnO nanoparticle thin film calculated in this work are also slightly smaller in magnitude than that in the work of Girish Lakhwani et al. [91], as shown in Figure 3. At higher temperatures, nanoparticle thin films behave like a bulk material.

3.3. Unpolarized Absorbance and Transmittance

The optical absorbance spectrum of synthesized ZnO nanoparticles dispersed in chloroform was obtained using UV/Vis/NIR absorption spectroscopy, as shown in Figure 4A. The spectrum displays the first exciton absorption peak at a 334.5 nm (3.7 eV) wavelength, much below the bulk bandgap wavelength of 368 nm [2]. The first exciton absorption peak attributes the approximate intrinsic bandgap of ZnO due to the electron transitions from the valence band to the conduction band ($O_{2p} \rightarrow Zn_{3d}$) [94]. In addition, the sharp absorption at the band edge and narrow peak position reveal that the size distribution of the nanoparticles is small. The absorbance peak position (334.5 nm) obtained in this work lies at smaller wavelength than from previous research, where it was in the range of 355 nm to 380 nm [44,95–99]. The weak absorption area range includes the whole visible spectrum region (400 nm to 700 nm) and some part of the infrared spectrum (>700 nm) in the measurement range. After the first exciton peak, a monotonical increase in the absorption of light continues until the effect of glass at the middle ultraviolet (MUV) region breaks at 248.5 nm, as shown in Figure 4A. Another experiment was conducted where all KOH solution was mixed into the $ZnAc_2$ solution simultaneously at once instead of via dropwise addition. The poly-dispersed ZnO nanoparticles obtained are shown in the Supplementary Information Figure S2.

At the absorbance edge, absorbance is contributed by the largest size nanoparticles, whereas at the region of absorbance maximum, absorbance contribution is from all particles. There must be a threshold value of wavelength where absorption just begins and is called threshold wavelength (λ_S). The threshold wavelength is 354.6 nm, only 20 nm higher than first exciton absorption peak, as shown in the Figure S3.

Transmittance % spectra of the ZnO nanoparticles of thin films of ~100 nm before and after annealing are shown in Figure 4B. In the Figure 4B, the red line represents transmission from as-synthesized film and other data represent transmission from annealed films. Five films of approximately similar thicknesses were prepared and four films were annealed at 300 °C for different times, as shown. The optimized annealing temperature of 300 °C was estimated based on the transmission measurement of four ZnO thin films at different temperatures, as shown in Figure 4C. When the film is annealed, traces of solvent and other byproducts evaporate, and defect states in the films are minimized or removed. The film becomes more crystalline which may be responsible for scattering the incident light and lowering the transmission. The small change in transmission in the visible range of the spectrum may be due to the slightly different thicknesses of the films. As shown in Figure 4B, in the visible range of spectrum, all films present a high average transmittance greater than 60%, indicating low absorbance and low reflectance, which confirms that these ZnO thin films can be used as antireflection coating in solar cells working mainly in the visible region, such as silicon solar cells [31,100]; in a large spectrum, such as GaN-based concentration solar cells [101]; or as a window layer in quantum dots solar cells [74].

The optical absorption measurement near the fundamental absorption edge is a standard method for the assessment of the bandgap energy. The variation of the optical absorption coefficient with incident wavelength (or energy) helps to explain the band structure and the type of transition of electrons. The optical absorption coefficient ($\alpha(\lambda)$) of the ZnO thin films were calculated using Beer–Lambert's relation [102]:

$$I = I_0 e^{-\alpha(\lambda)x} \rightarrow \alpha(\lambda) = 2.303\frac{A}{t} \quad (5)$$

where A represents the absorbance, and t represents film thickness. The basic procedure for absorption coefficient analysis is to obtain optical absorbance (A) data from transmittance of ZnO thin films taken from UV/Vis/NIR spectrophotometer by the relation: $A = 2 - log_{10}(T\%)$. As shown in Figure 4D, there is only a small change in optical absorption coefficient with respect to the temperature but at the band edge, the absorption coefficient is sharply increased with respect to as-synthesized film. This increase in α at the band edge may be attributed to the onset of interband transitions, which is more efficient in the case of annealed (crystalline) films than as-synthesized (less crystalline) ones. The absorption coefficient depends on the material's extinction coefficient (κ : how strongly a material absorbs light of a particular wavelength) and the wavelength of light (λ) being absorbed by the relation: $\alpha = \frac{4\pi\kappa}{\lambda}$. The κ value depends on the type of bandgap (direct vs. indirect) and the λ depends on the bandgap of the material.

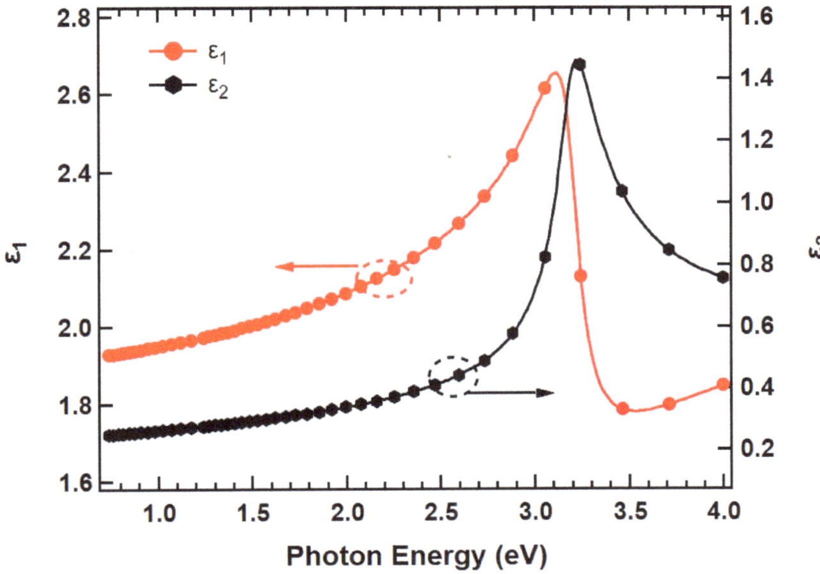

Figure 3. Dielectric function spectra for ZnO nanoparticle thin film measured by SE at room temperature. The red curve represents ε_1 plotted to the left y-axis and is represented by an arrow with a circle and black curve represents ε_2 as shown by a circle with an arrow to the right.

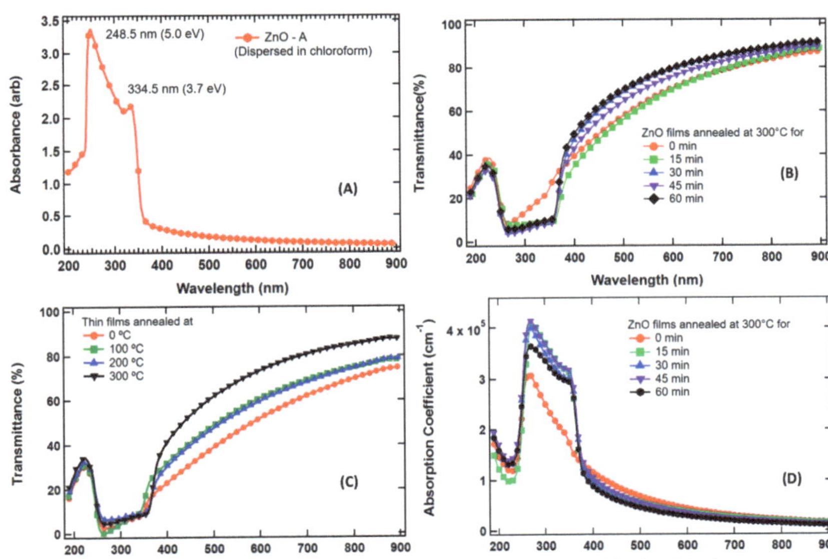

Figure 4. VU/Vis/NIR spectroscopy results: (**A**) Absorbance spectrum of as-obtained ZnO nanoparticles dispersed in chloroform. (**B**) Transmission % of ZnO thin film of ~100 nm before and after annealing the films. (**C**) Temperature-dependent transmission of ZnO thin films annealed for 30 min. When temperature is increased, transmission in infrared and visible regions is increased, whereas it is decreased in the band edge absorption regions. (**D**) Absorption coefficients (α, cm^{-1}) of thin films of ZnO with respect to wavelength in nm. Small changes in α were observed when the films were annealed for different time intervals.

The optical bandgap of the ZnO is estimated using Tauc's relation [103]:

$$\alpha h\nu = A\left(h\nu - E_g\right)^n \tag{6}$$

where A is a constant, $h\nu$ incident photon's energy, and E_g is the bandgap of the material to be determined. The factor n depends on the nature of the electron transition and is equal to ½ and 2 for the direct and indirect bandgap transitions, respectively.

Figure 5A gives the Tauc plot for ZnO, where α times the $h\nu$ to the second power is plotted versus the $h\nu$. The second power of $\alpha h\nu$ is used as ZnO is well known to have a direct allowed transition. The characteristic features of Tauc plot are evident: at low photon energies, the absorption approaches insignificant—the material is transparent; near the bandgap value, the absorption gets stronger and shows a region of linearity in this squared-exponent plot. The linear fit applied at the linear region is extrapolated to the x-axis intercept to find the bandgap value. The direct bandgap values decreased from 3.40 eV at room temperature to 3.33 eV when the film was annealed for 15 min at 300 °C. The bandgap continued decreasing and saturated after 30 min at 3.30 eV. Temperature variation of the energy gap in semiconductors is known to be due to the effects of lattice dilation (expansion) and electron–phonon interaction or phonon-induced atomic vibrations [104–106]. Both experimental and theoretical studies conducted by Zhang et al. [107] have shown that bandgap decreases as temperature increases. The phenomenon is more prevalent in ionic compounds, such as ZnO, than covalent compounds due to stronger lattice expansion with temperature. The bandgap of the annealed sample matches very well with some of the previous results [99,108].

An obvious redshift in the absorption edge was observed for the ZnO film with respect to the exciton peak observed in the solution, as shown in Figure 4A. This might be due to the increased dielectric constant and substantial electronic coupling when the nanoparticles come close together after the evaporation of the solvent. In addition, in thin films, effects due to interference might occur that not only lead to unexpected absorbance values but also to peak shifts.

Figure 5A displays lower and upper deviations from the region of linear behavior. On the lower energy side, the deviation from the linear region can be associated with defect absorption states located near the band edge. This phenomenon was first investigated by Urbach [85] in silver halides, and in subsequent years, was applied to a large number of other crystals, therefore, it is known as an "Urbach Tail". These defect absorption states exhibit an exponential energy dependence corresponding to a typical distribution of density of states. On the higher energy side, saturation of available density of states is responsible for a leveling out of absorbance strength [103].

Modification of band structure, which may be due to the introduction of defects in the material, can be recognized by measuring the material's Urbach energy. It was found that the absorption coefficient near the band edge shows an exponential dependence on photon energy given by [109,110]

$$\alpha(h\nu) = \alpha_0 \exp\left(\frac{h\nu - E_0}{E_U}\right) \tag{7}$$

where E_0 and α_0 are characteristic parameters of the material, and E_U, an inverse logarithmic slope of absorption coefficient, is the Urbach energy interpreted as the width of the tails of localized states, associated with the amorphous state, in the bandgap. To calculate the Urbach energy, a plot of $ln\alpha$ versus photon energy was made, as shown in Figure 5B. Then, the value of E_U was calculated by taking the reciprocal of the slopes of the linear portion in the lower photon energy region of the curve. The minimum Urbach energy of 0.91 eV was found from a film heated at 300 °C for 30 min and maximum Urbach energy of 1.24 eV was found from the as-synthesized film. The higher E_U from the as-synthesized film is due to the impure state of the material. Factors governing the Urbach energy are structural disorder, imperfection in stoichiometry and unpassivated surface state. Structural disorder may be higher in the as-synthesized nanoparticle thin film. Imperfection in stoichiometry may be caused by residual materials left after synthesis. In addition, dangling bonds disappear from the surface of the nanoparticles when the film is annealed.

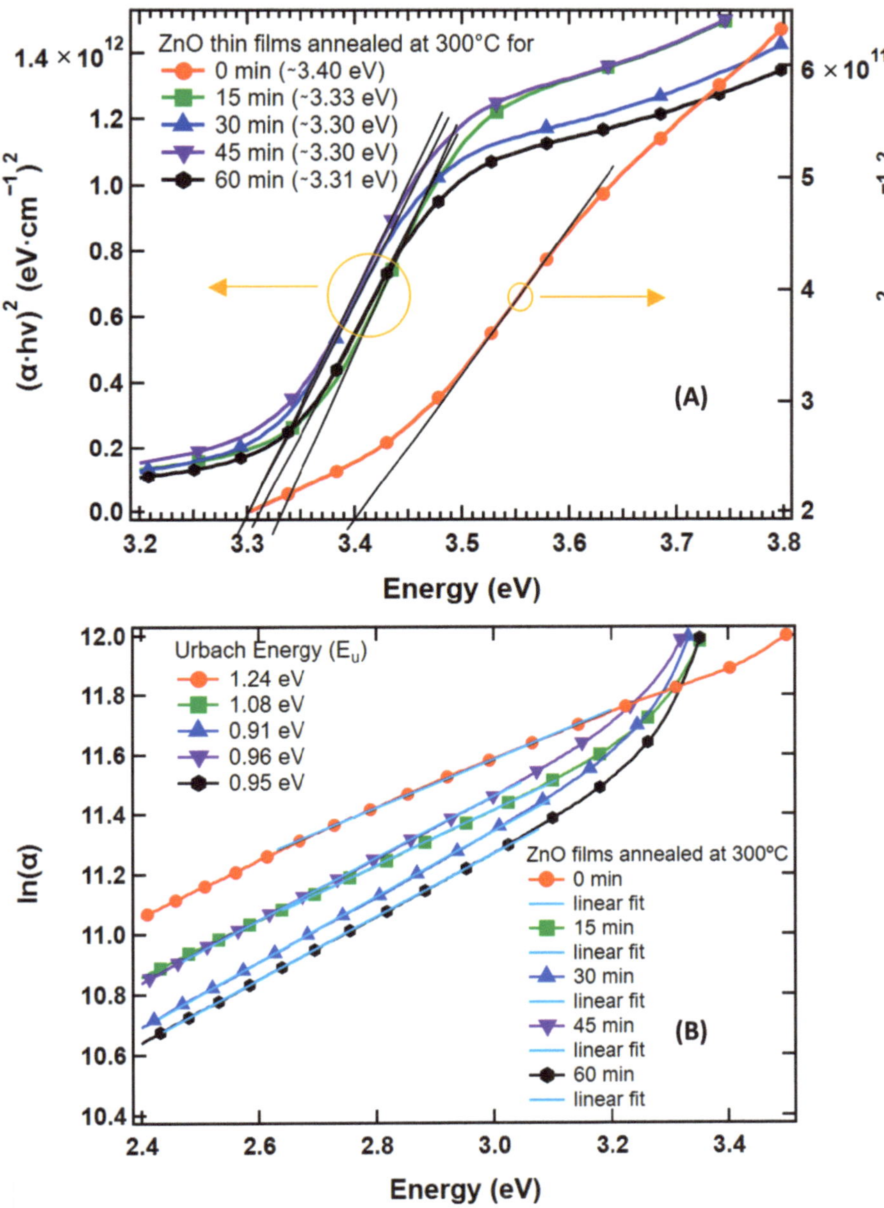

Figure 5. (**A**) Plot of $(\alpha h\nu)^2$ vs. $h\nu$ of ZnO nanoparticles' thin films. The red curve is from the as-synthesized film with y-axis to the right and the rest are from the heated films with y-axis to the left. Yellow circles and arrows represent directions of y-axes for respective curves. (**B**) Urbach energy calculation of ZnO nanoparticles. Low E_u indicates that annealed films are pure & crystalline. More defect states are present in the as-synthesized film, as depicted by higher E_u.

4. Conclusions

In this work, we have reported the synthesis of monodispersed ZnO nanoparticles with a hexagonal wurtzite structure via the solvothermal synthesis method using $ZnAc_2$ as zinc precursor and KOH as oxygen precursor. The nanoparticles were confirmed using XRD, SEM micrograph analysis, SP, and UV/Vis/NIR spectroscopy study. The SEM images showed identical morphology of ZnO nanoparticles with uniform particle size distribution. The XRD analyses proved the crystalline structure of the particles. The average crystallite size obtained from Scherrer's equation was ~9 nm, one of the smallest for ZnO nanoparticles. Optical and morphological properties of ZnO nanoparticle film were also studied using ellipsometry in the spectral range from 0.734 eV to 4.00 eV. The measured spectra revealed the distinct band edge structures at 3.22 eV, ~0.1 eV smaller than found in another method. The sharp first exciton peak revealed that the ZnO nanoparticle size in these samples were nearly monodisperse. The first exciton peak in solution measurement dispersed in chloroform was at 334.5 nm (~3.7 eV), with threshold absorption wavelength of 354.6 nm (~3.5 eV). The first exciton peak was found red shifted in thin film measurement relative to nanoparticles in solution with a value of 350 nm (~3.5 eV). The red shift in the first exciton peak was due to the increased dielectric constant and substantial electronic coupling of nanoparticles. As expected, the bandgap energy transition of ZnO nanoparticles was found to be directly allowed, with values of 3.40 eV and 3.30 eV, respectively, for as-synthesized and annealed films. Our investigation revealed that the nanoparticles fabricated in this work have a wide range of applications, especially in the electro-technological industries, such as photoelectronics, field emitter, sensors, UV lasers, and solar cells.

Supplementary Materials: The following supporting information can be downloaded at: https://www.mdpi.com/article/10.3390/nano13111795/s1, Figure S1: XRD patterns of as-synthesized ZnO thin film. The film was deposited by using a spin coating process. The material is less crystalline because the peak intensities are comparatively lower; Figure S2: Absorbance spectrum of ZnO nanoparticles dispersed in chloroform. Synthesis was done by mixing all KOH solution in methanol at once into all $ZnAc_2$ solution in methanol; Figure S3: $(A/\lambda)^2$ vs. $1/\lambda$ of ZnO nanoparticles dispersed in chloroform. References [111,112] are cited in the Supplementary Materials.

Author Contributions: K.P.B. prepared the manuscript with the support of D.R.S., M.K.J., Q.S. and R.W.C. All authors have read and agreed to the published version of the manuscript.

Funding: This research received no external funding.

Data Availability Statement: No new data were created.

Acknowledgments: This work was supported by the Dean's Research Lab Grants, Ohio Northern University. Also, all authors thank the editors and anonymous reviewers for their valuable suggestions to upgrade the manuscript.

Conflicts of Interest: The authors declare no conflict of interest.

References

1. Patil, L.; Bari, A.; Shinde, M.; Deo, V.; Kaushik, M. *Zinc Oxide: Fundamentals Materials and Device Technology*; John Wiley & Sons, Inc.: Hoboken, NJ, USA, 2009; p. 12.
2. Özgür, Ü.; Avrutin, V.; Morkoç, H. Chapter 16—Zinc Oxide Materials and Devices Grown by Molecular Beam Epitaxy. In *Molecular Beam Epitaxy*, 2nd ed.; Henini, M., Ed.; Elsevier: Amsterdam, The Netherlands, 2018; pp. 343–375. [CrossRef]
3. Olvera, M.; Maldonado, A.; Asomoza, R. Chemical stability of doped ZnO thin films. *J. Mater. Sci. Mater. Electron.* **2000**, *11*, 1–5. [CrossRef]
4. Lee, W.; Yeop, J.; Heo, J.; Yoon, Y.J.; Park, S.Y.; Jeong, J.; Shin, Y.S.; Kim, J.W.; An, N.G.; Kim, D.S.; et al. High colloidal stability ZnO nanoparticles independent on solvent polarity and their application in polymer solar cells. *Sci. Rep.* **2020**, *10*, 18055. [CrossRef]
5. Heinonen, S.; Nikkanen, J.-P.; Huttunen-Saarivirta, E.; Levänen, E. Investigation of long-term chemical stability of structured ZnO films in aqueous solutions of varying conditions. *Thin Solid Film.* **2017**, *638*, 410–419. [CrossRef]
6. Deschanvres, J.L.; Rey, P.; Delabouglise, G.; Labeau, M.; Joubert, J.C.; Peuzin, J.C. Characterization of piezoelectric properties of zinc oxide thin films deposited on silicon for sensors applications. *Sens. Actuators A Phys.* **1992**, *33*, 43–45. [CrossRef]
7. Membetsi, R.O.-N.Z.M.; Gnanga, H.; Omanda, H.; Foucaran, A. Microwave Analysis and Electrical Properties of ZnO thin Films Prepared by RF Magnetron Sputtering. *Int. J. Sci. Res. IJSR* **2015**, *4*, 1828–1833.

8. Segets, D.; Gradl, J.; Taylor, R.K.; Vassilev, V.; Peukert, W. Analysis of Optical Absorbance Spectra for the Determination of ZnO Nanoparticle Size Distribution, Solubility, and Surface Energy. *ACS Nano* **2009**, *3*, 1703–1710. [CrossRef]
9. Wahyuni, E.T.; Diantariani, N.P.; Kartini, I.; Kuncaka, A. Enhancement of the photostability and visible photoactivity of ZnO photocatalyst used for reduction of Cr(VI) ions. *Results Eng.* **2022**, *13*, 100351. [CrossRef]
10. Ayoub, I.; Kumar, V.; Abolhassani, R.; Sehgal, R.; Sharma, V.; Sehgal, R.; Swart, H.C.; Mishra, Y.K. Advances in ZnO: Manipulation of defects for enhancing their technological potentials. *Nanotechnol. Rev.* **2022**, *11*, 575–619. [CrossRef]
11. Kumar, V.; Ntwaeaborwa, O.; Soga, T.; Dutta, V.; Swart, H. Rare Earth Doped Zinc Oxide Nanophosphor Powder: A Future Material for Solid State Lighting and Solar Cells. *ACS Photonics* **2017**, *4*, 2613–2637. [CrossRef]
12. Capper, P.; Kasap, S.O.; Willoughby, A. *Zinc Oxide Materials for Electronic and Optoelectronic Device Applications*; Litton, C.W., Reynolds, D.C., Collins, T.C., Eds.; John Wiley & Sons: Hoboken, NJ, USA, 2011; Volume 35.
13. Singh, R.; Ichikawa, T.; Jain, A. 20—Zinc oxide as promising anode material for Li-ion battery. In *Nanostructured Zinc Oxide*; Awasthi, K., Ed.; Elsevier: Amsterdam, The Netherlands, 2021; pp. 575–603. [CrossRef]
14. Kim, H.; Jae, W.; Song, J.; Kim, J. Skein-shaped ZnO/N-doped carbon microstructures as a high performance anode material for lithium-ion batteries. *J. Alloys Compd.* **2019**, *772*, 507–515. [CrossRef]
15. McCluskey, M.D.; Jokela, S.J. Defects in ZnO. *J. Appl. Phys.* **2009**, *106*, 071101. [CrossRef]
16. Pearton, S.J.; Norton, D.P.; Ip, K.; Heo, Y.W.; Steiner, T. Recent advances in processing of ZnO. *J. Vac. Sci. Technol. B Microelectron. Nanometer Struct. Process. Meas. Phenom.* **2004**, *22*, 932–948. [CrossRef]
17. Klingshirn, C.; Hauschild, R.; Fallert, J.; Kalt, H. Room-temperature stimulated emission of ZnO: Alternatives to excitonic lasing. *Phys. Rev. B* **2007**, *75*, 115203. [CrossRef]
18. Tang, Z.K.; Wong, G.K.L.; Yu, P.; Kawasaki, M.; Ohtomo, A.; Koinuma, H.; Segawa, Y. Room-temperature ultraviolet laser emission from self-assembled ZnO microcrystallite thin films. *Appl. Phys. Lett.* **1998**, *72*, 3270–3272. [CrossRef]
19. Dong, H.; Zhou, B.; Li, J.; Zhan, J.; Zhang, L. Ultraviolet lasing behavior in ZnO optical microcavities. *J. Mater.* **2017**, *3*, 255–266. [CrossRef]
20. Egerton, E.J.; Sood, A.K.; Singh, R.; Puri, Y.R.; Davis, R.F.; Pierce, J.; Look, D.C.; Steiner, T. P-type doping utilizing nitrogen and Mn doping of ZnO using MOCVD for ultraviolet lasers and spintronic applications. *J. Electron. Mater.* **2005**, *34*, 949–952. [CrossRef]
21. Deng, G.; Zhang, Y.; Yu, Y.; Han, X.; Wang, Y.; Shi, Z.; Dong, X.; Zhang, B.; Du, G.; Liu, Y. High-Performance Ultraviolet Light-Emitting Diodes Using n-ZnO/p-hBN/p-GaN Contact Heterojunctions. *ACS Appl. Mater. Interfaces* **2020**, *12*, 6788–6792. [CrossRef]
22. Mohammad Reza, M. Metal Oxide ZnO-Based Varistor Ceramics. In *Advances in Ceramics*; Costas, S., Ed.; IntechOpen: Rijeka, Croatia, 2011; p. 564. [CrossRef]
23. Levinson, L.; Philipp, H. Zinc oxide varistors—A review. *Am. Ceram. Soc. Bull.* **1986**, *65*, 639–646.
24. Raghu, N.; Kutty, T.R.N. Varistors based on mixed-phase ceramics containing ZnO and negative temperature coefficient spinels. *Appl. Phys. Lett.* **1991**, *58*, 1050–1052. [CrossRef]
25. Park, K.; Lee, J.K. Mn–Ni–Co–Cu–Zn–O NTC thermistors with high thermal stability for low resistance applications. *Scr. Mater.* **2007**, *57*, 329–332. [CrossRef]
26. Cho, J.; Hwang, S.; Ko, D.-H.; Chung, S. Transparent ZnO Thin-Film Deposition by Spray Pyrolysis for High-Performance Metal-Oxide Field-Effect Transistors. *Materials* **2019**, *12*, 3423. [CrossRef] [PubMed]
27. Chen, Z.; Wang, J.; Wu, H.; Yang, J.; Wang, Y.; Zhang, J.; Bao, Q.; Wang, M.; Ma, Z.; Tress, W.; et al. A Transparent Electrode Based on Solution-Processed ZnO for Organic Optoelectronic Devices. *Nat. Commun.* **2022**, *13*, 4387. [CrossRef] [PubMed]
28. Burkov, S.I.; Zolotova, O.P.; Sorokin, B.P.; Turchin, P.P.; Talismanov, V.S. Features of acoustic wave propagation in the Me/ZnO/Me/diamond waveguide structure. *J. Acoust. Soc. Am.* **2018**, *143*, 16–22. [CrossRef] [PubMed]
29. Shiosaki, T.; Kawabata, A. Low-frequency piezoelectric-transducer applications of ZnO film. *Appl. Phys. Lett.* **1974**, *25*, 10–11. [CrossRef]
30. Galstyan, V.; Comini, E.; Baratto, C.; Faglia, G.; Sberveglieri, G. Nanostructured ZnO chemical gas sensors. *Ceram. Int.* **2015**, *41*, 14239–14244. [CrossRef]
31. Hussain, B.; Ebong, A.; Ferguson, I. Zinc oxide as an active n-layer and antireflection coating for silicon based heterojunction solar cell. *Sol. Energy Mater. Sol. Cells* **2015**, *139*, 95–100. [CrossRef]
32. Vittal, R.; Ho, K.-C. Zinc oxide based dye-sensitized solar cells: A review. *Renew. Sustain. Energy Rev.* **2017**, *70*, 920–935. [CrossRef]
33. Consonni, V.; Briscoe, J.; Kärber, E.; Li, X.; Cossuet, T. ZnO nanowires for solar cells: A comprehensive review. *Nanotechnology* **2019**, *30*, 362001. [CrossRef]
34. Le, A.T.; Ahmadipour, M.; Pung, S.-Y. A review on ZnO-based piezoelectric nanogenerators: Synthesis, characterization techniques, performance enhancement and applications. *J. Alloys Compd.* **2020**, *844*, 156172. [CrossRef]
35. Pearton, S.J.; Ren, F. Advances in ZnO-based materials for light emitting diodes. *Curr. Opin. Chem. Eng.* **2014**, *3*, 51–55. [CrossRef]
36. Žutić, I.; Fabian, J.; Das Sarma, S. Spin injection through the depletion layer: A theory of spin-polarized p-n junctions and solar cells. *Phys. Rev. B* **2001**, *64*, 121201. [CrossRef]
37. Miccoli, I.; Spampinato, R.; Marzo, F.; Prete, P.; Lovergine, N. DC-magnetron sputtering of ZnO: Al films on (00.1) Al_2O_3 substrates from slip-casting sintered ceramic targets. *Appl. Surf. Sci.* **2014**, *313*, 418–423. [CrossRef]

38. Liu, B.; Zeng, H.C. Hydrothermal Synthesis of ZnO Nanorods in the Diameter Regime of 50 nm. *J. Am. Chem. Soc.* **2003**, *125*, 4430–4431. [CrossRef] [PubMed]
39. Schlur, L.; Calado, J.R.; Spitzer, D. Synthesis of zinc oxide nanorods or nanotubes on one side of a microcantilever. *R. Soc. Open Sci.* **2018**, *5*, 180510. [CrossRef] [PubMed]
40. Somvanshi, D.; Jit, S. Analysis of Temperature-Dependent Electrical Characteristics of n-ZnO Nanowires (NWs)/p-Si Heterojunction Diodes. *IEEE Trans. Nanotechnol.* **2014**, *13*, 62–69. [CrossRef]
41. Wang, Y.; Li, X.; Wang, N.; Quan, X.; Chen, Y. Controllable synthesis of ZnO nanoflowers and their morphology-dependent photocatalytic activities. *Sep. Purif. Technol.* **2008**, *62*, 727–732. [CrossRef]
42. Zhang, N.; Yi, R.; Shi, R.; Gao, G.; Chen, G.; Liu, X. Novel rose-like ZnO nanoflowers synthesized by chemical vapor deposition. *Mater. Lett.* **2009**, *63*, 496–499. [CrossRef]
43. Meulenkamp, E.A. Synthesis and growth of ZnO nanoparticles. *J. Phys. Chem. B* **1998**, *102*, 5566–5572. [CrossRef]
44. Talam, S.; Karumuri, S.R.; Gunnam, N. Synthesis, characterization, and spectroscopic properties of ZnO nanoparticles. *Int. Sch. Res. Not.* **2012**, *2012*, 372505. [CrossRef]
45. Hong, R.; Pan, T.; Qian, J.; Li, H. Synthesis and surface modification of ZnO nanoparticles. *Chem. Eng. J.* **2006**, *119*, 71–81. [CrossRef]
46. Kattan, P. *Ratio of Surface Area to Volume in Nanotechnology and Nanoscience*; Petra Books: Kota Bandung, Indonesia, 2011.
47. Lu, H.; Liao, L.; Li, J.; Wang, D.; He, H.; Fu, Q.; Xu, L.; Tian, Y. High Surface-to-Volume Ratio ZnO Microberets: Low Temperature Synthesis, Characterization, and Photoluminescence. *J. Phys. Chem. B* **2006**, *110*, 23211–23214. [CrossRef]
48. Kumar, H.; Rani, R. Structural and optical characterization of ZnO nanoparticles synthesized by microemulsion route. *Int. Lett. Chem. Phys. Astron.* **2013**, *14*, 26–36. [CrossRef]
49. Yıldırım, Ö.A.; Durucan, C. Synthesis of zinc oxide nanoparticles elaborated by microemulsion method. *J. Alloys Compd.* **2010**, *506*, 944–949. [CrossRef]
50. Hasnidawani, J.; Azlina, H.; Norita, H.; Bonnia, N.; Ratim, S.; Ali, E. Synthesis of ZnO nanostructures using sol-gel method. *Procedia Chem.* **2016**, *19*, 211–216. [CrossRef]
51. Vafaee, M.; Ghamsari, M.S. Preparation and characterization of ZnO nanoparticles by a novel sol-gel route. *Mater. Lett.* **2007**, *61*, 3265–3268. [CrossRef]
52. Nagpal, K.; Rapenne, L.; Wragg, D.S.; Rauwel, E.; Rauwel, P. The role of CNT in surface defect passivation and UV emission intensification of ZnO nanoparticles. *Nanomater. Nanotechnol.* **2022**, *12*, 18479804221079419. [CrossRef]
53. Tsuzuki, T.; McCormick, P.G. Mechanochemical synthesis of nanoparticles. *J. Mater. Sci.* **2004**, *39*, 5143–5146. [CrossRef]
54. Tsuzuki, T.; McCormick, P.G. ZnO nanoparticles synthesised by mechanochemical processing. *Scr. Mater.* **2001**, *44*, 1731–1734. [CrossRef]
55. Tani, T.; Mädler, L.; Pratsinis, S.E. Homogeneous ZnO nanoparticles by flame spray pyrolysis. *J. Nanoparticle Res.* **2002**, *4*, 337–343. [CrossRef]
56. Ghaffarian, H.R.; Saiedi, M.; Sayyadnejad, M.A.; Rashidi, A.M. Synthesis of ZnO nanoparticles by spray pyrolysis method. *Iran. J. Chem. Chem. Eng. IJCCE* **2011**, *30*, 1–6.
57. Yang, Y.; Chen, H.; Zhao, B.; Bao, X. Size control of ZnO nanoparticles via thermal decomposition of zinc acetate coated on organic additives. *J. Cryst. Growth* **2004**, *263*, 447–453. [CrossRef]
58. Salavati-Niasari, M.; Davar, F.; Mazaheri, M. Preparation of ZnO nanoparticles from [bis (acetylacetonato) zinc (II)]-oleylamine complex by thermal decomposition. *Mater. Lett.* **2008**, *62*, 1890–1892. [CrossRef]
59. Bekermann, D.; Gasparotto, A.; Barreca, D.; Maccato, C.; Comini, E.; Sada, C.; Sberveglieri, G.; Devi, A.; Fischer, R. Co_3O_4/ZnO nanocomposites: From plasma synthesis to gas sensing applications. *ACS Appl. Mater. Interfaces* **2012**, *4*, 928–934. [CrossRef] [PubMed]
60. Hiragino, Y.; Tanaka, T.; Takeuchi, H.; Takeuchi, A.; Lin, J.; Yoshida, T.; Fujita, Y. Synthesis of nitrogen-doped ZnO nanoparticles by RF thermal plasma. *Solid State Electron.* **2016**, *118*, 41–45. [CrossRef]
61. Byrappa, K.; Ohara, S.; Adschiri, T. Nanoparticles synthesis using supercritical fluid technology—Towards biomedical applications. *Adv. Drug Deliv. Rev.* **2008**, *60*, 299–327. [CrossRef] [PubMed]
62. Viswanathan, R.; Gupta, R.B. Formation of zinc oxide nanoparticles in supercritical water. *J. Supercrit. Fluids* **2003**, *27*, 187–193. [CrossRef]
63. Kinge, S.; Crego-Calama, M.; Reinhoudt, D.N. Self-assembling nanoparticles at surfaces and interfaces. *ChemPhysChem* **2008**, *9*, 20–42. [CrossRef]
64. Baruwati, B.; Kumar, D.K.; Manorama, S.V. Hydrothermal synthesis of highly crystalline ZnO nanoparticles: A competitive sensor for LPG and EtOH. *Sens. Actuators B Chem.* **2006**, *119*, 676–682. [CrossRef]
65. Madathil, A.N.P.; Vanaja, K.; Jayaraj, M. Synthesis of ZnO nanoparticles by hydrothermal method. In Proceedings of the Nanophotonic Materials IV, San Diego, CA, USA, 26–27 August 2007; pp. 47–55.
66. Singh, L.; Sharma, R.; Singh, N.; Kumar, A.; Mahato, D.K.; Lee, Y.; Bechelany, M.; Mandal, K.D. Semi-wet growth and characterization of multi-functional nano-engineered mixed metal oxides for industrial application. *Prog. Cryst. Growth Charact. Mater.* **2021**, *67*, 100542. [CrossRef]
67. Huang, M.H.; Wu, Y.; Feick, H.; Tran, N.; Weber, E.; Yang, P. Catalytic growth of zinc oxide nanowires by vapor transport. *Adv. Mater.* **2001**, *13*, 113–116. [CrossRef]

68. Prete, P.; Lovergine, N.; Tapfer, L. Nanostructure size evolution during Au-catalysed growth by carbo-thermal evaporation of well-aligned ZnO nanowires on (100) Si. *Appl. Phys. A* **2007**, *88*, 21–26. [CrossRef]
69. Hasanpoor, M.; Aliofkhazraei, M.; Delavari, H. Microwave-assisted synthesis of zinc oxide nanoparticles. *Procedia Mater. Sci.* **2015**, *11*, 320–325. [CrossRef]
70. Bilecka, I.; Elser, P.; Niederberger, M. Kinetic and thermodynamic aspects in the microwave-assisted synthesis of ZnO nanoparticles in benzyl alcohol. *ACS Nano* **2009**, *3*, 467–477. [CrossRef]
71. Kahouli, M.; Barhoumi, A.; Bouzid, A.; Al-Hajry, A.; Guermazi, S. Structural and optical properties of ZnO nanoparticles prepared by direct precipitation method. *Superlattices Microstruct.* **2015**, *85*, 7–23. [CrossRef]
72. Sharma, R.K.; Ghose, R. Synthesis of zinc oxide nanoparticles by homogeneous precipitation method and its application in antifungal activity against *Candida albicans*. *Ceram. Int.* **2015**, *41*, 967–975. [CrossRef]
73. Lupan, O.; Emelchenko, G.A.; Ursaki, V.V.; Chai, G.; Redkin, A.N.; Gruzintsev, A.N.; Tiginyanu, I.M.; Chow, L.; Ono, L.K.; Roldan Cuenya, B.; et al. Synthesis and characterization of ZnO nanowires for nanosensor applications. *Mater. Res. Bull.* **2010**, *45*, 1026–1032. [CrossRef]
74. Luther, J.M.; Gao, J.; Lloyd, M.T.; Semonin, O.E.; Beard, M.C.; Nozik, A.J. Stability Assessment on a 3% Bilayer PbS/ZnO Quantum Dot Heterojunction Solar Cell. *Adv. Mater.* **2010**, *22*, 3704–3707. [CrossRef]
75. Choi, J.J.; Lim, Y.-F.; Santiago-Berrios, M.E.B.; Oh, M.; Hyun, B.-R.; Sun, L.; Bartnik, A.C.; Goedhart, A.; Malliaras, G.G.; Abruña, H.D.; et al. PbSe Nanocrystal Excitonic Solar Cells. *Nano Lett.* **2009**, *9*, 3749–3755. [CrossRef]
76. Beek, W.J.; Wienk, M.M.; Kemerink, M.; Yang, X.; Janssen, R.A. Hybrid zinc oxide conjugated polymer bulk heterojunction solar cells. *J. Phys. Chem. B* **2005**, *109*, 9505–9516. [CrossRef]
77. Cullity, B.; Stock, S. *Elements of X-ray Diffraction*, 3rd ed.; Prentice Hall: New York, NY, USA, 2001; pp. 174–177.
78. Bindu, P.; Thomas, S. Estimation of lattice strain in ZnO nanoparticles: X-ray peak profile analysis. *J. Theor. Appl. Phys.* **2014**, *8*, 123–134. [CrossRef]
79. Saleem, M.; Fang, L.; Ruan, H.; Wu, F.; Huang, Q.; Xu, C.; Kong, C. Effect of zinc acetate concentration on the structural and optical properties of ZnO thin films deposited by Sol-Gel method. *Int. J. Phys. Sci* **2012**, *7*, 2971. [CrossRef]
80. Ebin, B.; Arıg, E.; Özkal, B.; Gürmen, S. Production and characterization of ZnO nanoparticles and porous particles by ultrasonic spray pyrolysis using a zinc nitrate precursor. *Int. J. Miner. Metall. Mater.* **2012**, *19*, 651–656. [CrossRef]
81. Koutu, V.; Shastri, L.; Malik, M. Effect of NaOH concentration on optical properties of zinc oxide nanoparticles. *Mater. Sci. Pol.* **2016**, *34*, 819–827. [CrossRef]
82. Haque, M.J.; Bellah, M.M.; Hassan, M.R.; Rahman, S. Synthesis of ZnO nanoparticles by two different methods & comparison of their structural, antibacterial, photocatalytic and optical properties. *Nano Express* **2020**, *1*, 010007.
83. Mullin, J.W. *Crystallization*; Elsevier: Amsterdam, The Netherlands, 2001.
84. Habibi, R.; Daryan, J.T.; Rashidi, A.M. Shape and size-controlled fabrication of ZnO nanostructures using novel templates. *J. Exp. Nanosci.* **2009**, *4*, 35–45. [CrossRef]
85. Montero-Muñoz, M.; Ramos-Ibarra, J.; Rodríguez-Páez, J.; Ramirez, A.; Huamaní-Coaquira, J. Shape-control of Zinc Oxide nanoparticles: Enhancing photocatalytic activity under UV irradiation. In Proceedings of the VIII International Congress of Engineering Physics, Mérida, Mexico, 7–11 November 2016.
86. Jellison, G., Jr.; Modine, F. Parameterization of the optical functions of amorphous materials in the interband region. *Appl. Phys. Lett.* **1996**, *69*, 371–373. [CrossRef]
87. Sapkota, D.R. Characterization and Optimization of CuInSe2 Solar Cells Applicable for Tandem Devices. Ph.D. Thesis, University of Toledo, Toledo, OH, USA, 2022.
88. Aspnes, D.; Theeten, J.; Hottier, F. Investigation of effective-medium models of microscopic surface roughness by spectroscopic ellipsometry. *Phys. Rev. B* **1979**, *20*, 3292. [CrossRef]
89. Subedi, I.; Bhandari, K.P.; Ellingson, R.J.; Podraza, N.J. Near infrared to ultraviolet optical properties of bulk single crystal and nanocrystal thin film iron pyrite. *Nanotechnology* **2016**, *27*, 295702. [CrossRef]
90. Oldham, W. Numerical techniques for the analysis of lossy films. *Surf. Sci.* **1969**, *16*, 97–103. [CrossRef]
91. Lakhwani, G.; Roijmans, R.F.; Kronemeijer, A.J.; Gilot, J.; Janssen, R.A.; Meskers, S.C. Probing charge carrier density in a layer of photodoped ZnO nanoparticles by spectroscopic ellipsometry. *J. Phys. Chem. C* **2010**, *114*, 14804–14810. [CrossRef]
92. Yoshikawa, H.Y.H.; Adachi, S.A.S. Optical constants of ZnO. *Jpn. J. Appl. Phys.* **1997**, *36*, 6237. [CrossRef]
93. Malandrino, G.; Blandino, M.; Fragala, M.E.; Losurdo, M.; Bruno, G. Relationship between nanostructure and optical properties of ZnO thin films. *J. Phys. Chem. C* **2008**, *112*, 9595–9599. [CrossRef]
94. Zak, A.K.; Abrishami, M.E.; Majid, W.H.A.; Yousefi, R.; Hosseini, S.M. Effects of annealing temperature on some structural and optical properties of ZnO nanoparticles prepared by a modified sol-gel combustion method. *Ceram. Int.* **2011**, *37*, 393–398. [CrossRef]
95. Zak, A.K.; Razali, R.; Majid, W.H.; Darroudi, M. Synthesis and characterization of a narrow size distribution of zinc oxide nanoparticles. *Int. J. Nanomed.* **2011**, *6*, 1399–1403.
96. Bian, S.-W.; Mudunkotuwa, I.A.; Rupasinghe, T.; Grassian, V.H. Aggregation and Dissolution of 4 nm ZnO Nanoparticles in Aqueous Environments: Influence of pH, Ionic Strength, Size, and Adsorption of Humic Acid. *Langmuir* **2011**, *27*, 6059–6068. [CrossRef]

97. Ha, T.T.; Canh, T.D.; Tuyen, N.V. A Quick Process for Synthesis of ZnO Nanoparticles with the Aid of Microwave Irradiation. *ISRN Nanotechnol.* **2013**, *2013*, 497873. [CrossRef]
98. Kumbhakar, P.; Singh, D.; Tiwary, C.; Mitra, A. Chemical synthesis and visible photoluminescence emission from monodispersed ZnO nanoparticles. *Chem. Phys. Lett.* **2008**, *5*, 387–394.
99. Bindu, P.; Thomas, S. Optical Properties of ZnO Nanoparticles Synthesised from a Polysaccharide and $ZnCl_2$. *Acta Phys. Pol. A* **2017**, *131*, 1474–1478. [CrossRef]
100. Chen, J.; Sun, K.-W. Growth of vertically aligned ZnO nanorod arrays as antireflection layer on silicon solar cells. *Sol. Energy Mater. Sol. Cells* **2010**, *94*, 930–934. [CrossRef]
101. Nam, S.Y.; Choi, Y.; Song, Y.H.; Jung, M.H.; Kang, C.-M.; Kong, D.-J.; Park, S.J.; Lee, J.Y.; Namkoong, G.; Lee, D.-S. N-ZnO/i-InGaN/p-GaN heterostructure for solar cell application. *Phys. Status Solidi A* **2013**, *210*, 2214–2218. [CrossRef]
102. Kaplan, I. *Nuclear Physics*; Addison-Wesley Publishing: Boston, MA, USA, 1977.
103. Viezbicke, B.D.; Patel, S.; Davis, B.E.; Birnie, D. Evaluation of the Tauc Method for Optical Absorption Edge Determination: ZnO Thin Films as a Model System. *Phys. Status Solidi B* **2015**, *252*, 1700–1710. [CrossRef]
104. Varshni, Y.P. Temperature dependence of the energy gap in semiconductors. *Physica* **1967**, *34*, 149–154. [CrossRef]
105. Fan, H. Temperature dependence of the energy gap in semiconductors. *Phys. Rev.* **1951**, *82*, 900. [CrossRef]
106. Manoogian, A.; Woolley, J. Temperature dependence of the energy gap in semiconductors. *Can. J. Phys.* **1984**, *62*, 285–287. [CrossRef]
107. Zhang, Y.; Wang, Z.; Xi, J.; Yang, J. Temperature-dependent band gaps in several semiconductors: From the role of electron–phonon renormalization. *J. Phys. Condens. Matter* **2020**, *32*, 475503. [CrossRef]
108. Wang, M.-D.; Zhu, D.-Y.; Liu, Y.; Zhang, L.; Zheng, C.-X.; He, Z.-H.; Chen, D.-H.; Wen, L.-S. Determination of Thickness and Optical Constants of ZnO Thin Films Prepared by Filtered Cathode Vacuum Arc Deposition. *Chin. Phys. Lett.* **2008**, *25*, 743. [CrossRef]
109. Urbach, F. The long-wavelength edge of photographic sensitivity and of the electronic absorption of solids. *Phys. Rev.* **1953**, *92*, 1324. [CrossRef]
110. Kurik, M.V. Urbach rule. *Phys. Status Solidi A* **1971**, *8*, 9–45. [CrossRef]
111. Caponetti, E. Synthesis, size control, and passivation of CdS nanoparticles in water/AOT/n-heptane microemulsions. *Mater. Sci. Eng. C* **2003**, *23*, 531–539. [CrossRef]
112. Wang, Y.; Herron, N. Nanometer-sized semiconductor clusters: Materials synthesis, quantum size effects, and photophysical properties. *J. Phys. Chem.* **1991**, *95*, 525–532. [CrossRef]

applied sciences

MDPI

Review

Magnetic Nanoparticles as Mediators for Magnetic Hyperthermia Therapy Applications: A Status Review

Miloš Beković [1,*], Irena Ban [2], Miha Drofenik [2,3] and Janja Stergar [2,*]

1 Institute of Power Engineering, Faculty of Electrical Engineering and Computer Science, University of Maribor, Koroška cesta 46, 2000 Maribor, Slovenia
2 Laboratory of Inorganic Chemistry, Faculty of Chemistry and Chemical Engineering, University of Maribor, Smetanova ulica 17, 2000 Maribor, Slovenia; irena.ban@um.si (I.B.); miha.drofenik@um.si (M.D.)
3 Jožef Stefan Institute, Department for Materials Synthesis, Jamova 39, 1000 Ljubljana, Slovenia
* Correspondence: milos.bekovic@um.si (M.B.); janja.stergar@um.si (J.S.); Tel.: +386-2-220-71-72 (M.B.); +386-2-229-44-17 (J.S.)

Abstract: This concise review delves into the realm of superparamagnetic nanoparticles, specifically focusing on Fe_2O_3, $Mg_{1+x}Fe_{2-2x}Ti_xO_4$, $Ni_{1-x}Cu_x$, and Cr_xNi_{1-x}, along with their synthesis methods and applications in magnetic hyperthermia. Remarkable advancements have been made in controlling the size and shape of these nanoparticles, achieved through various synthesis techniques such as coprecipitation, mechanical milling, microemulsion, and sol–gel synthesis. Through this review, our objective is to present the outcomes of diverse synthesis methods, the surface treatment of superparamagnetic nanoparticles, their magnetic properties, and Curie temperature, and elucidate their impact on heating efficiency when subjected to high-frequency magnetic fields.

Keywords: magnetic nanoparticles; magnetic hyperthermia; specific absorption rate; calorimetric measurements

Citation: Beković, M.; Ban, I.; Drofenik, M.; Stergar, J. Magnetic Nanoparticles as Mediators for Magnetic Hyperthermia Therapy Applications: A Status Review. *Appl. Sci.* **2023**, *13*, 9548. https://doi.org/10.3390/app13179548

Academic Editors: Robert A. Varin and Geng Zhong

Received: 12 July 2023
Revised: 16 August 2023
Accepted: 18 August 2023
Published: 23 August 2023

1. Introduction

Magnetic nanoparticles (MNPs) are nanoscale materials with exceptional properties that find applications across diverse fields, including environmental, biomedical, and clinical domains [1–6]. The size range of MNPs is comparable to that of viruses (20–500 nm), proteins (5–50 nm), or genes (2 nm wide and 10–200 nm long). These nanoparticles possess magnetic characteristics, adhering to Coulomb's law, enabling their manipulation through an external magnetic field. Moreover, their substantial surface area can be effectively utilized for binding diverse biological agents [7–9].

The synthesis of magnetic nanoparticles (MNPs) plays a pivotal role in enhancing their properties and evaluating their potential applications [10–13]. It involves a series of carefully designed processes aimed at tailoring their size, shape, composition, and surface characteristics. In the synthesis of MNPs, the main problem is to control the particle size, which results from the high surface energy of these systems. The spontaneous reduction in surface area in magnetic nanoparticles (MNPs) is driven by interfacial tension, particularly during the initial stages of nucleation, growth, and Ostwald ripening. To ensure the stability of MNPs, it becomes crucial to maintain an appropriate surface area while simultaneously implementing effective protection measures [14–16]. The coating should improve the stability and solubility of MNPs, increase their biocompatibility and target specificity, and prevent agglomeration, oxidation, corrosion, and toxicity [17–23]. The MNPs can be synthesized through many different methods including coprecipitation [24–30], thermal decomposition [31–35], hydrothermal synthesis [36–40], microemulsion [41–44], polyol reduction [45–49], the sol–gel method [50–54], and others [55–63]. The synthesized MNPs are usually coated to ensure a proper surface coating and develop some effective protection strategies to maintain stability. Depending on the end use of the MNPs, specific surface modification processes are selected [2]. The applied coating strategies can roughly be divided into different groups: synthetic polymer [64,65], natural polymer [27,66], organic surfactants [67,68], inorganic components [69,70], and bioactive molecules and structures [71,72].

Magnetic materials serve as highly effective tools for the magnetic separation of small molecules, biomolecules, and cells. Their unique ability to respond to a magnetic field allows for efficient and precise manipulation in separation processes. Additionally, magnetic materials can be coated with various coatings simultaneously, further expanding their versatility and applicability [73]. In the biomedical field, magnetic particles and magnetic composites are utilized as drug carriers [74–81], as contrast agents for magnetic resonance imaging (MRI) [82–87], and in magnetic hyperthermia (MH) [26,55,77,80,88–98], which is also a focus of this short review.

So far, several methods have been discovered to heat the tissue and destroy tumor cells at the increased temperature. Hyperthermia is an alternative and promising method of treating cancer cells, in which the cancer cells die with minimal damage to healthy cells. Hyperthermia is very popularly used together with radiotherapy and chemotherapy, which is why it has attracted great interest from many researchers from different fields. They started with translational research materials arranged to protect the entire tumor. This method requires carefully planned surgery, but it was not possible to control the temperature.

The basis of MH is the increased heat sensitivity of cancer cells. It has been found that treatment of the cancer area is most effective at a therapeutic temperature between 41 and 46 °C, as it kills most cancer cells in the tissue of interest. This finding was based on the observation that the growth of cancer cells can be stopped at a temperature above 42 °C, while healthy cells survive at higher temperatures. Thermal energy is generated by magnetic particles adhering to diseased tissue exposed to an AMF. The basis of this method is the irradiation of cancerous tissue with implanted magnetic material in the form of magnetic particles or MNPs in an AMF. The amount of thermal energy released depends on the type of magnetic material and the parameters of the magnetic field.

MH is a therapy in which tissue temperature can be increased by exposing MNPs to an alternating magnetic field (AMF). The final heating temperature achieved depends on the Curie temperature (T_C) of the MNPs [99]. This method must overcome two problems: the temperature increase must be strictly limited only in the target region so that all other regions are not affected, and the temperature must be controlled inside and outside the target region. Self-regulation MH is a phenomenon in which the T_C changes with the changing chemical composition of the MNPs. Magnetic materials lose their magnetic properties above T_C, so magnetic heating is stopped. This type of therapy results in the death of cancer cells with minimal damage to surrounding healthy tissue [100–102].

The most important feature of magnetic hyperthermia is the fact that it is thermal energy generated in a well-defined space associated with the distribution of magnetic particles that we can design. In magnetic hyperthermia, in which magnetic nanoparticles are involved, four different mechanisms are mainly active, depending on the morphology of the magnetic particles and the frequencies of the alternating magnetic field: eddy losses, hysteresis losses, relaxation losses, and resonance losses, but these are small due to the relatively low frequencies. The main characteristic of magnetic particles in an alternating magnetic field is their heating and the conversion of magnetic energy into thermal energy, which heats their surroundings. In particular, the absorption rate of magnetic energy, which is then reflected in the heating of samples, is represented by the parameter "specific absorption rate" (SAR). This describes the heating ability of the corresponding material—MNPs—or the SAR that we consider to be the average absorbed power, not the unit mass of the material at the time it is exposed to a fluctuating magnetic field [103–105].

Despite all the biomedical applications of superparamagnetic nanoparticles, this short review mainly focuses on the superparamagnetic Fe_2O_3, $Mg_{1+x}Fe_{2-2x}Ti_xO_4$, $Ni_{1-x}Cu_x$, and Cr_xNi_{1-x} MNPs synthesized by various synthesis methods and their applications in MH. With this brief review, we have aimed to show the relationship between the synthesis routes, surface chemistry, magnetic properties, and T_C, which is near or in the therapeutic range of superparamagnetic nanoparticles. We wanted to explain their influence on the heating efficiency when they are only under the influence of an external alternating magnetic field.

2. Syntheses of MNPs

In our study, we employed four distinct synthesis methods to produce four different types of magnetic nanoparticles (MNPs) specifically designed for applications in MH treatments. For the synthesis of maghemite MNPs, we used $FeCl_2{\cdot}4H_2O$ and $FeCl_3{\cdot}6H_2O$. The precursors were prepared by the coprecipitation of Fe^{2+} and Fe^{3+} ions using NH_4OH and heated to 80 °C. Then, the CM-dextran solution was added for over 30 min. The suspension was then stored at 80 °C for 30 min and cooled to room temperature. The suspension was magnetically decanted and washed several times with deionized water [27].

$Mg(NO_3)_2{\cdot}6H_2O$, $Fe(NO_3)_3{\cdot}9H_2O$, and $Ti(C_3H_7O)_4$ were used for the synthesis of the $Mg_{1+x}Fe_{2-2x}Ti_xO_4$ complex ferrite. The precursors were prepared by coprecipitation of Fe^{3+}, Mg^{2+}, and Ti^{4+} ions in different stoichiometric ratios (x = 0.34; 0.37; 0.40) with a NaOH solution (6 M) and heated to 80 °C. Then, the suspensions were aged at 80 °C for 1 h. The product was washed several times with deionized water and dried [26].

NiCu MNPs were prepared by reducing a Ni, Cu-hydrazine complex using the microemulsion method [41], via reducing a Ni, Cu-oxide mixture in a silica matrix [50,103], obtained with a sol–gel method and by mechanical milling [55]. MNPs with a controlled T_C were prepared by reduction of a Ni, Cu-hydrazine complex synthesized in a compartmentalized state of reverse micelles. We used the phase diagram of water/N-cetyl-N,N,N-trimethylammonium bromide (CTAB), and n-butanol/isooctane as a starting point for the preparation of the microemulsion. We used the titration method to select a suitable composition range that would form a microemulsion in the mentioned system. Microemulsion A was prepared with a 0.3 M solution of Ni^{2+} and Cu^{2+} acidified with HCl and then heated to 60 °C in a water bath. Microemulsion B was then added to microemulsion A, using hydrazine as the metal-containing ligand. This mixture was heated at 60 °C for several hours. After the complex was formed, stirring was continued for another hour and NaOH was added. The color change in the microemulsion to a stable black suspension was indicative of the formation of MNPs. The suspension was then centrifuged and washed several times to sediment the particles [41]. MNPs of NiCu alloy with narrow size distribution were prepared by reduction of the Ni and Cu oxide mixture in a silica matrix obtained by the sol–gel method. $Ni(NO_3)_2{\cdot}6H_2O$, $Cu(NO_3)_2{\cdot}3H_2O$, and citric acid (CA) were dissolved in deionized water. After 15 min, absolute ethanol and tetraethyl orthosilicate (TEOS) were added to the solution with vigorous stirring. The molar ratio of all components, in the case of the Ni:Cu ratio of 67.5:32.5, was Ni:Cu:CA:TEOS:D.I.:ethanol = 0.675:0.325:1.1:2.9:40.6:11.6. The salt was dried at room temperature for 72 h and then calcined at 500 °C for 24 h (air atmosphere). We obtained a powder of nickel and copper oxides in a silica matrix, followed by a reduction in an Ar/H_2 atmosphere for one day at 850 °C. We obtained the final product $Ni_{1-x}Cu_x$ MNPs in a SiO_2 matrix. By etching, the solution was stirred for 1 day under an Ar atmosphere to remove the SiO_2 matrix. The final product was washed several times with etching solution and ethanol and redispersed in ethanol [50]. A series of MNPs from NiCu alloy, which had a T_C in the range of 51 and 63 °C, was prepared in the following article. Using the sol–gel method, we reduced a mixture of nickel and copper oxides in a silica matrix with some minor modifications [106].

$Ni_{1-x}Cu_x$ MNPs (x = 40, 30, 27.5, 27, 25, 20) were prepared by mechanical milling. We milled Cu (grain size < 63 μm) and Ni (grain size < 150 μm) in a SPEX (Metuchen, NJ, USA) 8000M and 8000D at 1425 rpm, using steel vials, while the balls-to-powder ratio was 20:1. We milled for 20 h under an inert N_2 atmosphere. We also added NaCl during milling to avoid agglomeration of MNPs [55]. Cr_xNi_{1-x} MNPs were prepared by water-in-oil microemulsions and mechanical milling. The titration method was used to determine the stability range of the microemulsion, with the phase diagram of the composition of water/CTAB and n-butanol/isooctane serving as a starting point. We prepared two microemulsions, the first was by dissolving aqueous Ni^{2+} (0.4 M) and Cr^{3+} (0.1 M) ions, and the second by adding $NaBH_4$ (0.8 M) to a mixture of CTAB, n-butanol, and isooctane. Both microemulsions were of the same volume; subsequently, after two hours of mixing in an inert N_2 atmosphere, a black solution was obtained. The mixtures were centrifuged to separate the black MNPs and washed several times with methanol. Finally, the as-prepared alloy powder was heat-treated at 200, 300, 400, and 600 °C. Cr_xNi_{1-x} MNPs were also prepared by mechanical milling. Cr (particle size < 74 μm) and Ni (particle size < 150 μm) were milled in a SPEX 8000M mill at 1425 rpm, where we used steel vials, while the ratio between balls and powder was 20:1. We milled for 20 h in an inert N_2 atmosphere [107].

3. Results and Discussion

The synthesized MNPs were characterized by various characteristic methods. We focused on and compared the results of X-ray diffraction analysis (XRD), thermogravimetric analysis (TGA/SDTA), modified thermogravimetric analysis, transmission electron microscopy (TEM), and magnetic and calorimetric measurements.

3.1. X-ray Diffraction Analysis

Figures 1–7 show the XRD patterns of as-prepared maghemite and maghemite particles covered with CM-dextran MNPs, (Mg, Ti)-ferrite MNPs, as-prepared NiCu MNPs, NiCu MNPs embedded in SiO_2 matrix, and as-prepared NiCr NPs heat-treated. XRD measurements were performed with a D5005 diffractometer (Bruker Siemens) and analyzed with Topas software (Bruker, AXS). All presented MNPs with selected compositions are well crystallized. Crystallite sizes were determined from XRD line broadening using the Sherrer equation (Table 1):

$$d_x = \frac{(0.94 \cdot \lambda)}{(\beta \cdot cos\theta)}, \quad (1)$$

where d_x is crystallite size; λ is the wavelength of the radiation; β is the line broadening at half the maximum intensity (FWHM) in radians and θ is the corresponding diffraction angle.

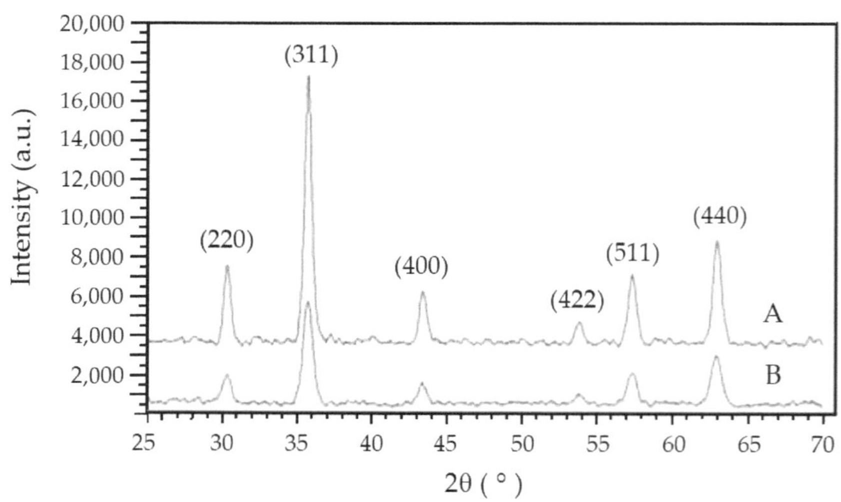

Figure 1. XRD spectra of as-prepared maghemite particles (A) and the particles covered with CM-dextran (B).

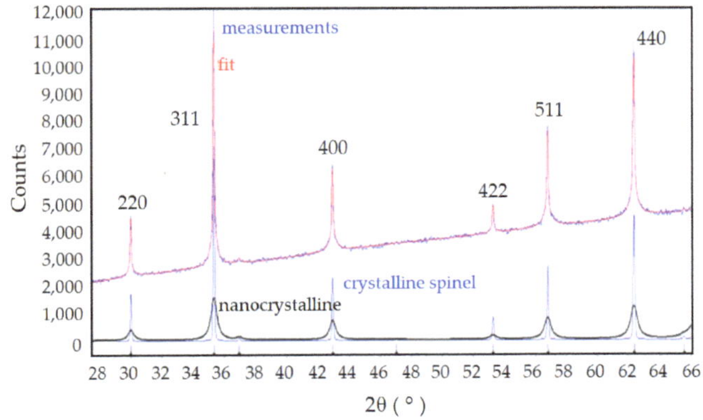

Figure 2. XRD spectra of the (Mg, Ti)-ferrite sample (x = 0.37). The curves of the crystalline and nanocrystalline spinel phases are shown below.

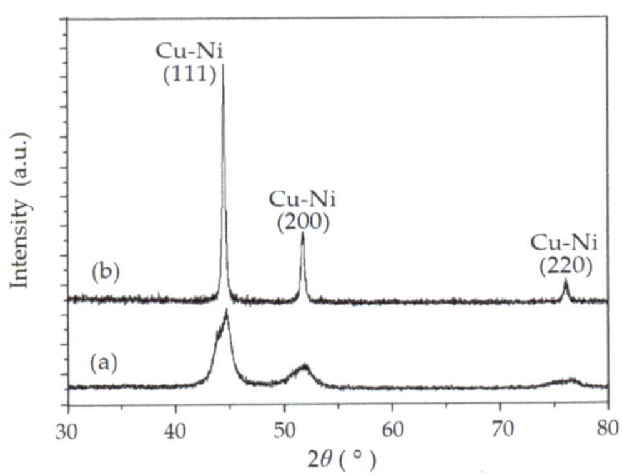

Figure 3. XRD spectra of (a) as-prepared $Ni_{0.725}Cu_{0.275}$ MNPs and (b) particles thermally homogenized in NaCl matrix (microemulsion method).

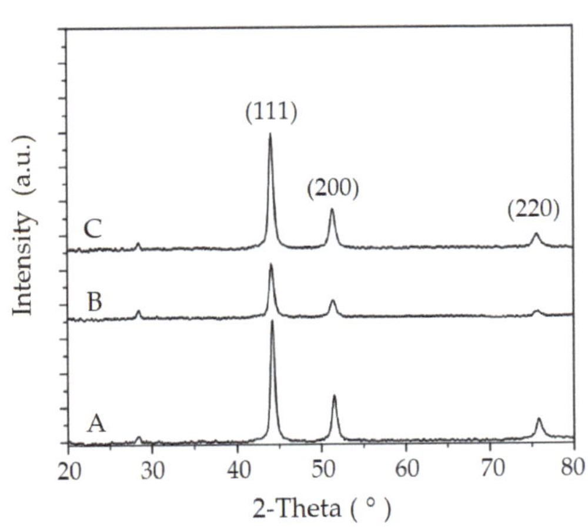

Figure 4. XRD powder diffraction patterns of $Ni_{1-x}Cu_x$ ($Ni_{67.5}Cu_{32.5}$ (A), $Ni_{62.5}Cu_{37.5}$ (B), $Ni_{60}Cu_{40}$ (C)) alloy MNPs embedded in the silica matrix (sol–gel method).

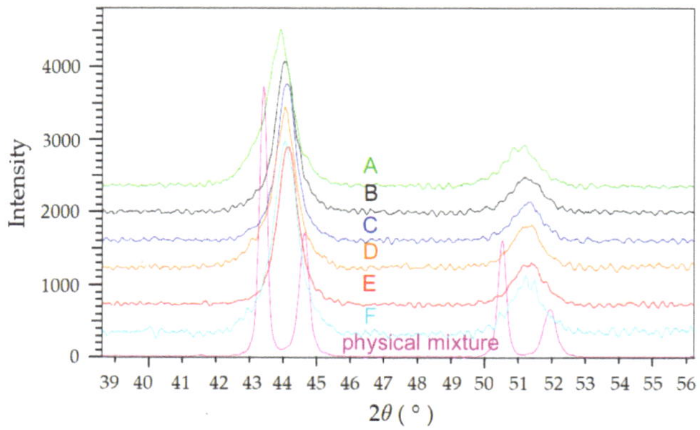

Figure 5. XRD patterns of $Cu_{1-x}Ni_x$ samples (milled 20 h) with various compositions. The patterns are compared with the pattern of a Cu–Ni physical mixture, where samples are: A = $Cu_{40}Ni_{60}$, B = $Cu_{30}Ni_{70}$, C = $Cu_{27.5}Ni_{72.5}$, D = $Cu_{27}Ni_{73}$, E = $Cu_{25}Ni_{75}$, F = $Cu_{20}Ni_{80}$.

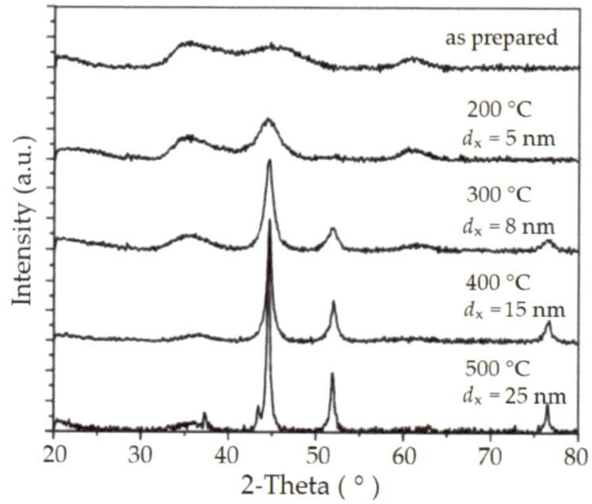

Figure 6. XRD patterns of synthesized $Cr_{20}Ni_{80}$ alloy, as-prepared and at different temperatures (microemulsion technique).

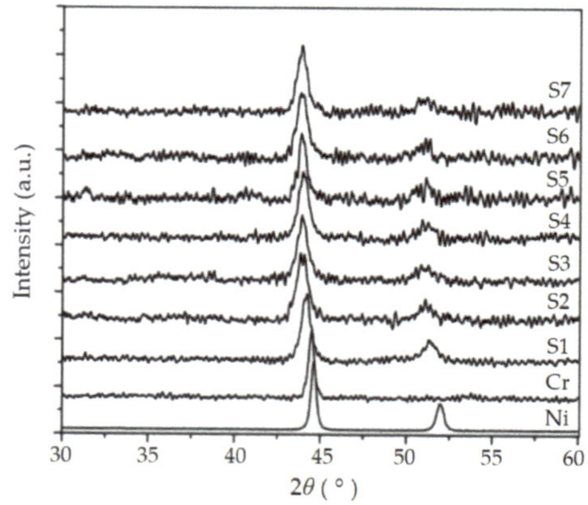

Figure 7. XRD patterns of the various Cr_xNi_{1-x} samples were obtained after 20 h of ball milling under an N_2 atmosphere. Samples are $Cr_{10}Ni_{90}$ (S1), $Cr_{15}Ni_{85}$ (S2), $Cr_{20}Ni_{80}$ (S3), $Cr_{26}Ni_{74}$ (S4), $Cr_{27}Ni_{73}$ (S5), $Cr_{28}Ni_{72}$ (S6), $Cr_{29}Ni_{71}$ (S7).

Table 1. Average particle diameters of different samples were obtained using the Sherrer equation. [108].

Sample Name	Figure	d_{XRD} [nm]	References
As prepared maghemite particles (A)	Figure 1	11.8	[27]
Maghemite MNPs covered with CM-dextran (B)	Figure 1	16.0	
(Mg, Ti)-ferrite MNPs (nanocrystallites)	Figure 2	20.0	[26]
(Mg, Ti)-ferrite MNPs (crystalline spinel)	Figure 2	200	
As-prepared $Ni_{0.725}Cu_{0.275}$ MNPs	Figure 3	7.0	[41]
$Ni_{0.725}Cu_{0.275}$ MNPs thermally homogenized in NaCl matrix	Figure 3	28.0	
$Ni_{67.5}Cu_{32.5}$ (A)	Figure 4	19.0	[50,103]
$Ni_{62.5}Cu_{37.5}$ (B)	Figure 4	17.0	
$Ni_{60}Cu_{40}$ (C)	Figure 4	17.0	
Cu_xNi_{1-x} (A–F)	Figure 5	10.0–12.0	[55]
$Cr_{20}Ni_{80}$	Figure 6	5.0–25.0	[107]
Cr_xNi_{1-x} (S1–S7)	Figure 7	12.0–18.0	

The X-ray diffraction patterns of the as-prepared maghemite MNPs (A) and the particles coated with CM-dextran (B) are shown in Figure 1. The diffraction peaks correspond to the cubic phase of maghemite (JCPDS No. 39-1346). Figure 1 clearly shows that the particles are crystalline. Using the Sherrer equation, we estimated the size of the maghemite nanoparticles to be around 16 nm (sample A) and the size of the nanoparticles coated with CM-dextran to be around 11.8 nm (sample B). The size of the nanoparticles is consistent with the analysis of TEM, but one of the main reasons that the coated particles are smaller than the uncoated ones is due to centrifugation, which removed the larger particles [27].

The X-ray diffraction patterns of milled (Mg, Ti)-ferrite sample (x = 0.37) are shown in Figure 2 [26]. The diffraction patterns are consistent with the peaks characteristic of $MgFe_2O_4$, but the diffraction peaks are slightly shifted. The diffraction peaks of $MgFe_2O_4$ (JPCDS 36-0398) have a lattice constant of a = 8.387 Å, those of the (Mg, Ti)-ferrite sample (x = 0.37) have a larger lattice constant of a = 9.418 Å. The XRD data were fitted with a nanoscale crystalline component in addition to a crystalline component, and we can see that the powder consists of two types of crystallites. The crystalline spinel phase with sharp XRD peaks is shown in blue in Figure 2, while the spinel phase is shown in black. The size of nanoparticles is also different in different phases. The size of the nanoparticles in the crystalline spinel phase is about 200 nm, while the size of the nanocrystallites is about 20 nm. According to the XRD analysis, the size of the individual unit cells also varies.

The 2θ diffraction angles in Figure 3 correspond to the (111), (200), and (220) planes of NiCu alloy crystallites, single FCC phase [41]. Sample (a) in Figure 3 shows as-prepared $Ni_{0.725}Cu_{0.275}$ MNPs (microemulsion method) and sample (b) in Figure 3 shows the particles thermally homogenized in the NaCl matrix. Figure 3 (a) shows the as-prepared nanoparticles synthesized by the microemulsion method. It can be seen that the peaks are broad and therefore were further homogenized under a reducing atmosphere (Ar/H_2). Figure 3 (b) clearly shows crystalline peaks that are extremely narrow. Homogenization in a reducing atmosphere also resulted in the growth of nanoparticles. As-prepared nanoparticles were about 7 nm in size, while the homogenized nanoparticles were about 28 nm in size. Thus, the average size of nanoparticles increased at the expense of elevated temperature (750 °C, 5 h) and reduced atmosphere.

The 2θ diffraction angles in Figure 4 correspond to the (111), (200), and (220) planes of the $Ni_{1-x}Cu_x$ alloy crystallites after reduction in a silica matrix, single-phase composition [103]. The samples are A ($Ni_{67.5}Cu_{32.5}$), B ($Ni_{62.5}Cu_{37.5}$), and C ($Ni_{60}Cu_{40}$). The silica matrix is manifested by a small peak of cristobalite at $2\theta = 28°$. Using sol–gel synthesis, we first achieved the formation of copper and nickel oxides [52], which were then heated to 850 °C. Namely, in the silica matrix, the nanoparticles could not agglomerate but only grew accordingly, and the reducing atmosphere (H_2/Ar) ensured the formation of $Ni_{1-x}Cu_x$ alloy nanoparticles. Compared with the microemulsion technique, the sol–gel method has the advantage that the homogenization of nanoparticles does not cause agglomeration and uncontrolled size of nanoparticles.

Figure 4 clearly shows that the nanoparticles are crystalline and have a size between 17 and 19 nm, which was estimated using the Sherrer equation. Moreover, the analysis of TEM shows that the particles are not agglomerated in the silica matrix using the sol–gel method and subsequent homogenization.

In Figure 5, we see the characteristic peaks corresponding to the (100) and (200) planes of bulk FCC metals, 43.47 °, 50.38 ° (Cu), and 44.6 °, and 51.91 ° (Ni) (PDF files 001-1241 and 001-1260). Using EDS analysis, we confirmed that the composition of Cu-Ni MNPs is consistent with the $[Cu^{2+}]$:$[Ni^{2+}]$ molar ratio used in the synthesis. Figure 5 shows samples that were milled for 20 h, as this time was found to be optimal for the formation of the CuNi alloy. It is clear from the X-ray diffraction patterns that as the copper content increases, the d_{111} spacing also increases. For samples A through F, we estimated particle size between 10 and 12 nm using Sherrer's equation, and the size was consistent with the size of nanoparticles estimated by magnetic measurements [55].

The XRD patterns in Figure 6 show three characteristic broad peaks at 2θ = 44.57°, 51.94°, and 76.51°. The as-prepared MNPs were amorphous, and upon heating with an elevated temperature, the size of the crystallites or crystallinity increased [107].

The composition of CrNi MNPs (microemulsion technique) was following the $[Cr^{3+}]$:$[Ni^{2+}]$ = 20:80 molar ratio used for the synthesis. The microemulsion technique for the synthesis of NiCr nanoparticles also proved not to be the best, as extremely amorphous nanoparticles were obtained without thermal homogenization. At the same time, the XRD results also show a peak at 37°, indicating the formation of NiO despite the inert atmosphere throughout the synthesis process. Only with the help of homogenization at elevated temperatures did we obtain crystalline nanoparticles that grew with increasing temperature. Unfortunately, homogenization leads to agglomeration of nanoparticles and at some locations also to larger grains in the form of platelets, which was confirmed by TEM analysis.

In Figure 7, we see the XRD results showing the formation of Cr_xNi_{1-x} alloy using mechanical milling [107]. With increasing Ni content (in %), a progressive diffraction angle was observed in both cases (111) and (200). NiCr samples prepared using mechanical milling were also milled for 20 h in the same manner as NiCu. Figure 7 shows that the Bragg peaks in both cases (111) and (200) shift to higher diffraction angles with increasing Ni content. The lattice constant and interplanar distance d_{111} also increase with increasing chromium content. In the microemulsion method, particles with a size of up to 25 nm were obtained, while in mechanical milling the particles were micrometer in size, which were comprised of nanocrystallites.

3.2. Thermogravimetric Analysis and Modified Thermogravimetric Analysis

Figures 8–13 show the TGA/SDTA analysis and thermomagnetic curves, respectively, and the corresponding T_C of CM-dextran-coated maghemite MNPs, (Mg, Ti)-ferrite MNPs, and thermally homogenized NiCu MNPs and ball-milled NiCu and NiCr MNPs. TGA/SDTA analysis was performed using the TGA/SDTA 851^e, Mettler Toledo (Columbus, OH, USA). The thermomagnetic (TM) curves and corresponding T_C were determined by thermal demagnetization using a modified TGA on a TGA/SDTA 851^e, Mettler Toledo. T_C was measured by placing a permanent magnet on top of the device.

Figure 8 shows the amount/concentration of CM-dextran absorbed into the maghemite MNPs [27]. During thermal heating from room temperature to 750 °C, there is a total weight loss of 21.2%. The weight loss of 16.4% due to the oxidation of the CM-dextran occurs in the first temperature range (150–500 °C), indicated by an exothermic peak of the simultaneous differential thermal analysis (SDTA). The last step, 500–750 °C, is associated with a weight loss of 3.1%, which is due to the removal of the CM-dextran covalently bound to the surface of the particles.

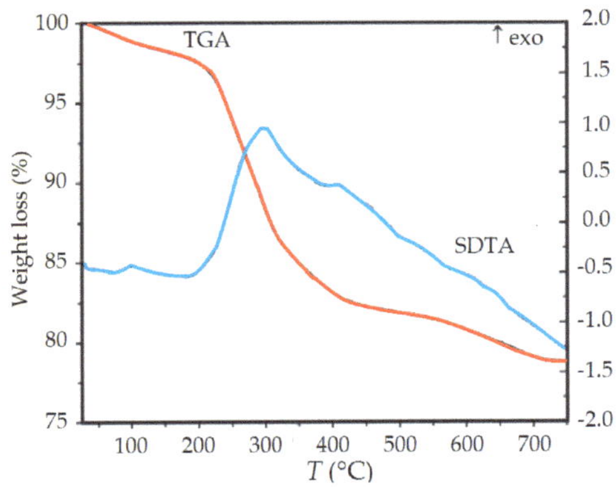

Figure 8. Thermal analysis of the maghemite MNPs coated with CM-dextran and the corresponding SDTA.

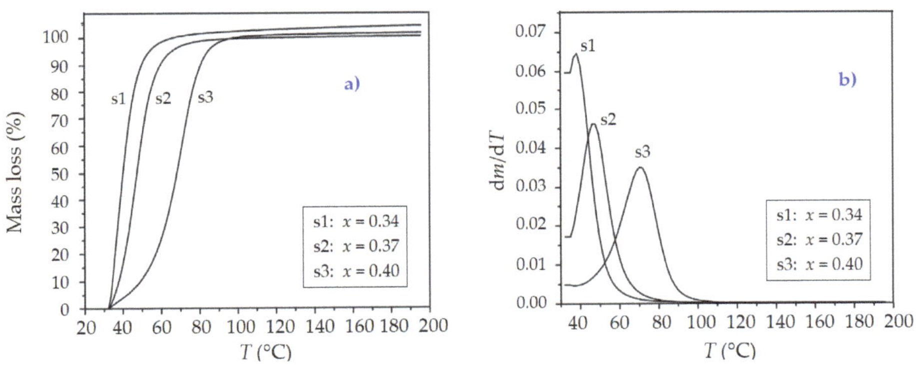

Figure 9. (**a**) TM curves of the ferrite particles s1, s2, s3 and (**b**) the corresponding T_C (the first derivative curve).

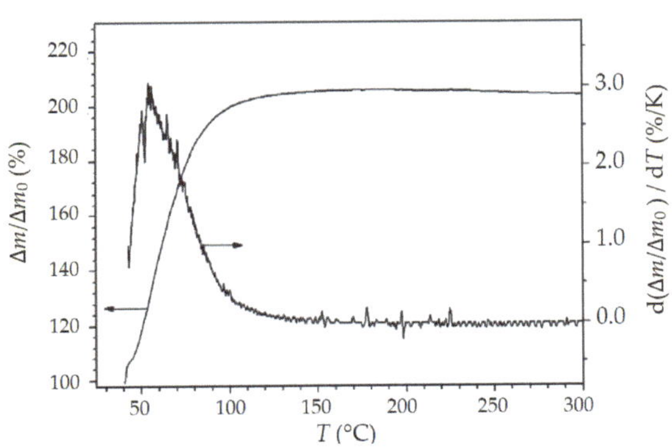

Figure 10. TM curve of the thermally homogenized particles $Ni_{0.725}Cu_{0.275}$.

Figure 9 shows the TM curves and the corresponding T_C of as-calcined (Mg, Ti)-ferrite powders (30 min, 1000 °C in the air) [26]. The T_C of MNPs was determined using differential thermomagnetic curves (DTMC), which show an apparent change in weight because of a decrease in magnetization due to an increase in temperature (Figure 9a). Figure 9b shows the first derivative in the TM curve of (Mg, Ti)-ferrite MNPs, the maximum of which is attributed to the corresponding T_C. Figure 7 shows the samples for three different compositions. The most promising sample was s2 (x = 0.37), which has a T_C of about 46 °C, which is ideal for use in MH. This is the temperature at which the nanoparticles transition from a ferromagnetic to a paramagnetic state, meaning that they no longer heat up under the influence of a magnetic field.

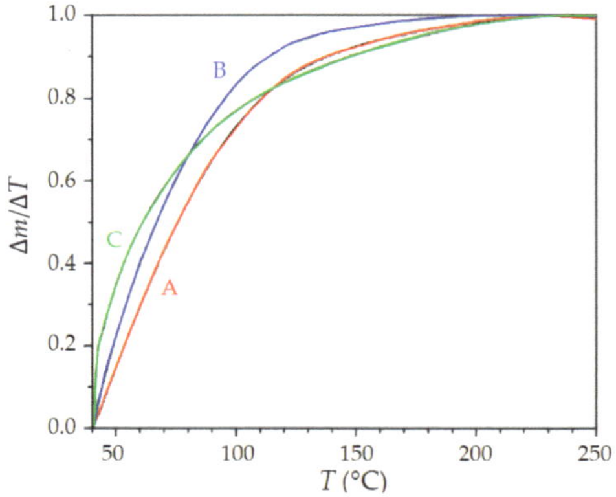

Figure 11. TM curves of analyzed samples (A) ($Ni_{67.5}Cu_{32.5}$), (B) ($Ni_{62.5}Cu_{37.5}$), and (C) ($Ni_{60}Cu_{40}$).

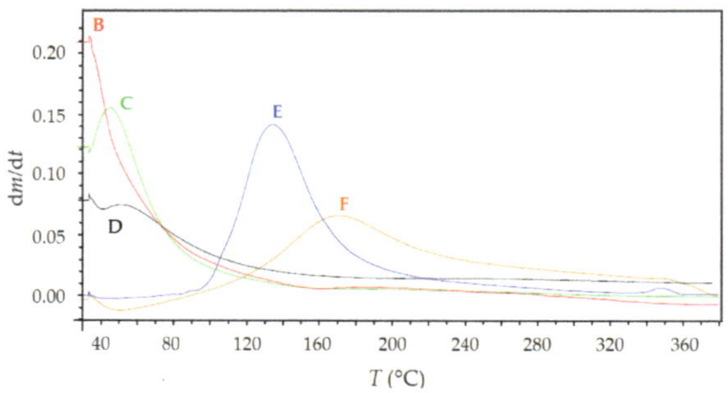

Figure 12. First derivative curves of TM curves for $Cu_{1-x}Ni_x$ (B–F) alloys.

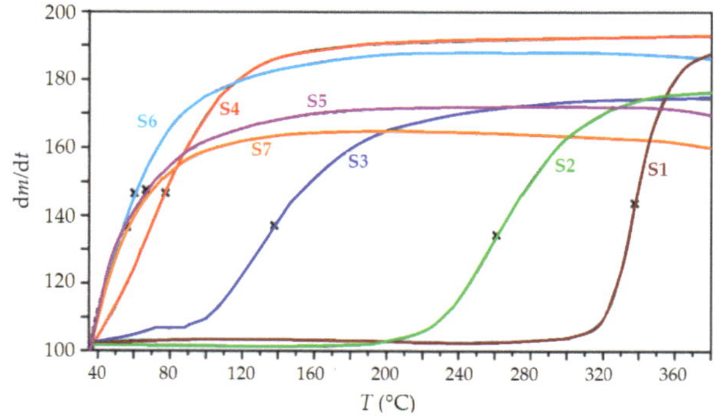

Figure 13. TM curves of Cr_xNi_{1-x} samples (S1–S7).

Figure 10 shows the TM curve and its derivative for the thermally homogenized for 5 h at 750 °C under a reducing atmosphere of Ar/H_2 alloy $Ni_{0.725}Cu_{0.275}$ MNPs (microemulsion method) [41]. For the as-prepared sample, the T_C was about 250 °C, close to the T_C of Ni, due to the different standard electrode potentials of Cu and Ni or to inhomogeneities in the sample itself. Using homogenization, we then determined the correct T_C for the $Cu_{27.5}Ni_{72.5}$ composition, which was about 45 °C, and which is in the therapeutic temperature range when we refer to MH.

Figure 11 shows the T_C as the temperature at which half of the magnetization weight gain/loss was recorded during heating [103]. Samples A ($Cu_{67.5}Cu_{32.5}$), B ($Ni_{62.5}Cu_{37.5}$), and C (Ni_{60}Cu40) were analyzed after heat treatment (850 °C in a reducing atmosphere, 6 h). Nanoparticles synthesized by the sol–gel method have a T_C in the therapeutic range for use in MH. However, homogenization in a reducing atmosphere is required. In Figure 11, the T_C of the samples follows each other, namely sample A has a T_C of 63 °C, sample B has a T_C of 54 °C, and sample C has a T_C of 51 °C. As the Ni content increases, so does the T_C.

Figure 12 shows the first derivative of TM curves of $Cu_{1-x}Ni_x$ alloys (ball-milling), the maximum of which is attributed to the corresponding T_C [55]. The TM curves have a slightly asymmetric shape, indicating inhomogeneity in the composition and size of the MNPs. Again, T_C increases with the increasing Ni content, so we can change the composition according to the desired T_C. T_C in the therapeutic range around 45 °C shows sample C with composition $Cu_{27.5}Ni_{72.5}$.

Figure 13 shows the DTMC of Cr_xNi_{1-x} alloys (ball-milling) [107]. The T_C of the samples was determined by the temperature, which belongs to 50% of the peaks of the demagnetization curves. The samples are heterogeneous in composition, as evidenced by asymmetric TM curves. Figure 13 also clearly shows how the T_C increases with increasing Ni content in each sample. Sample S1, which has the highest Ni content, also has the highest T_C, and conversely, S7 has the lowest T_C and the lowest Ni content.

The T_C for the synthesized (Mg, Ti)-ferrite MNPs, thermally homogenized NiCu MNPs, and NiCu and NiCr ball-milled MNPs were measured and are shown in Table 2.

Table 2. T_C of (Mg, Ti)-ferrite and NiCu, NiCr MNPs synthesized with different methods.

Sample Name	Figure	T_C [°C]	References
(Mg, Ti)-ferrite (x = 0.37)	Figure 9b	46	[26]
$Ni_{0.725}Cu_{0.275}$	Figure 10	45	[41]
$Ni_{67.5}Cu_{32.5}$ (A)	Figure 11	63	
$Ni_{62.5}Cu_{37.5}$ (B)	Figure 11	54	[103]
$Ni_{60}Cu_{40}$ (C)	Figure 11	51	
$Cu_{30}Ni_{70}$ (B)	Figure 12	24	
$Cu_{27.5}Ni_{72.5}$ (C)	Figure 12	45	
$Cu_{27}Ni_{73}$ (D)	Figure 12	53	[55]
$Cu_{25}Ni_{75}$ (E)	Figure 12	137	
$Cu_{20}Ni_{80}$ (F)	Figure 12	174	
$Cr_{10}Ni_{90}$ (S1)	Figure 13	340	
$Cr_{15}Ni_{85}$ (S2)	Figure 13	262	
$Cr_{20}Ni_{80}$ (S3)	Figure 13	138	
$Cr_{26}Ni_{74}$ (S4)	Figure 13	69	[107]
$Cr_{27}Ni_{73}$ (S5)	Figure 13	52	
$Cr_{28}Ni_{72}$ (S6)	Figure 13	44	
$Cr_{29}Ni_{71}$ (S7)	Figure 13	43	

The (Mg, Ti)-ferrite MNPs were synthesized by the coprecipitation method, thermally homogenized $Ni_{0.725}Cu_{0.275}$ were synthesized by a microemulsion method, and $Ni_{67.5}Cu_{32.5}$, $Ni_{62.5}Cu_{37.5}$, and $Ni_{60}Cu_{40}$ in a silica matrix were synthesized by a sol–gel method. The NiCu (B–F) (Figure 12) and NiCr (S1–S7) (Figure 13) MNPs were synthesized by mechanical milling. The transition from a ferrimagnetic to a paramagnetic state at T_C depends on energy exchange. The (Mg, Ti)-ferrite MNPs (sample B, x = 0.37) exhibit T_C close to 46 °C and thus meet the therapeutic requirement. To obtain the desired Tc, MNPs with composition $Cu_{27.5}Ni_{72.5}$ (microemulsion technique) were thermally homogenized at 750 °C for 5 h in a reducing Ar/H_2 atmosphere. The homogenized MNPs have T_C at 45 °C. For the MNPs ($Ni_{67.5}Cu_{32.5}$ (A), $Ni_{62.5}Cu_{37.5}$ (B), $Ni_{60}Cu_{40}$ (C)) synthesized by the sol–gel method, the determined T_C agreed with the selected nominal composition. In the case of mechanical milling, the measurements showed that the T_C of MNPs could be adjusted by changing the molar ratio of Cu/Ni. T_C increased with increasing nickel content. Similar results were obtained in the case of mechanical milling of the synthesized CrNi MNPs. For NiCu and NiCr nanoparticles, the table and individual figures clearly show that T_C increases with increasing Ni content. T_C for pure Ni is about 357 °C. T_C has a few samples in the therapeutic range for use in MH, but not all of them are suitable in terms of their shape or dispersion.

3.3. Transmission Electron Microscopy

Figures 14–17 show the TEM analysis of as-synthesized maghemite and the CM-dextran-coated maghemite MNPs, (Mg, Ti)-ferrite MNPs, as-prepared and thermally homogenized $Ni_{0.725}Cu_{0.275}$ MNPs, and of NiCu MNPs embedded in the silica matrix. The nanoparticle size and the crystallinity were characterized by TEM using a JEOL 2010F microscope. The MNPs were deposited on a cooper-grid-supported, perforated, transparent carbon film.

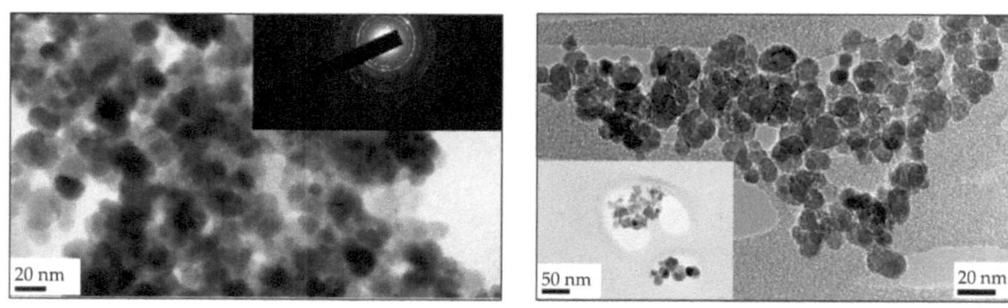

(**a**) (**b**)

Figure 14. TEM micrographs of (**a**) as-prepared maghemite particles (A), (**b**) coated maghemite with CM-dextran.

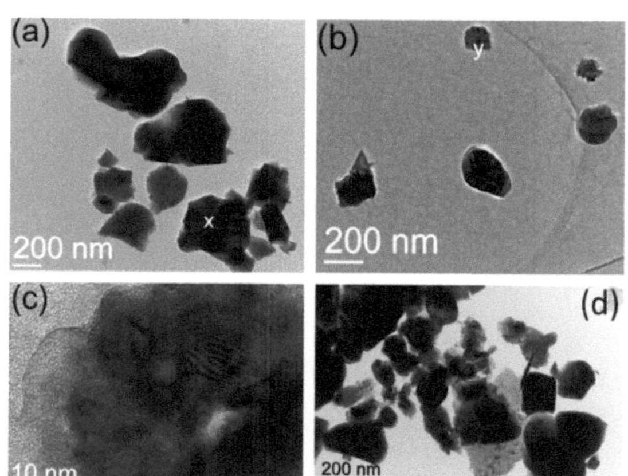

Figure 15. Particles of sample B with mechanically strained particle = x (**a**), consisting of nanocrystallites = y (**b**), enlarged view (**c**), and a typical image of the milled powder (**d**).

Figure 14 shows as-prepared maghemite MNPs (a) and the maghemite NP coated with CM-dextran (b) [27]. The main reason that the coated particles have a smaller average particle size is that they were centrifuged, which removes the larger particles from the magnetic fluid. Using TEM images, we also determined the size of the nanoparticles, which was about 14.5 nm for the maghemite nanoparticles and 12.0 nm for the maghemite nanoparticles coated with CM-dextran. When we compare the size estimated with Sherrer's equation, the results are consistent with or very similar to the XRD results.

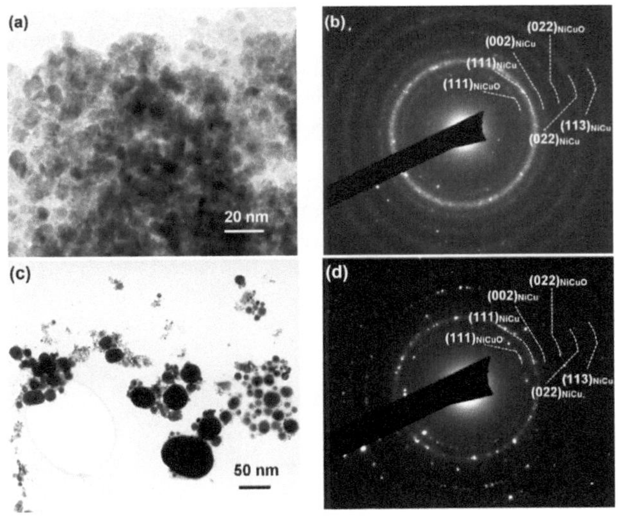

Figure 16. TEM image (**a**) and corresponding electron-diffraction pattern of the as-prepared MNPs (**b**) (microemulsion method). TEM image (**c**) and corresponding electron-diffraction pattern of the thermally homogenized particles (**d**).

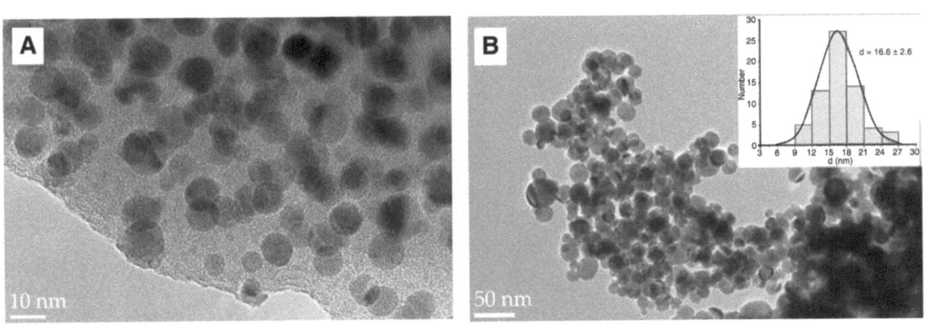

Figure 17. (**A**) Typical TEM images of MNPs of sample $Ni_{67.5}Cu_{32.5}$ embedded in a silica matrix and (**B**) bare alloy particles and the corresponding histogram.

Figure 15 shows the (Mg, Ti)-ferrite MNPs with a size of a few hundred nanometers [26]. The particle marked with x (b) shows contours that could have been formed by the mechanical strain. According to what was said or found by XRD analysis, TEM images confirm the formation of crystalline components and nanosized crystalline components. So, we can confirm with certainty that the sample consists of two types of crystallites. In Figure 15a, we see very large nanoparticles larger than a hundred nanometers, while in Figure 15b,c we see nanocrystallites.

In Figure 16 (above), we see a TEM image of prepared NiCu MNPs (microemulsion method) and the corresponding electron diffraction [41]. The corresponding electron diffraction shows a combination of two cubic FCC structures (Ni-Cu, NiCuO). Using TEM images, we were able to determine the size of the nanoparticles, which ranged from 3 to 10 nm. Figure 16 (below) shows a TEM image of thermally homogenized MNPs (750 °C, 5 h under H_2/Ar atmosphere) and their corresponding electron diffraction showing larger agglomerated particles of Ni-Cu and NiCuO structure. Homogenization resulted in a relatively broad distribution of nanoparticle sizes, ranging from a few tens of nm to several hundreds of nm. The samples were homogenized, but on the other hand, homogenization led to agglomeration, which we do not want in the case of use in MH.

Figure 17A shows the NiCu alloy MNPs embedded in the silica matrix (sol–gel method) [103]. Spherical, non-agglomerated grains with a relatively uniform size distribution were observed. The sol–gel method proved to be very promising as the nanoparticles were spherical despite homogenization and no agglomeration occurred due to the silica matrix. Figure 17B shows a TEM image of the bared MNPs after leaching. Using a TEM image, we also determined the size of the nanoparticles after leaching (Figure 17B), the nanoparticles are about 16.6 nm in size, which is consistent with the results of XRD analysis, or later, as we will see, with magnetic measurements.

In Figure 18, we see a TEM image of mechanically milled Cu-Ni, sample C ($Cu_{27.5}Ni_{72.5}$) [55]. The size distribution is broad, the MNPs are partially agglomerated, and longer platelets of 200 nm length and 5 nm thickness are visible. During mechanical milling we see that we do not obtain homogeneous nanoparticles, but nanoparticles with different shapes and sizes.

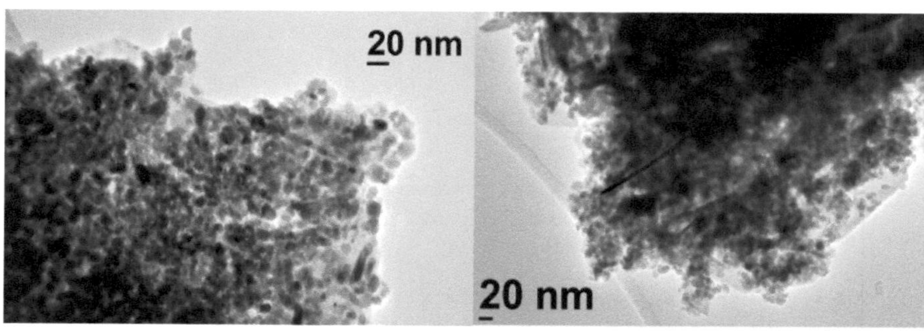

Figure 18. TEM image of milled $Cu_{27.5}Ni_{72.5}$ MNPs (sample C, sol–gel method).

The transmission electron micrograph of the $Cr_{20}Ni_{80}$ MNPs synthesized using the microemulsion method and heat treated at 400 °C is shown in Figure 19, left [107]. Like the previous figure, we see a broad size distribution, the particles are agglomerated, and larger grains are observed in some places. In this case of NiCr nanoparticles, mechanical milling also proved to be less than optimal when homogeneity is required, i.e., nanoparticles of the same size and shape.

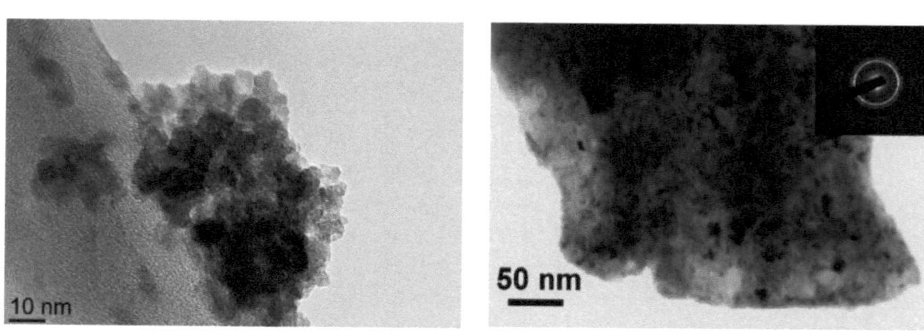

Figure 19. TEM micrograph of $Cr_{20}Ni_{80}$ MNPs synthesized using microemulsion method and heat-treated at 400 °C (**left**) and TEM image of mechanically alloyed $Cr_{29}Ni_{71}$ (**right**).

Figure 19, right, shows a TEM image of $Cr_{29}Ni_{71}$ MNPs synthesized by the ball milling method [107]; here, we see micrometer particles with nanocrystallites. The nanocrystallites are between 5 and 30 nm in size.

The particle size was also determined from the TEM analysis (Table 3).

Table 3. Average particle diameters of different samples were obtained using TEM.

Sample Name	Figure	d_M [nm]	References
As-prepared maghemite MNPs	Figure 14a	14.5	[27]
CM-dextran-coated maghemite MNPs	Figure 14b	12.0	
(Mg, Ti)-ferrite MNPs (x = 0.37)	Figure 15	a few hundred nm	[26]
$Ni_{0.725}Cu_{0.275}$	Figure 16a	3.0–10.0	[41]
$Ni_{0.725}Cu_{0.275}$	Figure 16c	a few tens to several hundreds of nm	
$Ni_{67.5}Cu_{32.5}$	Figure 17	16.6	[103]
$Cu_{27.5}Ni_{72.5}$ (C)	Figure 18	10	[55]
$Cr_{20}Ni_{80}$	Figure 19, left	5–10	[107]
$Cr_{29}Ni_{71}$	Figure 19, right	5–30	

Figure 14a,b show typical TEM images of the as-synthesized sample (A) and the CM-dextran-coated maghemite MNPs (B). The average particle diameters estimated from TEM images are 14.5 nm (sample A) and 12.0 nm (sample B), which is in good agreement with the XRD analysis. The (Mg, Ti)-ferrite MNPs are several hundred nanometers in size. The TEM analysis confirmed the XRD results. The MNPs synthesized by the microemulsion method (as-prepared, $Ni_{0.725}Cu_{0.275}$) are about 3 to 10 nm in size. The TEM analysis supports the XRD spectra. The TEM analysis agrees with the XRD analysis. The thermally homogenized MNPs ($Ni_{0.725}Cu_{0.275}$) prepared by the microemulsion method show a relatively broad size distribution (a few tens of nm to several hundred nm). The peaks of the thermally homogenized particles were much sharper, with an estimated value of 28 nm, as indicated by the XRD analysis in agreement with the TEM analysis. The MNPs synthesized by the sol–gel method (after leaching) have a relatively narrow distribution, with an average size of 16.6 nm. The size distribution of $Cu_{27.5}Ni_{72.5}$ (C) MNPs synthesized by mechanical milling is comparable to the XRD analysis and the Sherrer equation, and the average value is 10 nm. The MNPs size distribution of CrNi MNPs (heat-treated at 400 °C) synthesized by the microemulsion technique is relatively broad, with an average particle size comparable to XRD analysis and Sherrer's equation, between 5 and 10 nm. The nanocrystallites sizes of the MNPs ($Cr_{29}Ni_{71}$) synthesized by mechanical milling determined from TEM range from about 5 nm to 30 nm.

3.4. Magnetic Measurements

Figures 20–24 show the magnetic properties of maghemite MNPs and maghemite MNPs coated with CM-dextran, (Mg, Ti)-ferrite MNPs, as-prepared and thermally homogenized NiCu MNPs (microemulsion technique), NiCu MNPs (sol–gel method), and ball-milled NiCu MNPs. The magnetic properties of the MNPs were investigated using a Lake Shore 7307 vibrating sample magnetometer.

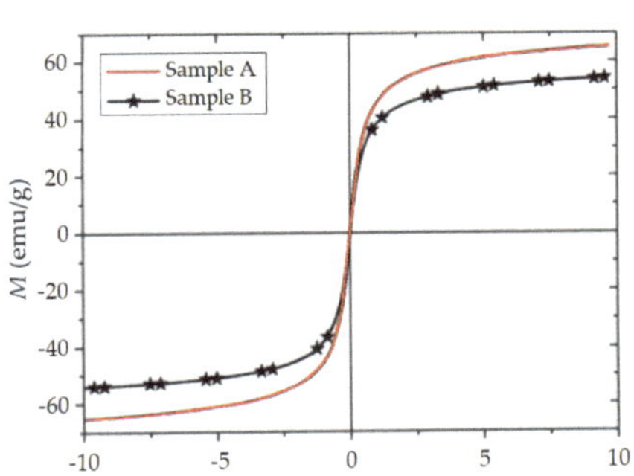

Figure 20. Magnetization of as-prepared maghemite MNPs (Sample A) and CM-dextran-coated MNPs (Sample B) at 300 K.

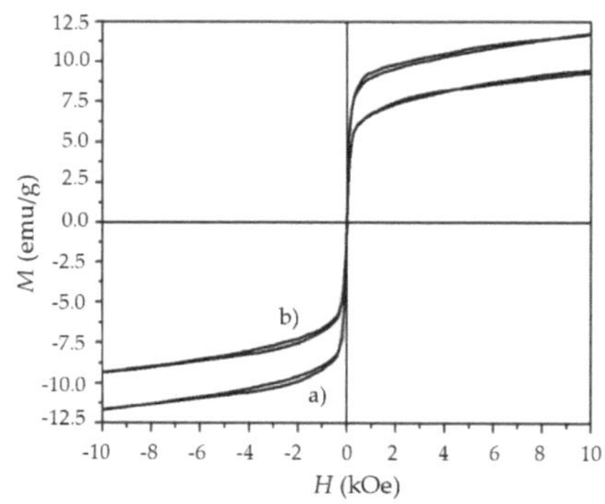

Figure 21. Hysteresis for (Mg, Ti)-ferrite as-synthesized (a) and milled (b) powder.

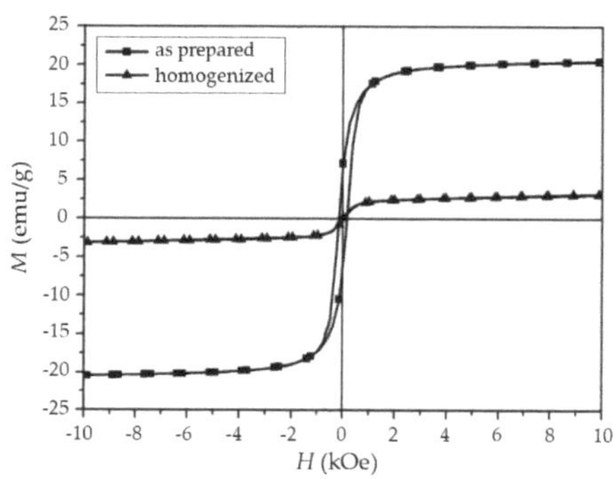

Figure 22. Magnetization vs. the AMF for the as-prepared and homogenized NiCu MNPs (microemulsion method).

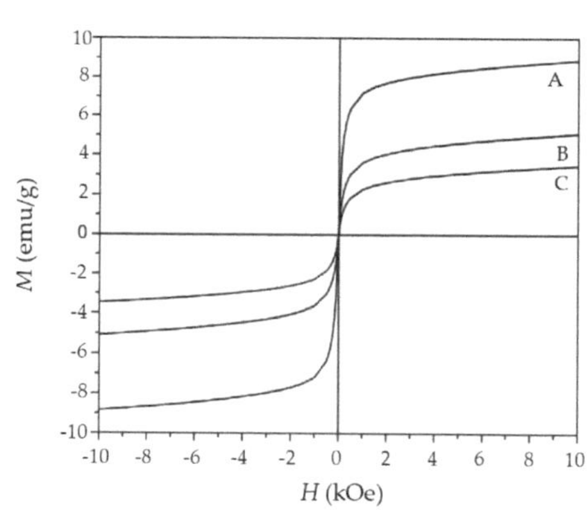

Figure 23. Magnetization of synthesized MNPs (sol–gel method) at 293 K: $Ni_{67.5}Cu_{32.5}$ (A), $Ni_{62.5}Cu_{37.5}$ (B), and $Ni_{60}u_{40}$ (C) vs. the magnetic field.

Figure 20 shows the magnetization vs. the magnetic field of the as-synthesized and CM-dextran-covered maghemite particles [27]. The uncoated particles show no hysteresis and have a saturation magnetization (M_s) of 65.2 emu/g. The magnetization of the coated MNPs was about 55 emu/g, which is lower than that of the uncoated sample. The average particle diameter of our particles determined by magnetic measurements was 8.20 nm (sample A) and 8.17 nm (sample B). Comparing the size with the XRD and the TEM analysis, certain sizes of nanoparticles are slightly smaller in the magnetic measurements, because the so-called "dead layer" is not considered in the magnetic measurements.

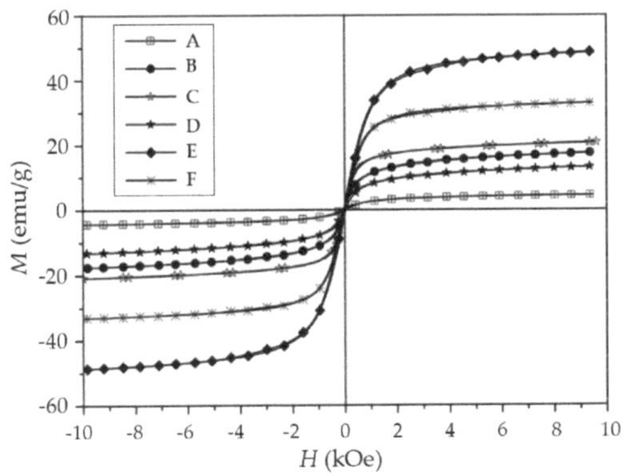

Figure 24. Magnetization vs. magnetic field of $Cu_{1-x}Ni_x$ alloy particles, samples (A–F) ball-milled for 20 h.

The magnetization vs. magnetic field curve for the (Mg, Ti)-ferrite particles measured at room temperature under an applied magnetic field of 1 T is shown in Figure 21 [26]. The magnetization of the as-prepared samples at 1 T is higher than the magnetization of the milled sample, which is 8 emu/g. This reduction in magnetization due to milling is a well-known phenomenon. It is caused by induced mechanical stresses and an increase in the specific surface area, consequently, an increase in the non-magnetic fraction ("dead layer"), which is usually located on the surface of the particles and is the result of imperfect coordination and, finally, a non-collinear spin configuration.

Figure 22 shows the magnetic measurements indicating that the magnetization is greatly reduced in the stabilized samples [41]. The MNPs in the as-prepared state exhibit M_s~20 emu/g, while it decreases to ~2.5 emu/g in the stabilized MNPs. To obtain homogenized nanoparticles that have a T_C in the therapeutic range, homogenization is necessary, which of course contributes to the drop in magnetization in homogenized nanoparticles.

Figure 23 shows the magnetization of the bare A, B, and C MNPs synthesized by a sol–gel method [103]. The magnetization was measured concerning the magnetic field at room temperature. It can be seen from the figure that the magnetizations measured at H = 10 kOe are good approximations for M_s since the slopes of the curves are low under the field H = 6 kOe. The considered samples have a similar size, which shows that the size does not influence the magnetization as much as the influence of the different compositions in the $Ni_{1-x}Cu_x$ alloy. The average particle size determined by magnetic measurements was 15.4 nm ($Ni_{67.5}Cu_{32.5}$ (A)), 16.4 nm ($Ni_{62.5}Cu_{37.5}$ (B)), and 17.6 nm ($Ni_{60}Cu_{40}$ (C)).

Figure 24 shows the magnetization vs. the magnetic field of the $Cu_{1-x}Ni_x$ alloy particles [55]. The measurements were performed at room temperature and under the influence of an AMF of 10 kOe. The hysteresis loop is not expressed in any sample, indicating the superparamagnetic properties of the $Cu_{1-x}Ni_x$ MNPs prepared by mechanical milling. The M_s of the samples ranges from 2 to 50 emu/g and depends on the nickel concentration of each sample.

3.5. Calorimetric Measurements

Figures 25–27 show the calorimetric measurements of maghemite MNPs coated with CM-dextran, (Mg, Ti)-ferrite MNPs, NiCu MNPs synthesized by microemulsion technique and sol–gel method, and ball-milled NiCu and NiCr MNPs. Magnetic heating effect measurements were performed in a conventionally built calorimetric system that generates an AMF with a nominal field strength of 2 kA/m up to 42 kA/m in the frequency range from 80 kHz to 800 kHz and equipped with a fiber optic temperature probe to measure the temperature required to determine the specific absorption rate (SAR).

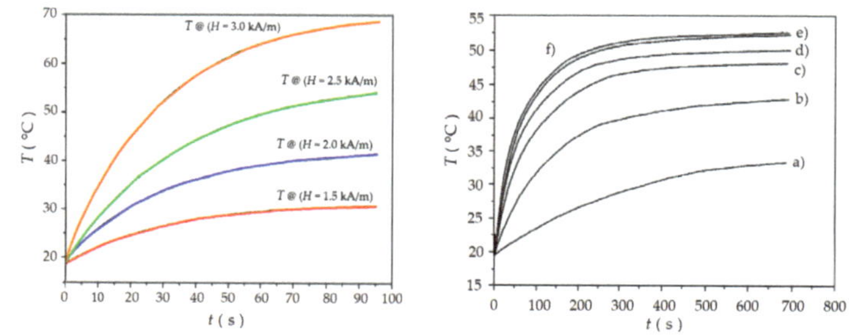

Figure 25. Time dependence of the heating of maghemite nanoparticles coated with CM-dextran (sample B) under the influence of different AMF intensities—(**left graph**)—and the calorimetric measurements of a sample (Mg, Ti)-ferrite (x = 0.37, sample B), at different magnitudes of AFM—(**right graph**) ((a) 8.4 kA/m; (b) 12,7 kA/m; (c) 16.9 kA/m; (d) 21.7 kA/m; (e) 25.4 kA/m; (f) 29.6 kA/m).

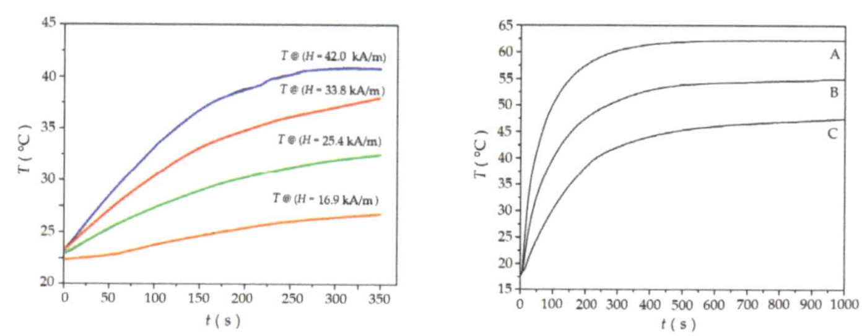

Figure 26. Calorimetric measurements for the $Ni_{0.725}Cu_{0.275}$ sample were performed under the influence of different AMF amplitudes—(**left graph**)—and the calorimetric measurements of samples A ($Ni_{67.5}Cu_{32.5}$), B ($Ni_{62.5}Cu_{37.5}$), and C ($Ni_{60}Cu_{40}$) at a magnetic field of 29.4 kA/m and a frequency of 100 kHz—(**right graph**).

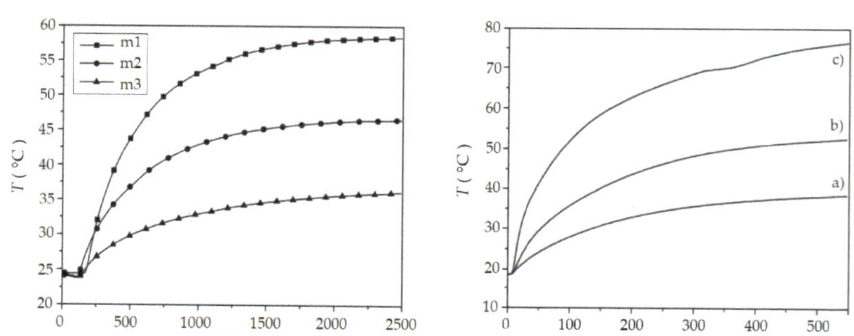

Figure 27. Time function of self-heating temperature under the influence of a magnetic field of varying strength for $Cu_{27.5}Ni_{72.5}$ (sample C) at AMF m1- H = 3.83 kA/m; m2- H = 3.13 kA/m; m3- H = 2.41 kA/m—(**left graph**)—and time function of self-heating temperature under the influence of a magnetic field of varying strength for $Cr_{28}Ni_{72}$ at AMF (a) H = 9.20 kA/m; (b) H = 12.52 kA/m; (c) H = 16.18 kA/m—(**right graph**).

In Figure 25, left graph, a time dependence of the heating of maghemite nanoparticles coated with CM-dextran (sample B) under the influence of different AMF intensities at 104 kHz [27] can be seen. The values of SAR were calculated based on the calorimetric response at a frequency of 104 kHz and in the range of 1.5 to 3 kA/m.

The SAR was estimated as:

$$\mathrm{SAR} = \left(\frac{C_{\mathrm{solvent}}}{x_{\mathrm{iron\ oxide}}}\right)\left(\frac{\Delta T}{\Delta t}\right), \tag{2}$$

where c_{solvent} is the heat capacity of the solvent (c = 4.18 J/Kg for water); $x_{\text{iron oxide}}$ is the weight fraction of iron oxide; and $(\Delta T/\Delta t)$ is the initial slope of the time dependence of the self-heating temperature. After a certain period of self-heating, the steady state was achieved when generated heat at a constant temperature was equal to that lost to the environment due to thermal convection. The maximum temperature of this steady state is important when iron oxide-based magnetic fluids (MFs) with a high T_C are used for in vivo applications, as this is the only way to use such an MF without damaging the surrounding tissue during MH.

Figure 25, right graph, shows tested MNPs and the corresponding MF for the SAR [26]. From the course, the SAR was estimated as:

$$\text{SAR} = \left(\frac{C_{solvent}}{x_{(Mg,\ Ti)-ferrite}}\right)\left(\frac{\Delta T}{\Delta t}\right), \tag{3}$$

where $c_{solvent}$ equals c_{water} 4.18 J/kg; $x_{(Mg,Ti)\text{-}ferrite}$ is the weight fraction of (*Mg, Ti*)-ferrite; and $(\Delta T/\Delta t)$ is the initial heating slope as a function of time. SAR values were estimated between 1.2 and 3.9 W/g (8.4 kA and 29.6 kA/m). The result of calorimetric measurements shows that the stationary temperature is higher than the planned T_C (sample B, x = 0.37).

In Figure 26, left graph, a calorimetric measurements for the $Ni_{0.725}Cu_{0.275}$ sample performed under the influence of AMF with an amplitude between 16.9 and 42.0 kA/m and a frequency of 104 kHz [41] can be seen. The SAR was estimated as:

$$\text{SAR} = (c_{\text{Ni}})\left(\frac{\Delta T}{\Delta t}\right), \tag{4}$$

where c_{Ni} represents the heat capacity of nickel and is 0.444 J/kg. $\Delta T/\Delta t$ represents the initial slope of the time dependence of the self-regulating temperature. After a certain time, the curve reaches a maximum corresponding to the temperature, T_C, already determined by thermomagnetic measurements. Table 4 lists the corresponding SAR values.

Table 4. SAR values as a function of AMF amplitude for $Ni_{0.725}Cu_{0.275}$ (microemulsion method).

H [kA/m]	SAR [mW/g]
16.9	4.3
25.4	21.8
33.8	35.8
42.0	41.6

In Figure 26, right graph [103], we determined the initial heating slopes $(\Delta T/\Delta t)$ that we used to determine the values of SAR:

$$\text{SAR} = \left(c_{\text{alloy}}\right)\left(\frac{\Delta T}{\Delta t}\right)\left(\frac{1}{x_{\text{Ni}}}\right), \tag{5}$$

where x_{Ni} is 0.67, the calculation was based on the mass fraction of nickel in each sample. Using the heat capacities of c_{Ni} (0.444 J/gK) and c_{Cu} (0.390 J/gK), we calculate the heat capacity of the alloy, at a temperature of 20 °C, since all measurements had started at this value, defined by the temperature of the surrounding thermal bath.

Table 5 lists the corresponding SAR values.

Table 5. Samples of composition $Ni_{1-x}Cu_x$ (sol–gel method) and data of SAR measurements.

Composition	SAR [W/g]
$Ni_{67.5}Cu_{32.5}$ (sample A)	0.60
$Ni_{62.5}Cu_{37.5}$ (sample B)	0.36
$Ni_{60}Cu_{40}$ (sample C)	0.12

To conduct calorimetric measurements, a solid powder sample was subjected to controlled heating. These measurements took place within a system capable of generating a nominal field strength of 2 kA/m and operating within a frequency range of 80 to 800 kHz. This setup allowed for precise characterization of the thermal properties and behavior of the sample under varying magnetic field conditions. Figure 27, left graph, shows the growth of the temperature of the powder sample in the calorimeter for different magnetic fields [55]. Remarkably, the $Cu_{27.5}Ni_{72.5}$ MNPs exhibited a significant heating effect during our experiments. Under an applied magnetic field strength of 3.13 kA/m, we were able to achieve temperatures surpassing 40 °C.

Figure 27, right graph, shows the temperature rise of the $Cr_{28}Ni_{72}$ sample in the calorimeter at different magnetic field strengths [107]. The magnetic measurements of the solid powdered $Cr_{28}Ni_{72}$ with a T_C of 44 °C were performed in a system that generates an AMF (9.2–16.2 kA/m, 800 kHz).

4. Conclusions

The potential of superparamagnetic nanoparticles in magnetic hyperthermia holds great promise. In this concise review, we explored different synthesis methods employed for the preparation of Fe_2O_3, $Mg_{1+x}Fe_{2-2x}Ti_xO_4$, $Ni_{1-x}Cu_x$, Cr_xNi_{1-x} MNPs, their resulting size, shape, functionalization, T_C, magnetic, and calorimetric properties. The results indicate that the described synthesis method of superparamagnetic nanoparticles with adjustable Curie temperature offers a convenient approach for producing controlled T_C nanoparticles. These synthesized particles demonstrate excellent suitability, particularly when optimized for MH applications.

Through the coprecipitation method, maghemite nanoparticles were successfully synthesized, and a supplementary CM-dextran coating was applied. The resulting nanoparticles displayed a distinct crystalline structure, as confirmed by both XRD and TEM analysis. The magnetization ranged from 55 to 65 emu/g, with the results of calorimetric measurements showing that the size of the nanoparticles and the thickness of the CM-dextran coating affect the SAR.

The (Mg, Ti)-ferrite nanoparticles were also synthesized by coprecipitation, obtaining nanoparticles of two crystallites, 20 and 200 nm. The TEM analysis revealed the presence of larger nanoparticles, displaying low magnetization. Nevertheless, these nanoparticles exhibited a favorable thermal response under the influence of alternating magnetic fields (AMF), with the measured T_C falling within the therapeutic range for magnetic hyperthermia (MH) applications.

The NiCu and NiCr nanoparticles were successfully synthesized using the microemulsion technique. However, in both cases, additional homogenization was required, resulting in partial agglomeration and the formation of larger nanoparticle clusters. Despite achieving a T_C of approximately 45 °C for NiCu nanoparticles, their dissipation and/or heating efficiency (SAR) proved inadequate. We also synthesized NiCu and NiCr nanoparticles of different compositions by mechanical milling. We obtained nanoparticles of different sizes and shapes, which in turn affected their magnetization and thermal response under the influence of AMF, but again the T_C were within the therapeutic temperature range for MH. The sol–gel method proved to be the most promising. We obtained nanoparticles with a size of about 18 nm, which could otherwise be further homogenized. Although the particles partially grew during homogenization, they did not agglomerate because the silica matrix prevented this. The only weakness of silica is that it affects magnetization, but when silica is successfully removed, nanoparticles with higher magnetization and good thermal response under the influence of AMF are obtained. Also, the T_C of the mentioned nanoparticles is in the therapeutic range for use in MH.

Through XRD measurements, we substantiated the successful completion of each synthesis process. This assessment not only validated the creation of magnetic nanoparticles (MNPs) with the intended composition but also allowed for the simultaneous estimation of the MNPs' size, employing Sherrer's equation. The size of the nanoparticles ranged from

5 to 25 nm, and in the case of homogenization, the particles were larger, or the growth of the nanoparticles proved to be a result of the aforementioned homogenization.

The TM curves and corresponding T_C were determined by thermal demagnetization. The T_C is very important in the field of MH. It is the temperature at which the material passes from the magnetic to the non-magnetic state, which means that it no longer gives off heat or heats up under the influence of an external magnetic field. The therapeutic temperature in MH is between 41 and 46 °C. Most of the synthesized nanoparticles have a T_C in the therapeutic range. For maghemite nanoparticles and magnetic NiCu nanoparticles synthesized by the microemulsion method in a silica matrix, this temperature is about 45 °C, and for magnetic NiCu nanoparticles synthesized by the sol–gel method, the T_C varies depending on the composition and ranged from 51 to 64 °C; similarly, for NiCu and NiCr nanoparticles synthesized by mechanical milling, in the case of the former T_C the temperature varies from 24 to 174 °C, while for NiCr nanoparticles it ranges from 69 to 340 °C. As the Ni content increases, so does the T_C.

Using TEM analysis, we determined the morphology, size, and size distribution of the synthesized nanoparticles. The size of the nanoparticles, which we estimated using Sherrer's equation, was largely confirmed. The morphology of the MNPs is different, they are mostly round nanoparticles, with the best distribution of the size of the nanoparticles in the case of the synthesis of NiCu nanoparticles by sol–gel synthesis; in these nanoparticles, no agglomeration can be detected. For other nanoparticles, homogenization leads to agglomeration and consequently to a larger size of nanoparticles, and the size distribution is also very wide.

The synthesized nanoparticles show magnetic properties, which we have confirmed by magnetic measurements. The highest magnetization is observed for the maghemite nanoparticles, but it decreases somewhat when they are coated with CM-dextran. The magnetization is much lower for the magnetic NiCu nanoparticles, and the magnetization increases with increasing nickel content. The difference is also evident in the sol–gel method of magnetic NiCu nanoparticles, where the magnetization is about 2 emu/g for particles in the silica matrix, while the value increases to 20 emu/g when the silica is leached.

Measurements of the magnetic heating effect were performed in a conventionally built calorimetric system generating an AMF with a nominal field strength of 2 kA/m up to 42 kA/m in the frequency range from 80 kHz to 800 kHz and equipped with a fiber-optic temperature probe to measure the temperature required to determine SAR. We monitored the temperature response of individually synthesized nanoparticles at different magnetic fields. Depending on the strength of the magnetic field, the nanoparticles showed different temperature responses; all of them mostly reached a heating temperature in the therapeutic range for use in MH, but it varied depending on the composition or strength of the magnetic field. In the case of NiCu nanoparticles synthesized by the microemulsion method, we also calculated the SAR values, which increased with increasing the magnetic field and ranged from 4.3 to 41.6 mW/g, while in the case of magnetic NiCu nanoparticles, the values ranged from 0.12 to 0.60 W/g, depending on the composition.

All synthesized MNPs have properties that are necessary for use in MH. A T_C, which is in the therapeutic temperature range, is particularly important because it means that heating in this range is automatically interrupted, which in the case of cancer cells means damage or destruction, while healthy cells remain undamaged. The magnetic NiCu nanoparticles synthesized by the sol–gel method were found to be most suitable for use in MH. They are about 10 nm in size and have a very narrow size distribution. The magnetization is very low in the case of silica, but only this can be removed. T_C is close to the therapeutic temperature, which can also be adjusted by simply changing the composition or the nickel content.

In this short review article, we have shown that it is possible to synthesize MNPs of different compositions with properties for use in MH using various synthetic methods. In addition, it will be necessary to think in the direction of biocompatibility tests, or the use of both in vitro and in vivo tests, especially of the synthesized nanoparticles that have shown the best properties for use in the field MH.

Funding: This brief review received no external funding.

Data Availability Statement: The data presented in this study are openly.

Conflicts of Interest: The authors declare no conflict of interest.

References

1. Angadi, J.V.; Manjunatha, K. 21-Present and future applications of magnetic nanoparticles in the field of medicine and bio-sensors. In *Fundamentals and Industrial Applications of Magnetic Nanoparticles*; Hussain, C.M., Patankar, K.K., Eds.; Woodhead Publishing: Sawston, UK, 2022; pp. 655–663.
2. Binandeh, M. Performance of unique magnetic nanoparticles in biomedicine. *Eur. J. Med. Chem. Rep.* **2022**, *6*, 100072. [CrossRef]
3. Ullah Khan, A.; Chen, L.; Ge, G. Recent development for biomedical applications of magnetic nanoparticles. *Inorg. Chem. Commun.* **2021**, *134*, 108995. [CrossRef] [PubMed]
4. Marzi, M.; Osanloo, M.; Vakil, M.K.; Mansoori, Y.; Ghasemian, A.; Dehghan, A.; Zarenezhad, E. Applications of Metallic Nanoparticles in the Skin Cancer Treatment. *BioMed Res. Int.* **2022**, *2022*, 2346941. [CrossRef] [PubMed]
5. Borghei, Y.-S.; Hosseinkhani, S.; Ganjali, M.R. Engineering in modern medicine using 'magnetic nanoparticles' in understanding physicochemical interactions at the nano–bio interfaces. *Mater. Today Chem.* **2022**, *23*, 100733. [CrossRef]
6. Hewlin, R.L.; Edwards, M.; Schultz, C. Design and Development of a Traveling Wave Ferro-Microfluidic Device and System Rig for Potential Magnetophoretic Cell Separation and Sorting in a Water-Based Ferrofluid. *Micromachines* **2023**, *14*, 889. [CrossRef] [PubMed]
7. Nikzamir, M.; Akbarzadeh, A.; Panahi, Y. An overview on nanoparticles used in biomedicine and their cytotoxicity. *J. Drug Deliv. Sci. Technol.* **2020**, *61*, 102316. [CrossRef]
8. Mehta, R. Synthesis of magnetic nanoparticles and their dispersions with special reference to applications in biomedicine and biotechnology. *Mater. Sci. Eng. C* **2017**, *79*, 901–916. [CrossRef] [PubMed]
9. Gao, Y.; Liu, Y.; Xu, C. Magnetic Nanoparticles for Biomedical Applications: From Diagnosis to Treatment to Regeneration. In *Engineering in Translational Medicine*; Cai, W., Ed.; Springer London: London, UK, 2014; pp. 567–583.
10. Wu, W.; He, Q.; Jiang, C. Magnetic Iron Oxide Nanoparticles: Synthesis and Surface Functionalization Strategies. *Nanoscale Res. Lett.* **2008**, *3*, 397–415. [CrossRef]
11. Ban, I.; Stergar, J.; Maver, U. NiCu magnetic nanoparticles: A review of synthesis methods, surface functionalization approaches, and biomedical applications. *Nanotechnol. Rev.* **2018**, *7*, 187–207. [CrossRef]
12. Chow, J.C.L. Synthesis and applications of functionalized nanoparticles in biomedicine and radiotherapy. In *Additive Manufacturing with Functionalized Nanomaterials*; Elsevier: Amsterdam, The Netherlands, 2021; pp. 193–218.
13. Zarenezhad, E.; Abdulabbas, H.T.; Marzi, M.; Ghazy, E.; Ekrahi, M.; Pezeshki, B.; Ghasemian, A.; Moawad, A.A. Nickel Nanoparticles: Applications and Antimicrobial Role against Methicillin-Resistant Staphy-lococcus aureus Infections. *Antibiotics* **2022**, *11*, 1208. [CrossRef]
14. Mohammed, L.; Gomaa, H.G.; Ragab, D.; Zhu, J. Magnetic nanoparticles for environmental and biomedical applications: A review. *Particuology* **2017**, *30*, 1–14. [CrossRef]
15. Kudr, J.; Haddad, Y.; Richtera, L.; Heger, Z.; Cernak, M.; Adam, V.; Zitka, O. Magnetic Nanoparticles: From Design and Synthesis to Real World Applications. *Nanomaterials* **2017**, *7*, 243. [CrossRef]
16. Selmani, A.; Kovačević, D.; Bohinc, K. Nanoparticles: From synthesis to applications and beyond. *Adv. Colloid Interface Sci.* **2022**, *303*, 102640. [CrossRef] [PubMed]
17. Catherine, C.B.; Curtis, A.S.G. Functionalisation of magnetic nanoparticles for applications in biomedicine. *J. Phys. D Appl. Phys.* **2003**, *36*, R198.
18. Issa, B.; Obaidat, I.M.; Albiss, B.A.; Haik, Y. Magnetic Nanoparticles: Surface Effects and Properties Related to Biomedicine Applications. *Int. J. Mol. Sci.* **2013**, *14*, 21266–21305. [CrossRef] [PubMed]
19. Hewlin, R.L.; Tindall, J.M. Computational Assessment of Magnetic Nanoparticle Targeting Efficiency in a Simplified Circle of Willis Arterial Model. *Int. J. Mol. Sci.* **2023**, *24*, 2545. [CrossRef] [PubMed]
20. Setia, A.; Mehata, A.K.; Vikas; Malik, A.K.; Viswanadh, M.K.; Muthu, M.S. Theranostic magnetic nanoparticles: Synthesis, properties, toxicity, and emerging trends for biomedical ap-plications. *J. Drug Deliv. Sci. Technol.* **2023**, *81*, 104295. [CrossRef]
21. Gambhir, R.P.; Rohiwal, S.S.; Tiwari, A.P. Multifunctional surface functionalized magnetic iron oxide nanoparticles for biomedical applications: A review. *Appl. Surf. Sci. Adv.* **2022**, *11*, 100303. [CrossRef]
22. Lin, H.; Yin, L.; Chen, B.; Ji, Y. Design of functionalized magnetic silica multi-core composite nanoparticles for synergistic magnetic hyper-thermia/radiotherapy in cancer cells. *Colloids Surf. B Biointerfaces* **2022**, *219*, 112814. [CrossRef]
23. Gupta, I.; Sirohi, S.; Roy, K. Strategies for functionalization of magnetic nanoparticles for biomedical applications. *Mater. Today Proc.* **2023**, *72*, 2757–2767. [CrossRef]
24. Yazdani, F.; Seddigh, M. Magnetite nanoparticles synthesized by co-precipitation method: The effects of various iron anions on specifications. *Mater. Chem. Phys.* **2016**, *184*, 318–323. [CrossRef]
25. Rashid, H.; Mansoor, M.A.; Haider, B.; Nasir, R.; Hamid, S.B.A.; Abdulrahman, A. Synthesis and characterization of magnetite nano particles with high selectivity using in-situ precipitation method. *Sep. Sci. Technol.* **2019**, *55*, 1207–1215. [CrossRef]

26. Ferk, G.; Drofenik, M.; Lisjak, D.; Hamler, A.; Jagličić, Z.; Makovec, D. Synthesis and characterization of $Mg_{1+x}Fe_{2-2x}Ti_xO_4$ nanoparticles with an adjustable Curie point. *J. Magn. Magn. Mater.* **2014**, *350*, 124–128. [CrossRef]
27. Ferk, G.; Ban, I.; Stergar, J.; Makovec, D.; Hamler, A.; Jagličić, Z.; Drofenik, M. A facile route to the synthesis of coated maghemite nanocomposites for hyperthermia applications. *Acta Chim. Slov.* **2012**, *59*, 366–374. [PubMed]
28. Senthilkumar, G.; Sakthivelu, A.; Rahman, M.A.; Parameswari, P. Enhancement of antibacterial and anticancer properties lanthanum insight into zinc oxide nanoparticles prepared via coprecipitation process. *Inorg. Chem. Commun.* **2023**, *155*, 111081. [CrossRef]
29. Ranga, R.; Kumar, K.; Kumar, A. Morphology, structural and magnetic study of superparamagnetic $Mg_{0.5}Zn_{0.5}Fe_{2-x}La_xO_4$ ($0 \leq x \leq 0.1$) ferrite nanoparticles synthesized by chemical coprecipitation method. *Ceram. Int.* **2023**, *49*, 2956–2966. [CrossRef]
30. Nkurikiyimfura, I.; Wang, Y.; Safari, B.; Nshingabigwi, E. Temperature-dependent magnetic properties of magnetite nanoparticles synthesized via co-precipitation method. *J. Alloy. Compd.* **2020**, *846*, 156344. [CrossRef]
31. Unni, M.; Uhl, A.M.; Savliwala, S.; Savitzky, B.H.; Dhavalikar, R.; Garraud, N.; Arnold, D.P.; Kourkoutis, L.F.; Andrew, J.S.; Rinaldi, C. Thermal Decomposition Synthesis of Iron Oxide Nanoparticles with Diminished Magnetic Dead Layer by Controlled Addition of Oxygen. *ACS Nano* **2017**, *11*, 2284–2303. [CrossRef]
32. Dixit, S.; Jeevanandam, P. Synthesis of Iron Oxide Nanoparticles by Thermal Decomposition Approach. *Adv. Mater. Res.* **2009**, *67*, 221–226. [CrossRef]
33. Tomar, D.; Jeevanandam, P. Synthesis of $ZnFe_2O_4$ nanoparticles with different morphologies via thermal decomposition approach and studies on their magnetic properties. *J. Magn. Magn. Mater.* **2022**, *564*, 170033. [CrossRef]
34. Heydaryan, K.; Kashi, M.A.; Montazer, A.H. Tuning specific loss power of $CoFe_2O_4$ nanoparticles by changing surfactant concentration in a combined co-precipitation and thermal decomposition method. *Ceram. Int.* **2022**, *48*, 16967–16976. [CrossRef]
35. Ivantsov, R.D.; Lin, C.-R.; Ivanova, O.S.; Altunin, R.R.; Knyazev, Y.V.; Molokeev, M.S.; Zharkov, S.M.; Chen, Y.-Z.; Lin, E.-S.; Chen, B.-Y.; et al. Mössbauer and MCD spectroscopy of the Fe_3S_4 nanoparticles synthesized by the thermal decomposition method with two different surfactants. *Curr. Appl. Phys.* **2021**, *25*, 55–61. [CrossRef]
36. Sreeja, V.; Joy, P.A. Microwave–hydrothermal synthesis of γ-Fe_2O_3 nanoparticles and their magnetic properties. *Mater. Res. Bull.* **2007**, *42*, 1570–1576. [CrossRef]
37. Chen, R.; Wang, W.; Zhao, X.; Zhang, Y.; Wu, S.; Li, F. Rapid hydrothermal synthesis of magnetic $CoxNi_{1-x}Fe_2O_4$ nanoparticles and their application on removal of Congo red. *Chem. Eng. J.* **2014**, *242*, 226–233. [CrossRef]
38. Wang, L.; Zhang, M. Study on synthesis and magnetic properties of Nd2Fe14B nanoparticles prepared by hydrothermal method. *J. Magn. Magn. Mater.* **2020**, *507*, 166841. [CrossRef]
39. Jesus, A.C.B.; Jesus, J.D.; Lima, R.J.S.; Moura, K.O.; Almeida, J.M.A.; Duque, J.G.S.; Meneses, C.T. Synthesis and magnetic interaction on concentrated Fe_3O_4 nanoparticles obtained by the co-precipitation and hydrothermal chemical methods. *Ceram. Int.* **2020**, *46*, 11149–11153. [CrossRef]
40. Silva, J.M.M.; Feuser, P.E.; Cercená, R.; Peterson, M.; Dal-Bó, A.G. Obtention of magnetite nanoparticles via the hydrothermal method and effect of synthesis parameters. *J. Magn. Magn. Mater.* **2023**, *580*, 170925. [CrossRef]
41. Stergar, J.; Ferk, G.; Ban, I.; Drofenik, M.; Hamler, A.; Jagodič, M.; Makovec, D. The synthesis and characterization of copper–nickel alloy nanoparticles with a therapeutic Curie point using the microemulsion method. *J. Alloy. Compd.* **2013**, *576*, 220–226. [CrossRef]
42. Pérez, J.A.L.; Quintela, M.A.L.; Mira, J.; Rivas, J.; Charles, S.W. Advances in the Preparation of Magnetic Nanoparticles by the Microemulsion Method. *J. Phys. Chem. B* **1997**, *101*, 8045–8047. [CrossRef]
43. Umair, H.M.; Bibi, I.; Majid, F.; Kamal, S.; Alwadai, N.; Arshad, M.I.; Ali, A.; Nouren, S.; Al Huwayz, M.; Iqbal, M. Ferroelectric, dielectric, magnetic and photocatalytic properties of Mn doped Ca-hexaferrite prepared via microemulsion route. *Mater. Chem. Phys.* **2023**, *307*, 128152. [CrossRef]
44. Martín, R.F.; Prietzel, C.; Koetz, J. Template-mediated self-assembly of magnetite-gold nanoparticle superstructures at the water-oil interface of AOT reverse microemulsions. *J. Colloid Interface Sci.* **2021**, *581*, 44–55. [CrossRef] [PubMed]
45. Hachani, R.; Lowdell, M.; Birchall, M.; Hervault, A.; Mertz, D.; Begin-Colin, S.; Thanh, N.T.K. Polyol synthesis, functionalisation, and biocompatibility studies of superparamagnetic iron oxide nanoparticles as potential MRI contrast agents. *Nanoscale* **2016**, *8*, 3278–3287. [CrossRef] [PubMed]
46. Kotoulas, A.; Samara, D.; Angelakeris, M.; Kalogirou, O. The Effect of Polyol Composition on the Structural and Magnetic Properties of Magnetite Nanoparticles for Magnetic Particle Hyperthermia. *Materials* **2019**, *12*, 2663. [CrossRef] [PubMed]
47. Khannanov, A.A.; Rossova, A.A.; Ignatyeva, K.A.; Ulakhovich, N.A.; Gerasimov, A.V.; Boldyrev, A.E.; Evtugyn, V.G.; Rogov, A.M.; Cherosov, M.A.; Gilmutdinov, I.F.; et al. Superparamagnetic cobalt nanoparticles in hyperbranched polyester polyol matrix with anti-protease activity. *J. Magn. Magn. Mater.* **2021**, *547*, 168808. [CrossRef]
48. Cheah, P.; Qu, J.; Li, Y.; Cao, D.; Zhu, X.; Zhao, Y. The key role of reaction temperature on a polyol synthesis of water-dispersible iron oxide nanoparticles. *J. Magn. Magn. Mater.* **2021**, *540*, 168481. [CrossRef]
49. Koventhan, C.; Kumar, N.K.R.; Chen, S.M.; Pandi, K.; Sangili, A. Polyol mediated synthesis of hexagonal manganese cobaltate nanoparticles for voltammetric determi-nation of thioridazine. *Colloids Surf. A Physicochem. Eng. Asp.* **2021**, *621*, 126625. [CrossRef]
50. Ferk, G.; Stergar, J.; Drofenik, M.; Makovec, D.; Hamler, A.; Jagličić, Z.; Ban, I. The synthesis and characterization of nickel–copper alloy nanoparticles with a narrow size distribution using sol–gel synthesis. *Mater. Lett.* **2014**, *124*, 39–42. [CrossRef]

51. Xu, J.; Yang, H.; Fu, W.; Du, K.; Sui, Y.; Chen, J.; Zeng, Y.; Li, M.; Zou, G. Preparation and magnetic properties of magnetite nanoparticles by sol–gel method. *J. Magn. Magn. Mater.* **2007**, *309*, 307–311. [CrossRef]
52. Tahir, M.; Fakhar-e-Alam, M.; Atif, M.; Mustafa, G.; Ali, Z. Investigation of optical, electrical and magnetic properties of hematite α-Fe2O3 nanoparticles via sol-gel and co-precipitation method. *J. King Saud Univ.-Sci.* **2023**, *35*, 102695. [CrossRef]
53. Jayakumar, T.; Rajeevgandhi, C.; Anand, P. Effect on magnetic behaviour of Ag and Cd doped cobalt ferrite nanoparticles prepared by glycine-assisted sol-gel auto combustion method. *J. Alloy. Compd.* **2023**, *962*, 171098. [CrossRef]
54. Castro-Lopes, S.; Guerra, Y.; Silva-Sousa, A.; Oliveira, D.; Gonçalves, L.; Franco, A.; Padrón-Hernández, E.; Peña-Garcia, R. Influence of pH on the structural and magnetic properties of Fe-doped ZnO nanoparticles synthesized by sol gel method. *Solid State Sci.* **2020**, *109*, 106438. [CrossRef]
55. Ban, I.; Stergar, J.; Drofenik, M.; Ferk, G.; Makovec, D. Synthesis of copper–nickel nanoparticles prepared by mechanical milling for use in magnetic hyperthermia. *J. Magn. Magn. Mater.* **2011**, *323*, 2254–2258. [CrossRef]
56. Karimzadeh, I.; Aghazadeh, M.; Doroudi, T.; Ganjali, M.R.; Kolivand, P.H. Superparamagnetic Iron Oxide (Fe_3O_4) Nanoparticles Coated with PEG/PEI for Biomedical Applications: A Facile and Scalable Preparation Route Based on the Cathodic Electrochemical Deposition Method. *Adv. Phys. Chem.* **2017**, *2017*, 9437487. [CrossRef]
57. Jamshidiyan, M.; Shirani, A.; Alahyarizadeh, G. Solvothermal synthesis and characterization of magnetic Fe_3O_4 nanoparticle by different sodium salt sources. *Mater. Sci.* **2017**, *35*, 50–57.
58. Younes, A.; Dilmi, N.; Bouamer, A. Effect of zinc oxide and alumina nanoparticles on Structural, magnetic and mechanical properties of the iron matrix synthesized by mechanical milling and thermal spraying. *Mater. Today Proc.* **2021**, *42*, 2990–2995. [CrossRef]
59. Chagas, E.F.; Ferreira, E.S. 6-Mechanical milling of ferrite nanoparticles. In *Ferrite Nanostructured Magnetic Materials*; Singh, J.P., Chae, K.H., Srivastava, R.C., Caltun, O.F., Eds.; Woodhead Publishing: Sawston, UK, 2023; pp. 85–102.
60. Ali, I.; Pan, Y.; Jamil, Y.; Chen, J.; Shah, A.A.; Imran, M.; Alvi, U.; Nasir, N.; Shen, Z. Hybrid Au/Co nanoparticles: Laser-assisted synthesis and applications in magnetic hyperthermia. *Phys. B Condens. Matter* **2023**, *657*, 414773. [CrossRef]
61. Kumar, A.; Gangawane, K.M. Synthesis and effect on the surface morphology & magnetic properties of ferrimagnetic nanoparticles by different wet chemical synthesis methods. *Powder Technol.* **2022**, *410*, 117867.
62. Zou, L.; Huang, B.; Zheng, X.; Pan, H.; Zhang, Q.; Xie, W.; Zhao, Z.; Li, X. Microfluidic synthesis of magnetic nanoparticles in droplet-based microreactors. *Mater. Chem. Phys.* **2021**, *276*, 125384. [CrossRef]
63. Fuentes-García, J.A.; Alavarse, A.C.; de Castro, C.E.; Giacomelli, F.C.; Ibarra, M.R.; Bonvent, J.J.; Goya, G.F. Sonochemical route for mesoporous silica-coated magnetic nanoparticles towards pH-triggered drug delivery system. *J. Mater. Res. Technol.* **2021**, *15*, 52–67. [CrossRef]
64. Yallapu, M.M.; Foy, S.P.; Jain, T.K.; Labhasetwar, V. PEG-Functionalized Magnetic Nanoparticles for Drug Delivery and Magnetic Resonance Imaging Applications. *Pharm. Res.* **2010**, *27*, 2283–2295. [CrossRef]
65. Li-Juan, A.; Zhao-Qiang, L.; Yan-Ping, W.; Bai, Y. Synthesis of Fe_3O_4/PMMA Nanocomposite Particles by Surface-Initiated ATRP and Characterization. *Chem. J. Chin. Univ.* **2006**, *27*, 1372–1375.
66. Díaz-Hernández, A.; Gracida, J.; García-Almendárez, B.E.; Regalado, C.; Núñez, R.; Amaro-Reyes, A. Characterization of Magnetic Nanoparticles Coated with Chitosan: A Potential Approach for Enzyme Immobilization. *J. Nanomater.* **2018**, *2018*, 9468574. [CrossRef]
67. Shete, P.; Patil, R.; Tiwale, B.; Pawar, S. Water dispersible oleic acid-coated Fe_3O_4 nanoparticles for biomedical applications. *J. Magn. Magn. Mater.* **2015**, *377*, 406–410. [CrossRef]
68. Ferk, G.; Drofenik, M.; Makovec, D.; Ban, I. Monodispersed water-soluble maghemite nanoparticles stabilized by a polymerized bilayer of 10-undecenoic acid. *Mater. Lett.* **2015**, *157*, 239–242. [CrossRef]
69. Stergar, J.; Ban, I.; Drofenik, M.; Ferk, G.; Makovec, D. Synthesis and Characterization of Silica-Coated Cu_{1-x}Nix Nanoparticles. *IEEE Trans. Magn.* **2012**, *48*, 1344–1347. [CrossRef]
70. Stergar, J.; Jirák, Z.; Veverka, P.; Kubíčková, L.; Vrba, T.; Kuličková, J.; Knížek, K.; Porcher, F.; Kohout, J.; Kaman, O. Mn-Zn ferrite nanoparticles coated with mesoporous silica as core material for heat-triggered release of therapeutic agents. *J. Magn. Magn. Mater.* **2019**, *475*, 429–435. [CrossRef]
71. Gupta, A.; Berry, C.; Gupta, M.; Curtis, A. Receptor-mediated targeting of magnetic nanoparticles using insulin as a surface ligand to prevent endocytosis. *IEEE Trans. NanoBiosci.* **2003**, *2*, 255–261. [CrossRef]
72. López-Abarrategui, C.; Figueroa-Espí, V.; Reyes-Acosta, O.; Reguera, E.; Otero-Gonzalez, A. Magnetic nanoparticles: New players in antimicrobial peptide therapeutics. *Curr. Protein Pept. Sci.* **2013**, *14*, 595–606. [CrossRef]
73. Anderson, S.D.; Gwenin, V.V.; Gwenin, C.D. Magnetic Functionalized Nanoparticles for Biomedical, Drug Delivery and Imaging Applications. *Nanoscale Res. Lett.* **2019**, *14*, 188. [CrossRef]
74. Mou, X.; Ali, Z.; Li, S.; He, N. Applications of Magnetic Nanoparticles in Targeted Drug Delivery System. *J. Nanosci. Nanotechnol.* **2015**, *15*, 54–62. [CrossRef]
75. Wilczewska, A.Z.; Niemirowicz, K.; Markiewicz, K.H.; Car, H. Nanoparticles as drug delivery systems. *Pharmacol. Rep.* **2012**, *64*, 1020–1037. [CrossRef] [PubMed]
76. Stergar, J.; Maver, U.; Bele, M.; Gradišnik, L.; Kristl, M.; Ban, I. NiCu-silica nanoparticles as a potential drug delivery system. *J. Sol-Gel Sci. Technol.* **2020**, *101*, 493–504. [CrossRef]

77. Ciofani, G.; Raffa, V.; Obata, Y.; Menciassi, A.; Dario, P.; Takeoka, S. Magnetic Driven Alginate Nanoparticles for Targeted Drug Delivery. *Curr. Nanosci.* **2008**, *4*, 212–218. [CrossRef]
78. Da, X.; Li, R.; Li, X.; Lu, Y.; Gu, F.; Liu, Y. Synthesis and characterization of PEG coated hollow Fe_3O_4 magnetic nanoparticles as a drug carrier. *Mater. Lett.* **2021**, *309*, 131357. [CrossRef]
79. Qiao, R.; Fu, C.; Forgham, H.; Javed, I.; Huang, X.; Zhu, J.; Whittaker, A.K.; Davis, T.P. Magnetic iron oxide nanoparticles for brain imaging and drug delivery. *Adv. Drug Deliv. Rev.* **2023**, *197*, 114822. [CrossRef] [PubMed]
80. Shahzad, K.; Mushtaq, S.; Rizwan, M.; Khalid, W.; Atif, M.; Din, F.U.; Ahmad, N.; Abbasi, R.; Ali, Z. Field-controlled magneto-electric core-shell $CoFe_2O_4$@$BaTiO_3$ nanoparticles as effective drug carriers and drug release in vitro. *Mater. Sci. Eng. C* **2020**, *119*, 111444. [CrossRef] [PubMed]
81. Gong, D.; Celi, N.; Zhang, D.; Cai, J. Magnetic Biohybrid Microrobot Multimers Based on *Chlorella* Cells for Enhanced Targeted Drug Delivery. *ACS Appl. Mater. Interfaces* **2022**, *14*, 6320–6330. [CrossRef] [PubMed]
82. Wang, Q.; Liang, Z.; Li, F.; Lee, J.; Low, L.E.; Ling, D. Dynamically switchable magnetic resonance imaging contrast agents. *Exploration* **2021**, *1*, e210. [CrossRef]
83. Gervits, N.E.; Gippius, A.A.; Tkachev, A.V.; Demikhov, E.I.; Starchikov, S.S.; Lyubutin, I.S.; Vasiliev, A.L.; Chekhonin, V.P.; Abakumov, M.A.; Semkina, A.S.; et al. Magnetic properties of biofunctionalized iron oxide nanoparticles as magnetic resonance imaging contrast agents. *Beilstein J. Nanotechnol.* **2019**, *10*, 1964–1972. [CrossRef]
84. Jin, R.; Lin, B.; Li, D.; Ai, H. Superparamagnetic iron oxide nanoparticles for MR imaging and therapy: Design considerations and clinical applications. *Curr. Opin. Pharmacol.* **2014**, *18*, 18–27. [CrossRef]
85. Xue, F.; Zhu, S.; Tian, Q.; Qin, R.; Wang, Z.; Huang, G.; Yang, S. Macrophage-mediated delivery of magnetic nanoparticles for enhanced magnetic resonance imaging and magnetothermal therapy of solid tumors. *J. Colloid Interface Sci.* **2023**, *629*, 554–562. [CrossRef] [PubMed]
86. Das, P.; Salvioni, L.; Malatesta, M.; Vurro, F.; Mannucci, S.; Gerosa, M.; Rizzuto, M.A.; Tullio, C.; Degrassi, A.; Colombo, M.; et al. Colloidal polymer-coated Zn-doped iron oxide nanoparticles with high relaxivity and specific absorption rate for efficient magnetic resonance imaging and magnetic hyperthermia. *J. Colloid Interface Sci.* **2020**, *579*, 186–194. [CrossRef] [PubMed]
87. Chow, J.C.L. 9-Magnetic nanoparticles as contrast agents in magnetic resonance imaging and radiosensitizers in radio-therapy. In *Fundamentals and Industrial Applications of Magnetic Nanoparticlesin*; Hussain, C.M., Patankar, K.K., Eds.; Woodhead Publishing: Sawston, UK, 2022; pp. 291–316.
88. Zhao, L.Y.; Liu, J.Y.; Ouyang, W.W.; Li, D.Y.; Li, L.; Li, L.Y.; Tang, J.T. Magnetic-mediated hyperthermia for cancer treatment: Research progress and clinical trials. *Chin. Phys. B* **2013**, *22*, 108104. [CrossRef]
89. Peiravi, M.; Eslami, H.; Ansari, M.; Zare-Zardini, H. Magnetic hyperthermia: Potentials and limitations. *J. Indian Chem. Soc.* **2022**, *99*, 100269. [CrossRef]
90. Bañobre-López, M.; Teijeiro, A.; Rivas, J. Magnetic nanoparticle-based hyperthermia for cancer treatment. *Rep. Pr. Oncol. Radiother.* **2013**, *18*, 397–400. [CrossRef] [PubMed]
91. Hergt, R.; Hiergeist, R.; Hilger, I.; Kaiser, W.; Lapatnikov, Y.; Margel, S.; Richter, U. Maghemite nanoparticles with very high AC-losses for application in RF-magnetic hyperthermia. *J. Magn. Magn. Mater.* **2004**, *270*, 345–357. [CrossRef]
92. Kuznetsov, A.A.; Leontiev, V.G.; Brukvin, V.A.; Vorozhtsov, G.N.; Kogan, B.Y.; Shlyakhtin, O.A.; Yunin, A.M.; Tsybin, O.I.; Kuznetsov, O.A. Local radiofrequency-induced hyperthermia using CuNi nanoparticles with therapeutically suitable Curie temperature. *J. Magn. Magn. Mater.* **2007**, *311*, 197–203. [CrossRef]
93. Zarenezhad, E.; Kanaan, M.H.G.; Abdollah, S.S.; Vakil, M.K.; Marzi, M.; Mazarzaei, A.; Ghasemian, A. Metallic Nanoparticles: Their Potential Role in Breast Cancer Immunotherapy via Trained Immunity Provocation. *Biomedicines* **2023**, *11*, 1245. [CrossRef]
94. Hedayatnasab, Z.; Abnisa, F.; Daud, W.M.A.W. Review on magnetic nanoparticles for magnetic nanofluid hyperthermia application. *Mater. Des.* **2017**, *123*, 174–196. [CrossRef]
95. Nori, Z.Z.; Bahadori, M.; Moghadam, M.; Tangestaninejad, S.; Mirkhani, V.; Mohammadpoor-Baltork, I.; Jafari, S.S.; Emamzadeh, R.; Alem, H. Synthesis and characterization of a new gold-coated magnetic nanoparticle decorated with a thiol-containing dendrimer for targeted drug delivery, hyperthermia treatment and enhancement of MRI contrast agent. *J. Drug Deliv. Sci. Technol.* **2023**, *81*, 104216. [CrossRef]
96. Ferreira, I.; Isikawa, M.; Nunes, L.; Micheletto, M.; Guidelli, E. Magnetic nanoparticles covered with polycyclic aromatic hydrocarbons as singlet oxygen carriers for combining photodynamic therapy and magnetic hyperthermia. *J. Photochem. Photobiol. A Chem.* **2023**, *444*, 114902. [CrossRef]
97. Gasser, A.; Ramadan, W.; Getahun, Y.; Garcia, M.; Karim, M.; El-Gendy, A.A. Feasibility of superparamagnetic $NiFe_2O_4$ and GO-$NiFe_2O_4$ nanoparticles for magnetic hyperthermia. *Mater. Sci. Eng. B* **2023**, *297*, 116721. [CrossRef]
98. Hajalilou, A.; Ferreira, L.; Jorge, M.M.; Reis, C.; Cruz, M. Superparamagnetic Ag-Fe_3O_4 composites nanoparticles for magnetic fluid hyperthermia. *J. Magn. Magn. Mater.* **2021**, *537*, 168242. [CrossRef]
99. Tartaj, P.; del Puerto Morales, M.; Veintemillas-Verdaguer, S.; González-Carreño, T.; Serna, C.J. The preparation of magnetic nanoparticles for applications in biomedicine. *J. Phys. D Appl. Phys.* **2003**, *36*, R182–R197. [CrossRef]
100. Andrä, W.; Nowak, H. *Magnetism in Medicine: A Handbook*, 2nd ed.; completely rev. and enl. ed; Wiley-VCH xxiii: Weinheim, Germany, 2007; p. 629.
101. Häfeli, U.; Schütt, W.; Teller, J.; Zborowski, M. *Scientific and Clinical Applications of Magnetic Carriers*; Springer: Berlin/Heidelberg, Germany, 1997.

102. Moroz, P.; Jones, S.K.; Gray, B.N. Magnetically mediated hyperthermia: Current status and future directions. *Int. J. Hyperth.* **2002**, *18*, 267–284. [CrossRef] [PubMed]
103. Nain, S.; Kumar, N.; Chudasama, B.; Avti, P.K. The SLP estimation of the nanoparticle systems using size-dependent magnetic properties for the magnetic hyperthermia therapy. *J. Magn. Magn. Mater.* **2023**, *565*, 170219. [CrossRef]
104. Deb, P.K.; Odetallah, H.M.A.; Al-Jaidi, B.; Akkinepalli, R.R.; Al-Aboudi, A.; Tekade, R.K. Chapter 11-Biomaterials and Nanoparticles for Hyperthermia Therapy. In *Biomaterials and Bionanotechnology*; Tekade, R.K., Ed.; Academic Press: Cambridge, MA, USA, 2019; pp. 375–413.
105. Farzin, L.; Saber, R.; Sadjadi, S.; Mohagheghpour, E.; Sheini, A. Nanomaterials-based hyperthermia: A literature review from concept to applications in chemistry and bi-omedicine. *J. Therm. Biol.* **2022**, *104*, 103201. [CrossRef]
106. Ferk, G.; Stergar, J.; Makovec, D.; Hamler, A.; Jagličić, Z.; Drofenik, M.; Ban, I. Synthesis and characterization of Ni–Cu alloy nanoparticles with a tunable Curie temperature. *J. Alloy. Compd.* **2015**, *648*, 53–58. [CrossRef]
107. Ban, I.; Stergar, J.; Drofenik, M.; Ferk, G.; Makovec, D. Synthesis of chromium-nickel nanoparticles prepared by a microemulsion method and mechanical milling. *Acta Chim. Slov.* **2013**, *60*, 750–755.
108. Houssaini, J.; Bennani, M.N.; Ziyat, H.; Arhzaf, S.; Qabaqous, O.; Amhoud, A. Study of the Catalytic Activity of the Compounds Hydrotalcite Type Treated by Microwave in the Self-Condensation of Acetone. *Int. J. Anal. Chem.* **2021**, *2021*, 1551586. [CrossRef]

applied sciences

MDPI

Article

Towards the Development of a Z-Scheme FeOx/g-C_3N_4 Thin Film and Perspectives for Ciprofloxacin Visible Light-Driven Photocatalytic Degradation

Murilo Fendrich [1], Om Prakash Bajpai [1], Raju Edla [1], Alessandra Molinari [2], Paola Ragonese [3], Chiara Maurizio [3], Michele Orlandi [1] and Antonio Miotello [1,*]

[1] Physics Department, University of Trento, Via Sommarive 14, 38123 Trento, Italy; murilo.fendrich@unitn.it (M.F.); michele.orlandi@unitn.it (M.O.)
[2] Department of Chemical, Pharmaceutical and Agricultural Sciences, University of Ferrara, Via L. Borsari 46, 44121 Ferrara, Italy
[3] Physics and Astronomy Department and CNISM, University of Padova, Via Marzolo 8, 35131 Padova, Italy
* Correspondence: antonio.miotello@unitn.it

Abstract: Thermally synthesized graphitic carbon nitride (g-C_3N_4) over pulsed laser deposition (PLD) produced urchin-like iron oxide (FeOx) thin films were fabricated via in situ and ex situ processes. Materials characterisation revealed the formation of the graphitic allotrope of C_3N_4 and a bandgap E_g for the combined FeOx/g-C_3N_4 of 1.87 and 1.95 eV for each of the different fabrication strategies. The in situ method permitted to develop a novel petal-like morphology, whereas for the ex situ method, a morphological mixture between FeOx bulk and g-C3N4 was observed. Given the improved optical and morphological properties of the in situ film, it was employed as a proof of concept for the direct photocatalysis and photo-Fenton removal of ciprofloxacin antibiotic (CIP) under visible light irradiation. Improved photocatalytic activity (rate constant k = 8.28×10^{-4} min^{-1}) was observed, with further enhancement under photo-Fenton conditions (k = 2.6×10^{-3} min^{-1}), in comparison with FeOx + H_2O_2 (k = 1.6×10^{-3} min^{-1}) and H_2O_2 only (k = 1.3×10^{-4} min^{-1}). These effects demonstrate the in situ methodology as a viable route to obtain working heterojunctions for solar photocatalysis in thin-film materials, rather than the more common powder materials.

Keywords: FeOx/g-C_3N_4 thin films; petal-like morphologies; Z-scheme; visible light photocatalysis; ciprofloxacin degradation

Citation: Fendrich, M.; Bajpai, O.P.; Edla, R.; Molinari, A.; Ragonese, P.; Maurizio, C.; Orlandi, M.; Miotello, A. Towards the Development of a Z-Scheme FeOx/g-C_3N_4 Thin Film and Perspectives for Ciprofloxacin Visible Light-Driven Photocatalytic Degradation. *Appl. Sci.* **2023**, *13*, 10591. https://doi.org/10.3390/app131910591

Academic Editors: Hicham Idriss and Hyungjun Kim

Received: 27 July 2023
Revised: 25 August 2023
Accepted: 19 September 2023
Published: 22 September 2023

1. Introduction

Direct Z-scheme photocatalysis mechanisms, obtained by combining two different photocatalysts (PC), have been proven to be a promising option for enhancing photocatalytic reactions efficiency [1].

In this configuration, inspired by the study of natural photosynthesis, photogenerated electrons in the conduction band (CB) of a PC I can recombine with photogenerated holes in the valence band (VB) of a PC II: this results in a charge-separated state where holes are left on the VB of PC I and electrons on the CB of PC II, separated in space. This inhibits recombination, thus enhancing lifetimes and boosting the yield of redox processes employing them. Z-schemes can be implemented using two semiconductors with opportunely aligned band edges and offers the possibility of using materials with different bandgaps, with a double advantage: (1) materials that are good oxidants but poor reductants can be used as PC I and the reverse can be used for PC II, thus maximizing the activity of their redox potentials; and (2) a broader portion of sunlight spectrum can be exploited.

For these reasons, Z-schemes are increasingly popular in photoactive materials and are increasingly studied for a variety of applications, from water splitting to wastewater remediation [2,3].

Graphitic carbon nitride (g-C_3N_4)-based materials recently attracted great attention due to, among other things, useful optical properties, high chemical stability, and ease of fabrication [4]. g-C_3N_4-based materials could also be used as Fenton-like catalysts, adsorbents, disinfectants, and membrane materials [5–7]. Despite these advantages, g-C_3N_4 also has significant drawbacks, such as poor UV–visible absorption (<450 nm), low surface area, and fast recombination kinetics for photogenerated electron–hole couples, limiting its photocatalytic performance.

Different methodologies for modifying g-C_3N_4 have been investigated in order to improve its performance, for example by doping g-C_3N_4 with different elements: TiO_2 [8–12], Cu [13], SiC [14], Ag [15,16], ZnO [17], graphitized polyacrylonitrile [18], CdO [19], and WO_3 [20]. However, most of these dopings comprise nanomaterials in powdered form and are used for colloidal photocatalysis.

Much less investigated is the possibility of utilizing g-C_3N_4 (either alone or in composites) as thin films over substrate surfaces, for example FTO [21–24]. In water purification applications, the use of colloidal suspensions still represents the main research field due the available high surface areas. However, the recovery of the material, after use, constitutes an additional problem [25]. From an industrial perspective, working with immobilized photocatalysts, such as thin films, allows the facile removal at the end of the process and displays enhanced compatibility with flow reactor installations [26,27]. On the other hand, the application of Z-schemes or other useful heterojunction-based design principles is quite straightforward and well established for nanopowders but much more difficult for thin-films. Indeed, for such a material to be effective in photocatalysis, both the components need to have a solid–liquid interface at which redox reactions with substrates can occur, and this is not possible with the layered composites generally obtained with most thin-film fabrication methods.

In photocatalysis, g-C_3N_4 composites have been recently investigated for the degradation of emerging pharmaceutical contaminants (like sulphonamides, acyclovir) as well as of pesticides [28–33]. Only a few papers deal with the synthesis of FeOx/C_3N_4 composites for oxidation processes, water splitting, electro-catalysis, dye degradation, or artificial photosynthesis [34–36]. Note that both materials are industrially scalable, economically viable, abundant (in the case of iron oxides), and environmentally compatible [37], so that a composite, based on these materials, which incorporates heterojunction effects into a thin-film design would be a breakthrough towards application. Regarding pharmaceutical contaminants specifically, CIP, a fluoroquinolone antibiotic used in medicine and veterinary practice both, is of particular interest: it is one of the most widely used antibiotics worldwide and standard wastewater treatments are largely ineffective for its removal, leading to its frequent detection in surface waters in many regions. It is thus considered a potential source of antimicrobial resistance (AMR), an increasingly severe threat to public health [38]. Note that, while the final target of solar-wastewater treatments is generally the complete removal of pollutants via mineralisation, avoiding AMR in principle only requires the inactivation of antibiotic residues in water instead [39]. To this end, solar-driven photocatalysis would be a potentially cost-effective, high-efficiency treatment process.

In this paper, we investigate the production of Z-scheme composite thin films based on the combination of pulsed laser deposition (PLD)-produced urchin-like iron oxide (FeOx) nanostructured film with g-C_3N_4 to be employed for ciprofloxacin (CIP) degradation. The urchin-like FeOx was recently developed in our lab and has been selected as a platform for the composite synthesis because it has good solar spectrum compatibility, with an energy bandgap E_g = 2.2 eV; consists of hierarchical structures, which are beneficial for applications requiring high surface area [40,41]; and finally, it is photo-Fenton active [42,43]. Two different strategies were used to prepare FeOx/g-C_3N_4 composite films. In the first case, we grew the petal-like g-C_3N_4 over iron urchins by directly drop-casting the melamine solution followed by thermal treatment (in situ). In the second case, previously synthesized g-C_3N_4 powder was spin-coated over iron urchins (ex situ). The photocatalytic performance of the coatings has been evaluated via the degradation of ciprofloxacin (CIP) under visible light. CIP is indeed an important emerging contaminant, but at the same time it is well-studied, its degradation mechanism well reported, and it can easily be followed by simple UV–visible methods; therefore, it can be used to evaluate and tune the performances of our new materials [44–49].

2. Materials and Methods

2.1. Materials and Reagents

Starting elements were metallic iron powder (Ventron GmBh (Karlsruhe, Germany), purity 97%, particle size mesh 325, 56 g/mol), boric acid (H_3BO_3 Sigma-Aldrich (St. Louis, MO, USA) BioXtra, purity > 99.5%, 61.83 g/mol), and melamine (C3H6N6, Sigma-Aldrich, purity > 99%, 126.12 g/mol). Ethanol absolute (VWR, purity 97%, 46.08 g/mol) was used as a dispersing agent for melamine and synthesized g-C_3N_4 powder. Ciprofloxacin ($C_{17}H_{18}FN_3O_3$, Standard PHR1044 Sigma-Aldrich, purity 99%, 385.82 g/mol) was employed to test the efficiency of photocatalysis. H_2O_2 solution (Merck (Rahway, NJ, USA), EMSURE ISO, concentration 30%, 34.01 g/mol) was employed in the photo-Fenton reactions.

2.2. Synthesis of Iron Oxide Urchins

Urchin-like FeOx nanostructured thin films were produced via PLD following the protocol and apparatus published elsewhere [40,50]. Briefly, the films were obtained using a KrF Lambda Physik excimer laser with a 248 nm wavelength, 25 ns pulse duration, and 20 Hz repetition rate. Targets were prepared as pellets of cold-pressed iron and H_3BO_3-mixed powders (1:1 weight). The deposition of films was carried out on both Si slides (for the purpose of characterisation) and glass slides (75 mm × 26 mm) in O_2 atmosphere (4.5 Pa) at room temperature by keeping them at distance of 4.5 cm from the target. During the deposition, a laser fluence of 3 J/cm^2 was applied using optical focusing, and 10,000 laser pulses were used for each single substrate glass slide. The prepared films were thermally annealed at 550 °C for 4 h in air atmosphere with a heating rate of 5 °C/min. Deposited films weight was measured using a balance with 10^{-5} g resolution.

2.3. Synthesis of g-C_3N_4 Powder

Melamine was used as the precursor for g-C_3N_4 synthesis. A total of 10 g of melamine powder was added into a porcelain crucible (200 mL) with a closed cap. Before the synthesis, the porcelain crucible was well polished with an abrasive sheet. In the synthesis process, the crucible containing melamine was heated inside a muffle furnace at 550 °C for 4 h with the heating rate of 9 °C/min [51]. This protocol was used to produce bulk g-C_3N_4 powder for characterisation and as part of the process to produce ex situ film.

2.4. Preparation of FeOx/g-C_3N_4 Composite Thin Films

The FeOx/g-C_3N_4 composite materials were prepared by employing the iron oxide urchin film described in Section 2.2 as a support for the g-C_3N_4, following two distinct procedures as described below and schematically presented in Figure 1:

(a) The g-C_3N_4 powder was synthesized ex situ by thermally heating melamine for 4 h (protocol from Section 2.3). The synthesized g-C_3N_4 was dispersed in ethanol with concentration of 40 mg/mL. The prepared suspension was spin-coated on top of the iron urchins with a rotational speed of 800 rpm for 60 s. The composite material was dried at 60 °C for 4 h to evaporate the extra solvent. The final composite sample is referred to from here onwards as (FeOx/g-C_3N_4)*ex*.

(b) Starting from melamine, g-C_3N_4 was synthesized in situ on the surface of iron oxide urchin. Specifically, the melamine powder was dispersed in ethanol with a concentration of 100 mg/mL. The prepared suspension was used to spread a controlled amount of melamine precursor on top of the iron urchin film via the drop-casting method. Subsequently, films were heated in a muffle furnace at 550 °C for 1 h to obtain the g-C_3N_4 on top of the iron oxide urchins. The prepared composite is referred to from here onwards as (FeOx/g-C_3N_4)*in*.

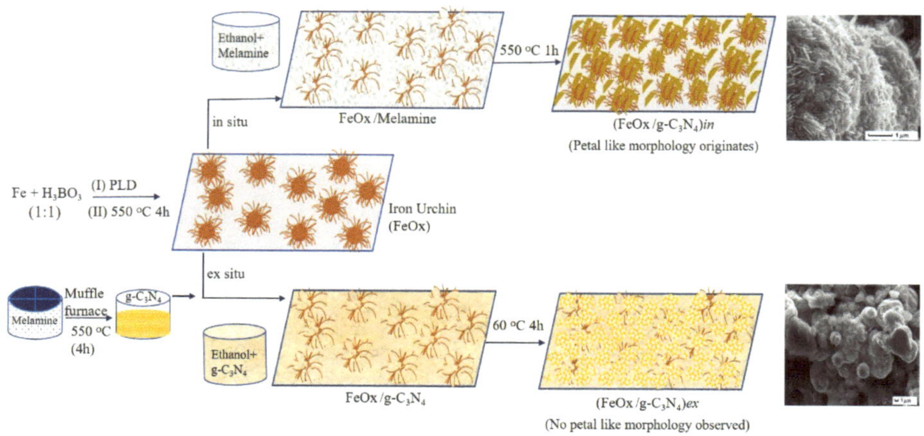

Figure 1. Schematic showing the synthesis of iron urchin (FeOx), g-C_3N_4, and FeOx/g-C_3N_4 composites via different procedures.

Here, we selected 4 h and 1 h durations of annealing for ex situ and in situ methods, respectively, based on the optimisation described in the Supplementary Materials (Section S1).

2.5. Materials Characterisation

The properties of the obtained materials were investigated via different characterisation techniques. A LabRAM Aramis Jobin-Yvon Horiba µ-Raman equipped with a He-Ne laser source (785 nm) was used for Raman spectroscopy measurements. FT-IR was performed using a JASCO model 4600 instrument (Easton, PA, USA). Grazing incidence XRD (GIXRD) was employed to analyse the crystallinity of samples using a X'Pert Pro diffractometer, working with the Cu Kα radiation. A field emission scanning electron microscope (FEG-SEM, JSM-7001F, JEOL (Tokyo, Japan)) equipped with an Energy-Dispersive X-ray Spectroscopy (EDXS-INCA PentaFETx3, Oxford Instruments (Abingdon, UK)) analyser was used to examine morphology and composition, whereas TEM analysis was performed in order to add structural information at a microscopic scale (S/TEM TALOS F200S, ThermoFisher, Waltham, MA, USA). UV and visible (UV–Vis) absorption spectra were measured using a VARIAN Cary 5000 UV-Vis-NIR spectrophotometer over a wavelength range of 200 to 800 nm. The Kubelka–Munk function (Equation (1)) and the Tauc equation were used to evaluate the direct bandgap of the synthesized materials:

$$F(R) = \frac{(1-R)^2}{2R} \quad (1)$$

where R is the reflectance of the sample.

The Tauc equation for direct electronic transitions is described in Equation (2):

$$(\alpha h\nu)^2 = A(h\nu - E_g) \quad (2)$$

where α is the optical absorption coefficient; $h\nu$ is the incident photon energy in eV; A is the material-dependent absorption constant; and Eg is the energy bandgap.

X-ray photoemission spectroscopy (XPS) measurements were performed on the thin film coatings to understand the surface chemical states of the elements. The X-ray source was a monochromatized Al Kα (1486.6 eV) source, and the spectra were collected from an electrostatic hemispherical electron energy analyser with $\pm 8°$ of acceptance angle. The X-ray beam incidence was at an angle of 45° on the surface of the samples and all spectra were acquired in normal emission conditions with an initial base pressure of 2×10^{-9} mbar.

The calibration of the data was performed using the C 1 s native peak present on the surface of the samples.

2.6. Photocatalysis Experiments: Degradation of Ciprofloxacin

In a typical experiment, the photocatalyst film was positioned in the centre of a borosilicate glass beaker and immersed in an aqueous solution (50 mL) containing CIP (C_0 = 10 ppm) on. The beaker was covered with aluminium foil and kept in the dark for 30 min while stirring (150 rpm) to achieve adsorption–desorption equilibrium. After 30 min of initial dark, the foil was removed and the beaker was irradiated for 180 min. When required, H_2O_2 was added to the starting solution to reach a (5×10^{-4} M) concentration. Aliquots (1 mL) of irradiated solution were taken at fixed time intervals and UV–Vis spectra were recorded to evaluate the extent of CIP disappearance. The absorbance values at λ_{max} = 272 nm were collected and the degradation of CIP was reported as:

$$\text{Degradation efficiency } (\%) = \left(\frac{C_o - C_t}{C_o}\right) \times 100 = \left(\frac{A_o - A_t}{A_o}\right) \times 100 \quad (3)$$

where C_o and A_o are the starting molar concentration and absorbance of CIP while C_t and A_t are the molar concentration and absorbance of CIP at time t. Control experiments were carried out by irradiating the starting solution of CIP in the absence of any photocatalyst.

The irradiation was provided by three halogen lamps (Osram (Berlin, Germany) HALOPAR 30, 75W, 30° aperture angle, luminous flux of 350 lumens) positioned in a vertical support at 40 cm height from the reactor, and they were slightly inclined to direct the focus towards the centre of the beaker containing the catalyst and the CIP solution. The spectral irradiance of the light source is reported in Figure S1. Cooling was provided by adding a constant supply of ice around the reactor with a monitored temperature (35–20 °C). UV–Visible absorption spectra were recorded using a VARIAN Cary 5000 UV–Vis–NIR spectrophotometer. When working in photo-Fenton mode, a 5×10^{-4} M of H_2O_2 is employed, which is high enough to ensure a large excess (ca. 20 times with respect to the 2.59×10^{-5} M CIP) and low enough for the absorbance of H_2O_2 not to interfere with the spectrophotometric determination of CIP.

3. Results and Discussion

3.1. g-C_3N_4 Characterisation

Synthesized g-C_3N_4 powder and melamine precursor were characterized in terms of structure and morphology (Supplementary Materials, Figures S2–S7). A UV–Vis spectrum was acquired to characterize the melamine precursor before synthesis to g-C_3N_4, and the spectrum shows significant absorbance only in the UV region (<275 nm), as seen in Figure S2. After the melamine synthesis, a powder-like morphology with agglomerated particles under 100 nm in size was observed in the SEM analysis, and the presence of only C and N elements was confirmed from corresponding EDXS spectra (Figure S3). In the FTIR (Figure S4), the peaks that appear in the 1200–1700 cm^{-1} region are related to the stretching modes of C=N and C–N hetero-cycles, and the peaks at 875 and 804 cm^{-1} indicate the presence of the stretching mode of the triazine ring [51]. Micro-Raman analysis proves the presence of g-C_3N_4 from the peaks at 211, 359, 471, 588, 706, 768, 984, 1117, 1151, 1232, and 1312 cm^{-1}, respectively (Figure S5 and Table S1). Also, the peaks 1360 cm^{-1} (D band) and 1580 cm^{-1} (G band) confirm the formation of C–N [52,53]. XRD analysis further confirmed the presence of the graphitic-like phase of C_3N_4 lattice system given by the (002) peak (Figure S6). An increase in UV absorbance was observed in the 400–500 nm range with a maximum at 400 nm, and the Tauc plot demonstrates that the direct bandgap of the synthesized g-C_3N_4 powder is 2.92 eV (Figure S7).

3.2. FeOx/g-C_3N_4 Composite Characterisation

Figure 2 presents the FTIR spectra comparison between the FeOx iron urchins film only, (FeOx/g-C_3N_4)*in* and (FeOx/g-C_3N_4)*ex* composites. Both the composite materials present the C-N and C=N vibration states in the wavenumbers range between 1200 and 1700 cm^{-1}. A triazine peak is also observed at 804 cm^{-1}. Sharper peak formation in the mentioned range is more evident in the ($FeOx/g\text{-}C_3N_4$)*ex* composite in comparison to ($FeOx/g\text{-}C_3N_4$)*in*. This can possibly be attributed to the higher mass of $g\text{-}C_3N_4$ on top of the FeOx supporting material. The corresponding FeOx spectrum involves the unresolved peaks below 1000 cm^{-1}, which are related to Fe-O stretching vibrations [54]. In Figure 2, a cluster of weak bands found at around 1500 cm^{-1} correspond to the vibration modes of carbonates resulting from interaction of the surface with CO_2 from the air. This is commonly observed in iron oxide surfaces exposed to air humidity, which leads to the formation of hydroxides that are, in turn, susceptible to carbonate formation [55].

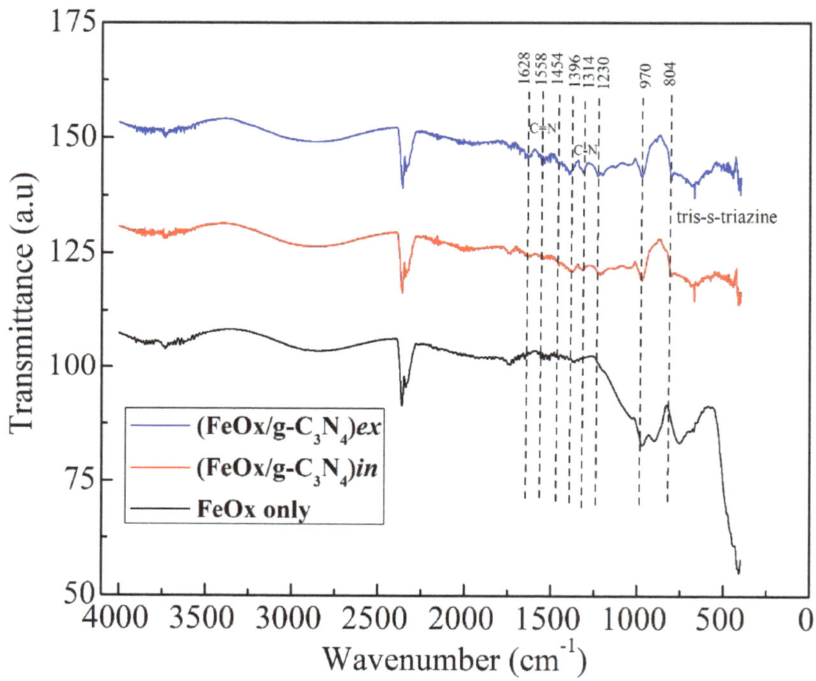

Figure 2. FTIR spectra comparisons between FeOx and the composites $FeOx/g\text{-}C_3N_4$.

Micro-Raman analysis was also employed to investigate the properties of the ($FeOx/g\text{-}C_3N_4$)*in* and ($FeOx/g\text{-}C_3N_4$)*ex* as well as the reference FeOx support material. The Raman spectra (Figure 3) show the peaks of $g\text{-}C_3N_4$ identified at 211, 359, 471, 588, 706, 768, 984, 1117, 1151, 1232, and 1312 cm^{-1}, respectively. For the ($FeOx/g\text{-}C_3N_4$)*ex*, spectrum shows essentially the $g\text{-}C_3N_4$ bulk nature due to the complete coverage of the FeOx thin film. The ($FeOx/g\text{-}C_3N_4$)*in* spectrum presents the $g\text{-}C_3N_4$ with the characteristic peaks shifted with respect to the bulk powder $g\text{-}C_3N_4$, such as the 521, 749, 810, 999, 1156, 1266, and 1346 cm^{-1} and small traces of FeOx at 298 and 1386 cm^{-1} being the other peaks in overlap with $g\text{-}C_3N_4$. The shifting on the $FeO_x/g\text{-}C_3N_4$ (in situ) sample is attributed to the atomic lattice compressive stress because of synthesis procedure. In particular, in situ synthesis induces stress modifications on the FeOx because of the melamine thermal conversion to C3N4 on its surface that indeed favours strong adhesion with the substrate. On the other hand, in the ex situ preparation condition, the interaction between the already prepared g-C3N4 powder and the FeOx surface is milder, as proved by the poor adhesion [56]. Peaks assignments are presented in Supplementary Table S1.

(a)

(b)

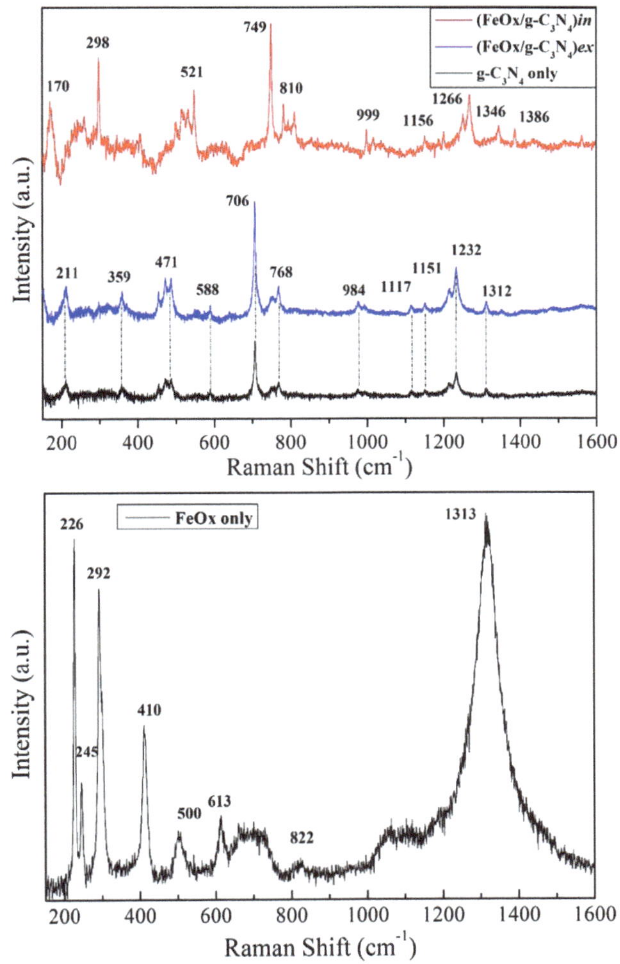

Figure 3. Raman spectra of g-C_3N_4 compared to composites (FeOx/g-C_3N_4)*in* and (FeOx/g-C_3N_4)*ex* (**a**) and FeOx supporting material (**b**).

Figure 4 presents the GIXRD patterns for both (FeOx/g-C_3N_4)*in* and (FeOx/g-C_3N_4)*ex*. (FeOx/g-C_3N_4)*ex* clearly presents the peaks at 13.1° and 27.5° corresponding to the (100) and (002) crystal planes of the g-C_3N_4 (JCPDS No. 87-1526). The diffraction peaks typical of hematite are labelled (012), (104), (113), (024), (116), and (122) (JCPDS No. 33-0664), and the observed traces of magnetite phase are labelled (111), (220), (311), (222), and (400) (JCPDS No. 19-0629). According to the growth mechanism of the urchins during the oxidizing annealing, the most oxidized iron phase (Fe_2O_3) is expected to most likely be present on the surface of the nanostructured film, in contact with the solution and with the g-C_3N_4.

On the other hand, in the (FeOx/g-C_3N_4)*in* there is clear evidence of hematite and traces of magnetite. However, in comparison to (FeOx/g-C_3N_4)*ex*, the (002) peak of g-C_3N_4 is observed slightly shifted toward lower diffraction angle with considerably reduced intensity and also the (100) peak is apparently supressed. This suggests the intercalation of the layered structure of g-C_3N_4 by the abundant urchin-like structures (size ~10–20 nm) present on the surface of FeOx contributing to diminish the peak [57,58]. According to the Bragg equation ($\lambda = 2d\sin\theta$), the expanded interlayer distance of the d-spacing due to the intercalation of iron urchins into g-C_3N_4 layers may lead to the shift of the diffraction peak toward low angle. Moreover, the reduced intensity of the peak shows the possibility of less dense lamellar structure of the g-C_3N_4 in the composite due to the presence of small amounts of g-C_3N_4 (in the form of thin petals) on the FeOx surface [59,60]. These structural changes likely imply that in the in situ method of the formation of composite films, the g-C_3N_4 interacts and reconfigures the surface of FeOx differently than ex situ method.

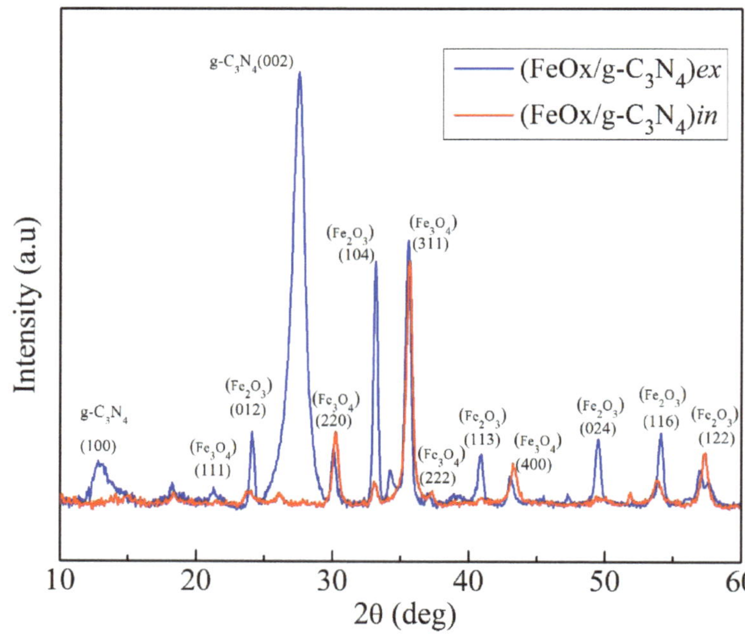

Figure 4. GIXRD patterns for the films (FeOx/g-C_3N_4)*in* and (FeOx/g-C_3N_4)*ex* recorded at incidence angle = 0.5 deg.

The absorption spectra and Tauc plot of the FeOx/g-C_3N_4 composites are presented in Figure 5. An absorption region can be observed from the visible range (<600 nm) while a constant increase up to the UV range of the spectrum is detected for both composite materials. Interestingly, the (FeOx/g-C_3N_4)*in* composite film shows lower bandgap (~1.87 eV) than the (FeOx/g-C_3N_4)*ex* composite (~1.95 eV). Also, the observed bandgaps (1.87 and 1.95 eV) are in agreement with the literature [61]. For both synthesis methods, due to the presence of FeOx as the supporting material, the absorbance is shifted towards the visible part of the spectrum in comparison to the bulk g-C_3N_4 presented in Figure S7.

Morphology and composition of the synthesized FeOx iron urchins and of the composites (FeOx/g-C_3N_4)*in* and (FeOx/g-C_3N_4)*ex* FeOx were investigated using SEM and EDXS techniques. Figure 6 presents the images of the surface of the FeOx urchin-like structure produced by PLD and annealed at 550 °C for 4 h in a tubular furnace used as the support material. EDXS results essentially confirm the presence of Fe and O on the microstructures (Figure S8). This agrees with previously published reports on this material [40,42]. The mass of the FeOx films was 21 ± 2 mg before the composite preparation steps.

SEM images in Figure 7 show the morphology of the (FeOx/g-C_3N_4)*in* composite. The formation of a petal-like microstructure (Figure 7a,b) is now observed as compared to the bare FeOx urchin. From Figure 7c, it is also observable that urchins present on the FeOx surface possibly intercalate the petal structure. A special formation of petal-like morphology growth is highlighted in (Figure 7d) as a single petal close-up (magnified image in the inset). Moreover, EDXS (Figure S9) confirms the presence of the C, N, Fe, and O elements in the composite film. The mass of the composite is about 10 mg more than the 21 ± 2 mg of the initial FeOx. This corresponds to 5–10% conversion yield of melamine.

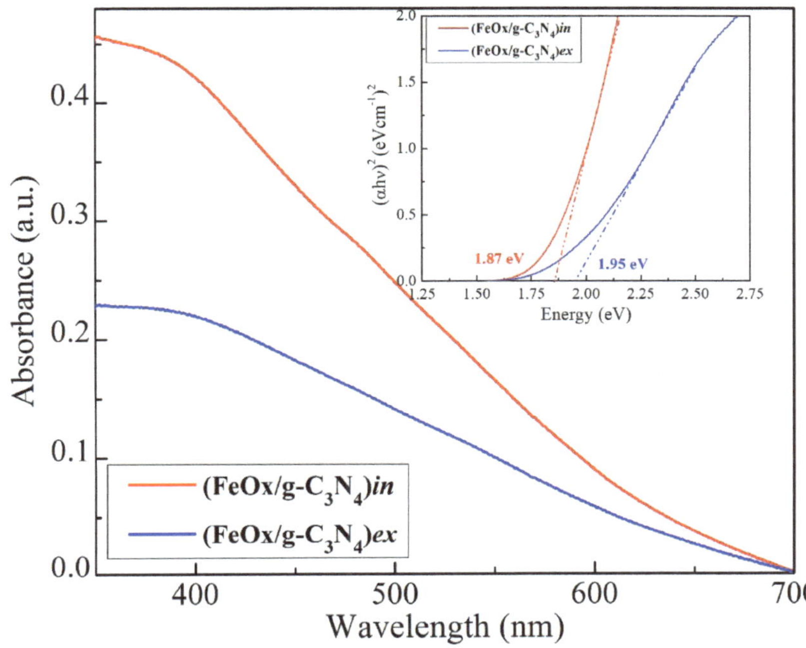

Figure 5. UV–Vis absorption spectra and corresponding Tauc plot (inset) of the synthesized (FeOx/g-C_3N_4)*in* and (FeOx/g-C_3N_4)*ex* composites materials.

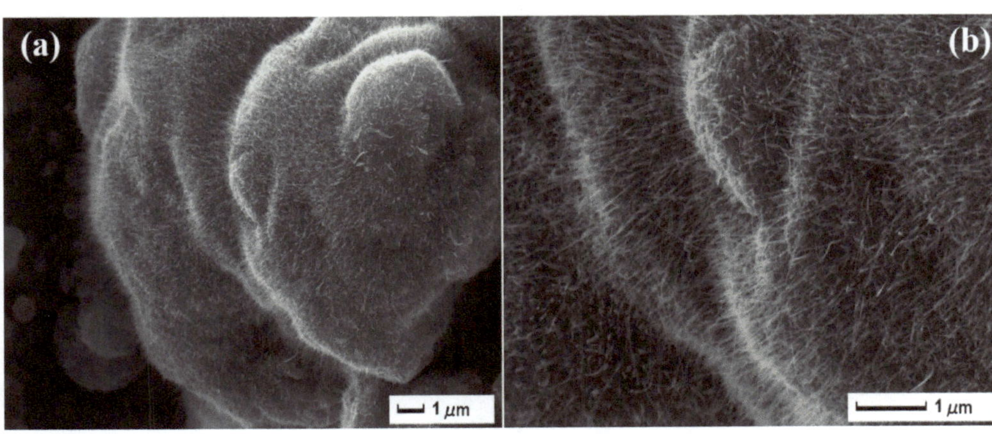

Figure 6. SEM images (**a**,**b**) showing the bare surface of the FeOx urchin-like structures employed in the composite with g-C_3N_4.

Morphology of the synthesized composite (FeOx/g-C_3N_4)*ex* are presented in Figure 8a–c. The visual inspection of the images demonstrates the absence of petal-like morphology and island separation (Figure 8d) between the FeOx particles and g-C_3N_4 deposited on the surface of the iron urchins. In this case, the mass increase was about 40 mg (about 60 mg total composite mass), due to the drying of 1mL solution after the spin-coating process. EDXS (Figure S10) confirms the distinct separation of particles in the composite materials.

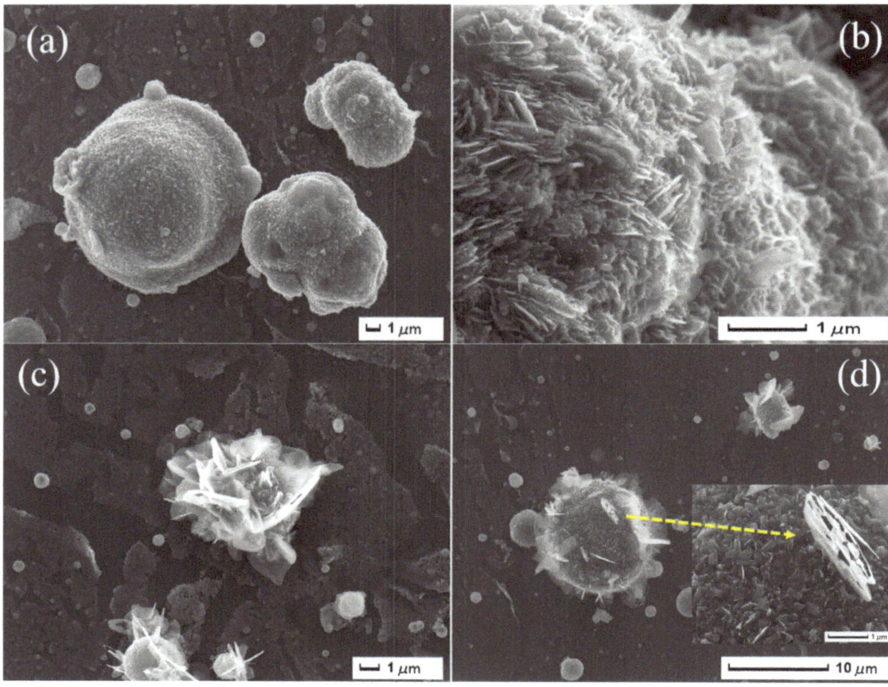

Figure 7. SEM images (**a–d**) of $(FeOx/g\text{-}C_3N_4)$*in* composite thin films.

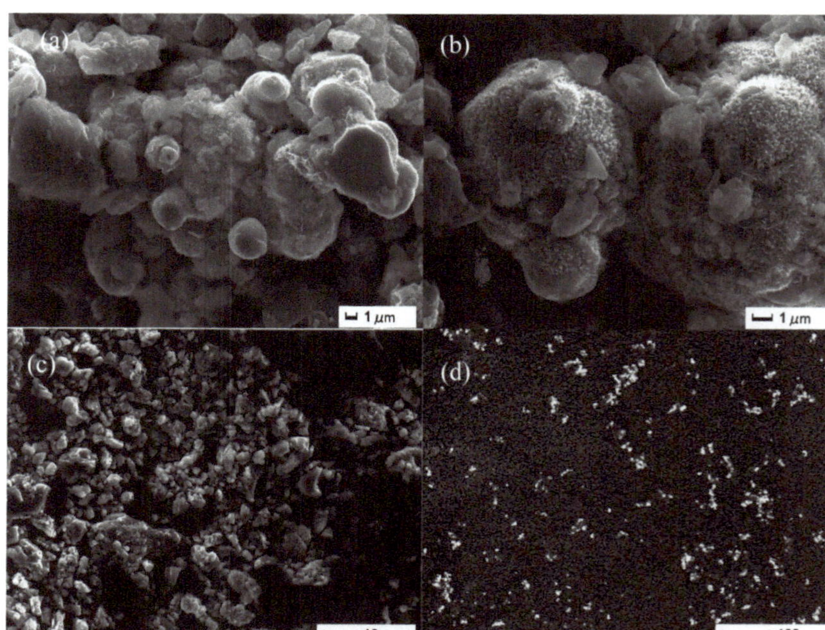

Figure 8. SEM images (**a–d**) of $(FeOx/g\text{-}C_3N_4)$*ex* composite thin films.

The morphology, composition, and crystalline structures of the composite $(FeOx/g\text{-}C_3N_4)$*in* were also investigated using HRTEM. Figure 9a–c presents the image of a single petal partially covered by $g\text{-}C_3N_4$ (darker areas). The magnification of Figure 9a shows lattice fringes 0.252 and 0.295 nm and the white arrowheads define intervals corresponding to 10 lattice planes. The SAED spectra of clean single crystal return hematite or magnetite spot diffraction patterns (Figure S11), whereas larger area SAED spectra containing crystalline structure and darker regions show the presence of both Fe_2O_3 and Fe_3O_4 as well as the (002) peak of g-C_3N_4, which supports the XRD results (SAED patterns were identified with the identical XRD JCPDS card database).

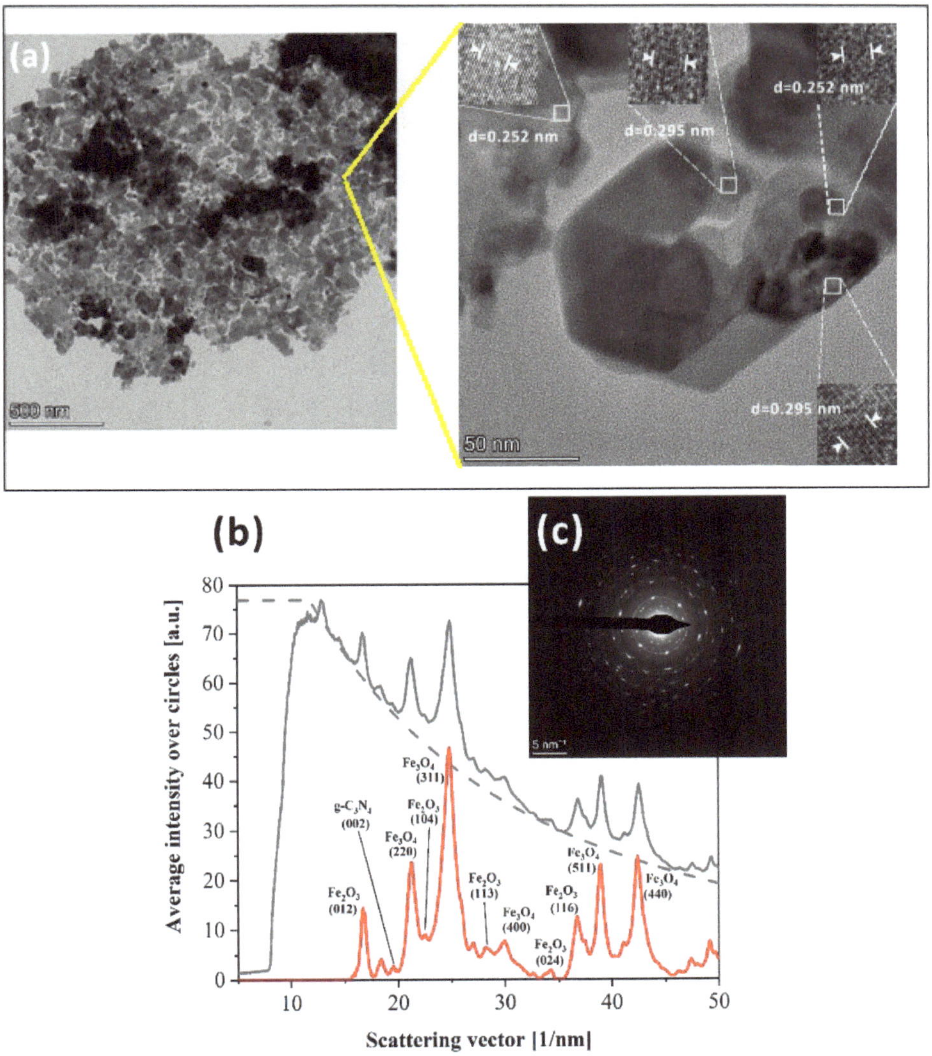

Figure 9. (FeOx/g-C_3N_4)*in* TEM and HRTEM analysis (**a**). Bright field TEM image of a crystalline petal whose SAED diffraction pattern is given in (**c**). The intensity profile obtained from (**c**) is shown in (**b**): gray lines indicate gross intensity (solid line) and the subtracted background (dashed line), and the red line is the resulting net intensity.

3.3. XPS Analysis

XPS was used to analyse the chemical bonding and composition of all the prepared materials. Figure 10 presents the full survey scan of the FeOx iron urchins, (FeOx/g-C_3N_4)*in*, and (FeOx/g-C_3N_4)*ex* composites that are compared to the FeOx as a support material only. The XPS survey spectra reveals presence of Fe, O, N, and C.

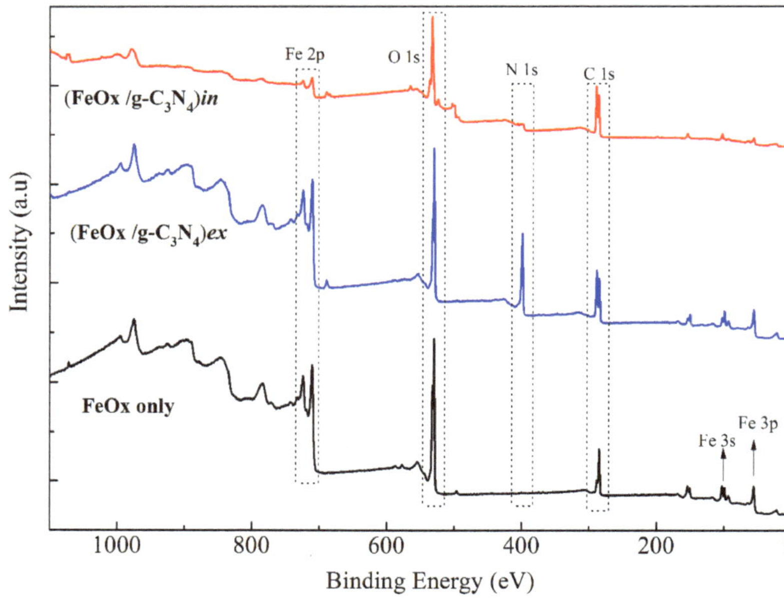

Figure 10. XPS survey scan for all elements in the FeOx urchin and its composites with g-C_3N_4.

Figure 11 shows Fe $2p_{3/2}$ core level spectra of FeOx and its composites, the Fe 2p survey can be found in the Supplementary Materials (Figure S12a). The Fe $2p_{3/2}$ core level main peak position (without curve fitting) is located at around 711 eV binding energy (BE) with spin-orbit splitting and a satellite at 719 eV. The peak at 711 eV was further deconvoluted into three peaks, namely 709.8 eV, 710.8 eV, and 712 eV, which correspond to +3 oxidation state of alpha hematite [62–64]. This clearly indicates the presence of dominating alpha Fe_2O_3 phase (hematite) with the +3 oxidation state on the surface of FeOx and (FeOx/g-C_3N_4)*ex* samples. For the XPS spectra of (FeOx/g-C_3N_4)*in*, the main BE positions at 711.5 eV with spin-orbit splitting indicate the presence of a Fe_3O_4 phase. Therefore, the presence of both hematite and Fe_3O_4 cannot be excluded in the samples, thus further supporting our previously reported (Figure 4) XRD results. In Figure 12a, the C1s spectrum of the composite materials contains two peaks at 285 eV and 288.9 eV for (FeOx/g-C_3N_4)*in*, which arise from C–N and sp^2-hybridized C in N=C–N. However, for (FeOx/g-C_3N_4)*ex*, the peaks originate at 285.1 eV and 289.3 eV [65,66]. Thus, the C 1s core level spectra of both composite films, i.e., (FeOx/g-C_3N_4)*ex* and (FeOx/g-C_3N_4)*in*, clearly show the presence of the C–N bonding structure coming from g-C_3N_4 with specific contributions as shown in adjacent peaks resulting from the fitted components. A slight modification of peak positions related to g-C_3N_4 was observed in case of the (FeOx/g-C_3N_4)*in* composite film, which could be attributed to interface effects with FeOx [65–70].

The N 1s spectrum (Figure 12b) related to (FeOx/g-C_3N_4)*in* is deconvoluted into three different peaks at 399.8, 401.5, and 402.1 eV, which are attributed, respectively, to sp^2-hybridized N in C=N–C, sp^3-hybridized N in N–C_3, and C-N-H from amino groups and π excitations in heterocycles. The N 1s core level spectra also clearly indicate the respective bonding contributions (reported with fitted components) originating from the g-C_3N_4 chemical structure. It is also interesting to see that the fitted components with respective chemical structures show modifications in the intensity as well. This is not fully understood yet but probably results from minor changes in the atomic lattice arrangement. Also, the O 1s spectra reported in the Figure S12b, identifies the presence of Fe-O chemical bonding as in Fe oxides with two peaks at 530 eV and 532.6 eV arising from the lattice oxygen in FeOx and the related hydroxide-oxyhydroxide system [71,72]. The small peak around 532 eV is associated with the presence of adsorbed water on the surface of the samples [73].

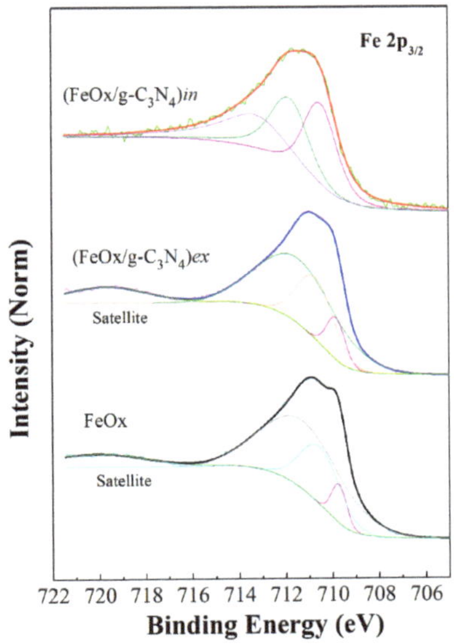

Figure 11. XPS spectra of the Fe $2p_{3/2}$ core level and deconvolution peaks of the iron urchin and its composites with g-C_3N_4. All the spectra were normalized to the maximum of the peak and curve fitting was achieved using Shirley background and Voigt profiles.

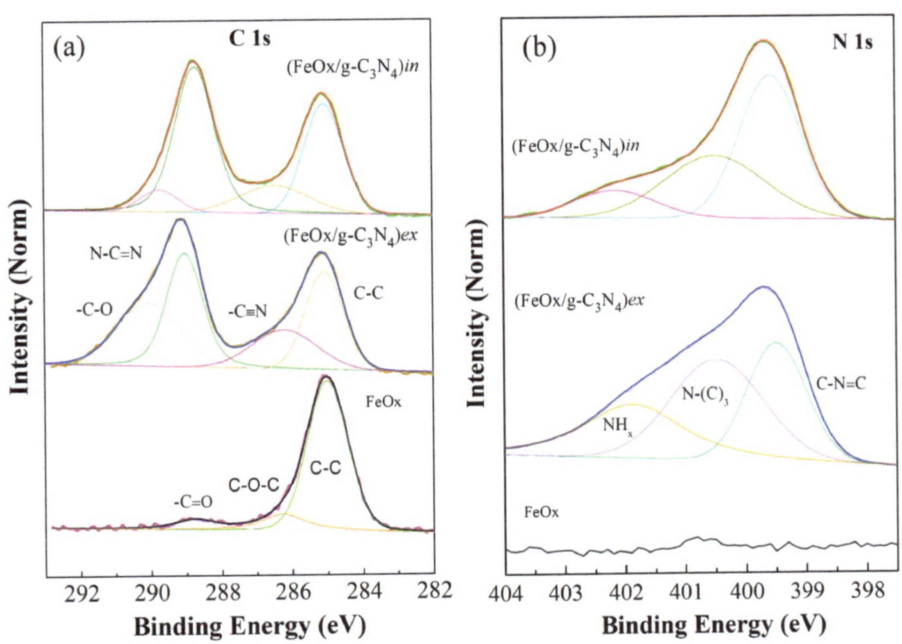

Figure 12. XPS spectra of (**a**) C 1s and (**b**) N 1s core levels and the corresponding deconvolution peaks of the iron urchin and its composites with g-C_3N_4. All the spectra were normalized to the maximum of the peak and curve fitting was achieved using Shirley background and Voigt profiles.

Moreover, looking at the lower bandgap (inset in Figure 5) of the ($FeOx/g$-C_3N_4)*in* thin film with respect to ($FeOx/g$-C_3N_4)*ex*, we can suggest that the detected minor differences (1.87 vs. 1.95 eV) could also be related to slight changes in atomic lattice arrangement in the in situ-produced g-C_3N_4 and FeOx likely promoting charge carrier separation and mobility.

3.4. Photocatalytic Activity

The photocatalytic activity of the prepared materials was evaluated at the proof-of-concept level in CIP degradation. First the g-C_3N_4 dispersed powder and coating samples were tested. The adsorption of powders and the poor photocatalytic degradation of CIP in films and powders was observed (Figure S13). Also, the ($FeOx/g$-C_3N_4)*ex* thin films show only 5% CIP disappearance after 180 min irradiation (Figure S14). The observed inferior CIP removal of ($FeOx/g$-C_3N_4)*ex* was due to the weak g-C_3N_4 adhesion with consequent partial detachment during the experiment, as pointed out by SEM. Therefore, we have not performed further in-depth analysis of these samples.

The photocatalytic activity of the ($FeOx/g$-C_3N_4)*in* thin films were compared in the photocatalytic removal of CIP through visible irradiation. Results are reported in Figure 13, and the corresponding UV–Vis spectra obtained are reported in Supplementary Figure S16. Low adsorption (less than 5%) is observed in the dark during the equilibrium period. Upon visible illumination, the system ($FeOx/g$-C_3N_4)*in* shows a better performance and leads to a 14% of CIP disappearance after 180 min of irradiation with a kinetic constant of 8.28×10^{-4} min^{-1}. The results obtained via ($FeOx/g$-C_3N_4)*in* are of relevance if one considers the control experiments reported in Figure S15. It is observed that both thermolysis at 50 °C and visible irradiation in the absence of a photocatalyst (photolysis) do not cause any decrease in CIP concentration. This demonstrates that the presence of the composite ($FeOx/g$-C_3N_4)*in* is mandatory for achieving CIP degradation and that degradation is not a consequence of the temperature increase from the heat developed by the radiation system or photolysis.

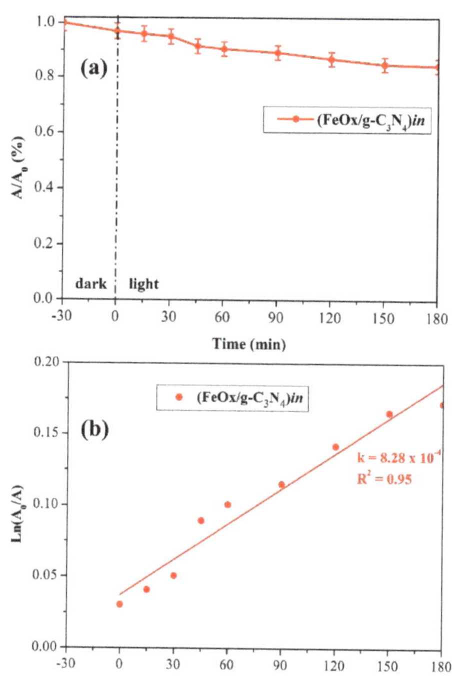

Figure 13. (**a**) removal yields and (**b**) corresponding reaction kinetics of the CIP upon irradiation of ($FeOx/g$-C_3N_4)*in* immersed in aqueous solutions containing CIP (C_0 = 10 ppm).

Given the results obtained from (FeOx/g-C_3N_4)*in*, the photo-Fenton reaction was tested with the addition of H_2O_2. Results are reported in Figure 14, where experiments performed with FeOx alone and in the presence of H_2O_2 5×10^{-4} M are reported for comparison. The irradiation of FeOx alone (black) caused a disappearance of CIP of about 15%. When H_2O_2 (5×10^{-4} M) is present, the material showed a higher CIP disappearance in the dark (17%), possibly due to an initial Fenton reaction, and upon irradiation a disappearance of 35% of CIP is observed with moderate improvements of the degradation after the initial 60 min of illumination on both photo-Fenton cases (FeOx + H_2O_2 and (FeOx/g-C_3N_4)*in* + H_2O_2). This moderate enhancement of degradation is attributed to slight temperature changes and is consistent with the literature [74]. The calculated kinetic constant is $k = 1.6 \times 10^{-3}$ min^{-1}. A comparison of the results of FeOx (Figure 14) with (FeOx/g-C_3N_4)*in* (Figure 13) shows that their photocatalytic activities are very similar. However, the addition of H_2O_2 (5×10^{-4} M) to the aqueous solution of CIP in which (FeOx/g-C_3N_4)*in* is present had a booster effect: in fact, 60% of CIP disappearance is achieved after irradiation, a value significantly higher than that observed with FeOx and (5×10^{-4} M) H_2O_2 alone (6%).

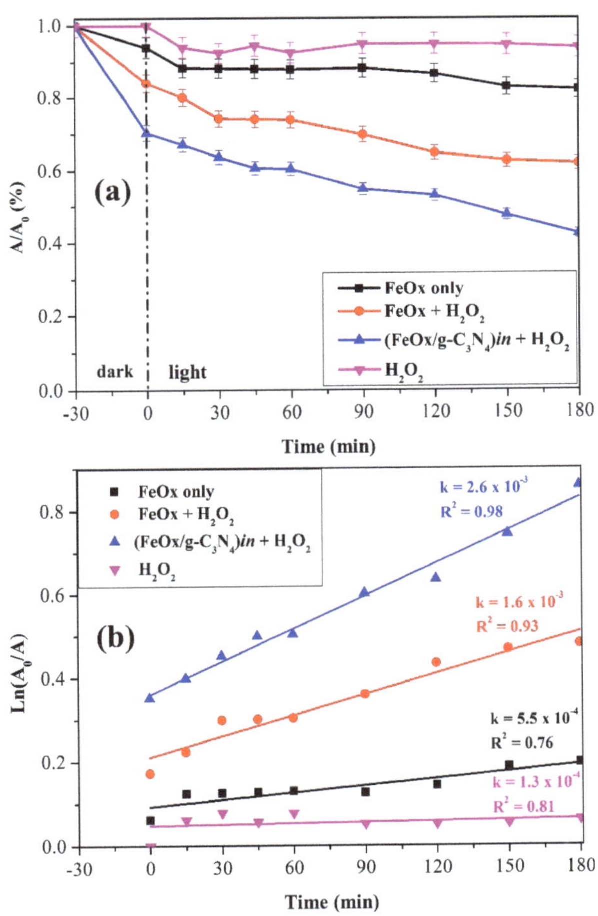

Figure 14. Removal yields (**a**) and reaction kinetics (**b**) of the CIP disappearance upon the irradiation of FeOx (black), FeOx + H_2O_2 (5×10^{-4} M) (red), and FeOx/g-C_3N_4 composites + H_2O_2 (5×10^{-4} M) (blue). The starting concentration of CIP is 10 ppm.

Therefore, the photo-Fenton process further enhances the degradation efficiency of the (FeOx/g-C_3N_4)*in* composite material compared with the two parts alone. In addition, good adhesion over the glass slides was observed both during photocatalysis and the photo-Fenton reaction.

An overview of the results obtained during the photocatalytic and photo-Fenton experiments are provided in Table 1.

Table 1. Summary of the testing conditions and results for the photocatalysts.

Experimental Condition	Catalyst Form	CIP Adsorption	CIP Removal Yield	Kinetic Constant (k, min^{-1})
H_2O_2 (5×10^{-4} M)	No catalyst	0%	6%	1.3×10^{-4}
(FeOx/g-C_3N_4)*in*		5%	13%	8.28×10^{-4}
FeOx		5%	15%	5.5×10^{-4}
FeOx + H_2O_2 (5×10^{-4} M)		17%	35%	1.6×10^{-3}
(FeOx/g-C_3N_4)*in* + H_2O_2 (5×10^{-4} M)		30%	60%	2.6×10^{-3}

It has been observed that the (FeOx/g-C_3N_4)*in* composite material exhibits very important photoactivity during CIP removal under photo-Fenton conditions, showing an increase of one order of magnitude of the relative kinetic constant. This activity is higher than that of each counterpart alone. This picture is compatible with the arising of a heterojunction effect, such as a complete (where the solid–liquid interfaces of both materials are redox active) or partial (where only one, i.e., the FeOx conduction band is redox active) Z-scheme in the composite. Another possible explanation for the enhanced performance in the composite is that g-C_3N_4 is simply acting as an electron acceptor, thus temporarily trapping electrons from the FeOx conduction band and inhibiting charge recombination [75]. We consider this last scenario unlikely, as the bandgap of g-C_3N_4 is also compatible with visible absorption.

A likely photocatalysis mechanism for the (FeOx/g-C_3N_4)*in* composite, presented in Figure 15, can be hypothesized based on our data and the current literature. It is known that the degradation of CIP in the presence of iron oxides and g-C_3N_4 occurs mainly through the formation of active species, i.e., superoxide radicals (O_2^-), hydroxyl radical ($\cdot$OH), and holes (h+) [30,48,76–80]. In the case of (FeOx/g-C_3N_4)*in*, the degradation of CIP is then expected to proceed via two different pathways depending on whether H_2O_2 is present or not. Without H_2O_2 (Figure 15a), the CIP degradation probably proceeds through the formation of a Z-scheme between FeOx and g-C_3N_4 where primarily the (O_2^-) and ($\cdot$OH) active species are produced. Note that in this case, $\cdot$OH can only be produced through the reaction between h+ and water [78–82] and that the superoxide radicals are not very efficient in reacting with CIP [76,83]. When H_2O_2 is added, $\cdot$OH can also be produced through the reactions between CB electrons and H_2O_2 and between superoxide and H_2O_2, (Figure 15b) [76,83–87], resulting in a greater availability of highly active $\cdot$OH, which would explain the higher activity in the photo-Fenton mode. A detailed investigation of the composite materials working mechanism via transient UV–visible spectroscopy and using selective radical inhibitors will be undertaken in future investigations to confirm this mechanism hypothesis. Nevertheless, we consider the results reported here as a proof-of-concept that the in situ approach is a viable route for achieving heterojunction effects on thin-film photocatalysts for application in pollutant degradation. This is proven by data in Table 2 where a comparison is reported between the best results we obtained with the (FeOx/g-C_3N_4)*in* +H_2O_2 (5×10^{-4} M) samples and literature data (90−97) for the removal of ciprofloxacin in water and additional information can also be found from literature [88–91].

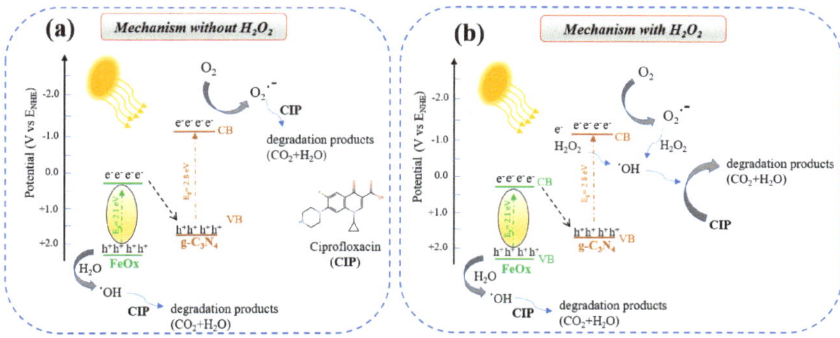

Figure 15. Schematic showing the (FeOx/g-C_3N_4)*in* photocatalyst degradation mechanism (**a**) without H_2O_2 and (**b**) with H_2O_2.

Table 2. Comparison for CIP removal with photocatalysts.

Ref	Catalyst	CIP Concentration (ppm)	Tested Volume (mL)	Experimental Time (min)	Illumination Source	Removal Yield
[92]	N,S-CQDs/Bi_2MoO_6@TiO_2	20	15	270	Visible light ca. 200 mW/cm^2	90%
[93]	FTO/$BaTiO_3$/SnO_2	5	75	180	100 W Xe lamp + Ultrasound 24 KHz	60%
[94]	Bi_2MoO_6@Bi_2MoO_{6-x}	20	100	150	6.2 W LEDs between 400 and 700 nm	95%
[95]	TiO_2 MOCVD	20	100	480	UV LED 365 nm	100%
[96]	(nc-TiO_2) from Degussa P25	10	100	600	UV lamp 276 nm	100%
[97]	TiO_2 films grown electrochemically	40	190	180	UV lamp (100–280 nm)	70%
[98]	Ag_2MoO_4	20	100	40	UV lamp (200–400 nm)	90%
[99]	CeO_2/CuO/ZnO/Zn	10	100	120	Mercury lamp 120 W—UV light	70%
This work	(FeOx/g-C_3N_4)*in* + H_2O_2 (5×10^{-4} M)	10	50	180	Visible Halogen lamps, 225 W	60%

4. Conclusions

The development of composites produced by combining iron oxide urchins (FeOx) with g-C_3N_4 following two different procedures (FeOx/g-C_3N_4)*in* and (FeOx/g-C_3N_4)*ex* was reported along with its application in the photocatalytic degradation of ciprofloxacin pharmaceuticals.

The PLD procedure to obtain annealed hematite iron urchins followed by the drop-casting of the melamine precursor to thermally condense the g-C_3N_4 (FeOx/g-C_3N_4)*in* composite thin film evolved into to a petal-like morphology with a high surface area featuring good absorption in the visible range (<600 nm). This composite material exhibits an improved photocatalytic performance when compared to the single components, with the optimal condition for ciprofloxacin degradation (from 20 to nearly 60%) observed under the photo-Fenton condition. These results are compatible with a heterojunction effect in the composite, likely of the Z-scheme type. The advantages with (FeOx/g-C_3N_4)*in* rely on the mechanical stability of the coating on the glass surface, which allows us to overcome the problems related to the use of nanoparticles in suspension. The research presented in this paper will be further developed as we consider the reported results as a significant starting point for obtaining a Z-scheme strategy specifically for thin-films. Indeed, further activities aimed at elucidating the working mechanism of these composites under light are ongoing.

Supplementary Materials: The following supporting information can be downloaded at: https://www.mdpi.com/article/10.3390/app131910591/s1, Section S1: Different aspects regarding ex situ and in situ materials synthesis methods, Section S2: light source for photocatalysis experiments, Section S3: characterisation of g-C_3N_4 bulk and Section S4: FeOx composite films, Section S5: supporting information for photocatalytic tests, Section S6: Raman peak assignments , and Section S7: UV–Vis spectra from the photocatalysis experiments. Figure S1: Measured and normalized spectrum of visible light lamps (75 W each) employed in this work (Osram HALOPAR 30, 75 W, 30° aperture angle, luminous flux of 350 lumens); Figure S2: Absorption spectrum of melamine used as precursor to g-C_3N_4 synthesis; Figure S3: SEM image of g-C_3N_4 powder synthesized at 550 °C 4h and its corresponding EDXS spectra; Figure S4: FTIR spectra of the melamine and g-C_3N_4 synthesized at 550 °C 4 h; Figure S5: Raman spectra of melamine and g-C_3N_4 synthesized at 550 °C 4 h; Figure S6: GIXRD pattern of the synthesized g-C_3N_4 powder at 550 °C; Figure S7: UV–Vis absorption spectra and Tauc plot of the g-C_3N_4 synthesized at 550 °C 4h; Figure S8: EDXS spectrum of the surface of the FeOx urchin like structures used as support in the composite with g-C_3N_4; Figure S9: EDXS spectra from 4 different regions of the (FeOx/g-C_3N_4)*in* composite thin films; Figure S10: EDXS spectra from two different regions of the (FeOx/g-C_3N_4)*ex* composite thin films; Figure S11 (FeOx/g-C_3N_4)*in* TEM image and corresponding larger area SAED pattern; Figure S12 XPS spectra of (a) Fe 2p (top left) and specifically Fe $2p_{3/2}$ (top right) and (b) O1s core levels without curve fitting of the iron urchin and its composites with g-C_3N_4; Figure S13: Removal yields of the CIP disappearance upon irradiation of g-C_3N_4 4h in powder (black) and in the film (blue) in aqueous solutions containing CIP (C_0 = 10 ppm); Figure S14: Removal yields of the CIP disappearance upon irradiation of (FeOx/g-C_3N_4)*ex* thin films immersed in aqueous solutions containing CIP (C_0 = 10 ppm); Figure S15: Control experiments: aerated aqueous solution of CIP (10 ppm) maintained at 50 °C in the dark (red) or visible irradiated (blue) in the absence of any photocatalyst; Table S1: The modes of every peak identified in the Raman measurements. The peaks identifications are based on the literature [100–103]; Figure S16: UV–Vis spectra from obtained from the photocatalysis and photo-Fenton experiments.

Author Contributions: Conceptualisation: M.F., O.P.B. and M.O.; sample preparation and analysis: M.F., O.P.B., R.E., A.M. (Alessandra Molinari), P.R. and C.M.; original draft preparation: M.F. and O.P.B.; writing—Review and editing: M.O., A.M. (Antonio Miotello), A.M. (Alessandra Molinari), R.E. and C.M.; supervision: M.O. and A.M. (Antonio Miotello); funding acquisition: A.M. (Antonio Miotello) and M.O. All authors have read and agreed to the published version of the manuscript.

Funding: This work was supported by the ERICSOL project, funded by the University of Trento, and by the FarSol project, funded by the Caritro Foundation (Research & Development 2020 Grant). We also acknowledge project "Produrre Idrogeno in Trentino—H2@TN" (PAT-Trento). Chiara Maurizio and Paola Ragonese acknowledge project FUN-FACE (SID_2019) of the Physics and Astronomy Department of the University of Padova (Italy).

Institutional Review Board Statement: Not applicable.

Informed Consent Statement: Not applicable.

Data Availability Statement: Raw data or datasets not included here are available from the authors upon request.

Acknowledgments: We thank Nicola Bazzanella for SEM/TEM measurements and Mario Rossi for FT-IR support.

Conflicts of Interest: The authors declare no conflict of interest.

References

1. Xu, Q.; Zhang, L.; Yu, J.; Wageh, S.; Al-Ghamdi, A.A.; Jaroniec, M. Direct Z-scheme photocatalysts: Principles, synthesis, and applications. *Mater. Today* **2018**, *21*, 1042–1063. [CrossRef]
2. Ghorai, K.; Panda, A.; Hossain, A.; Bhattacharjee, M.; Chakraborty, M.; Bhattacharya, S.K.; Show, B.; Sarkar, A.; Bera, P.; Kim, H.; et al. $LaNiO_3$/g-C_3N_4 nanocomposite: An efficient Z-scheme photocatalyst for wastewater treatment using direct sunlight. *J. Rare Earths* **2022**, *40*, 725–736. [CrossRef]
3. Xu, C.; Zhou, Q.; Huang, W.-Y.; Yang, K.; Zhang, Y.-C.; Liang, T.-X.; Liu, Z.-Q. Constructing Z-scheme β-Bi_2O_3/ZrO_2 heterojunctions with 3D mesoporous SiO_2 nanospheres for efficient antibiotic remediation via synergistic adsorption and photocatalysis. *Rare Met.* **2022**, *41*, 2094–2107. [CrossRef]
4. Guo, X.; Duan, J.; Li, C.; Zhang, Z.; Wang, W. Highly efficient Z-scheme g-C_3N_4/ZnO photocatalysts constructed by co-melting-recrystallizing mixed precursors for wastewater treatment. *J. Mater. Sci.* **2020**, *55*, 2018–2031. [CrossRef]
5. Yuan, Y.; Zhang, L.; Xing, J.; Utama, M.I.B.; Lu, X.; Du, K.; Li, Y.; Hu, X.; Wang, S.; Genç, A.; et al. High-yield synthesis and optical properties of g-C_3N_4. *Nanoscale* **2015**, *7*, 12343–12350. [CrossRef] [PubMed]
6. Chen, Z.; Zhang, S.; Liu, Y.; Alharbi, N.S.; Rabah, S.O.; Wang, S.; Wang, X. Synthesis and fabrication of g-C_3N_4-based materials and their application in elimination of pollutants. *Sci. Total Environ.* **2020**, *731*, 139054. [CrossRef] [PubMed]
7. Schwarzer, A.; Saplinova, T.; Kroke, E. Tri-s-triazines (s-heptazines)-From a "mystery molecule" to industrially relevant carbon nitride materials. *Coord. Chem. Rev.* **2013**, *257*, 2032–2062. [CrossRef]
8. Song, G.; Chu, Z.; Jin, W.; Sun, H. Enhanced performance of g-C_3N_4/TiO_2 photocatalysts for degradation of organic pollutants under visible light. *Chin. J. Chem. Eng.* **2015**, *23*, 1326–1334. [CrossRef]
9. Jo, W.K.; Natarajan, T.S. Influence of TiO_2 morphology on the photocatalytic efficiency of direct Z-scheme g-C_3N_4/TiO_2 photocatalysts for isoniazid degradation. *Chem. Eng. J.* **2015**, *281*, 549–565. [CrossRef]
10. Hu, K.; Li, R.; Ye, C.; Wang, A.; Wei, W.; Hu, D.; Qiu, R.; Yan, K. Facile synthesis of Z-scheme composite of TiO_2 nanorod/g-C_3N_4 nanosheet efficient for photocatalytic degradation of ciprofloxacin. *J. Clean. Prod.* **2020**, *253*, 120055. [CrossRef]
11. Lettieri, S.; Pavone, M.; Fioravanti, A.; Amato, L.S.; Maddalena, P. Charge carrier processes and optical properties in TiO_2 and TiO_2-based heterojunction photocatalysts: A review. *Materials* **2021**, *14*, 1645. [CrossRef] [PubMed]
12. Zhou, D.; Chen, Z.; Yang, Q.; Shen, C.; Tang, G.; Zhao, S.; Zhang, J.; Chen, D.; Wei, Q.; Dong, X. Facile Construction of g-C_3N_4 Nanosheets/TiO_2 Nanotube Arrays as Z-Scheme Photocatalyst with Enhanced Visible-Light Performance. *ChemCatChem* **2016**, *8*, 3064–3073. [CrossRef]
13. Li, X.; Wang, B.; Yin, W.; Di, J.; Xia, J.; Zhu, W.; Li, H. Cu^{2+} modified g-C_3N_4 photocatalysts for visible light photocatalytic properties. *Acta Phys. Chim. Sin.* **2020**, *36*, 1902001. [CrossRef]
14. Zhu, H.; Yang, B.; Yang, J.; Yuan, Y.; Zhang, J. Persulfate-enhanced degradation of ciprofloxacin with SiC/g-C_3N_4 photocatalyst under visible light irradiation. *Chemosphere* **2021**, *276*, 130217. [CrossRef] [PubMed]
15. Li, H.; Jing, Y.; Ma, X.; Liu, T.; Yang, L.; Liu, B.; Yin, S.; Wei, Y.; Wang, Y. Construction of a well-dispersed Ag/graphene-like g-C_3N_4 photocatalyst and enhanced visible light photocatalytic activity. *RSC Adv.* **2017**, *7*, 8688–8693. [CrossRef]
16. Zhu, Z.; Lu, Z.; Wang, D.; Tang, X.; Yan, Y.; Shi, W.; Wang, Y.; Gao, N.; Yao, X.; Dong, H. Construction of high-dispersed Ag/Fe_3O_4/g-C_3N_4 photocatalyst by selective photo-deposition and improved photocatalytic activity. *Appl. Catal. B Environ.* **2016**, *182*, 115–122. [CrossRef]
17. Wang, F.; Zhu, Z.; Guo, J. 2D-2D ZnO/N doped g-C_3N_4 composite photocatalyst for antibiotics degradation under visible light. *RSC Adv.* **2021**, *11*, 35663–35672. [CrossRef]
18. He, F.; Chen, G.; Yu, Y.; Hao, S.; Zhou, Y.; Zheng, Y. Facile approach to synthesize g-PAN/g-C_3N_4 composites with enhanced photocatalytic H_2 evolution activity. *ACS Appl. Mater. Interfaces* **2014**, *6*, 7171–7179. [CrossRef]
19. Munusamy, T.D.; Yee, C.S.; Khan, M.M.R. Construction of hybrid g-C_3N_4/CdO nanocomposite with improved photodegradation activity of RhB dye under visible light irradiation. *Adv. Powder Technol.* **2020**, *31*, 2921–2931. [CrossRef]
20. Yoon, M.; Oh, Y.; Hong, S.; Lee, J.S.; Boppella, R.; Kim, S.H.; Marques Mota, F.; Kim, S.O.; Kim, D.H. Synergistically enhanced photocatalytic activity of graphitic carbon nitride and WO_3 nanohybrids mediated by photo-Fenton reaction and H_2O_2. *Appl. Catal. B Environ.* **2017**, *206*, 263–270. [CrossRef]
21. Safaei, J.; Mohamed, N.A.; Noh, M.F.M.; Soh, M.F.; Riza, M.A.; Mustakim, N.S.M.; Ludin, N.A.; Ibrahim, M.A.; Isahak, W.N.R.W.; Teridi, M.A.M. Facile fabrication of graphitic carbon nitride, (g-C_3N_4) thin film. *J. Alloys Compd.* **2018**, *769*, 130–135. [CrossRef]
22. Mohamed, N.A.; Safaei, J.; Ismail, A.F.; Jailani, M.F.A.M.; Khalid, M.N.; Noh, M.F.M.; Aadenan, A.; Nasir, S.N.S.; Sagu, J.S.; Teridi, M.A.M. The influences of post-annealing temperatures on fabrication graphitic carbon nitride, (g-C_3N_4) thin film. *Appl. Surf. Sci.* **2019**, *489*, 92–100. [CrossRef]
23. Zhang, Y.; Zhao, S.-M.; Su, Q.-W.; Xu, J.-L. Visible light response ZnO–C_3N_4 thin film photocatalyst. *Rare Met.* **2021**, *40*, 96–104. [CrossRef]
24. Zhao, W.; Yang, X.; Liu, C.; Qian, X.; Wen, Y.; Yang, Q.; Sun, T.; Chang, W.; Liu, X.; Chen, Z. Facile Construction of All-Solid-State Z-Scheme g-C_3N_4/TiO_2 Thin Film for the Efficient Visible-Light Degradation of Organic Pollutant. *Nanomaterials* **2020**, *10*, 600. [CrossRef] [PubMed]
25. Nandy, S.; Savant, S.A.; Haussener, S. Prospects and challenges in designing photocatalytic particle suspension reactors for solar fuel processing. *Chem. Sci.* **2021**, *12*, 9866–9884. [CrossRef]

26. Fendrich, M.; Popat, Y.; Orlandi, M.; Quaranta, A.; Miotello, A. Pulsed laser deposition of nanostructured tungsten oxide films: A catalyst for water remediation with concentrated sunlight. *Mater. Sci. Semicond. Process.* **2020**, *119*, 105237. [CrossRef]
27. El Golli, A.; Fendrich, M.; Bazzanella, N.; Dridi, C.; Miotello, A.; Orlandi, M. Wastewater remediation with ZnO photocatalysts: Green synthesis and solar concentration as an economically and environmentally viable route to application. *J. Environ. Manag.* **2021**, *286*, 112226. [CrossRef]
28. Sun, Z.; Zhang, X.; Zhu, R.; Dong, X.; Xu, J.; Wang, B. Facile synthesis of visible light-induced g-C_3N_4/rectorite composite for efficient photodegradation of ciprofloxacin. *Materials* **2018**, *11*, 2452. [CrossRef]
29. Song, Y.; Tian, J.; Gao, S.; Shao, P.; Qi, J.; Cui, F. Photodegradation of sulfonamides by g-C_3N_4 under visible light irradiation: Effectiveness, mechanism and pathways. *Appl. Catal. B Environ.* **2017**, *210*, 88–96. [CrossRef]
30. Suyana, P.; Ganguly, P.; Nair, B.N.; Pillai, S.C.; Hareesh, U.S. Structural and compositional tuning in g-C_3N_4 based systems for photocatalytic antibiotic degradation. *Chem. Eng. J. Adv.* **2021**, *8*, 100148. [CrossRef]
31. Nguyen, T.B.; Huang, C.P.; Doong, R.-A.; Chen, C.W.; Dong, C. Di Visible-light photodegradation of sulfamethoxazole (SMX) over Ag-P-codoped g-C_3N_4 (Ag-P@UCN) photocatalyst in water. *Chem. Eng. J.* **2020**, *384*, 123383. [CrossRef]
32. Li, G.; Nie, X.; Gao, Y.; An, T. Can environmental pharmaceuticals be photocatalytically degraded and completely mineralized in water using g-C_3N_4/TiO_2 under visible light irradiation?—Implications of persistent toxic intermediates. *Appl. Catal. B Environ.* **2016**, *180*, 726–732. [CrossRef]
33. Mohamed, R.M. Synthesis and characterization of AgCl@graphitic carbon nitride hybrid materials for the photocatalytic degradation of atrazine. *Ceram. Int.* **2015**, *41*, 1197–1204. [CrossRef]
34. Geng, Y.; Chen, D.; Li, N.; Xu, Q.; Li, H.; He, J.; Lu, J. Z-Scheme 2D/2D α-Fe_2O_3/g-C_3N_4 heterojunction for photocatalytic oxidation of nitric oxide. *Appl. Catal. B Environ.* **2021**, *280*, 119409. [CrossRef]
35. Alduhaish, O.; Ubaidullah, M.; Al-Enizi, A.M.; Alhokbany, N.; Alshehri, S.M.; Ahmed, J. Facile Synthesis of Mesoporous α-Fe_2O_3@g-C_3N_4-NCs for Efficient Bifunctional Electro-catalytic Activity (OER/ORR). *Sci. Rep.* **2019**, *9*, 14139. [CrossRef] [PubMed]
36. Wang, Z.; Fan, Y.; Wu, R.; Huo, Y.; Wu, H.; Wang, F.; Xu, X. Novel magnetic g-C_3N_4/α-Fe_2O_3/Fe_3O_4 composite for the very effective visible-light-fenton degradation of Orange II. *RSC Adv.* **2018**, *8*, 5180–5188. [CrossRef] [PubMed]
37. Vesborg, P.C.K.; Jaramillo, T.F. Addressing the terawatt challenge: Scalability in the supply of chemical elements for renewable energy. *RSC Adv.* **2012**, *2*, 7933–7947. [CrossRef]
38. aus der Beek, T.; Weber, F.-A.; Bergmann, A.; Hickmann, S.; Ebert, I.; Hein, A.; Küster, A. Pharmaceuticals in the environment—Global occurrences and perspectives. *Environ. Toxicol. Chem.* **2016**, *35*, 823–835. [CrossRef]
39. Ribeiro, A.R.; Nunes, O.C.; Pereira, M.F.R.; Silva, A.M.T. An overview on the advanced oxidation processes applied for the treatment of water pollutants defined in the recently launched Directive 2013/39/EU. *Environ. Int.* **2015**, *75*, 33–51. [CrossRef]
40. Edla, R.; Tonezzer, A.; Orlandi, M.; Patel, N.; Fernandes, R.; Bazzanella, N.; Date, K.; Kothari, D.C.; Miotello, A. 3D hierarchical nanostructures of iron oxides coatings prepared by pulsed laser deposition for photocatalytic water purification. *Appl. Catal. B Environ.* **2017**, *219*, 401–411. [CrossRef]
41. Popat, Y.; Orlandi, M.; Patel, N.; Edla, R.; Bazzanella, N.; Gupta, S.; Yadav, M.; Pillai, S.; Patel, M.K.; Miotello, A. Pulsed laser deposition of $CoFe_2O_4$/CoO hierarchical-type nanostructured heterojuction forming a Z-scheme for efficient spatial separation of photoinduced electron-hole pairs and highly active surface area. *Appl. Surf. Sci.* **2019**, *489*, 584–594. [CrossRef]
42. Chieregato, F.; Molinari, A.; Milani, M.; Fendrich, M.; Orlandi, M.; Miotello, A. Colloids and Surfaces A: Physicochemical and Engineering Aspects an immobilized iron-oxides catalytic platform for photocatalysis and photosynthesis: Visible light induced hydroxylation reactions. *Colloids Surf. A Physicochem. Eng. Asp.* **2022**, *648*, 129428. [CrossRef]
43. Orlandi, M.; Filosa, N.; Bettonte, M.; Fendrich, M.; Girardini, M.; Battistini, T.; Miotello, A. Treatment of surfactant-rich industrial wastewaters with concentrated sunlight: Toward solar wastewater remediation. *Int. J. Environ. Sci. Technol.* **2019**, *16*, 2109–2114. [CrossRef]
44. Phillips, G.; Johnson, B.E.; Ferguson, J. The loss of antibiotic activity of ciprofloxacin by photodegradation. *J. Antimicrob. Chemother.* **1990**, *26*, 783–789. [CrossRef]
45. Karim, A.V.; Shriwastav, A. Degradation of ciprofloxacin using photo, sono, and sonophotocatalytic oxidation with visible light and low-frequency ultrasound: Degradation kinetics and pathways. *Chem. Eng. J.* **2020**, *392*, 124853. [CrossRef]
46. Rashid, J.; Abbas, A.; Chang, L.C.; Iqbal, A.; Haq, I.U.; Rehman, A.; Awan, S.U.; Arshad, M.; Rafique, M.; Barakat, M.A. Butterfly cluster like lamellar BiOBr/TiO_2 nanocomposite for enhanced sunlight photocatalytic mineralization of aqueous ciprofloxacin. *Sci. Total Environ.* **2019**, *665*, 668–677. [CrossRef]
47. Wagner, D.R.; Ament, K.; Mayr, L.; Martin, T.; Bloesser, A.; Schmalz, H.; Marschall, R.; Wagner, F.E.; Breu, J. Terrestrial solar radiation driven photodecomposition of ciprofloxacin in clinical wastewater applying mesostructured iron(III) oxide. *Environ. Sci. Pollut. Res.* **2021**, *28*, 6222–6231. [CrossRef]
48. Rajiv, P.; Mengelizadeh, N.; McKay, G.; Balarak, D. Photocatalytic degradation of ciprofloxacin with Fe_2O_3 nanoparticles loaded on graphitic carbon nitride: Mineralisation, degradation mechanism and toxicity assessment. *Int. J. Environ. Anal. Chem.* **2023**, *103*, 2193–2207. [CrossRef]
49. Wang, Y.; Li, X.; Lei, W.; Zhu, B.; Yang, J. Novel carbon quantum dot modified g-C_3N_4 nanotubes on carbon cloth for efficient degradation of ciprofloxacin. *Appl. Surf. Sci.* **2021**, *559*, 149967. [CrossRef]

50. Bonelli, M.; Cestari, C.; Miotello, A. Pulsed laser deposition apparatus for applied research. *Meas. Sci. Technol.* **1999**, *10*, 27–30. [CrossRef]
51. Svoboda, L.; Praus, P.; Lima, M.J.; Sampaio, M.J.; Matýsek, D.; Ritz, M.; Dvorský, R.; Faria, J.L.; Silva, C.G. Graphitic carbon nitride nanosheets as highly efficient photocatalysts for phenol degradation under high-power visible LED irradiation. *Mater. Res. Bull.* **2018**, *100*, 322–332. [CrossRef]
52. Zhang, Y.; Pan, Q.; Chai, G.; Liang, M.; Dong, G.; Zhang, Q.; Qiu, J. Synthesis and luminescence mechanism of multicolor-emitting g-C_3N_4 nanopowders by low temperature thermal condensation of melamine. *Sci. Rep.* **2013**, *3*, 1943. [CrossRef] [PubMed]
53. Jiang, J.; Zou, J.; Wee, A.T.S.; Zhang, W. Use of single-layer g-C_3N_4/Ag hybrids for surface-enhanced raman scattering (SERS). *Sci. Rep.* **2016**, *6*, 34599. [CrossRef] [PubMed]
54. Mishra, D.; Arora, R.; Lahiri, S.; Amritphale, S.S.; Chandra, N. Synthesis and characterization of iron oxide nanoparticles by solvothermal method. *Prot. Met. Phys. Chem. Surf.* **2014**, *50*, 628–631. [CrossRef]
55. Ristić, M.; De Grave, E.; Musić, S.; Popović, S.; Orehovec, Z. Transformation of low crystalline ferrihydrite to α-Fe_2O_3 in the solid state. *J. Mol. Struct.* **2007**, *834–836*, 454–460. [CrossRef]
56. Soler, M.A.G.; Qu, F. *Raman Spectroscopy of Iron Oxide Nanoparticles BT—Raman Spectroscopy for Nanomaterials Characterization;* Kumar, C.S.S.R., Ed.; Springer: Berlin/Heidelberg, Germany, 2012; pp. 379–416. ISBN 978-3-642-20620-7.
57. Zhang, Z.; Huang, J.; Yuan, Q.; Dong, B. Intercalated graphitic carbon nitride: A fascinating two-dimensional nanomaterial for an ultra-sensitive humidity nanosensor. *Nanoscale* **2014**, *6*, 9250–9256. [CrossRef]
58. Shi, Q.; Huang, J.; Yang, Y.; Wu, J.; Shen, J.; Liu, X.; Sun, A.; Liu, Z. In-Situ construction of urchin-like hierarchical g-C_3N_4/NiAl-LDH hybrid for efficient photoreduction of CO_2. *Mater. Lett.* **2020**, *268*, 127560. [CrossRef]
59. Ding, Q.; Lam, F.L.Y.; Hu, X. Complete degradation of ciprofloxacin over g-C_3N_4-iron oxide composite via heterogeneous dark Fenton reaction. *J. Environ. Manag.* **2019**, *244*, 23–32. [CrossRef]
60. Liu, L.; Wang, M.; Wang, C. In-Situ synthesis of graphitic carbon nitride/iron oxide−copper composites and their application in the electrochemical detection of glucose. *Electrochim. Acta* **2018**, *265*, 275–283. [CrossRef]
61. Wang, X.; Han, D.; Ding, Y.; Liu, J.; Cai, H.; Jia, L.; Cheng, X.; Wang, J.; Fan, X. A low-cost and high-yield approach for preparing g-C_3N_4 with a large specific surface area and enhanced photocatalytic activity by using formaldehyde-treated melamine. *J. Alloys Compd.* **2020**, *845*, 156293. [CrossRef]
62. Wang, S.; Meng, C.; Bai, Y.; Wang, Y.; Liu, P.; Pan, L.; Zhang, L.; Yin, Z.; Tang, N. Synergy Promotion of Elemental Doping and Oxygen Vacancies in Fe_2O_3 Nanorods for Photoelectrochemical Water Splitting. *ACS Appl. Nano Mater.* **2022**, *5*, 6781–6791. [CrossRef]
63. Yamashita, T.; Hayes, P. Analysis of XPS spectra of Fe^{2+} and Fe^{3+} ions in oxide materials. *Appl. Surf. Sci.* **2008**, *254*, 2441–2449. [CrossRef]
64. Guo, C.; Xia, F.; Wang, Z.; Zhang, L.; Xi, L.; Zuo, Y. Flowerlike iron oxide nanostructures and their application in microwave absorption. *J. Alloys Compd.* **2015**, *631*, 183–191. [CrossRef]
65. Wang, W.; Zhang, H.; Zhang, S.; Liu, Y.; Wang, G.; Sun, C.; Zhao, H. Potassium-Ion-Assisted Regeneration of Active Cyano Groups in Carbon Nitride Nanoribbons: Visible-Light-Driven Photocatalytic Nitrogen Reduction. *Angew. Chem. Int. Ed.* **2019**, *58*, 16644–16650. [CrossRef] [PubMed]
66. Zhang, J.-R.; Ma, Y.; Wang, S.-Y.; Ding, J.; Gao, B.; Kan, E.; Hua, W. Accurate K-edge X-ray photoelectron and absorption spectra of g-C_3N_4 nanosheets by first-principles simulations and reinterpretations. *Phys. Chem. Chem. Phys.* **2019**, *21*, 22819–22830. [CrossRef]
67. Nikolaev, V.; Sladkevich, S.; Divina, U.; Prikhodchenko, P.V.; Gasser, G.; Falciola, L.; Longhi, M.; Lev, O. LC-MS analysis of nitroguanidine compounds by catalytic reduction using palladium modified graphitic carbon nitride catalyst. *Microchim. Acta* **2021**, *188*, 152. [CrossRef]
68. Shen, L.; Xing, Z.; Zou, J.; Li, Z.; Wu, X.; Zhang, Y.; Zhu, Q.; Yang, S.; Zhou, W. Black TiO_2 nanobelts/g-C_3N_4 nanosheets Laminated Heterojunctions with Efficient Visible-Light-Driven Photocatalytic Performance. *Sci. Rep.* **2017**, *7*, 41978. [CrossRef]
69. Xu, Y.; Huang, S.; Xie, M.; Li, Y.; Xu, H.; Huang, L.; Zhang, Q.; Li, H. Magnetically separable Fe_2O_3/g-C_3N_4 catalyst with enhanced photocatalytic activity. *RSC Adv.* **2015**, *5*, 95727–95735. [CrossRef]
70. Yi, S.S.; Wulan, B.R.; Yan, J.M.; Jiang, Q. Highly Efficient Photoelectrochemical Water Splitting: Surface Modification of Cobalt-Phosphate-Loaded Co_3O_4/Fe_2O_3 p–n Heterojunction Nanorod Arrays. *Adv. Funct. Mater.* **2019**, *29*, 1801902. [CrossRef]
71. McIntyre, N.S.; Zetaruk, D.G. X-ray photoelectron spectroscopic studies of iron oxides. *Anal. Chem.* **1977**, *49*, 1521–1529. [CrossRef]
72. Grosvenor, A.P.; Kobe, B.A.; Biesinger, M.C.; McIntyre, N.S. Investigation of multiplet splitting of Fe 2p XPS spectra and bonding in iron compounds. *Surf. Interface Anal.* **2004**, *36*, 1564–1574. [CrossRef]
73. Zhao, L.; Guo, L.; Tang, Y.; Zhou, J.; Shi, B. Novel g-C_3N_4/C/Fe_2O_3 Composite for Efficient Photocatalytic Reduction of Aqueous Cr(VI) under Light Irradiation. *Ind. Eng. Chem. Res.* **2021**, *60*, 13594–13603. [CrossRef]
74. Carbajo, J.; Silveira, J.E.; Pliego, G.; Zazo, J.A.; Casas, J.A. Increasing Photo-Fenton process Efficiency: The effect of high temperatures. *Sep. Purif. Technol.* **2021**, *271*, 118876. [CrossRef]
75. Yadav, M.; Yadav, A.; Fernandes, R.; Popat, Y.; Orlandi, M.; Dashora, A.; Kothari, D.C.; Miotello, A.; Ahuja, B.L.; Patel, N. Tungsten-doped TiO_2/reduced Graphene Oxide nano-composite photocatalyst for degradation of phenol: A system to reduce surface and bulk electron-hole recombination. *J. Environ. Manag.* **2017**, *203*, 364–374. [CrossRef]

76. He, W.; Jia, H.; Li, Z.; Miao, C.Q.; Lu, R.; Zhang, S.; Zhang, Z. Magnetic recyclable g-C_3N_4/Fe_3O_4@MIL-100(Fe) ternary catalyst for photo-Fenton degradation of ciprofloxacin. *J. Environ. Chem. Eng.* **2022**, *10*, 108698. [CrossRef]
77. Liu, Y.; Yu, Y.X.; Zhang, W. De Photoelectrochemical study on charge transfer properties of nanostructured Fe_2O_3 modified by g-C_3N_4. *Int. J. Hydrogen Energy* **2014**, *39*, 9105–9113. [CrossRef]
78. Van Pham, V.; Truong, T.K.; Hai, L.V.; La, H.P.P.; Nguyen, H.T.; Lam, V.Q.; Tong, H.D.; Nguyen, T.Q.; Sabbah, A.; Chen, K.-H.; et al. S-Scheme α-Fe_2O_3/g-C_3N_4 Nanocomposites as Heterojunction Photocatalysts for Antibiotic Degradation. *ACS Appl. Nano Mater.* **2022**, *5*, 4506–4514. [CrossRef]
79. Danish, M.; Saud Athar, M.; Ahmad, I.; Warshagha, M.Z.A.; Rasool, Z.; Muneer, M. Highly efficient and stable Fe_2O_3/g-C_3N_4/GO nanocomposite with Z-scheme electron transfer pathway: Role of photocatalytic activity and adsorption isotherm of organic pollutants in wastewater. *Appl. Surf. Sci.* **2022**, *604*, 154604. [CrossRef]
80. Wang, J.; Zuo, X.; Cai, W.; Sun, J.; Ge, X.; Zhao, H. Facile fabrication of direct solid-state Z-scheme g-C_3N_4/Fe_2O_3 heterojunction: A cost-effective photocatalyst with high efficiency for the degradation of aqueous organic pollutants. *Dalton Trans.* **2018**, *47*, 15382–15390. [CrossRef]
81. Li, X.; Qiu, Y.; Zhu, Z.; Chen, T.; Zhang, H.; Yin, D. Construction of magnetically separable dual Z-scheme g-C_3N_4/α-Fe_2O_3/Bi_3TaO_7 photocatalyst for effective degradation of ciprofloxacin under visible light. *Chem. Eng. J.* **2022**, *440*, 135840. [CrossRef]
82. Shen, Y.; Han, Q.; Hu, J.; Gao, W.; Wang, L.; Yang, L.; Gao, C.; Shen, Q.; Wu, C.; Wang, X.; et al. Artificial Trees for Artificial Photosynthesis: Construction of Dendrite-Structured α-Fe_2O_3/g-C_3N_4 Z-Scheme System for Efficient CO_2 Reduction into Solar Fuels. *ACS Appl. Energy Mater.* **2020**, *3*, 6561–6572. [CrossRef]
83. Zhou, L.; Wang, L.; Zhang, J.; Lei, J.; Liu, Y. Well-Dispersed Fe_2O_3 Nanoparticles on g-C_3N_4 for Efficient and Stable Photo-Fenton Photocatalysis under Visible-Light Irradiation. *Eur. J. Inorg. Chem.* **2016**, *2016*, 5387–5392. [CrossRef]
84. Liu, D.; Li, C.; Ni, T.; Gao, R.; Ge, J.; Zhang, F.; Wu, W.; Li, J.; Zhao, Q. 3D interconnected porous g-C_3N_4 hybridized with Fe_2O_3 quantum dots for enhanced photo-Fenton performance. *Appl. Surf. Sci.* **2021**, *555*, 149677. [CrossRef]
85. Zhao, H.; Tian, C.; Mei, J.; Yang, S.; Wong, P.K. Faster electron injection and higher interface reactivity in g-C_3N_4/Fe_2O_3 nanohybrid for efficient photo-Fenton-like activity toward antibiotics degradation. *Environ. Res.* **2021**, *195*, 110842. [CrossRef] [PubMed]
86. Guo, T.; Wang, K.; Zhang, G.; Wu, X. A novel α-Fe_2O_3@g-C_3N_4 catalyst: Synthesis derived from Fe-based MOF and its superior photo-Fenton performance. *Appl. Surf. Sci.* **2019**, *469*, 331–339. [CrossRef]
87. Ge, F.; Li, X.; Wu, M.; Ding, H.; Li, X. A type II heterojunction α-Fe_2O_3/g-C_3N_4 for the heterogeneous photo-Fenton degradation of phenol. *RSC Adv.* **2022**, *12*, 8300–8309. [CrossRef]
88. Huang, Y.; Chen, B.; Duan, J.; Yang, F.; Wang, T.; Wang, Z.; Yang, W.; Hu, C.; Luo, W.; Huang, Y. Graphitic Carbon Nitride (g-C3N4): An Interface Enabler for Solid-State Lithium Metal Batteries. *Angew. Chemie Int. Ed.* **2020**, *59*, 3699–3704. [CrossRef]
89. Liu, X.; Yang, W.; Chen, L.; Liu, Z.; Long, L.; Wang, S.; Liu, C.; Dong, S.; Jia, J. Graphitic Carbon Nitride (g-C3N4)-Derived Bamboo-Like Carbon Nanotubes/Co Nanoparticles Hybrids for Highly Efficient Electrocatalytic Oxygen Reduction. *ACS Appl. Mater. Interfaces* **2020**, *12*, 4463–4472. [CrossRef]
90. Ong, W.-J.; Tan, L.-L.; Ng, Y.H.; Yong, S.-T.; Chai, S.-P. Graphitic Carbon Nitride (g-C3N4)-Based Photocatalysts for Artificial Photosynthesis and Environmental Remediation: Are We a Step Closer to Achieving Sustainability? *Chem. Rev.* **2016**, *116*, 7159–7329. [CrossRef]
91. Wang, Y.; Zhang, X.; Ding, X.; Li, Y.; Wu, B.; Zhang, P.; Zeng, X.; Zhang, Q.; Du, Y.; Gong, Y.; et al. Stitching Graphene Sheets with Graphitic Carbon Nitride: Constructing a Highly Thermally Conductive rGO/g-C3N4 Film with Excellent Heating Capability. *ACS Appl. Mater. Interfaces* **2021**, *13*, 6699–6709. [CrossRef]
92. Qu, Y.; Li, X.; Zhang, H.; Huang, R.; Qi, W.; Su, R.; He, Z. Controllable synthesis of a sponge-like Z-scheme N,S-CQDs/Bi_2MoO_6@TiO_2 film with enhanced photocatalytic and antimicrobial activity under visible/NIR light irradiation. *J. Hazard. Mater.* **2022**, *429*, 128310. [CrossRef] [PubMed]
93. Masekela, D.; Hintsho-Mbita, N.C.; Mabuba, N. Application of a piezo-photocatalytic thin film (FTO/$BaTiO_3/SnO_2$) for enhanced degradation of organic pollutants and disinfection of wastewater. *Ceram. Int.* **2023**, *49*, 7566–7579. [CrossRef]
94. Gómez, E.; Cestaro, R.; Philippe, L.; Serrà, A. Electrodeposition of nanostructured Bi_2MoO_6@Bi_2MoO_{6-x} homojunction films for the enhanced visible-light-driven photocatalytic degradation of antibiotics. *Appl. Catal. B Environ.* **2022**, *317*, 121703. [CrossRef]
95. Triquet, T.; Tendero, C.; Latapie, L.; Manero, M.H.; Richard, R.; Andriantsiferana, C. TiO_2 MOCVD coating for photocatalytic degradation of ciprofloxacin using 365 nm UV LEDs—Kinetics and mechanisms. *J. Environ. Chem. Eng.* **2020**, *8*, 104544. [CrossRef]
96. Tantis, I.; Bousiakou, L.; Karikas, G.A.; Lianos, P. Photocatalytic and photoelectrocatalytic degradation of the antibacterial agent ciprofloxacin. *Photochem. Photobiol. Sci.* **2015**, *14*, 603–607. [CrossRef]
97. Lima, A.S.; Rocha, R.D.C.; Pereira, E.C.; Sikora, M.d.S. Photodegradation of Ciprofloxacin antibiotic over TiO_2 grown by PEO: Ecotoxicity response in *Lactuca sativa* L. and *Lemna minor*. *Int. J. Environ. Sci. Technol.* **2022**, *19*, 2771–2780. [CrossRef]
98. Kumar, J.V.; Karthik, R.; Chen, S.M.; Muthuraj, V.; Karuppiah, C. Fabrication of potato-like silver molybdate microstructures for photocatalytic degradation of chronic toxicity ciprofloxacin and highly selective electrochemical detection of H_2O_2. *Sci. Rep.* **2016**, *6*, 34149. [CrossRef]
99. Mirzai, M.; Akhlaghian, F.; Rahmani, F. Photodegradation of ciprofloxacin in water using photocatalyst of zinc oxide nanowires doped with copper and cerium oxides. *Water Environ. J.* **2020**, *34*, 420–431. [CrossRef]

100. Marshall, C.P.; Dufresne, W.J.B.; Rufledt, C.J. Polarized Raman spectra of hematite and assignment of external modes. *J. Raman Spectrosc.* **2020**, *51*, 1522–1529. [CrossRef]
101. Tonda, S.; Kumar, S.; Kandula, S.; Shanker, V. Fe-doped and -mediated graphitic carbon nitride nanosheets for enhanced photocatalytic performance under natural sunlight. *J. Mater. Chem. A* **2014**, *2*, 6772–6780. [CrossRef]
102. Meier, R.J.; Maple, J.R.; Hwang, M.-J.; Hagler, A.T. Molecular Modeling Urea- and Melamine-Formaldehyde Resins. 1. A Force Field for Urea and Melamine. *J. Phys. Chem.* **1995**, *99*, 5445–5456. [CrossRef]
103. Jelı, M.; Zemek, J.; Trchová, M.; Vorlı, V.; Lančok, J.; Tomov, R.; Šimečková, M. CNx films created by combined laser deposition and r.f. discharge: XPS, FTIR and Raman analysis. *Thin Solid Film.* **2000**, *366*, 69–76. [CrossRef]

applied sciences

MDPI

Article

Development of Polymer-Cored Akaganeite Adsorbent for Phosphate Adsorption

Jiyeol Bae [1,2,†], Hyobin We [3,†], Suho Kim [1], Sungjik Oh [1] and Soyoung Baek [1,*]

1 Department of Environmental Research, Korea Institute of Civil Engineering and Building Technology (KICT), Goyang 10223, Republic of Korea; baejiyeol@kict.re.kr (J.B.)
2 Civil and Environmental Engineering, University of Science and Technology (UST), Daejeon 34113, Republic of Korea
3 Atmospheric Environment Team, Gyeonggi Province Institute of Health and Environment Northern Support Institute, Uijeongbu-si 11780, Republic of Korea
* Correspondence: soyoung@kict.re.kr
† These authors contributed equally to this work.

Abstract: Environmental issues related to phosphate and resource depletion have recently emerged as serious problems. This study focuses on solving the problems of phosphate removal and recovery using synthesized granular akaganeite (GAK). This study identified that akaganeite, which possesses an FeOOH structure in iron oxyhydroxide, can be synthesized and used as a reusable material. Immobilization with the core–shell method using polyethersulfone was applied as a strategy to recover phosphate anions from a trace of phosphate solution. GAK was successfully analyzed using SEM/TGA/BET to understand its physical properties. XRD and SAD pattern analyses suggested that the GAK powder form was amorphous in nature. The powdered akaganeite had a surface area of 231 mg^2/g and a maximum adsorption capacity of 21.27 mg/g. To prevent the dispersion of powder during granulation, polyethersulfone was used as a scaffold since akaganeite particles can be effectively immobilized onto PES polymer scaffolds, as substantiated by the SEM/EDS results. Moreover, a lack of changes in the pore sizes suggested that physical properties remained unchanged. Furthermore, compared to the granular akaganeite, the surface area of powdered akaganeite decreased 4–5-fold. The adsorption kinetic of granular akaganeite fit the pseudo-second-order model. The powdered form displayed high removal efficiency, intimate with phosphate anions, when $n > 1.0$, instead of lower K_F. On the other hand, granular akaganeite showed lower affinity when $n < 1.0$, but appeared positive for an adsorbate with higher KF. This implies that the granulation of akaganeite with the PES polymer did not change its adsorption property, with the maximum adsorption capacity for granular akaganeite being 3.65 mg/g.

Keywords: polymer core granulation; phosphate removal; akaganeite adsorbent

Citation: Bae, J.; We, H.; Kim, S.; Oh, S.; Baek, S. Development of Polymer-Cored Akaganeite Adsorbent for Phosphate Adsorption. *Appl. Sci.* **2024**, *14*, 146. https://doi.org/10.3390/app14010146

Academic Editor: Domenico Lombardo

Received: 8 November 2023
Revised: 18 December 2023
Accepted: 21 December 2023
Published: 23 December 2023

1. Introduction

Phosphorus is commonly found in many terrestrial and aquatic areas where it acts as an essential nutrient for the growth of plants, animals, bacteria, and soil-based organisms. Owing to this, fertilizers containing phosphate have been used in agricultural production for many years. Moreover, inorganic phosphate-based products have been used in practical applications, including toothpaste, cosmetics, synthetic detergents, bone cement, and medical products [1–6].

The increasing consumption of phosphates has led to two significant problems. First, phosphate rock, which contains high amounts of phosphate, is a limited global resource. With the increasing global phosphate consumption, the prices of phosphate rock have been climbing in the mineral market. In fact, global phosphorus reserves likely peaked, causing greater derived demand for phosphate rock [7]. The Food and Agricultural Organization of the United Nations (FAO) reported that demands for fertilizers increased by approximately

2.6% each year from 2010 to 2014 [8]. Based on this price inflation, further investigations have revealed that the rising demand for phosphate rock will continue to deplete phosphate resources in the future. In addition, investigators predict that given the current phosphate usage in America and China, their resources are expected to be depleted by 2070 [9]. Furthermore, an overload of waste phosphorus from sewage treatment plants directly damages aquatic systems, causing excessive microbial growth in developed cities. It is difficult to safely remove this wastewater, including phosphate, from sewage treatment plants; typical plants employ only biological processes, making it hard to remove traces of phosphate.

The overconsumption of phosphates creates two main problems: on a smaller scale, it influences aquatic systems, and on a larger scale, it depletes phosphate reserves. These increasingly problematic issues have garnered international attention. A more gradual use of this resource will put pressure on conservation and environmental protection efforts.

Recently, municipal sewage treatment plants have started making an effort to remove phosphates by incorporating a variety of chemical and precipitation methods. The most commonly used method is the use of coagulants, such as inorganic coagulants of Al^{+3} and Fe^{+2}, to settle the precipitation [10]. However, such chemical dosing generates by-products and disposal problems given the nonrecovery of phosphorus. Thus, to address this problem, a crystallization technique with magnesium dosing has been introduced for removing ammonium from wastewater and obtaining the nutrient elements struvite and apatite [11]. Though this technique has improved the recovery of coagulated phosphorus, it has failed to recover the phosphate resources. However, it is still difficult to remove the precipitate containing the phosphorus from the bottom of the grit chamber. The powdered chemical products used in most of these techniques cause fine particles to be dissolved in water, which are released from effluent sewage treatment plants. These secondary pollutants were first reported in Korean rivers by the National Institute of Environmental Research of Korea [12]. Three factors are essential for overcoming these problems: developing high removal capacity with economics, developing reusable materials with recovered phosphorus, and immobilizing fine particles to prevent secondary pollution.

Iron-based adsorbents are particularly useful because they are safe and economical. Iron-based adsorption studies have received significant, sustained attention for many years now. Many long-term studies have verified that iron-based adsorbents are capable of removing various heavy metals [13–18]. They exist in various forms and have varied properties, and they can be classified as iron oxides, iron hydroxides, or iron oxyhydroxides, depending on their characteristics [19,20]. Among them, iron oxyhydroxide is well known for adsorbing anionic substances. However, the reusability of the material depends on its specific characteristics or adsorption mechanism impacts [21,22]. Thus, this study used iron oxyhydroxide adsorbents to recover reusable phosphorus.

In previous studies, powder forms of akaganeite were evaluated for the removal of oxyanions such as phosphate [19,20]. However, the use of powdered adsorbents in water treatment is problematic, e.g., due to the need for posterior removal of fine adsorbent particles from the purified water. Applications where powdered adsorbents are encased in the filtration columns require a high working pressure. Thus, in order to solve this problem fundamentally, the powdered adsorbents need to be immobilized, and immobilizing a powdered adsorbent onto a polymer scaffold can be a good way. Various immobilization techniques have been introduced and are being widely used for adsorption, acting as a catalyst, in the medical field [23–26]. Granulation is an important component of the adsorption process where nanoparticles are immobilized from the scaffold material and are directly put in contact with targeted anion materials on the edge of the beads. Polyethersulfone (PES) polymers, as organic compounds, have been found to be suitable for immobilizing fine particles; they provide an optimal material that is non-toxic and freely forms a shape given the high viscosity of the liquid. Moreover, the material is highly resistant to acidic and alkaline conditions and has superior strength in the membrane field [27]. In terms of recovering phosphorus from sewage treatment plants, these two strategies are extremely close to being successful and have the potential to relieve the environmental burden.

In this study, phosphate removal and recovery were accomplished by using akaganeite, which is an iron oxyhydroxide polymorph. This study allowed the identification of adsorbents that could be used to recycle phosphate. Granular iron oxyhydroxides were synthesized using a simple core–shell technique using the PES polymer as a scaffold. Immobilizing the nanoparticles on the surface of the beads is important because it increases the effective area for removing phosphate anions on granular adsorbents. Therefore, we attempted immobilizing nanoparticles on the outer surface using granular adsorbents. First, the characteristics of powdered iron oxyhydroxide were analyzed. To determine its adsorption, tests were conducted on the phosphate anion (PO_4^{3-}; as HPO_4^{2-} and $H_2PO_4^-$ at near-neutral pH) using isotherm adsorption, kinetic adsorption and desorption, the effects of pH, and regeneration experiments for granular forms. Based on these results, this study investigated the formation of struvite using eluted phosphorus and evaluated the application of granular iron oxyhydroxide to water treatment technology.

2. Materials and Methods

2.1. Preparation of Powdered Iron Oxyhydroxide

2.1.1. Materials

All chemicals in this study were of reagent grade and used without further purification. Sodium hydroxide (NaOH) and ammonium carbonate (NH_4OH) were purchased from Sigma-Aldrich. Iron species reagents purchased were as follows: Iron (III) sulfate pentahydrate ($Fe_2(SO_4)_3{\cdot}5H_2O$) from Acros Organics in Germany; Iron (III) chloride hexahydrate ($FeCl_3{\cdot}6H_2O$) from Fluka in Germany; and Iron(II) nitrate 9-hydrate ($Fe(NO_3)_3{\cdot}9H_2O$) from Yakuri Pure Chemical Ltd. in Kyoto, Japan. Urea (H_2NCONH_2) and ethylene glycol ($C_2H_6O_2$) were purchased from Daejung in the Republic of Korea. Polyethersulfone (Gafone 3000) was obtained from Solvay Advanced polymers, and Dimethylformamide (DMF) was purchased from Sigma-Aldrich (St. Louis, MO, USA).

2.1.2. Synthesis of Iron Oxyhydroxide Adsorbents

A total of 0.1 M of the ammonium carbonate solution (2000 mL) was added dropwise with constant vigorous stirring to 0.1 M ferric chloride solution (2000 mL). The generated brown precipitate was then washed thrice with 2000 mL of deionized (DI) water and separated by centrifugation at 2000 rpm for 10 min. Since the radius of the centrifuge rotor used in the centrifugation process is 5 cm, the relative centrifugal force applied in this process is 224× g. The final products were stored in an oven at 55 °C for 48 h and then washed with DI water.

2.2. Granulation of Powdered Adsorbents

To immobilize particles, 20% PES polymer solution was prepared by dissolving 6 g of polyethersulfone in 24 mL DMF solution. Later, the PES polymer solution was stored at 150 rpm for 24 h in a shaking incubator at room temperature. The PES polymer solution was dropped onto powdered materials via a syringe. A drop of translucent PES polymer on the plate was shaken until powdered akaganeite was covered with PES (Figure 1). Each drop that fell in DI water was indurated. After separation from DI water, sphere-shaped akaganeite was obtained (Figure 1).

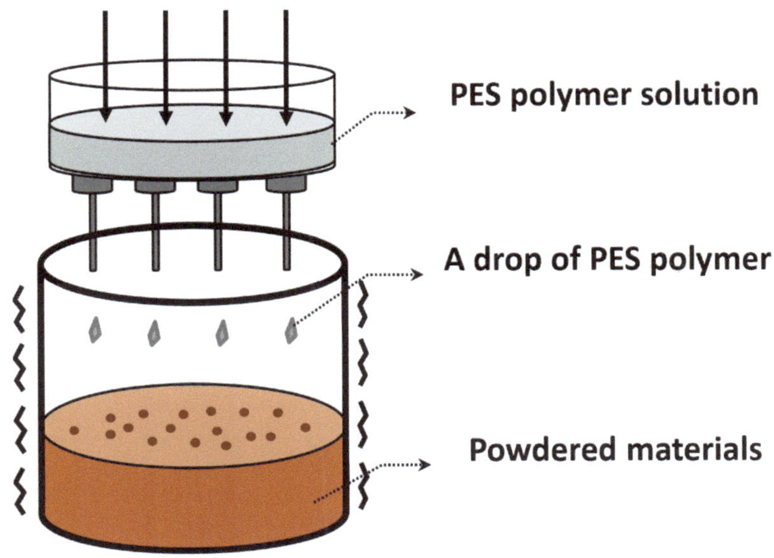

Figure 1. Schematic diagram of granulation process using PES polymer.

2.3. Characterization

2.3.1. X-ray Diffraction (XRD) Analysis

XRD patterns were measured using an X-ray diffractometer (X'Pert PRO Multi-Purpose X-ray Diffractometer, PANalytical) using Cu Ka radiation at 40 kV and 25 mA, with a wavelength equal to 1.5405980 Å. A sample of goethite, which has an orthorhombic structure, was recorded at scan speed of 1°/min at low 2θ angles from 20 to 80°. A sample of akaganeite, which has a tetragonal structure, was recorded at a scan speed of 1°/min at low 2θ angles from 10 to 40°. Finally, a sample of lepidocrocite, which has an orthorhombic structure, was recorded at a scan speed of 1°/min at low 2θ angles from 20 to 80°. The samples were then analyzed using selected area diffraction (SAD) patterns. The standard data were obtained from the International Centre for Diffraction Data (Goethite, ICDD No. 98-015-9962; Akaganeite, ICDD No. 96-900-1320; Lepidocrocite, ICDD No. 98-010-8876).

2.3.2. Transmission Electron Microscopy (TEM) Analysis

Powdered samples were prepared by dispersing them in ethanol and dried on a Cu grid coated with carbon film. Field emission transmission electron microscopy (JEM-2100F, JEOL, Peabody, MA, USA) was performed to obtain TEM images.

2.3.3. Nitrogen Adsorption–Desorption Measurement

Nitrogen adsorption–desorption measurements were performed at 77 K on a Micromeritics ASAP 2020 analyzer. A sample of powdered goethite was degassed for 24 h at 150 °C, while a sample of powdered akaganeite was degassed for 16 h at room temperature; a sample of powdered lepidocrocite was degassed for 20 h at 80 °C. Each sample of granular iron oxyhydroxide was degassed under the same conditions as those for each powdered iron oxyhydroxide sample. Then, the samples were analyzed using the Barett–Joyner–Halenda (BJH) method. The adsorption pore size distribution (PSD), which is based on nitrogen adsorption isotherms, was calculated. Additionally, the pore volume was obtained at the $P/P_0 \leq 0.976$ single point, and BET surface area was calculated from adsorption data.

2.3.4. Isoelectric Point Measurement and Zeta Potential Measurement

The point of zero charge pH (pH_{pzc}) for each powdered iron oxyhydroxide was determined using a zeta potential analyzer (Nano ZS, Malvern, Chester County, PA, USA).

Each iron hydroxide sample of 0.5 g/L was suspended in 0.01 M $NaNO_3$ solution (50 mL) as an electrolyte, and the aqueous suspension was equilibrated at different pH values using the HCl and NaOH solutions. The average zeta potential at a certain pH value was plotted on pH in the range from 3 to 11.

2.3.5. Scanning Electron Microscopy (SEM) Analysis

The cross-section of the granular iron oxyhydroxide was observed using a field emission scanning electron microscope (S-4700, Hitachi, Tokyo, Japan). Each sample was divided into blank, 30, 60, 120, 300, and 1440 s exposure times in air. All measurements were performed at 10 kV. An energy-dispersive X-ray spectrometer (EDS) was also used to confirm the chemical constituents of the surface. EDS analysis and the analysis of the distribution of Fe on the PES were conducted with a spot with an approximate diameter of 20 μm in selected areas on the sample.

2.3.6. Thermogravimetric Analysis (TGA)

The weights of each powdered and granular iron oxyhydroxide required for immobilization on the PES polymer were analyzed with increasing temperature using a thermogravimetric analyzer (TGA 2050, TA instruments, New castle, DE, USA) under a nitrogen flow of 100 mL/min.

2.4. Adsorption/Desorption Experiments

Batch Experiment

A series of batch experiments were conducted to evaluate the performance of the new material in terms of regeneration and removal of aqueous phosphate. Isothermal adsorption was conducted by adjusting the concentration of phosphate for powdered/granular adsorbents. Additionally, the kinetic experiment of adsorption/desorption was conducted with increasing time for powdered/granular adsorbents. The pH effect experiment was performed by adjusting the pH using HCl and NaOH solutions for the granular adsorbent. The desorption experiment was performed using NaOH 0.5 M solution for 6 h by injecting the adsorbent adsorbed from phosphate for granular adsorbents. Finally, 10-cycle regeneration was conducted through continuous adsorption/desorption experiments. During the regeneration experiments, the adsorbents were washed with DI water following the desorption, before the adsorption step was started again. All of the adsorption experiments were carried out in a 50 mL conical tube containing 50 mL of phosphate solution (initial concentration: 5 mg/L, except isothermal adsorption experiment) and 0.05 g of the powdered/granular adsorbents. All of the samples were then shaken at 150 rpm/min in a shake incubator at 25. Samples were filtered through a 0.2 μm cellulose acetate membrane filter, extracted by solid-phase extraction, and analyzed using the UV spectrum (HS-3300, HUMAS, Daejeon, Republic of Korea).

3. Results and Discussion

3.1. Characteristic Analysis of Iron Oxyhydroxide Material

3.1.1. XRD Analysis of Powdered Iron Oxyhydroxide

The XRD patterns of the synthesized iron oxyhydroxide nanoparticles are presented in Figure 2. The diffraction peaks obtained from the XRD analysis of FeOOH nanoparticles were compared with the standard database (The International Centre for Diffraction Data (ICDD), 1997) using PCPDFWIN Version 2.00 software. The red color lines indicate reference peaks applicable to each nanoparticle, as provided by ICDD. For synthesized akaganeite data, the 203 hkl peak displayed weakening intensity against the reference data, indicating that the lattice plane of 203, 200, 001 hkl in a unit cell showed amorphous structure responding to the weak diffraction from constructive interference by an irregular structure. A peak of 312 hkl with relatively strong intensities can suggest that the akaganeite phase formed at a length perpendicular to 312 hkl.

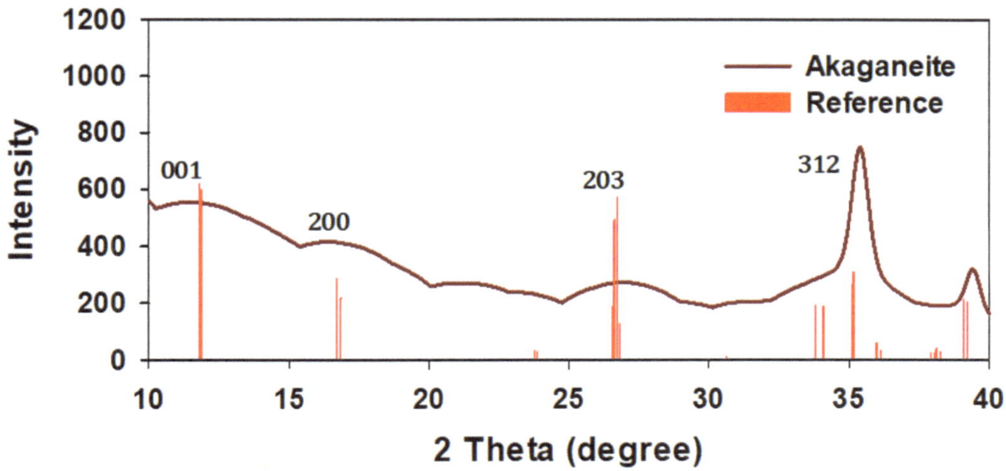

Figure 2. The diffraction peaks obtained from each iron akaganeite with those of the standard database (red).

3.1.2. Transmission Electron Microscopy (TEM) Analysis of Powdered Iron Oxyhydroxide

The morphology and crystallinity of each iron oxyhydroxide were investigated through TEM, as shown in Figure 3. The shape of goethite resembled a long stick, as mentioned in the XRD analysis. The goethite particle was over 200 nm in length and was confirmed to have crystallinity through SAD. For akaganeite, XRD showed particles similar to aggregated millet grains having an average size of 5 nm. SAD patterns for akaganeite indicated an amorphous phase, as seen in the XRD data of akaganeite. Overall, the study suggests that akaganeite appears to have a different phase and has a rod shape, which is a result of a blunt shape due to a noncrystalline structure. Lepidocrocite develops in a flower shape (Figure 3). The phase appears as flakes with very sharp tips, and the particles were observed in aggregation. The crystallinity of lepidocrocite was confirmed through SAD patterns. The lepidocrocite particles had a size of approximately 5–10 nm, and the size of the extended stem on the particle was less than 200 nm in length.

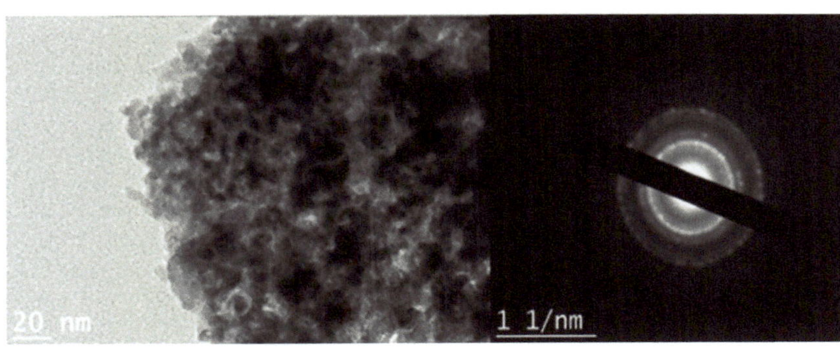

Figure 3. TEM images and selected area diffraction (SAD) patterns of akaganeite.

3.1.3. BET Analysis and Zeta-Potential Analysis of Powdered Iron Oxyhydroxide

The akaganeite had a higher BET surface area (231 m^2/g) than other materials. When considering the pore size and volume, the effective area of akaganeite is advantageous for adsorbing polluted substances with regard to contact between anions and those surfaces. This may be attributed to akaganeite forming a tunnel structure in a unit cell [28]. Meanwhile, akaganeite PZC was observed to have a higher pH (pH 10) than general oxyhydroxide (pH 7~9) [29]; lepidocrocite PZC had a pH of 9.5. Phosphate has a natural pH of

5~10 mg/L; thus, the zeta potential of akaganeite in the range of pH 6–7 is advantageous for adsorbing phosphate anions by attracting electrostatic force. Iron oxyhydroxide is known for adsorption via electrostatic forces or chemisorption. Chemisorption implies that it does not return to its original state on bonding with other anions. In the absence of this, it is acted upon by electrostatic forces from anions of hydroxyl groups, thus helping the desorption of the anion by switching the O^- on the FeOOH surface [21]. Further investigations are required to understand interfacial behavior in order to distinguish the types of bonds. This study attempted to understand the sorption behavior by conducting repetition tests on these powders.

3.2. Granulation of Akaganeite

3.2.1. Optimum Granulation Condition for Iron Oxyhydroxide Powder

Time variation with PES polymer exposure to the air during the process of granulation was monitored, and corresponding SEM images of cross-sections were obtained. Only granular PES without powdered akaganeite showed that bulky pores were located in internal granules and fine pores were located at the edge (Figure 4a). SEM images (b)–(f) suggest that powdered akaganeite was immobilized around the edge of the PES polymer. Though (a) and (b) appear similar, the latter has a rough surface on the edge when compared to blank (a), where the powdered akaganeite attaches to the PES polymer. With increasing the exposure time, a pore in an internal granule was converted to a huge hole, and the akaganeite layer formed a thin layer (Figure 4b–f). The reason for phase inversion has been attributed to the non-solvent/solvent exchange rate, with DI water as the non-solvent and the DMF as the solvent; this parameter plays the role of drying phase inversion phenomena that control the phase of the PES polymer in a membrane field [30]. The DMF solvent dissolved in water when immersed, and the position was filled with water by allowing water to form empty pores in the internal PES. Moreover, images (b)~(f) explain why more pores were produced with decreasing exposure time and the non-solvent/solvent exchange rate of water was faster than that of natural volatilization. These results suggest that a granule with much higher exposure time had a big hole, not a pore, and that the drying phase inversion did occur, which did not allow PES enough time to form pores (f); thus, PES polymer formed a much denser non-porous structure. Because of this, a granule of (f) must form a hole in the internal granule. As a result, granular akaganeite at 30 s of exposure time, which does not float on water, was selected and carefully analyzed.

To analyze the components of the granule surface, an edge was analyzed through field emission scanning electron microscopy with energy-dispersive X-ray spectroscopy (FESEM-EDS). Akaganeite formed the chemical constituents of $FeOOH \cdot H_2O$ (Fe 52.2%, H_2O 25.0%) [22], and polyethersulfone was used to immobilize particle forms ($C_{12}H_8O_3S$) [31].

According to the results obtained from qualitative analysis (Table 1), the S element existed only in the PES polymer, not in akaganeite. Thus, the S element can be a reference for the existence of PES. In order to measure the distribution of akaganeite powder in the akaganeite granule, EDS mapping was performed on the surface of the granule where the akaganeite was immobilized (#1 in Figure 5b) and the inside of the granule (#2 in Figure 5b). The O and Fe contents in #1 were higher than those in #2 (Figure 5a,b). Moreover, the EDS mapping results (red dot in (b) and white dot in (c) in Figure 5) confirmed that the Fe and O elements were situated close to the edge (left side in boundary). In contrast, the S element ((white dot) in (d)) was detected far from the edge. Thus, these data suggest that akaganeite was doped onto the PES polymer scaffolds and that the Fe was gradually distributed from the inner area to the edge of the granularity. These data verify the immobilization of the akaganeite particles from the PES polymer scaffolds.

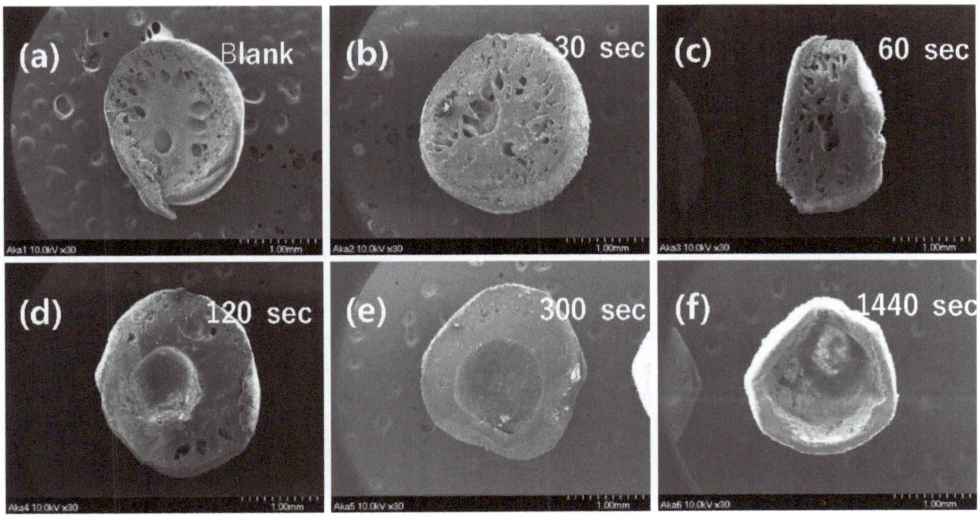

Figure 4. SEM images of cross-section area for synthesized granular akaganeite depending on the exposure time in the air. The samples are named as follows: blank (**a**), 30 s (**b**), 60 s (**c**), 120 s (**d**), 300 s (**e**), and 1440 s (**f**).

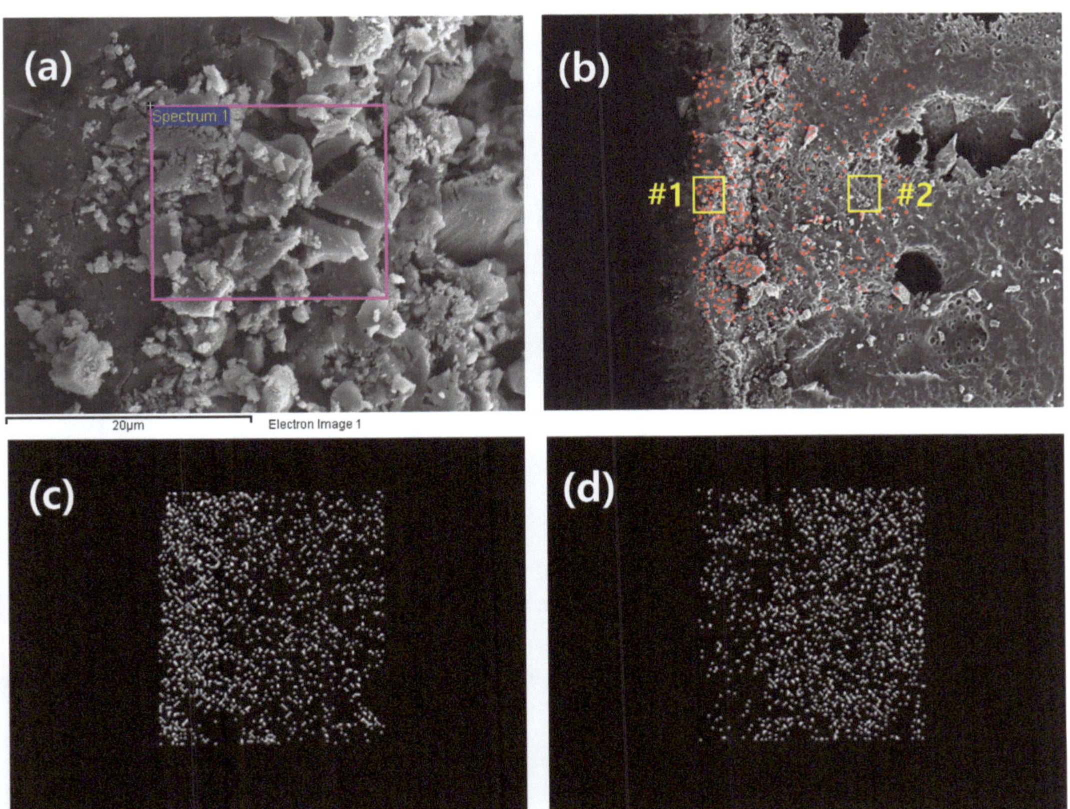

Figure 5. SEM images of cross-sections of granular akaganeite (**a**), EDS mapping of Fe (**b**), EDS mapping of O (**c**) and EDS mapping of S (**d**).

Table 1. The constituent elements from qualitative analysis of the PES polymer and cross-section area on the edge part of granular akaganeite as shown in Figure 5b.

Only PES			#1			#2	
Element	**Weight (%)**	**Atomic (%)**	**Element**	**Weight (%)**	**Atomic (%)**	**Weight (%)**	**Atomic (%)**
C K	53.60	66.36	O K	35.81	64.05	37.50	62.20
O K	26.03	24.20	S K	8.03	7.16	22.97	19.01
S K	20.37	9.45	Fe K	56.17	28.78	39.53	18.78
Total	100.00		Total	100.00		100.00	

3.2.2. TGA Analysis of Granular Iron Oxyhydroxide

TGA profiles of the thermal decomposition were analyzed to determine the PES polymer retention on a granule. First, powdered akaganeite showed weight loss by hydroxyl group decomposition up to 300 °C (Figure 6 (red line)). According to previous studies, akaganeite heated at temperatures above 300 °C transforms into hematite and magnetite [32,33]. Secondly, granular akaganeite showed a substantial weight loss at around 550 °C (Figure 6 (black line)); a substantial amount of PES polymer was volatilized due to the combination of carbon dioxide with the oxygen and carbon atoms. From this result, it is clear that the remaining PES in the granule accounted for approximately a 55% difference, with 86.05% for powdered akaganeite and 31.03% for granule akaganeite, suggesting that 55% of PES polymer was in granule form while the rest in powder form.

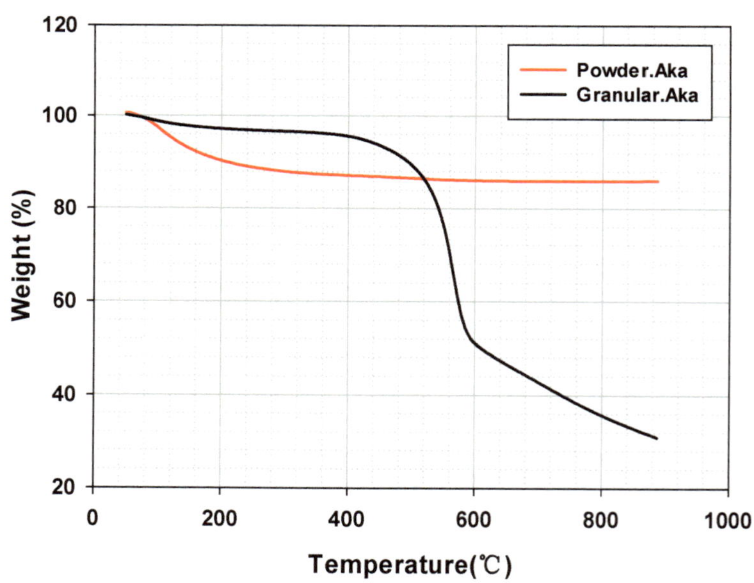

Figure 6. Thermogravimetric analysis curves of powdered/granular akaganeite.

3.2.3. BET Analysis of Granular Iron Oxyhydroxide

The hysteresis loop obtained from granular akaganeite showed type V with mesoporosity. This indicates that the pore shape of akaganeite immobilized onto PES polymer suggests ink-bottle-shaped pores (Figure 7 (blue line)) [29]. Moreover, no difference was identified between the pore sizes of powdered/granular akaganeite (Table 2), implying no structural changes after granulation. However, the BET surface area of granular akaganeite was reduced by 4 times compared with the powdered akaganeite, as suggested by SEM/TGA analysis. Preferentially, the TGA analysis confirmed that half-weight loss of beads was observed in each powdered akaganeite/PES polymer. The outer surface of the bead has an affinity for nitrogen adsorption (Figure 8a), but inner akaganeite immobilized onto the PES polymer does not contribute towards adsorption due to loss by fixation (Figure 8b). This means that the akaganeite particles were reduced to half. Thus, consideration of the amount of akaganeite in the bead and the effective area on the bead surface could explain why BET surface area had reduced to one-fourth of the adsorption capacity; the analysis accurately corresponded to the 4-fold reduction in surface area (Table 2).

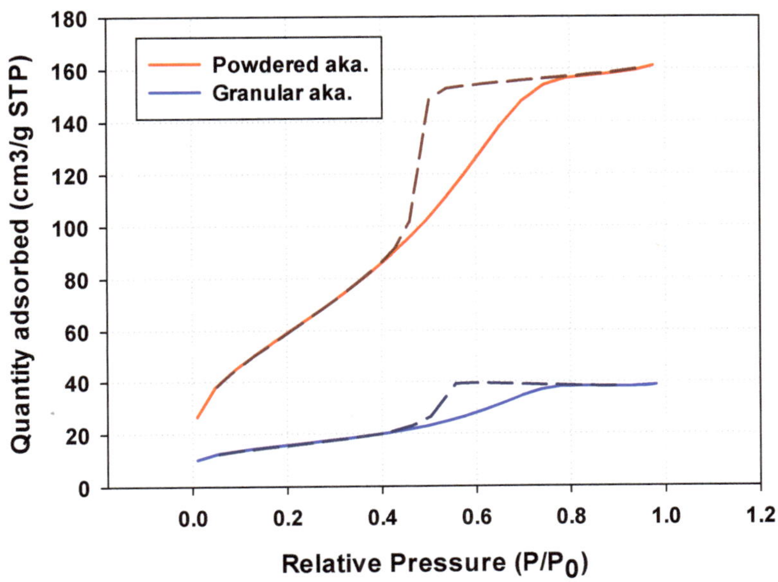

Figure 7. Nitrogen adsorption/desorption isotherm obtained for powdered/granular iron oxyhydroxide.

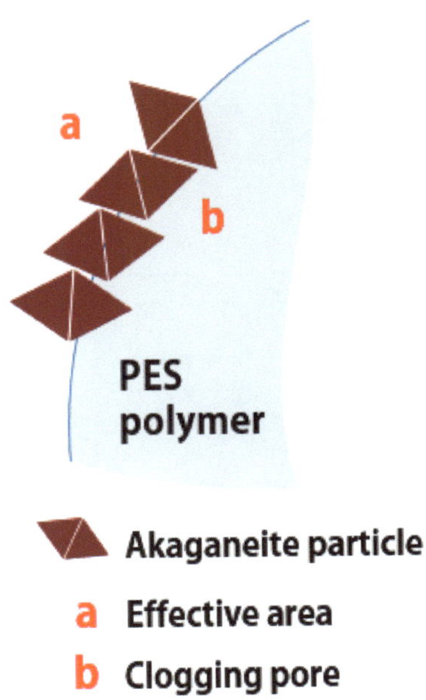

Figure 8. Schematic diagram of pore clogging.

Table 2. Comparison of physical properties of the powdered/granular akaganeite.

	Powdered Akaganeite	Granular Akaganeite
BET surface area (m^3/g)	231	54
Pore volume (cm^3/g)	0.249	0.060
Pore size (nm)	4.3	4.4

3.3. Adsorption Experiments

3.3.1. Kinetic Adsorption/Desorption Experiment

Kinetic adsorption/desorption of phosphate anions by GAK was measured using phosphate 5 mg/L as initial phosphate concentration, followed by desorption of adsorbed granular akaganeite using 50 mL of NaOH 0.5 M. The adsorption kinetic experiment data (Figure 9) were fitted into a pseudo-second-order rate kinetic model, and the rate constants and determination coefficients were tabulated (Figure 9). The adsorption/desorption kinetics indicate that the desorption rate was faster than the adsorption rate due to the strong base of NaOH 0.5 M with a pH range of 13~14. This is supported by the rate constant (k_2). Accordingly, each equilibrium time at 5 mg/L of phosphate was estimated as 12 h for adsorption and 6 h for desorption. Furthermore, the pseudo-second-order kinetic model calculated the equilibrium adsorption capacity (q_e) to be 2.71 mg/g and the desorption capacity to be 2.53 mg/g. The desorption rate was thus estimated to be 93%.

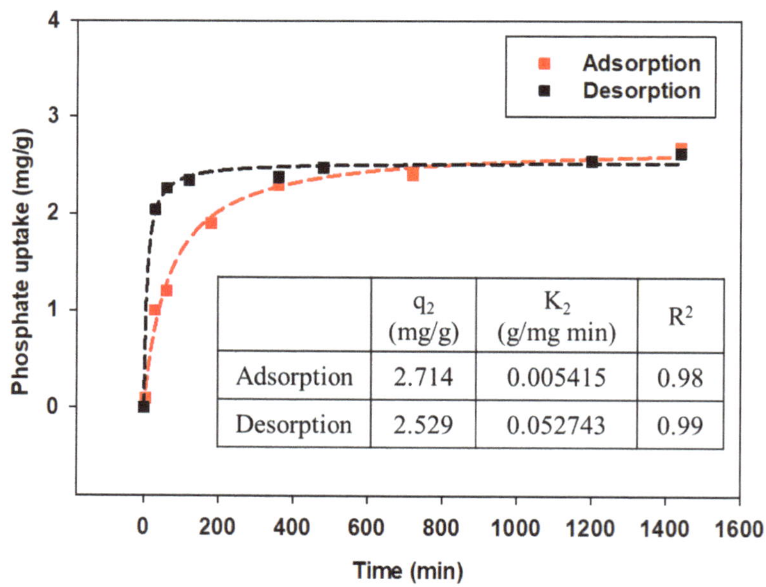

Figure 9. Effect of contact time on the adsorption of phosphate onto granular akaganeite applied by the pseudo-second-order kinetic model.

3.3.2. Isothermal Adsorption Experiment

Experimental data were expressed by q_e (amount of phosphate anions attached to adsorbents) versus Ce (equilibrium concentration of phosphate anions in the solution) plots, and the parameters were calculated by applying Freundlich and Langmuir isotherm models for each powdered/granular akaganeite (Figures 10 and 11). The adsorption isotherm model of powdered/granular akaganeite followed the Freundlich isotherm model with $R^2 > 0.88$ for both powdered and granular akaganeite, using the Sigma plot 12.0 program. This result was analyzed using the Freundlich model with a high R^2. It showed that powdered akaganeite has an affinity for phosphate anions, as shown by $n > 1.0$; conversely, granular akaganeite showed a lower affinity, as shown by $n < 1.0$. The maximum adsorption capacity was calculated to be 21.27 mg/g for powdered akaganeite and 3.65 mg/g for granular akaganeite. The reason for the lower adsorption capacity of granular akaganeite was that the BET surface area was decreased by the immobilization of the particles from the PES polymer, and only the outer surface had effective adsorption. Thus, it is possible to remove phosphate at a concentration of 5 mg/L, which a biological treatment process cannot remove.

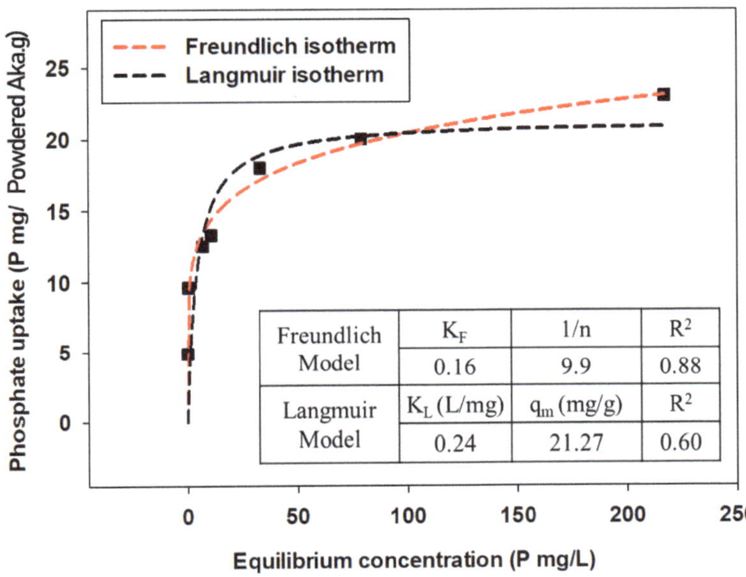

Figure 10. Adsorption isotherms for powdered akaganeite obtained using the Freundlich model (red dash line) and Langmuir model (black dash line).

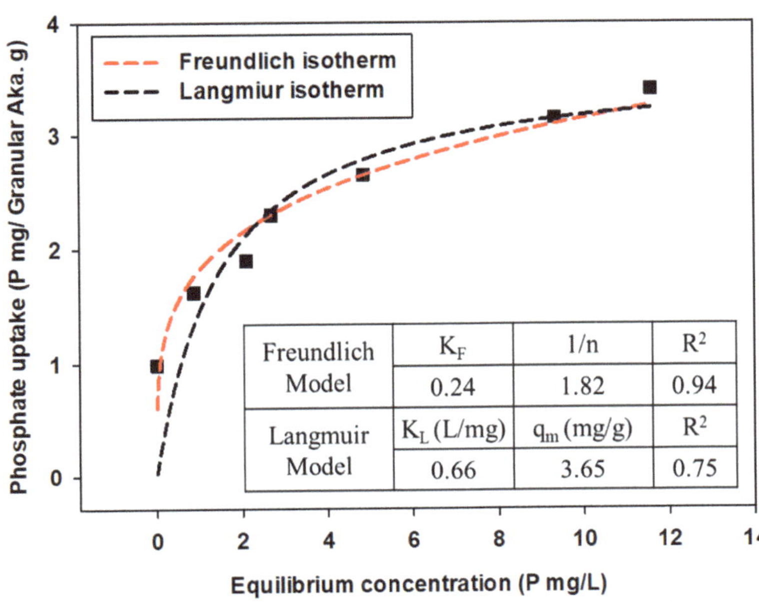

Figure 11. Adsorption isotherms for granular akaganeite obtained using the Freundlich model (red dash line) and Langmuir model (black dash line).

4. Conclusions

In this study, akaganeite with an FeOOH structure in iron oxyhydroxide was synthesized, and reusable materials were developed. Immobilization with the core–shell method using polyethersulfone was applied as a strategy to recover phosphate anions from a trace of phosphate solution. The synthesized granular akaganeite was successfully investigated through various analyses, such as SEM/TGA/BET, to understand its physical properties.

The XRD analysis of the powder form of the synthesized akaganeite showed a blunt peak, while TEM showed an aggregated millet grain shape with an average size of 5 nm. The SAD patterns of akaganeite verified the amorphous form owing to scattering patterns; these data were supported by the results from XRD analysis. Powdered akaganeite had a surface area of 231 mg^2/g and was estimated as a superior material given that the zeta potential of akaganeite was higher than that of goethite and lepidocrocite at pH 6–7, with 5 mg/L of phosphate. In the granulation process, to prevent the dispersion of the powder, granular akaganeite was synthesized using polyethersulfone as a scaffold. The granulation of akaganeite was carried out to improve the field applicability of intrinsic akaganeite. In the analysis of its characteristics, the result of the SEM images of the cross-section area of granular akaganeite depending on the exposure time was explained by the dry phase inversion due to DMF volatility participating in forming holes in an internal granule. SEM/EDS analyses of the edge surface of granular akaganeite suggest that akaganeite particles are well immobilized onto PES polymer scaffolds. Moreover, based on the pore size data, the physical properties were found to have remained unchanged. Finally, the surface area of granular akaganeite decreased by 4–5 times compared to powdered akaganeite. The adsorption kinetic of granular akaganeite was well fitted by the pseudo-second-order model ($R^2 > 0.99$ for adsorption and $R^2 > 0.99$ for desorption). The adsorption equilibrium time for granular akaganeite reached 12 h, while the desorption equilibrium time reached 6 h. Furthermore, the equilibrium adsorption capacity (q_e) was calculated as 2.71 mg/g using the pseudo-second-order kinetic model, while the desorption capacity was calculated as 2.53 mg/g. The Freundlich and the Langmuir isotherm models were applied using granular akaganeite as well.

Powdered akaganeite showed a high removal efficiency in affinity with phosphate anions. On the other hand, granular akaganeite showed a lower affinity. This suggests that the granulation of akaganeite with the PES polymer did not change its adsorption property. As a result, the maximum adsorption capacities calculated were 21.27 mg/g for powdered akaganeite and 3.65 mg/g for granular akaganeite.

Author Contributions: Conceptualization, J.B.; methodology, H.W. and S.K.; validation, S.B. and J.B.; investigation, S.O.; data curation, J.B.; writing—original draft preparation, H.W. and J.B.; writing—review and editing, S.O., S.K. and S.B.; supervision, J.B.; project administration, S.B.; funding acquisition, J.B. All authors have read and agreed to the published version of the manuscript.

Funding: This research was supported by the Korea Ministry of Environment under the Aquatic Ecosystem Conservation Research Program (No. 20230206-001). The authors are very grateful for the funds (Project #20230160-001) provided by the "Korea Institute of Civil Engineering and Building Technology" (KICT), Republic of Korea.

Institutional Review Board Statement: Not applicable for studies not involving humans or animals.

Informed Consent Statement: Not applicable for studies not involving humans.

Data Availability Statement: The data presented in this study are available on request from the corresponding author. The data are not publicly available due to privacy.

Conflicts of Interest: The authors declare no conflicts of interest.

References

1. Bajammal, S.S.; Zlowodzki, M.; Lelwica, A.; Tornetta, P., 3rd; Einhorn, T.A.; Buckley, R.; Leighton, R.; Russell, T.A.; Larsson, S.; Bhandari, M. The use of calcium phosphate bone cement in fracture in treatment: A meta-analysis of randomized trials. *J. Bone Jt. Surg.* **2008**, *90*, 1186–1196. [CrossRef] [PubMed]

2. Putt, M.S.; Milleman, K.R.; Milleman, J.L.; Ghassemi, A. Comparison of a dual-phase fluoride thoothpaste containing calcium, phosphate, and sodium bicarbonate with a regular fluoride thoothpaste on calculus formation. *Compend. Contin. Educ. Dent.* **2004**, *25*, 44–51. [PubMed]
3. Dal'Belo, S.E.; Gaspar, L.R.; Maia Campos, P.M. Moisturinzing effect of cosmetic formulations containing Aloe vera extract in different concentrations assessed by skin bioengineering techniques. *Ski. Res. Technol.* **2006**, *12*, 241–246. [CrossRef] [PubMed]
4. Xu, S.; Zhu, S.; Xiong, H. phosphate adsorption removals by five synthesized isomeric α-, β-, γ-FeOOH. *Water Air Soil Pollut.* **2022**, *233*, 454. [CrossRef]
5. Kalaitzidou, K.; Zouboulis, A.; Mitrakas, M. Thermodynamic study of phosphate adsorption and removal from water using iron oxyhydroxides. *Water* **2022**, *14*, 1163. [CrossRef]
6. Ginebra, M.P.; Espanol, M.; Montufar, E.B.; Perez, R.A.; Mestres, G. New processing approaches in calcium phosphate cements and their applications in regenerative medicine. *Acta Biomater.* **2010**, *6*, 2863–2873. [CrossRef]
7. de Ridder, M.; de Jong, S.; Polchar, J.; Lingemann, S. *Risks and Opportunities in the Global Phosphate Rock Market*; The Hague Centre for Strategic Studies: Hague, The Netherlands, 2012.
8. Jasinski, S.M. *Mineral Commodity Summaries 2011*; U.S. Geological Survey: Reston, VA, USA, 2011.
9. Cooper, J.; Lombardi, R.; Boardman, D.; Carliell-Marquet, C. The future distribution and production of global phosphate rock reserves. *Resour. Conserv. Recycl.* **2011**, *57*, 78–86. [CrossRef]
10. Abu-Obaid, S.; Aktij, S.A.; Tabe, S.; Sadrzadeh, M.; Farnood, R.R. Surfactant-modified adsorptive electrospun nanofiber membrane impregnated with akageneite for phosphorus recovery from wastewater. *J. Environ. Chem. Eng.* **2022**, *10*, 108786. [CrossRef]
11. Wilsenach, J.A.; Schuurbiers, C.A.; van Loosdrecht, M.C. Phosphate and potassium recovery from source separated urine through struvite precipitation. *Water Res.* **2007**, *41*, 458–466. [CrossRef]
12. Kim, S.-J.; Park, T.-J.; Kim, S.-H.; Yang, H.-J. *Characteristics of Residual Metals from Phosphorus Removal in Sewage Treatment Plants around Paldang Lake*; National Institute of Environmental Research in Korea: Incheon, Republic of Korea, 2012.
13. Cheng, W.; Li, J.; Sun, J.; Luo, T.; Marsac, R.; Boily, J.F.; Hanna, K. Nalidixic Acid and Fe (II)/Cu (II) Coadsorption at Goethite and Akaganéite Surfaces. *Environ. Sci. Technol.* **2023**, *57*, 15680–15692. [CrossRef]
14. Boparai, H.K.; Joseph, M.; O'Carroll, D.M. Kinetics and thermodynamics of cadmium ion removal by adsorption onto nano zerovalent iron particles. *J. Hazard. Mater.* **2011**, *186*, 458–465. [CrossRef] [PubMed]
15. Altundogan, H.S. Cr (VI) removal from aqueous solution by iron (III) hydroxide-loaded sugar beet pulp. *Process Biochmistry* **2005**, *40*, 1443–1452. [CrossRef]
16. Phuengprasop, T.; Sittiwong, J.; Unob, F. Removal of heavy metal ions by iron oxide coated sewage sludge. *J. Hazard. Mater.* **2011**, *186*, 502–507. [CrossRef] [PubMed]
17. Gong, J.-L.; Wang, X.-Y.; Zeng, G.-M.; Chen, L.; Deng, J.-H.; Zhang, X.-R.; Niu, Q.-Y. Copper (II) by pectin-iron oxide magnetic nanocomposite adsorbent. *Chem. Eng. J.* **2012**, *185–186*, 100–107. [CrossRef]
18. Gu, B.; Lian, L.; Dickey, J.; Yin, X.; Dai, S. Reductive precipitation of uranium (VI) by zero-valent iron. *Environ. Sci. Technol.* **1998**, *32*, 3366–3373. [CrossRef]
19. Hanesch, M. Raman spectroscopy of iron oxides and (oxy)hydroxides at low laser power and possible application in environmental magnetic studies. *Geophys. J. Int.* **2009**, *177*, 941–948. [CrossRef]
20. Navrotsky, A.; Mazeina, L.; Majzlan, J. Size-driven structural and thermodynamic complecity in iron oxides. *Science* **2008**, *319*, 1635–1638. [CrossRef]
21. Kim, J.; Li, W.; Philipsc, B.L.; Grey, C.P. Phosphate adsorption on the iron oxyhydroxides goethite (α-FeOOH), akaganeite (β-FeOOH), and lepidocrocite (γ-FeOOH): P NMR study. *Energy Environ. Sci.* **2011**, *4*, 4298–4305. [CrossRef]
22. Chitrakar, R.; Tezuka, S.; Sonoda, A.; Sakane, K.; Ooi, K.; Hirotsu, T. Phosphate adsorption on synthetic goethite and akaganeite. *J. Colloid Interface Sci.* **2006**, *298*, 602–608. [CrossRef]
23. Li, N.; Bai, R. Copper adsorption on chitosan-cellulose hydrogel beads: Behaviors and mechanisms. *Sep. Purif. Technol.* **2005**, *42*, 237–247. [CrossRef]
24. Chang, J.; Woo, H.; Ko, M.S.; Lee, J.; Lee, S.; Yun, S.T.; Lee, S. Targeted removal of trichlorophenol in water by oleic acid-coated nanoscale palladium/zero-valent iron alginate beads. *J. Hazard. Mater.* **2015**, *293*, 30–36. [CrossRef] [PubMed]
25. Grisdanurak, N.; Akewaranugulsiri, S.; Futalan, C.; Tsai, W.-C. The study of copper adsorption from aqueous solution using crosslinked chitosan immobilized on bentonite. *J. Appl. Polym. Sci.* **2012**, *125*, 132–142. [CrossRef]
26. Liu, R.; Guo, Y.; Odusote, G.; Qu, F.; Priestley, R.D. Core-shell Fe_3O_4 polydopamine nanoparticles serve multipurpose as drug carrier, catalyst support and carbon adsorbent. *ACS Appl. Mater. Interfaces* **2013**, *5*, 9167–9171. [CrossRef] [PubMed]
27. Li, J.-F.; Xu, Z.-L. Effect of TiO_2 nanoparticles on the surface morphology performance of microporous PES membrane. *Appl. Surf. Sci.* **2009**, *255*, 4725–4732. [CrossRef]
28. Song, X. Surface and Bulk Reactivity of Iron Oxyhydroxides: A Molecular Perspective. Ph.D. Thesis, Umeå University, Umeå, Sweden, 2013.
29. Jemutai-Kimosop, S.; Okello, V.A.; Shikuku, V.O.; Orata, F.; Getenga, Z.M. Synthesis of mesoporous akaganeite functionalized maize cob biochar for adsorptive abatement of carbamazepine: Kinetics, isotherms, and thermodynamics. *Clean. Mater.* **2022**, *5*, 100104. [CrossRef]
30. Klaysom, C.; Moon, S.-H.; Ladewig, B.; Lu, G.Q.; Wang, L. Preparation of porous ion-exchange membranes (IEMs) and their characterizaotns. *J. Membr. Sci.* **2011**, *371*, 37–44. [CrossRef]

31. Wen, S.; Lu, Y.; Dai, J.; Huang, X.; An, S.; Liu, J.; Liu, Z.; Du, Y.; Zhang, Y. Stability of organic matter-iron-phosphate associations during abiotic reduction of iron. *J. Hazard. Mater.* **2023**, *449*, 131016. [CrossRef]
32. Asenath-Smith, E.; Estroff, L.A. Role of Akaganeite (β-FeOOH) in the Growth of Hematite (α-Fe_2O_3) in an inorganic silica hydrogel. *Cryst. Growth Des.* **2015**, *15*, 3388–3398. [CrossRef]
33. Choi, J.; Cha, J.; Lee, J.-K. Synthesis of various magnetite nanoparticles through simple phase transformation and their shape dependent magnetic properties. *RSC Adv.* **2013**, *3*, 8365–8371. [CrossRef]

energies

Article

An Experimental Study on the Effect of Nanofluids on the Thermal Conductivity and Rheological Properties of a Coolant for Liquids

Le Sun [1,*], Jiafeng Geng [2], Kaijun Dong [1] and Qin Sun [1]

[1] Guangzhou Institute of Energy Conversion, Chinese Academy of Sciences, Guangzhou 510640, China; dongkj@ms.giec.ac.cn (K.D.); sunqinsq@hotmail.com (Q.S.)
[2] School of Water and Environment, Chang'an University, Xi'an 710064, China; gengjf@chd.edu.cn
* Correspondence: sunyuexuzhou@hotmail.com; Tel.: +86-020-38455713

Abstract: Thermal conductivity and viscosity are important properties for nanofluids as they significantly affect the flow and heat transfer process. To date, the rheological properties of water-based nanofluids have been well studied, while the results are scarce for non-aqueous nanofluids. In this study, the thermal conductivity and rheological properties of two different kinds of oxide nanofluids (CuO and Al_2O_3) in a typical commercial data center focusing on liquid coolants were systematically investigated at different mass fractions and temperatures. The results showed that the addition of nanoparticles can significantly improve the heat conduction capacity of mineral oil coolants. There is an average increase in thermal conductivity of up to 20–25%. The shear rate–shear stress and shear rate–viscosity curves all showed that mineral oil coolant-based oxide nanofluids behaved as Newtonian fluids and that nanoparticles did not cause the increment in viscosity. The effect of temperature on rheological properties was also studied, and the result showed that high temperatures resulted in low viscosity and shear stress. Finally, the effect of particle type was investigated, and it was found that no matter what kind of nanoparticles were added, their effects on the rheological behaviors were the same.

Keywords: liquid cooling; nanofluids; heat conductivity; viscosity; rheology

Citation: Sun, L.; Geng, J.; Dong, K.; Sun, Q. An Experimental Study on the Effect of Nanofluids on the Thermal Conductivity and Rheological Properties of a Coolant for Liquids. *Energies* **2024**, *17*, 1313. https://doi.org/10.3390/en17061313

Academic Editors: Robert A. Varin and Geng Zhong

Received: 19 January 2024
Revised: 29 February 2024
Accepted: 7 March 2024
Published: 8 March 2024

1. Introduction

Nanofluids have attracted much attention in the past several years thanks to the small sizes and large specific surface areas of nanoparticles [1–9]. In general, nanofluids are defined as a new kind of heat transfer fluid with nanosized particles dispersing into conventional working fluids, such as water, methanol, ethanol, glycol and oil [10,11]. Various metallic or nonmetallic nanoparticles can be applied to form nanofluids, for example, Cu, CuO, Al_2O_3, SiO_2, TiO_2, Fe_3O_4, etc. [12–24]. Many researchers have found that compared to fluids without nanoparticles, nanofluids have many superior features, such as enhanced solar energy absorption, high heat transfer efficiency, an enhanced rate of oil and gas recovery and other properties related to spreading, wetting and antibacterial activity [25].

To date, most informed research studies regarding nanofluids have only focused on the behaviors during the thermal conduction, convective heat transfer and phase change heat transfer processes. Although very little attention has been paid to the rheological properties of nanofluids, especially for nanofluids applied in the data center cooling process, rheological properties play equally significant roles as thermal parameters in the engineering applications and parameters of nanofluids. For example, the viscosity of working fluids has a tight relation with pressure drop, which determines the power of the pump. Overall, from the point view of engineering applications, it is valuable to find ideal nanofluids that not only have high thermal conductivity, but also possess low viscosity.

The most widely used metal nanoparticle in the literature is Cu. Leong [18] investigated the heat transfer enhancement of ethylene glycol-based copper nanofluids in an automotive cooling system. The volume concentration of nanoparticles they used ranged from 0% to 2%, and they studied the heat transfer rate in different air and coolant conditions with different Reynolds numbers. They found that with the increase in nanoparticles, the heat transfer rate can increase by up to 3.8%. The increase in the Reynolds number also has an effect on the heat enhancement of Cu-ethylene glycol nanofluids. When the air's Reynolds number was increased from 4000 to 6000, the heat transfer rate showed a 45.2% enhancement. But when the Reynolds number of the coolant was increased from 5000 to 7000, the enhancement of the heat transfer rate was not obvious (only 0.4%).

Another commonly used metal nanoparticle is Ag. Behrangzade and Heyhat [26] compared the effect of using nano-silver–water nanofluid and pure water for energy efficiency enhancement at the same flow rate and same Reynolds number, respectively. By using 100 ppm of nano-silver dispersed water-based nanofluid, they obtained a 16.79% enhancement in the overall heat transfer coefficient, while the pressure drop value did not exhibit an obvious change.

Al_2O_3 nanofluids are the most widely investigated type of nanofluids in the literature. Hassani et al. [27] used different volume fractions of Al_2O_3 nanoparticles to synthesized nanofluids. Al_2O_3 nanofluids were used as nanofluid coolants for an electronic heat sink. The results disclosed that Al_2O_3 nanofluids with 0.5% and 1.0% volume fractions can enhance the performance index of the heat sink by averages of 14.7% and 28.3%.

Al_2O_3–water nanofluids were used by Nguyen [28] in an electronic liquid cooling system. Among the different volume concentrations of Al_2O_3–water nanofluids tested in their experiments, it was found that 6.8% of the Al_2O_3 particle can enhance the heat transfer coefficient by 40% compared to pure water. And the particle size also has an effect on heat transfer. A 36 nm Al_2O_3 particle is better at enhancing heat transfer coefficients than a 47 nm particle.

CuO is another commonly studied nanosized metal oxide material. Chein and Chuang [29] designed a series of experiments to investigate the effect of using a CuO-water nanofluid as a coolant. The CuO particle volume fraction ranges from 0.2% to 0.4% in nanofluids. The authors found that a CuO-water nanofluid can take away more heat than pure water in a microchannel heat sink at a low flow rate. But when the flow rate is high, the nanofluid has little effect on extra heat transfer and the heat transfer is mainly affected by the volume flow rate.

The performance of a CuO-water nanofluid in a heat sink was also studied by Sarafraz [30]. A CuO-water nanofluid, gallium liquid metal and water were used to transfer the heat of a CPU when it worked separately at three states (standby, normal and overload). Compared to the other two working fluids, a CuO-water nanofluid can provide higher thermal performance than water and lower pressure drop and pumping power than gallium liquid metal when the heat flux is not very high.

The heat transfer performance and rheological properties of a coolant are important for the practical application of liquid cooling in data centers. To date, the heat transfer and rheological properties of water-based nanofluids have been sufficiently investigated. However, both the thermal conductivity and viscosity of nanofluids based on data center coolants have rarely been studied. Thus, in the present work, we report on the synthesis of nanofluids based on a data center coolant containing CuO and Al_2O_3 nanoparticles separately. We also measure and analyze their thermal conductivity and rheological behaviors, including shear stress and viscosity. Additionally, the effects of temperature, particle type and volume fraction are fully investigated and discussed. Our experimental work will be useful for the optimal design of various electronic components in cooling processes that use nanofluids as working fluids.

2. Materials and Methods

2.1. Materials and Preparation

Commercially available data center coolant (YT198, purchased from Amer Technology Co., Ltd., Dongguan, China) was used as base fluid. The basic parameters of the coolant are listed below (see Table 1). Various volume fractions of CuO (40 nm, Macklin Biochemical Cooperation, Shanghai, China) and Al_2O_3 (80 nm, Sinopharm Chemical Reagent Cooperation, Shanghai, China) nanofluids, respectively, were formulated. And then the rheological properties of the nanofluids were investigated with a rotating rheometer (Haake Mar 60, Thermo Fisher Scientific Inc., Waltham, MA, USA) over a shear rate ranging from 0 s^{-1} to 100 s^{-1}.

Table 1. Typical properties of coolant (YT198) used for this study.

Density (20 °C)	Pour Point	Flash Point	Acidity	Specific Heat (40 °C)
804.3 kg/m^3	−38 °C	198 °C	0.03 mgKOH/g	2.089 KJ/kg·K
Breakdown Voltage	**Relative Permittivity (90 °C)**	**Volume Resistivity (20 °C)**	**Surface Tension**	**Global Warming Potential**
62 KV	2.039	1.9×10^{10} Ω·cm	16 mN/m	0

Table 2 shows the typical properties of the nanoparticles employed for this study; the properties were measured with reliable accuracy by reagent manufacturers.

Table 2. Typical properties of nanoparticles employed for this study.

Nanoparticle	Particle Size	Density (20 °C)	Melting Point	Boiling Point
CuO	40 nm	6.315 g/cm^3	1326 °C	1026 °C
Al_2O_3	80 nm	1.06 g/cm^3	2000 °C	2977 °C

In order to disperse nanoparticles into the base fluid fully and uniformly, an ultrasonic homogenizer was applied in this experiment. The ultrasonic homogenizer (YT-JY99-IIDN, Shanghai Yetuo Technology Co., Ltd., Shanghai, China) has a max power of 1800 W, and the diameter of its ultrasonic horn is 22 mm. A picture of the ultrasonic homogenizer used in this work is shown in Figure 1. The main reasons for our choice of these kinds of nanoparticles are that they can be easily obtained in industry and their chemical properties are very stable but not harmful to human beings.

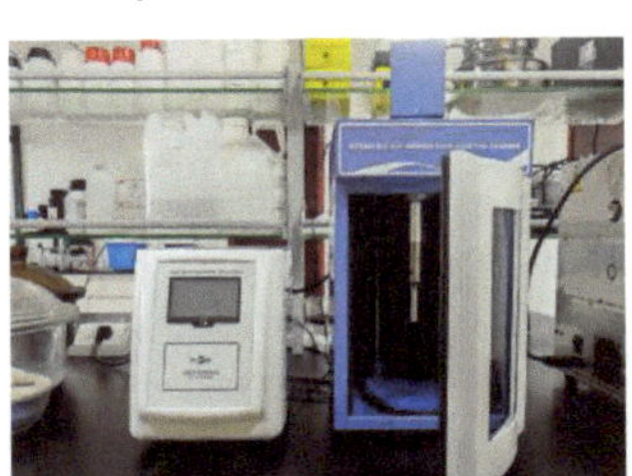

Figure 1. The ultrasonic homogenizer used in this work.

To start the experiment, the two kinds of nanoparticles with different volume fractions were added into the coolant using the following two-step method. The relationship between mass fraction and volume fraction is listed hereinafter (see Equation (1)):

$$\varphi = \frac{\omega \cdot \rho_{bf}}{(1 - \frac{\omega}{100})\rho_p + \frac{\omega}{100}\rho_{bf}} \tag{1}$$

In this equation, φ and ω stand for volume fraction and mass fraction, respectively, while ρ represents the density of the two different kinds of nanoparticles, and subscripts p and bf stand for particles and base fluid, respectively.

Firstly, the different kinds of nanoparticles were accurately weighed using an electronic balance and were then added in a methanol base fluid. The mass fractions investigated in our study are 0.01%, 0.05%, 0.1% and 0.15%, respectively. After that, the mixtures of the nanoparticles and base fluid were placed into an ultra-sonic homogenizer so that the nanoparticles could be fully and uniformly dispersed in the coolant. After 3 h of ultrasonic treatment, the methanol-based CuO and Al_2O_3 nanofluids were ready for use. And then a certain volume of nanofluids (19 mL) was taken out as a sample, and the rheological investigation was conducted.

2.2. *Thermal Conductivity and Rheological Investigation*

Thermal conductivity was measured using HotDisk TPS2500S (Hotdisk AB, Co., Ltd., Gothenburg, Sweden). A photo of the thermal conductivity measure system is shown in Figure 2. More details about the instrument can be found in reference [31].

Figure 2. HotDisk TPS2500S, used to measure thermal conductivity.

A Haake Mars 60 rheometer (See Figure 3) was mainly used for the rheological measurement. The temperature was controlled by a refrigerated circulating bath connected with the rheometer. A schematic diagram of the apparatus is given in Figure 3. More details about this rheometer can be found in reference [32].

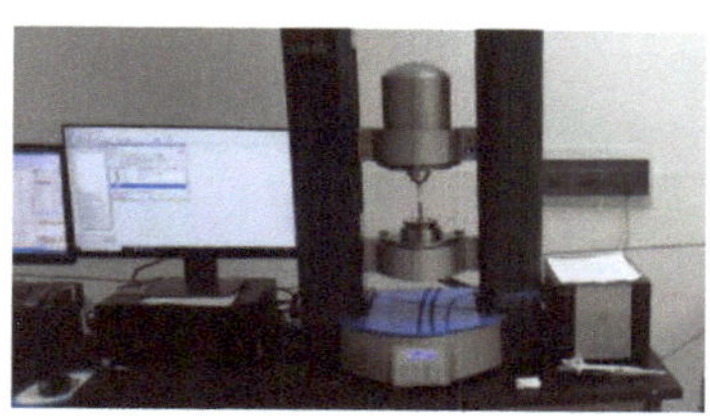

Figure 3. Haake Mars 60 rheometer used to study the rheological behavior.

The shear stress and viscosity of CuO-YT198 and Al_2O_3-YT198 at various volume fractions (0.01%, 0.05%, 0.1% and 0.15%) were measured through two programed modes, respectively: shear rate sweep and temperature sweep. In shear rate sweep mode, the shear rates were increased from 0 s^{-1} to 100 s^{-1} at 25 °C; at the same time, the shear stress and viscosity were measured. For temperature sweep, the temperature rose from 25 to 70 °C at a heating rate of 1 °C/min. At the same time, shear stress and viscosity were measured, while the shear rate was fixed at 50 s^{-1}.

2.3. *Experimental Uncertainty*

The thermal conductivity (k) was calculated according to Fourier's Law [33] using Equation (2):

$$k = \frac{q \times s}{\Delta T} \tag{2}$$

where q is the measured heat flux (W/m^2) on the face of the wall specimen, s is the thickness of the wall specimen (meters) and ΔT is the temperature difference between the two wall specimens' faces (°C).

The sensors were calibrated, and the uncertainty of each of sensor was ±0.01 °C (temperature sensor), ±0.1 W/m^2 (heat flux sensor) and ±0.01 mm (caliper for specimen thickness). The uncertainty of the final thermal conductivity result was associated with the propagation of uncertainty of indirect experimental measurements [34]. The calculated uncertainty propagation for thermal conductivity is ±0.00005 W/m·K.

The error value of the rheological measurements depended on the rheological measurement device used and its settings. The whole set of the determined characteristics was defined by repeating experiments under identical initial conditions. Many factors can result in random errors, for example, uncontrolled vibrations during measurements [35]. To process the results, including the identification and elimination of gross errors, standard approaches [36] were used, including calculating the mathematical expectation (Equation (4)), variance of a random variable (Equation (5)), and standard deviation for each series (Equation (6)).

$$M_x = \frac{1}{n}\sum_{i=1}^{n} X_i \tag{3}$$

$$V = \frac{1}{n-1}\sum_{i=1}^{n}(X_i - M_x)^2 \tag{4}$$

$$\sigma = V^{1/2} \tag{5}$$

where M_x is the mathematical expectation; X_i is the measurement result; n is the number of measurements; V is the variance; and σ is the standard deviation. The calculated uncertainty propagation for the viscosity is ±0.017 mPa·s.

3. Results and Discussion

3.1. The Basic Properties of the Investigated Nanofluids

The particle size distribution and morphology of the particles not only affect the dispersion stability of the nanofluid, but also the heat transfer characteristics [37]. A large number of studies have shown that the optimization of the effective thermal conductivity is a function of particle size reduction. This effect is even more obvious when the fluid temperature and particle concentration increase.

When the particle size is less than 100 nm, the thermal conductivity decreases with the increase in particle size. Beyond this particle size, further increases in particle size do not have effects on the effectiveness of thermal conductivity. However, even for the same nanofluid, the rate of change of thermal conductivity with nanoparticle size distribution at the same particle concentration and fluid temperature has different conclusions in different studies.

Figure 4a,b exhibit SEM images of CuO and Al_2O_3, respectively. The shapes and morphologies of the employed nanoparticles in our experiments can be seen in this figure. In Figure 4, one can see that the average size of a CuO nanoparticle is 40 nm, which conforms to the declared diameter, while the average size of an Al_2O_3 nanoparticle is 35 nm, which is much less than its declared diameter.

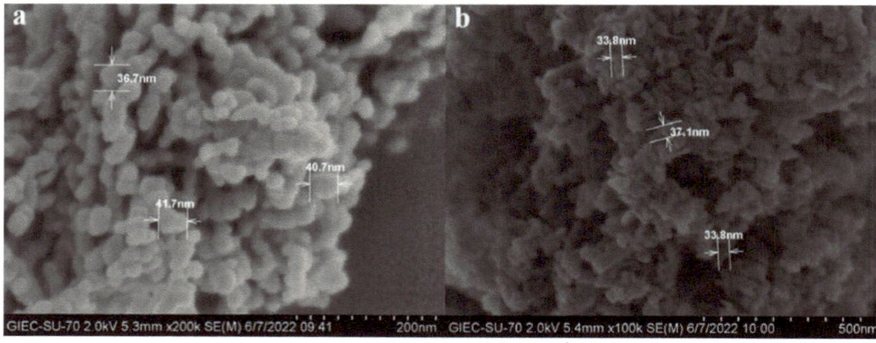

Figure 4. SEM images of (**a**) CuO and (**b**) Al_2O_3 nanoparticles.

3.2. Thermal Conductivity Enhancement of Nanofluids

In the following study, the effects of adding different volume fractions of nanoparticles are discussed. Before measuring the thermal conductivity of the nanoparticle coolant system, the thermal conductivity was tested, and the result is shown in Figure 5. As we can see from Figure 5, the thermal conductivity of pure YT198 was 0.1390 and 0.1450 W/(m·K) at 25 °C and 55 °C, respectively. After the addition of CuO and Al_2O_3 nanoparticles, the thermal conductivity increased obviously, regardless of the amount and kind of nanoparticles and the temperature.

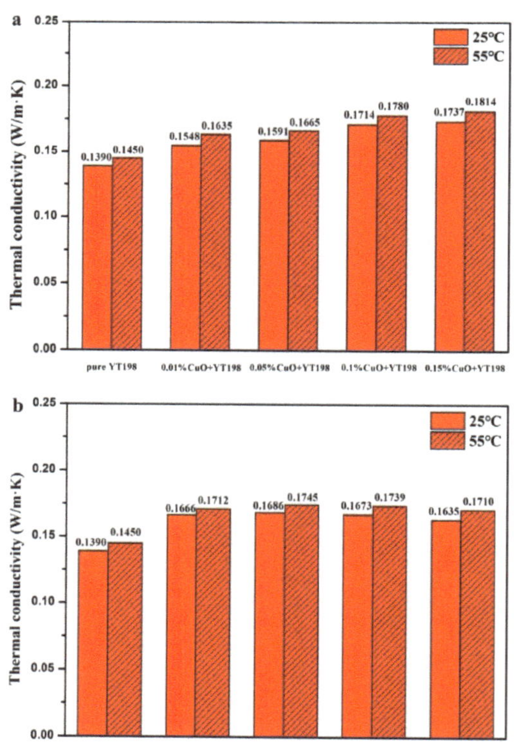

Figure 5. Thermal conductivity of pure coolant and coolant with different volume fractions: (**a**) CuO and (**b**) Al_2O_3 nanoparticles.

In Figure 5a, one can see that with the increment in the CuO nanoparticle's volume fraction, the thermal conductivity also increased with monotony; when the volume concentration of CuO nanoparticles was 0.15%, the thermal conductivity reached 0.1737 W/(m·K) at 25 °C and 0.1814 W/(m·K) at 55 °C. Compared with pure YT198, these growths in thermal conductivity represent growths of 20% and 25%, respectively.

Temperature variations affect the dispersion of nanoparticles in base fluids. High temperatures can result in more intense Brownian motion. Thus, the aggregation and sedimentation of nanoparticles can be hindered.

Similar laboratory findings can also be seen in Figure 5b; the addition of very few Al_2O_3 nanoparticles (0.01%) can cause a great enhancement in thermal conductivity from 0.1390 and 0.1450 W/(m·K) to 0.1666 and 0.1712 W/(m·K) at 25 °C and 55 °C, respectively, and the increasing percentages can be as large as 20% and 18%. But the growth trend is not synonymous with the volume of nanoparticles. The largest enhancement in thermal conductivity occurred when the volume fraction of Al_2O_3 + YT198 nanofluids was 0.05%; the increasing percentage reached 20% and 21% compared with the pure coolant. When comparing Figure 5a,b, it can be seen that the CuO nanoparticles had a slightly more significant improvement in thermal conductivity than the Al_2O_3 nanoparticles overall.

A possible explanation for the better thermal conductivity enhancement of the CuO nanoparticle in comparison to that of the Al_2O_3 nanoparticle is that CuO nanoparticles have greater conductivity [38]. Furthermore, as the number of nanoparticles increases, there may be more rapid interactions between the nanoparticles, and the Brownian motion of the nanoparticles may increase accordingly.

Durga Bhavani et al. observed that Al_2O_3 nanofluids had high compressibility at a high concentration [39]. When the concentration of Al_2O_3 nanofluids increased, the bulk modulus also increased, and the adiabatic compressibility decreased, which led to a strong cohesive interaction force among the molecules and atoms. Al_2O_3 particles tend to move less freely, and the Brownian motion of these nanoparticles also decreases. This may explain the decrease in thermal conductivity when the concentration of Al_2O_3 nanofluids surpassed 0.1%.

Surface modifications of nanoparticles contribute to the enhanced properties of nanofluids through the anti-aggregation effect produced by surface-modified nanoparticles. As studied by Linyang Dan. et al. [40], due to the Coulombic energy from the heteroatoms of SiO_2 nanoparticles, nanoparticles lacking alkyl chains can form aggregates and show a strong tendency to approach each other. By grafting alkyl chains on the surfaces of SiO_2 nanoparticles, Coulombic interactions can be shielded effectively, and the van der Waals interaction with natural esters can increase. Thus, the aggregation tendency of nanoparticles can be reduced. Furthermore, surface modification can reduce the diffusion ability of nanoparticles under the steric hindrance effect, thereby inhibiting the collision and aggregation of nanoparticles.

The pH mainly affects nanofluids in the following two aspects: the stability of nanofluids and the surface tension of nanofluids [41]. By adjusting the pH, the surface tension of a nanofluid can be changed. A pH value that is too high or too low can lead to corrosion and rusting [42].

Surfactants can increase the stability of nanofluids. The addition of surfactants does not have a great effect on the thermal conductivity, but it can increase the viscosity of the nanofluid.

The thermal conductivity of the base fluid has a great influence on the thermal conductivity of the nanofluid. The thermal conductivity of water-based nanofluids is generally higher than that of other nanofluids. The thermal conductivity of synthetic oil-based nanofluids is greater than that of water-based nanofluids mainly because synthetic oil-based nanofluids are more stable.

3.3. Rheological Properties of Various Nanofluids at Different Shear Rates

As typical experimental results, Figure 6 shows the shear stress vs. shear rate curve of a pure coolant and various nanofluids with different volume fractions at room temperature. In Figure 6, we can see that regardless of whether nanoparticles were added or not, pure

YT198 and nanofluids all exhibited properties of Newtonian fluids. The shear stress of pure YT198 and nanofluids showed linear monotonic growth as the shear rate increased. For example, in the case of CuO nanofluids, there is a small increase in the shear stress after adding CuO nanoparticles compared to pure coolant. But there is little difference between various volume fractions of CuO nanofluids. This trend is consistent with the findings of our previous works [43,44]. In our previous work [43], we found that with the mass fractions of CuO, Cu_2O and TiO_2 being increased from 0.01% to 0.15%, the viscosity showed almost no obvious change.

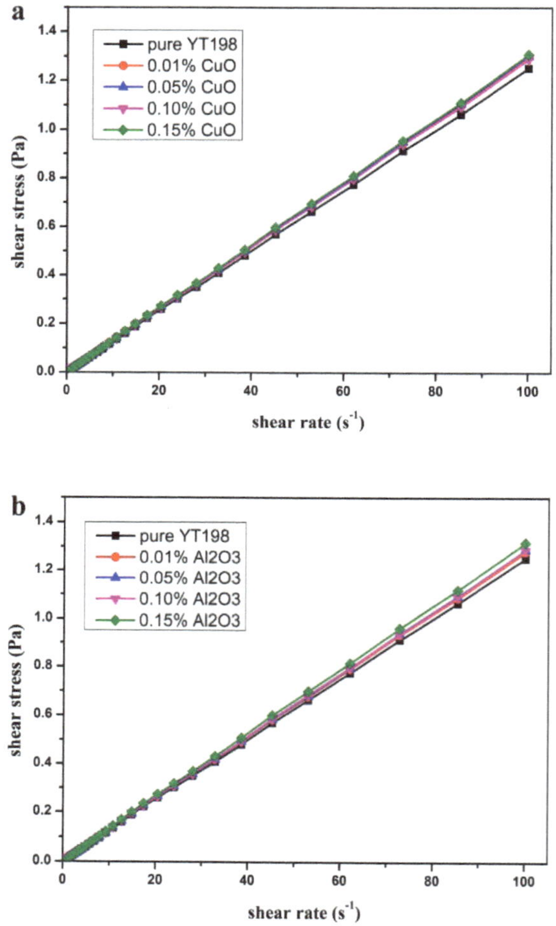

Figure 6. Shear stress vs. shear rate for various nanofluids with different volume fractions of (**a**) CuO and (**b**) Al_2O_3 nanoparticles at 25 °C.

The Al_2O_3 nanofluids also showed similar trends, and 0.15% Al_2O_3 nanofluids showed the highest shear stress. One explanation for this phenomenon may be the low levels of nanoparticle addition. The shear stress of the base fluids cannot be changed if the concentration of nanoparticles is too low. Another important factor influencing the shear stress–shear rate curve is temperature. When the temperature is low (25 °C, as is shown in Figure 6), the interactions between the molecules and the Brownian motion are not strong. Thus, the shear stress does not show an obvious change despite the concentration being increased.

The comparison of viscosity between pure YT198 and nanofluids is also shown in Figure 7. In Figure 7, one can see that the viscosity increased with the increasing shear rate before the shear rate reached 20 s^{-1}, and then, after a slight decline, the viscosity of both pure YT198 and nanofluids tended to achieve stable values. For the CuO nanofluids (Figure 7a), the descending order of viscosity values at the same shear rate is 0.15% > 0.05% > 0.01% > 0.10% > pure YT198; the maximum increase in viscosity occurred between pure YT198 and 0.15% CuO at 20 s^{-1} (5.34%). And the various numbers of nanoparticles added did not show obvious differences.

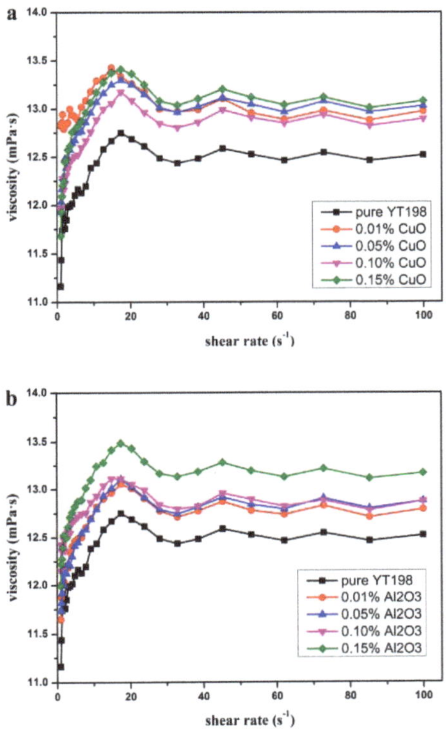

Figure 7. Viscosity vs. shear rate for various nanofluids with different volume fractions of (**a**) CuO and (**b**) Al_2O_3 nanoparticles at 25 °C.

As shown in Figure 7b, the addition of Al_2O_3 nanoparticles also caused the viscosity to increase. The greater the concentration of nanoparticles added, the more obvious the increase in viscosity as follows: 0.15% > 0.01% = 0.05% = 0.10% > pure YT198.

3.4. Effects of Temperature on Rheological Properties of Nanofluids

In this section, the different volume fractions (0.01%, 0.05%, 0.1% and 0.15%) of nanoparticles will be discussed in order to study the effect of temperature on their rheological properties. It was first found, as shown in Figure 8, that with the increment in temperature, the shear stress decreased exponentially. And one can see that various volume fractions of nanoparticles have little effect on the shear stress. This result is different from that obtained by Amir Yousuf Bhat and Adnan Qayoum [45]. They measured the viscosity of CuO nanofluids and reported a maximum increase of 0.5% in viscosity. The main reason why the results of our experiments differ from those in the literature is that the volume concentrations they used were between 1% and 4%, which are much larger than those used in this work.

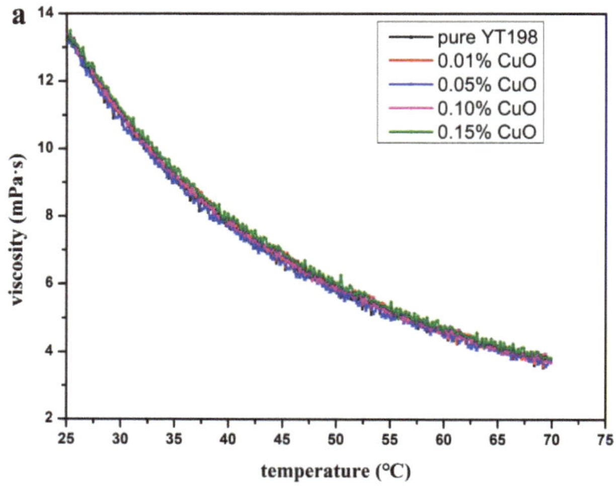

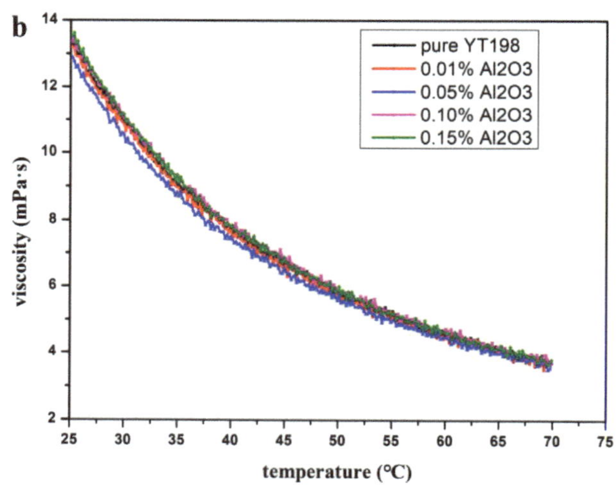

Figure 8. Effect of temperature on shear stress of pure YT198 and nanofluids at shear rate of 50 s^{-1}. (**a**) CuO; (**b**) Al_2O_3.

Additionally, the viscosity of CuO-YT198 and Al_2O_3-YT198 at different volume concentrations can be analyzed in Figure 9. As shown in Figure 9a,b, the results are consistent with the trend found in the study by Madhusree Kole and T.K. Dey [46]. Their results show that the viscosity of a nanofluid increases with an increasing nanoparticle concentration and decreases with an increase in temperature. The viscosity of these nanofluids decreased sharply with the increase in temperature, and the decreasing trend was also non-linear, which also illustrated that these mineral oil-based nanofluids are fluids whose rheological properties are strongly temperature-dependent. By comparing Figures 8 and 9 with the work of Amir Yousuf Bhat and Adnan Qayoum, we can see that a small number of nanoparticles can enhance heat transfer without causing a significant increase in viscosity.

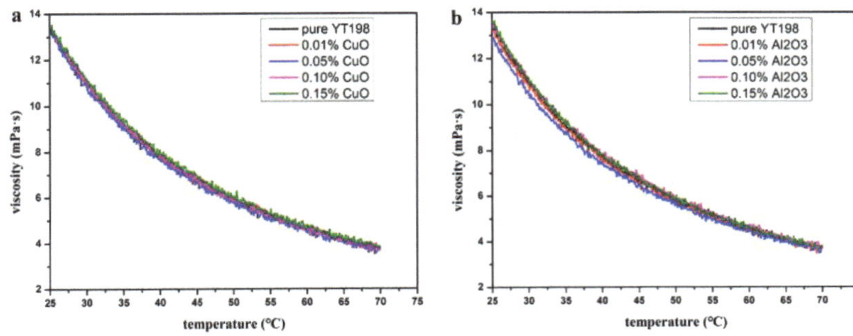

Figure 9. Effect of temperature on viscosity of pure YT198 and nanofluids at shear rate of 50 s^{-1}. (**a**) CuO; (**b**) Al_2O_3.

The stability of nanofluids is affected by temperature. Temperature variations affect the dispersion of nanoparticles in base fluids. High temperatures can result in more intense Brownian motion. Thus, the aggregation and sedimentation of nanoparticles can be hindered [47].

The nanofluid does not contain nitrogen, sulfur or aromatic hydrocarbons. So, over the course of long-term use, the color of the coolant will not turn yellow due to oxidation, which is a very common occurrence in other traditional cooling media. It will not form paint film and oil scale on electronic components. All of these characteristics ensure that the cooling medium has a long service life of more than 6 years and greatly decrease the cost of the cooling medium.

The cooling medium is made of extremely stable synthetic materials. It has a very low acid value and does not contain mineral oil or aromatic hydrocarbons. Long-term experiments have proven that nanofluids will not produce hardening, cracking, deformation, discoloration, corrosion, adhesion or other phenomena. They are harmless to the common plastics, resins, non-ferrous metals, plating and other components of the server circuit board, hard disk, central processing unit, etc. Using nanoparticles can ensure the stable performance of the parts and long-term trouble-free operation.

3.5. Derivation of Empirical Correlation

In this subsection, a mathematical model used to predict thermal conductivity and four famous models used to predict viscosity were used to fit and compare our experimental values of viscosity.

The Maxwell model [48] was suggested to predict the thermal conductivity of a nanofluid, and it is shown in the following Equation (6):

$$\frac{k_{nf}}{k_{bf}} = \frac{2k_{bf} + k_{np} + 2\varphi\left(k_{np} - k_{bf}\right)}{2k_{bf} + k_{np} - \varphi\left(k_{np} - k_{bf}\right)} \tag{6}$$

The suggested model is best suited for a nanofluid with nanoparticles with spherical shapes and lower volume fractions; a comparison of the theoretical predictions with the experimental results is shown in Figure 10.

As can be seen in Figure 10a, the Maxwell model was fixed well with the experimental results of thermal conductivity. The error between the experimental value and the predicted value of the model for the thermal conductivity of CuO nanofluids is no more than 6.1%. At concentrations of 0.01% and 0.05%, the predicted values of the model were greater than those measured in the experiments. At concentrations of 0.1% and 0.15%, the model predicted values were smaller than those from the experimental measurements. When comparing Figure 10a,b, it can be seen that the model agrees better with Al_2O_3.

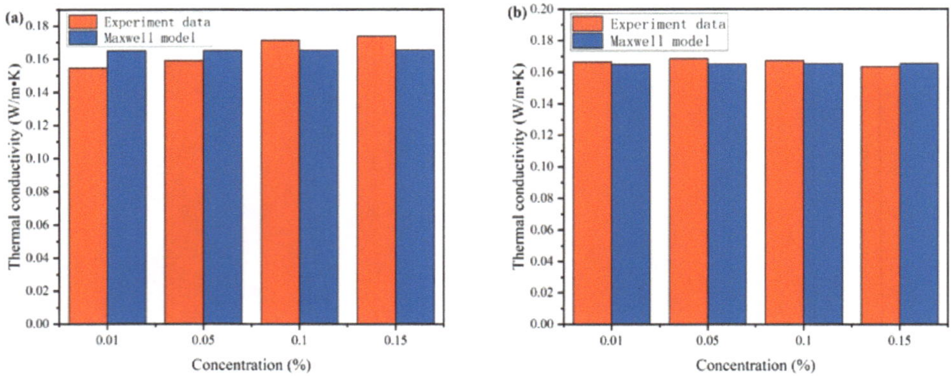

Figure 10. Comparison of theoretical predictions with experimental results of thermal conductivity at 25 °C for (**a**) CuO + YT198 and (**b**) 0.01% Al_2O_3 + YT198.

The four famous rheological models we used to fit and compare with our experimental values were Einstein's equation [49], Brinkman's equation [50], Batchelor's equation [51] and Wang et al.'s equation [52]. By using these four models, we could derive empirical correlations and then obtain quantitively valuable insights into the YT198 coolant-based metal oxide nanofluid system. The equations of the four models are listed and briefly discussed in the following sections.

As the first theoretical model created to calculate the viscosity of mixtures and composites, the Einstein model has been widely proven to be valid in the case of mixtures with a volume fraction (φ) of less than 0.02. The key assumption of the Einstein model is that the mixtures were simplified as linear viscous fluids and spherical particles. The equation for this model is as follows:

$$\mu_{nf} = \mu_{bf}(1 + 2.5\varphi) \tag{7}$$

Currently, the Brinkman-modified Einstein's equation can be used for suspensions with volume fractions of up to 4%.

$$\mu_{nf} = \mu_{bf}(1 - \varphi)^{-2.5} \tag{8}$$

As a further modification, the exitance of the Brownian motion effect was considered in the model proposed by Batchelor.

$$\mu_{nf} = \mu_{bf}\left(1 + 2.5\varphi + 6.5\varphi^2\right) \tag{9}$$

In 1999, Wang et al. [52] proposed another novel model to forecast the viscosity of nanofluids.

$$\mu_{nf} = \mu_{bf}\left(1 + 7.3\varphi + 123\varphi^2\right) \tag{10}$$

For all of the above equations, μ represents the viscosity, and the subscripts *nf* and *bf* represent nanofluids and base fluids, respectively. φ represents the volume fraction of nanoparticles.

A comparison of the theoretical predictions with the experimental results at 20 °C is shown in Figure 11.

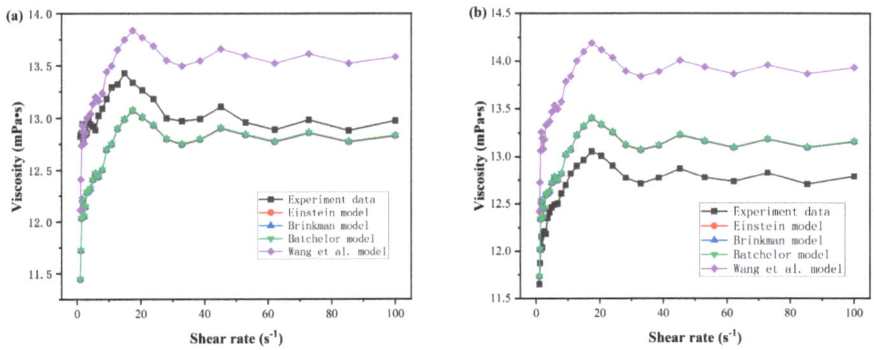

Figure 11. A comparison of the theoretical predictions with the experimental results at 20 °C for (**a**) 0.01% CuO + YT198 and (**b**) 0.01% Al_2O_3 + YT198.

As shown in Figure 11, the predictions of the Brinkman model and the Batchelor model almost exactly coincide, mainly because the volume fraction of the nanofluid is too small. For CuO, the experimental prediction is made between the Wang model and the Brinkman model. As can be seen in (b), for Al_2O_3, all model predictions are higher than the experimental values. The biggest difference between the predicted value of the model and the experimental value is 20 s^{-1}, and the errors are about 3.8% and 8.5%. Overall, the predicted and experimental values of the model are in good agreement. One possible reason for the model and experimental errors is that nanoparticles are not usually spherical.

4. Conclusions

The thermal conductivity and rheological properties of two different kinds of oxide nanofluids (CuO and Al_2O_3) in YT198, a typical mineral oil coolant used for data center liquid cooling, were measured at different volume fractions and temperatures. The thermal conductivity measurement results showed that even adding a small number of nanoparticles can significantly improve the heat conduction capacity of mineral oil coolants. The average increase in thermal conductivity is 20–25%. The temperature of nanoparticles also has an effect on the rheological properties of methanol-based nanofluids. The higher the temperature, the lower the viscosity and shear stress. However, it was found that the type of nanoparticles has little effect on the rheological properties. The difference in the effect of adding copper oxide and aluminum oxide on the viscosity is not very obvious.

More experimental research studies are needed to obtain the best types of nanoparticle additives which have the best rheological parameters and the best thermal conductivity enhancement effects. A good rheological parameter can guarantee minimum flow energy consumption, and high thermal conductivity provides optimum heat transfer. Additionally, more characteristics such as the sizes, shapes and types of nanoparticles should be taken into consideration because they are essential to establish more numerical models for further studies of heat transfer and rheology.

Based on the outcomes of this study, several suggestions for future research are listed below:

1. There are several parameters that have influences on the effect of a nanofluid coolant, including but not limited to particle size and size distribution, particle shape and morphology, the concentration of nanoparticles and temperature. These factors must be comprehensively and adequately studied in future works.
2. The stability of nanofluids is critical for their use in practical applications. Thus, more investigations about the surface modifications of nanoparticles and the use of surfactants should be conducted in order to determine an appropriate surfactant with remarkable stability, moderate viscosity and thermal conductivity.

3. The long-term effects of using these nanofluids in cooling systems, such as corrosion or clogging, deserve more attention. A life cycle analysis (LCA) of cooling systems using nanofluids may be necessary to evaluate whether a nanofluid is suitable for long-term use.
4. There are various models that can predict the thermal conductivity and viscosity of nanofluids separately. In future works, more mechanisms such as Brownian diffusion, particle aggregation, thermomigration and nanolayer formation can be taken into consideration. Therefore, models that will be developed in the future can be more accurate.

Author Contributions: Conceptualization, L.S. and J.G.; methodology, L.S.; validation, L.S.; formal analysis, L.S.; investigation, L.S. and J.G.; resources, Q.S.; data curation, L.S.; writing—original draft preparation, L.S.; writing—review and editing, L.S.; visualization, L.S.; supervision, K.D.; project administration, K.D.; funding acquisition, K.D. and Q.S. All authors have read and agreed to the published version of the manuscript.

Funding: This research was funded by the Shenzhen Science and Technology Program, grant number KCXST20221021111216038, and Guangzhou Development Zone International Science and Technology Cooperation Project Funding, grant number 2021GH07.

Data Availability Statement: Data are contained within the article.

Conflicts of Interest: The authors declare no conflicts of interest.

References

1. Patra, A.K.; Nayak, M.K.; Misra, A. Viscosity of nanofluids—A review. *Int. J. Thermofluid Sci. Technol.* **2020**, *7*, 070202. [CrossRef]
2. Nabil, M.F.; Azmi, W.H.; Hamid, K.A.; Mamat, R.; Hagos, F.Y. An experimental study on the thermal conductivity and dynamic viscosity of TiO_2-SiO_2 nanofluids in water: Ethylene glycol mixture. *Int. Commun. Heat Mass Transf.* **2017**, *86*, 181–189. [CrossRef]
3. Khodadadi, H.; Toghraie, D.; Karimipour, A. Effects of nanoparticles to present a statistical model for the viscosity of MgO-Water nanofluid. *Powder Technol.* **2019**, *342*, 166–180. [CrossRef]
4. Saxena, G.; Raj, J. A critical review on applications of nano-fluid as coolant. *Int. J. Eng. Manag. Res.* **2017**, *7*, 304–311.
5. Saravanakumar, P.T.; Surya, M.; Vijay, D.; Santhoshkumar, G. Improving performance in engine cooling system using nano fluids. *Int. Res. J. Eng. Technol.* **2017**, *4*, 1495–1499.
6. Junior, L.C.C.; Nogueira, E. Influence of the coolant flow containing silver nanoparticles (Ag) from an aqueous solution based on ethylene glycol (EG50%) on the thermal-hydraulic performance of an automotive radiator. *World J. Nano Sci. Eng.* **2020**, *10*, 14. [CrossRef]
7. Sandhya, M.; Ramasamy, D.; Sudhakar, K.; Kadirgama, K.; Harun, W.S.W. Hybrid nano-coolants in automotive heat transfer–an updated report. *Maejo Int. J. Energy Environ. Commun.* **2023**, *2*, 43–57. [CrossRef]
8. Erkan, A.; Tüccar, G.; Tosun, E.; Özgür, T. Comparison of effects of nanofluid utilization (Al_2O_3, SiO_2, TiO_2) with reference water in automotive radiators on exergetic properties of diesel engines. *SN Appl. Sci.* **2021**, *3*, 365. [CrossRef]
9. Sundari, K.G.; Asirvatham, L.G.; Ninolin, E.; Surekha, B. Feasibility of Glycerin/Al_2O_3 nanofluid for automotive cooling applications. *J. Therm. Eng.* **2020**, *6*, 619–632. [CrossRef]
10. Halelfadl, S.; Mare, T.; Estelle, P. Efficiency of carbon nanotubes water based nanofluids as coolants. *Exp. Therm. Fluid Sci.* **2014**, *53*, 104–110. [CrossRef]
11. Suganthi, K.S.; Rajan, K.S. Improved transient heat transfer performance of ZnO propylene glycol nanofluids for energy management. *Energy Convers Manag.* **2015**, *96*, 115–123. [CrossRef]
12. Asadi, A.; Pourfattah, F. Heat transfer performance of two oil-based nanofluids containing ZnO and MgO nanoparticles: A comparative experimental investigation. *Powder Technol.* **2019**, *343*, 296–308. [CrossRef]
13. Guo, W.; Li, G.; Zheng, Y.; Dong, C. Measurement of the thermal conductivity of SiO_2 nanofluids with an optimized transient hot wire method. *Thermochim. Acta* **2018**, *661*, 84–97. [CrossRef]
14. Abdolbaqi, M.K.; Sidik, N.A.C.; Rahim, M.F.A.; Mamat, R.; Azmi, W.H. Experimental investigation and development of new correlation for thermal conductivity and viscosity of BioGlycol/water based SiO_2 nanofluids. *Int. Commun. Heat Mass Transf.* **2016**, *77*, 54–63. [CrossRef]
15. Krishnakumar, T.S.; Sheeba, A.; Mahesh, V.; Jose, P.M. Heat transfer studies on ethylene glycol/water nanofluid containing TiO_2 nanoparticles. *Int. J. Refrig.* **2019**, *102*, 55–61. [CrossRef]
16. Perez-Tavernier, J.; Vallejo, J.P.; Cabaleiro, D.; Fernandez-Seara, J.; Lugo, L. Heat transfer performance of a nano-enhanced propylene glycol: Water mixture. *Int. J. Therm. Sci.* **2019**, *139*, 413–423. [CrossRef]
17. Yashawantha, K.M.; Vinod, A.V. ANN modelling and experimental investigation on effective thermal conductivity of ethylene glycol: Water nanofluids. *J. Therm. Anal. Calorim.* **2020**, *145*, 609–630. [CrossRef]

18. Leong, K.Y.; Saidur, R.; Kazi, S.N.; Mamun, A.H. Performance investigation of an automotive car radiator operated with nanofluid-based coolants (nanofluid as a coolant in a radiator). *Appl. Therm. Eng.* **2010**, *30*, 2685–2692. [CrossRef]
19. Subhedar, D.G.; Ramani, B.M.; Gupta, A. Experimental investigation of heat transfer potential of Al_2O_3/Water-Mono Ethylene Glycol nanofluids as a car radiator coolant. *Case Stud. Therm. Eng.* **2018**, *11*, 26–34. [CrossRef]
20. Mhamed, B.; Sidik, N.A.; Akhbar, M.F.A.; Mamat, R.; Najafi, G. Experimental study on thermal performance of MWCNT nanocoolant in Perodua Kelisa 1000cc radiator system. *Int. Commun. Heat Mass Transf.* **2016**, *76*, 156–161. [CrossRef]
21. Goudarzi, K.; Jamali, H. Heat transfer enhancement of Al_2O_3-EG nanofluid in a car radiator with wire coil inserts. *Appl. Therm. Eng.* **2017**, *118*, 510–517. [CrossRef]
22. Kumar, A.; Hassan, M.A.; Chand, P. Heat transport in nanofluid coolant car radiator with louvered fins. *Powder Technol.* **2020**, *376*, 631–642. [CrossRef]
23. Said, Z.; El Haj Assad, M.; Hachicha, A.A.; Bellos, E.; Abdelkareem, M.A.; Alazaizeh, D.Z. Enhancing the performance of automotive radiators using nanofluids. *Renew Sustain. Energy Rev.* **2019**, *112*, 183–194. [CrossRef]
24. Tijani, A.S.; Sudirman, A.S. Thermos-physical properties and heat transfer characteristics of water/anti-freezing and Al_2O_3/CuO based nanofluid as a coolant for car radiator. *Int. J. Heat Mass Transf.* **2018**, *118*, 48–57. [CrossRef]
25. Muruganandam, M.; Mukesh Kumar, P.C. Experimental analysis on internal combustion engine using MWCNT/water nanofluid as a coolant. *Mater. Today Proc.* **2020**, *21*, 248–252. [CrossRef]
26. Behrangzade, A.; Heyhat, M.M. The effect of using nano-silver dispersed water based nanofluid as a passive method for energy efficiency enhancement in a plate heat exchanger. *Appl. Therm. Eng.* **2016**, *102*, 311–317. [CrossRef]
27. Hassani, S.M.; Khoshvaght, A.M.; Mazloumi, S.H. Influence of chevron fin interruption on thermo-fluidic transport character-istics of nanofluid-cooled electronic heat sink. *Chem. Eng. Sci.* **2018**, *191*, 436–447. [CrossRef]
28. Nguyen, C.T.; Roy, G.; Gauthier, C.; Galanis, N. Heat transfer enhancement using Al_2O_3–water nanofluid for an electronic liquid cooling system. *Appl. Therm. Eng.* **2007**, *27*, 1501–1506. [CrossRef]
29. Chein, R.; Chuang, J. Experimental microchannel heat sink performance studies using nanofluids. *Int. J. Therm. Sci.* **2007**, *46*, 157–166. [CrossRef]
30. Sarafraz, M.M.; Arya, A.; Hormozi, F.; Nikkhah, V. On the convective thermal performance of a CPU cooler working with liquid gallium and CuO/water nanofluid: A comparative study. *Appl. Therm. Eng.* **2017**, *112*, 1373–1381. [CrossRef]
31. Sun, S.; Jin, X.; Liu, C.; Meng, Q.; Zhang, Y. Study on in-situ measurement method of THF hydrate thermal conductivity. *Int. J. Heat Mass Transf.* **2018**, *127*, 88–96. [CrossRef]
32. Gu, C.; Lv, Y.; Fan, X.; Zhao, C.; Dai, C.; Zhao, G. Study on rheology and microstructure of phenolic resin cross-linked nonionic polyacrylamide (NPAM) gel for profile control and water shutoff treatments. *J. Pet. Sci. Eng.* **2018**, *169*, 546–552. [CrossRef]
33. Amaral, C.; Pinto, S.; Silva, T. Development of polyurethane foam incorporating phase change material for thermal energy storage. *J. Energy Storage* **2020**, *28*, 101177. [CrossRef]
34. Alex, R.; Alexander, M.T.; Gabriel, F.; Gaurav, S.; Laurent, P. Thermal conductivity of cementitious composites containing microencapsulated phase change materials. *Int. J. Heat Mass Transf.* **2017**, *104*, 71–82.
35. Maxim, P.; Daniil, R.; Pavel, S. Stability and rheology of carbon-containing composite liquid fuels under subambient temperatures. *Energy* **2023**, *278*, 127912.
36. Theodorsson, E. Uncertainty in measurement and total error: Tools for coping with diagnostic uncertainty. *Clin. Lab. Med.* **2017**, *37*, 15–34. [CrossRef] [PubMed]
37. Tehmina, A.; Man-Hoe, K. Influence of particle size on the effective thermal conductivity of nanofluids: A critical review. *Appl. Energy* **2020**, *264*, 114684.
38. Nishant, K.; Shiriram, S. Experimental study of thermal conductivity and convective heat transfer enhancement using CuO and TiO_2 nanoparticles. *Int. Commun. Heat Mass* **2016**, *76*, 98–107.
39. Durga, B.J.; Tami, S.G.; Subashini, G.; Saravanan, P.; Muthumareeswaran, M.; Abdullah, N.A.; Nirmala, G.A. Ultrasonic Interferometry and Physiothermal properties of Al_2O_3/CuO nanofluids. *Case Stud. Therm. Eng.* **2024**, *55*, 104120.
40. Dan, L.; Zhang, K.; Wang, Q.; Liu, N. Surface modification boosts dispersion stability of nanoparticles in dielectric fluids. *J. Ind. Eng. Chem.* **2024**, *132*, 518–528. [CrossRef]
41. Younes, H.; Mao, M.; Murshed, S.S.; Lou, D.; Hong, H.; Peterson, G.P. Peterson. Nanofluids: Key parameters to enhance thermal conductivity and its applications. *Appl. Therm. Eng.* **2022**, *10*, 118202. [CrossRef]
42. Yılmaz, D.A.; Gürü, M. Nanofluids: Preparation, stability, properties, and thermal performance in terms of thermo-hydraulic, thermodynamics and thermo-economic analysis. *J. Therm. Anal. Calorim.* **2022**, *147*, 7631–7664. [CrossRef]
43. Sun, L.; Zhang, Y.; Gao, W.; Li, X.; Jing, D. Rheology of methanol-based metal oxide nanofluids: The effect of temperature and particle type and mass fraction. *Interfacial Phenom. Heat Transf.* **2020**, *8*, 165–181. [CrossRef]
44. Sun, L.; Zhao, Q.; Zhang, Y.; Gao, W.; Jing, D. Insights into the rheological behavior of ethanol-based metal oxide nanofluids. *J. Mol. Liq.* **2021**, *323*, 115006. [CrossRef]
45. Amir, Y.B.; Adnan, Q. Viscosity of CuO nanofluids: Experimental investigation and modelling with FFBP-ANN. *Thermochim. Acta* **2022**, *714*, 179267.
46. Kole, M.; Dey, T.K. Viscosity of alumina nanoparticles dispersed in car engine coolant. *Exp. Therm. Fluid Sci.* **2010**, *34*, 677–683. [CrossRef]

47. Chen, Y.; Huang, Y.; Li, K. Temperature effect on the aggregation kinetics of CeO_2 nanoparticles in monovalent and divalent electrolytes. *J. Environ. Anal. Toxicol.* **2012**, *2*, 158–162.
48. Maxwell, J. A physical treatise on electricity and magnetism. *J. Frankl. Inst.* **1881**, *111*, 386–389.
49. Einstein, A. Eine neue bestimmung der moleküldimensionen. *Ann. Phys.* **1906**, *324*, 289–306. [CrossRef]
50. Brinkman, H. The viscosity of concentrated suspensions and solutions. *J. Chem. Phys.* **1952**, *20*, 571. [CrossRef]
51. Batchelor, G.K. The effect of Brownian motion on the bulk stress in a suspension of spherical particles. *J. Fluid Mech.* **1977**, *83*, 97. [CrossRef]
52. Wang, X.; Xu, X.; Choi, S.U. Thermal conductivity of nanoparticle-fluid mixture. *J. Thermophys. Heat Transf.* **1999**, *13*, 474–480. [CrossRef]

applied sciences

MDPI

Article

Lotus Root Type Nickel Oxide-Carbon Nanofibers: A Hybrid Supercapacitor Electrode Material

Juhyeong Kim [1,†], Junho Lee [1,†], Hwajeong Ryu [1], Yoon Hwa [2], Aneel Pervez [1], Taeyong Choi [1], Suyong Nam [3] and Yoonkook Son [1,*]

1 Department of Electrical Engineering, Chosun University, 309, Pilmun-daero, Dong-gu, Gwangju 61452, Republic of Korea; juhyeong@chosun.kr (J.K.); kodume@hanmail.net (J.L.); ghkwjdfb7620@chosun.kr (H.R.); aneelpervez@chosun.ac.kr (A.P.); o304@chosun.ac.kr (T.C.)
2 School of Electrical, Computer and Energy Engineering, Arizona State University, Tempe, AZ 85287, USA; yoon.hwa@asu.edu
3 Department of Nanotechnology Engineering, Pukyong National University, 45, Yongso-ro, Nam-gu, Busan 48547, Republic of Korea; suynam@pknu.ac.kr
* Correspondence: y_son@chosun.ac.kr
† These authors contributed equally to this work.

Abstract: In this study, we investigate the electrochemical properties of a nickel oxide-carbon (NiO/C) material, synthesized in the form of highly porous carbon nanofibers through the electrospinning of polymers such as polyacrylonitrile (PAN) and polystyrene (PS) followed by a carbonization process. The primary focus of this work is to determine the optimal mixing ratio for the hybrid material composed of NiO and carbon. While it is widely acknowledged that supercapacitor materials benefit from having a high specific surface area, our findings reveal that hybrid carbon nanofibers with a 45% specific carbon-to-nickel oxide ratio exhibit significantly enhanced capacitance (39.9 F g^{-1}). This outcome suggests the promising potential of our materials as an energy storage material for hybrid supercapacitors, combining the advantages of electric double-layer capacitors (EDLC) and Pseudo capacitors (Pseudo).

Keywords: supercapacitor; NiO; hybrid; electrospinning; porous material

Citation: Kim, J.; Lee, J.; Ryu, H.; Hwa, Y.; Pervez, A.; Choi, T.; Nam, S.; Son, Y. Lotus Root Type Nickel Oxide-Carbon Nanofibers: A Hybrid Supercapacitor Electrode Material. *Appl. Sci.* **2024**, *14*, 2977. https://doi.org/10.3390/app14072977

Academic Editors: Robert A. Varin and Geng Zhong

Received: 3 March 2024
Revised: 28 March 2024
Accepted: 30 March 2024
Published: 1 April 2024

1. Introduction

In the realm of electrochemical energy storage and conversion technology, the supercapacitor is one of the renowned systems for its practicality and versatility. Supercapacitors are distinguished from other electrochemical energy storage systems by offering a unique balance of high-power capabilities, exceeding that of batteries, along with enhanced energy density compared to traditional capacitors [1]. Based on the charge storage mechanism, supercapacitors are categorized as electric double-layer capacitors (EDLCs), pseudo-capacitors, and hybrid supercapacitors.

EDLCs exhibit high power density and cycle life, which are attributed to the efficient separation of charges through electrostatic attraction within the electrochemical double layer. EDLCs generally employ activated carbon-based nanomaterials with high surface area and electrical conductivity in their electrodes to facilitate the rapid movement of ions and electrons [2]. The nanometer-scale pore size plays a pivotal role in reducing charge separation distances, consequently leading to increased capacitance [3]. Consequently, a common strategy involves the fabrication of nanostructured electrodes to bolster electrode material conductivity [4] while simultaneously reducing diffusion pathways [5], with specific surface area serving as a direct determinant of specific capacitance. In contrast, Pseudo capacitors employ electrode materials consisting of conductive polymers and transition metal oxides to store charge through surface-controlled faradaic reactions taking place at the electrode–electrolyte interface at rates comparable to those of electrochemical double-layer formation in EDLCs. The architecture and composition of electrode materials have a significant impact on the performance of supercapacitors. Various electrode materials with different dimensional structures such as 0D, 1D, 2D, and 3D have been designed to enhance their performance. Among them, 1D nanomaterials synthesized by the electrospinning method show promising potential due to their easy preparation, high aspect ratio, which facilitates efficient electron and mass transportation, and strong plasticity. The composite materials synthesized using carbon materials, carbide materials, conductive polymers, sulfides, nitrides, metal oxides, etc., show better electrochemical performance due to their high stability, conductivity, and electron storage capacity [6,7]. In general, transition metal oxides offer higher capacitance than carbon materials and superior charge-discharge cycle stability compared to polymers, despite their lower electronic conductivity [8]. Among transition metal oxides, NiO is a popular choice as a pseudo-capacitor electrode material due to its high theoretical capacitance, large surface area, cycling stability, multi-oxidation states, and reversible redox reactions [7]. However, NiO grapples with challenges such as low electronic conductivity, limited redox active sites, and a low charge/discharge ratio, which inevitably impact capacitance and cycling performance. To address these issues, strategies have been explored, including the nanostructuring of NiO electrodes to enhance the effective surface area, hybridization with highly conductive carbon-based materials, and doping with foreign elements [9]. Various studies have been reported to enhance the capacitance of capacitors by using NiO-based composite materials. For instance, R. D. Noce and coworkers achieved 150 F g^{-1} capacitance using NiO/CNF with Na_2SO_4 as electrolyte [10], T. Hussain et al. reported 157.9 F g^{-1} with CNF-NiO [11], and 742.2 F g^{-1} capacitance was reported by S. Shin and coworkers [12].

In our study, we have developed a hybrid Nickel oxide-carbon nanofiber (NiO-CNF) hybrid capacitor electrode, notably for its rectangular shape observed in cyclic voltammetry (CV), stable operating voltage, high energy, low cost, self-balancing, and close resemblance to the ideal characteristics of a capacitor [13]. Moreover, the NiO-CNFs hybrid capacitor exhibited significantly faster charge and discharge rates compared to bare NiO. This notable improvement can be attributed to enhancements in both the redox reaction kinetics and the electrode's resistivity. The superior electronic conductivity of CNF in comparison to NiO played an important role in achieving this enhanced performance. Specifically, the inclusion of Polystyrene (PS) in the synthesis process led to the development of internal lotus root-like pores in CNF during the subsequent heat treatment process, which encompassed oxidation stabilization and carbonization. Additionally, this process resulted in the formation of numerous surface curvatures, akin to a chemical etching effect. This enhancement in specific surface area was strategically designed to mitigate the internal resistance of the electrode, consequently boosting the overall capacitance of the supercapacitor. The porous structure of carbon materials was found to exert a profound influence on the mobility of ions within pores, with small pores hampering ion transport [14]. Hybrid capacitors combine the strengths of EDLC and pseudo-capacitors, utilizing different electrochemical materials with varying working potentials for higher energy and power densities. They can be created using various electrode material combinations [15]. The investigation demonstrates that the superior performance of micropores (<2 nm) persists over mesopores (2–50 nm), even in instances where mesopores undergo fine-tuning to achieve a carbon structure. Improved capacitance can be facilitated through the process of partial desolvation, accessed by the pores measuring less than 2 nm within the microporous structure [16]. We synthesized Polyacrylonitrile (PAN)-based nanofibers containing Ni and PS through electrospinning, with subsequent heat treatment to remove the PS polymer and form nanopores, thereby augmenting the specific surface area. The process results in a lotus root-type NiO-CNFs nanocomposite. The utilization of PS proved to be effective in enhancing the specific surface area of CNFs, a fact corroborated by various analytical techniques [17]. This facilitated a comprehensive comparison of material properties between NiO and carbon after the carbonization process [18].

2. Materials and Methods

2.1. Material Preparation

The first step in the experimental procedure was to dissolve the solute in a solvent to synthesize a polymer solution for electrospinning, and the following materials were used to prepare the active material. To prepare the carbonized material, polyacrylonitrile (PAN) (150,000 g/mol, Sigma-Aldrich, St. Louis, MO, USA), polystyrene (PS) (192,000 g/mol, Sigma-Aldrich), and nickel(II) acetate tetrahydrate (99.995%, Sigma-Aldrich) were used as solutes, dimethylformamide (N,N-Dimethylformamide, DMF) (99.5%, JUNSEI, Tokyo, Japan) was used as a solution, and nanofibers were manufactured using an electrospinning/spraying system (eS-robot, NanoNC, Seoul, Republic of Korea).

During the experiment, the operation was conducted in an environment where the temperature and relative humidity were kept constant at 25 °C and 50%, respectively. The nozzle tip through which the polymer solution was discharged was 0.33 mm, the discharge rate of the solution was 30 μL min^{-1}, and the voltage applied to the nozzle tip at this time was DC 20.0 kV. The polymer fibers manufactured through electrospinning are heated to 250 °C at 6 °C min^{-1} in normal air through a sample-sintering electric furnace (Box Furnace, K-TEC, Incheon, Republic of Korea), and then undergo an oxidation stabilization step and are thermally decomposed in a carbonization process at 1000 °C [19].

In the case of high-elasticity fibers, a higher temperature is required [20]. The carbon fiber manufacturing process involves a carbonization process in which the PAN polymer fiber is graphitized at a high temperature and converted into a carbon structure with a purity of at least 99% [21]. The secondary heat treatment for the carbonization process was performed at 1000 °C for 1 h in a hydrogen gas atmosphere at a flow rate of 500 mL min^{-1} in a tube furnace (ScienTech, Suwon, Republic of Korea). Figure 1 is a schematic diagram of the material manufacturing process.

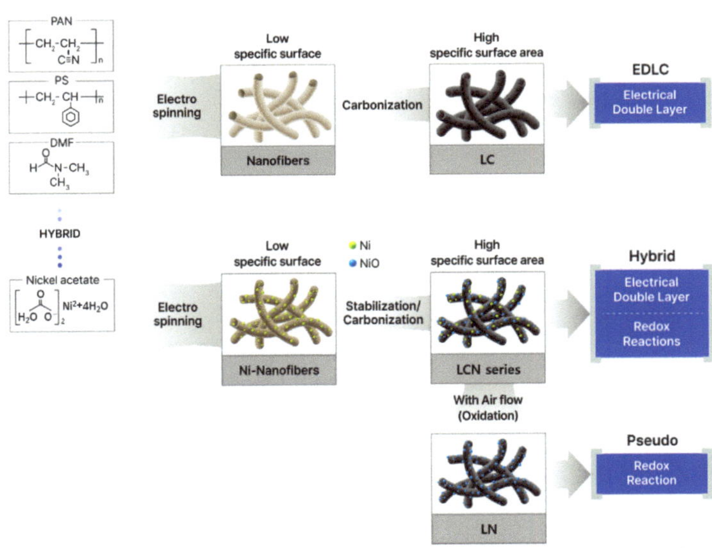

Figure 1. Schematic representation illustrating the reaction mechanism associated with the manufacturing method and material of CNF used in supercapacitors.

Heat treatment for the oxidation of nickel metal was carried out at 400 °C and 450 °C for 30 min at a heating rate of 6 °C min^{-1}, respectively, to obtain NiO/C samples. The powder produced by PAN/PS was named Lotus-root-typed Carbon nanofiber (LC). The sample types were divided into 0.5 g and 0.7 g depending on the nickel content with 8 mL of DMF and 1.0 g of PAN, and the heat treatment temperature was set at 400 °C and 450 °C to classify the sample types. The nomenclature of the Lotus-root-typed Carbon Nickel oxide nanofiber (LCN) series is based on the nickel content remaining after the removal of carbon by TGA and is divided into LCN-35 (35.6%), LCN-40 (40.3%), LCN-45 (45.5%), and LCN-48 (48.6%). In addition, a sample containing nickel content is additionally heat-treated in a normal atmospheric atmosphere, leaving only a trace amount of carbon necessary to maintain the shape of the lotus root structure, and is named lotus-root-typed nickel oxide nanofiber (LN).

2.2. Material Characterization

The morphology of samples was investigated through the use of a field emission scanning electron microscope (FE-SEM) (S-4800, Hitachi, Tokyo, Japan). High-resolution X-ray diffraction (XRD) patterns were acquired employing a state-of-the-art X-ray diffractometer (HRXRD, X'pert Pro MRD, PANalytical, Tokyo, Japan) and analyzed using Cu-Kα radiation within the range of $10° \leq 2\theta \leq 90°$ at a scanning rate of 2° min^{-1}. Thermogravimetric Analysis (TGA) (TGA 2050, TA Instruments, New Castle, DE, USA) was used to analyze the quantitative content of carbon and nickel/nickel oxide, and each sample was heated from room temperature to 800 °C at a rate of 6 min^{-1} under normal air atmosphere. N_2 adsorption/desorption isotherms were obtained using an automated specific surface area and pore size distribution analyzer (BELSORP-mini II, MicrotracBEL, Osaka, Japan). The specific surface area and pore properties of the samples were assessed employing the Brunauer–Emmett–Teller (BET) and Barrett–Joyner–Halenda (BJH) methods.

2.3. Electrochemical Characterization

The supercapacitor electrode is made by spreading a slurry composed of NiO-CNFs nanocomposites, a conductive material (super-P), styrene-butadiene rubber (SBR), and carboxymethyl cellulose (CMC) binder in a weight ratio of 80:10:10 on a 1 × 1 cm^2 nickel metal foil. The mass of active material within the slurry was 0.92 mg, 0.51 mg, 0.53 mg, 0.36 mg, 0.48 mg, and 0.90 mg after drying for LC, LCN-35, LCN-40, LCN-45, LCN-48, and LN, respectively. It was manufactured by casting and the casted electrode was placed in a vacuum oven and dried at 100 °C for 1 h. In addition to the manufactured working electrode, a counter electrode and reference electrode were used to form a three-electrode system with platinum and Hg/HgO, respectively. When evaluating the electrochemical properties, 6 M KOH was applied as the electrolyte.

Cyclic voltammetry (CV) was measured in the voltage range of 0.25 to 0.5 V. Galvanostatic charge/discharge (GCD) characteristics of the electrode were performed between 0 and 0.45 V. The scan rates for CV and GCD were 2 to 100 mV s^{-1} and 0.2–3.0 A g^{-1}, respectively. Long-term cycle life was performed with GCD measurements for 1000 cycles with a current density of 1.0 A g^{-1}.

The capacitance value (C_m) determined as the area of the cyclic voltammetric curve was calculated for each scanning speed according to Equation (1).

$$C_m = \frac{1}{mv(E_2 - E_1)} \int_{E_1}^{E_2} i(E)dE \; [F/g] \quad (1)$$

Here, C_m is the mass of the active electrode material [g], v is the scan rate [V/s], $i(E)$ is the respective current [A], and $E_2 - E_1$ is ΔV [V] the voltage of one scanning portion. From the time potential difference curve of the constant current charge/discharge curve, the weight-based capacitance (C_m) is calculated according to Equation (2).

$$C_m = \frac{I \times \Delta t}{m \times \Delta V} \; [F/g] \quad (2)$$

Here, C_m is the total capacitance [F], I is the discharge current value [A], Δt is the discharge time [s], m is the mass of the working electrode active material [g], and ΔV [V] is the potential change excluding the IR drop within Δt [22]. Additionally, to determine the internal resistance that impedes the flow of electrons, EIS was measured in the frequency range of 100,000 to 0.1 Hz with an amplitude voltage of 5 mV using IVUM STAT (HS Technologies, Plano, TX, USA).

3. Results and Discussion

The color of the nanofiber produced by electrospinning was initially white and later changed to yellow or brown through oxidation stabilization heat treatment, and the sample after final carbonization was black. The morphological characteristics of the final carbonization samples were observed using a field emission-scanning electron microscope (FE-SEM), and the corresponding images are presented in Figure 2. When viewed at a low magnification, the samples showed long nanofibers with diameters ranging from 700 nm to 1.5 μm. Additionally, these fiber samples displayed numerous internal pores and noticeable waviness on the fiber surface. This suggests that in contrast to CNFs fabricated solely from PAN [23], the portion of fibers containing PS is removed during the heat treatment, as shown in Figure 2a. PS is removed as it decomposes, and during this process, it plays an important role in forming the lotus root shape and additionally forms the internal pores and external etched surface of the electrospun nanofiber. This removal process results in the formation of internal pores and surface curvature in the CNFs. Interestingly, Figure 2b–e, illustrating various Ni content levels and temperature differences during the third heat treatment process, did not exhibit significant morphological distinctions. However, Figure 2f, where carbon was entirely removed, leaving only NiO, exhibited intensified fiber destruction and surface curvature. Nonetheless, it was apparent that the fiber's morphology, including internal pores, was retained, similar to the CNF sample. These findings suggest that CNFs acted as a morphological template.

Figure 3a displays the results of the TGA analysis, which was performed to assess the thermal decomposition characteristics and NiO content of LC, LCN series, and LN samples. To investigate the changes occurring during the heat treatment process, thermal gravimetric analysis (TGA) was conducted under the condition of a constant temperature increase of 6 °C per minute in an air atmosphere using a TGA 2050 (TA Instruments) instrument, with temperatures rising up to 800 °C. The NiO content was determined by comparing the remaining mass to the initial mass, considering the reduction in carbon and moisture loss due to thermal decomposition within the measured temperature range. The contents of NiO were determined to be 0, 35.6, 40.3, 45.5, 48.6, and 94.3 wt% for LC, LCN-35, LCN-40, LCN-45, LCN-48, and LN, respectively. In the case of LC samples without NiO, the thermal decomposition behavior indicated a change due to the moisture removal from the sample between 60 °C to 100 °C. Subsequently, carbon decomposition began at 450 °C, culminating in complete carbon removal at 750 °C. In contrast, the LCN series samples exhibited a steady mass decrease from around 450 °C until approximately 600 °C, signifying carbon decomposition. No further mass change was observed above 650 °C.

Figure 3b shows the results of XRD analysis, confirming the crystalline properties of the samples. In the case of LC, the XRD pattern demonstrated an amorphous carbon, while LN, where most of the carbon had disappeared, exhibited a distinct NiO structure. For the LCN series, diffraction peaks associated with graphite and NiO were evident in all samples after undergoing three heat treatment processes. As the NiO content increased, carbon content decreased, and the peak at 26.02° corresponding to carbon gradually diminished. Notably, with the increase in NiO content from the LCN series to LN, the peaks related to NiO gradually intensified, relative to the proportion of NiO in the composite. Therefore, these XRD results confirmed the formation of a NiO-CNF composite in the LCN series samples.

The specific surface area and pore properties of the prepared samples were analyzed using Brunauer, Emmett, and Teller (BET) and Barrett, Joyner, and Halenda (BJH) measurements. Figure S1a depict the adsorption isotherm, pore distribution, and BET plots for each sample, while Figure S2 summarizes the data in Table S1. The specific surface area for LC, LCN-35, LCN-40, LCN 45, LCN-48, and LN is 124.71 $m^2\ g^{-1}$, 209.89 $m^2\ g^{-1}$, 122.91 $m^2\ g^{-1}$, 133.34 $m^2\ g^{-1}$, and 205.68 $m^2\ g^{-1}$, respectiviely. In the LCN series, the specific surface area equaled or exceeded that of the LC sample. This increase can be attributed to the deepening of surface curvature resulting from the heat treatment process of the sample containing the nickel source. Figure S1a presents the adsorption isotherm curve, indicating the pore size and volume distribution within each sample. All samples, including the LC, LCN series, and LN, exhibited a sharp increase in the latter, with the isotherm line rising above 0.9, indicative of a porous structure [24]. Figure S1b illustrates the BJH analysis results, portraying the pore size distribution graph for each sample, spanning a range of 2 to 185 nm. Overall, pore sizes exhibited non-uniformity, with various pore sizes present. LC primarily featured pores around 80 nm, the LCN series displayed an even distribution around 90 nm, and LN exhibited an even distribution around 100 nm. Mostly, it has been reported that a large specific surface area can enhance capacitance performance by facilitating electrolyte diffusion to the redox reaction's active site. Meanwhile, it has been recently reported that specific surface area alone does not exclusively dictate a capacitance increase [25].

Figure 2. SEM images of (**a**) LC, (**b**) LCN-35, (**c**) LCN-40, (**d**) LCN-45, (**e**) LCN-48, and (**f**) LN.

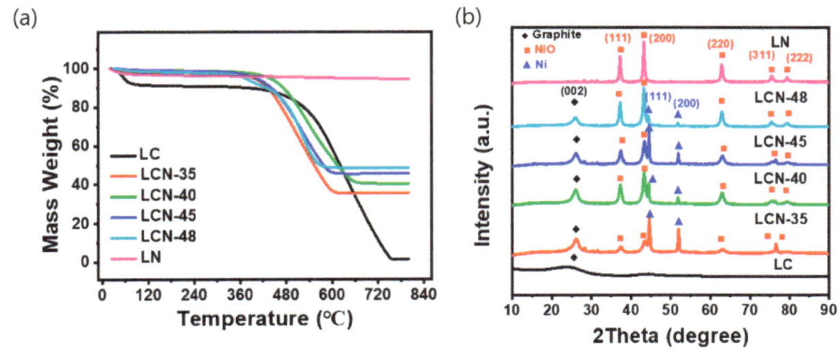

Figure 3. (**a**) TGA measurements and (**b**) XRD patterns of LC, LCN series, and LN.

The electrochemical properties of the LC, LCN series, and LN samples, classified based on NiO content through TGA and XRD analysis, were analyzed through Cyclic Voltammetry (CV), Galvanostatic Charge/Discharge (GCD), and Electric Impedance Spectroscopy (EIS). CV was conducted using a 6 M KOH electrolyte and a three-electrode system. All electrodes were analyzed by selecting a voltage range (0.25–0.5 V) wherein oxidation/reduction reactions occur at a fixed scanning speed of 2 mV s^{-1}. As the curve area increased, so did the capacitance values [26]. In Figure 4a, the CV curve for LC exhibits a distorted rectangular shape with an unclear redox reaction, whereas the CV curve for LN displayed a distinct redox peak attributed to the reversible Faradaic transition of nickel, indicating typical pseudocapacitive behavior. In Figure 4b, the CV curves for LCN series samples exhibited a distorted rectangular shape coupled to a distinct pseudo-capacitance behavior attributed to the Faradaic redox reaction. The reason for the different positioning of redox peaks among LCN series might be attributed to the synthetic electrochemical interference between NiO and carbon. Among the LCN series, the highest capacitance was calculated for the LCN-45 sample. Figure 4c shows the CV curves within the voltage range of 0.25 to 0.5 V for the LCN-45 sample. CV was also used to measure chemical changes in the electrode at the scanning speed ranging from 2 to 100 mV s^{-1}. An increase in the scanning speed resulted in an expanded curve area, demonstrating the potential for improved charge storage capability. Figure 4d shows specific capacitance values derived from CV conducted at various voltage scan rates for LCN-45. This highlights the charge storage mechanism of the NiO/C electrode, resulting from faradic redox reactions [27]. The highest capacitance for the LCN-45, measured at 2 mV s^{-1}, reached 92.52 F g^{-1} (0.18504 C g^{-1}), with corresponding energy and power density of 2.89 Wh kg^{-1} and 46.05 W kg^{-1}, respectively. As the scan rate increases, capacitance decreases, with values of 90.72 F g^{-1}, 85.91 F g^{-1}, 80.09 F g^{-1}, 70.65 F g^{-1}, and 61.11 F g^{-1} (0.4536 C g^{-1}, 0.8591 C g^{-1}, 1.6018 C g^{-1}, 3.5325 C g^{-1}, and 6.111 C g^{-1}) observed for 5 mV, 10 mV, 20 mV, 50 mV, and 100 mV, respectively. The phenomenon of pseudo-redox, which can result in the irreversibility of the redox process, is often associated with factors such as the generation of voltage polarization and ohmic resistance during the faradic redox reaction and ion diffusion limitation within the porous electrode during the redox reaction process [28]. The electrochemical Faradaic reaction of NiO, responsible for the pseudo reaction, involves a simple adsorption/desorption reaction of OH^-. The positive peak corresponds to the oxidation of NiO to NiOOH, while the negative peak corresponds to the reverse reaction, involving the reversible conversion between Ni^{2+} and Ni^{3+}. This redox reaction occurs on the surface of the NiO electrode and can be expressed as the following reaction (3) [29,30].

$$NiO + OH^- \leftrightarrow NiOOH + e^- \quad (3)$$

For galvanostatic charge–discharge (GCD) experiments, electrodes were prepared from each sample and the prepared electrode was employed in the three-electrode cell, with the working electrode immersed in a 6 M KOH aqueous solution. Figure 5a shows the charge–discharge curves of the electrodes for each sample within the voltage range between 0 and 0.45 V and at a current density of 0.2 A g^{-1}. In the low potential range of 0 to 0.25 V, no significant Faradaic reactions, such as forming an electric double layer. However, as the current density increased, the electric double layer showed a shape close to an asymmetric triangle due to the non-Faradaic current. At the beginning of the discharge, it shows a completely asymmetric graph with a fairly large IR voltage drop. It exhibited 0.14 V, 0.16 V, 0.17 V, 0.07 V, 0.13 V, and 0.14 V for LC, LCN-35, LCN-40, LCN-45, LCN-48, and LN, respectively. Figure 5b shows the discharge curve behavior at different current densities for each sample and shows a weight-based capacitance graph in the time potential difference curve, which is a constant current discharge curve with a voltage range of 0.45 V. In Figure 5c, bar graphs representing the specific capacitance of each sample calculated at a fixed current density of 0.2 A g^{-1} are displayed. The calculated values for LC, LCN-35, LCN-40, LCN 45, LCN-48, and LN are 14.2 F g^{-1}, 23.6 F g^{-1}, 21.7 F g^{-1}, 39.9 F g^{-1}, 24.5 F g^{-1}, and 18 F g^{-1} (3.195 C g^{-1}, 5.31 C g^{-1}, 4.8825 C g^{-1}, 8.9775 C g^{-1}, 5.5125 C g^{-1}, and 4.05 C g^{-1}), respectively.

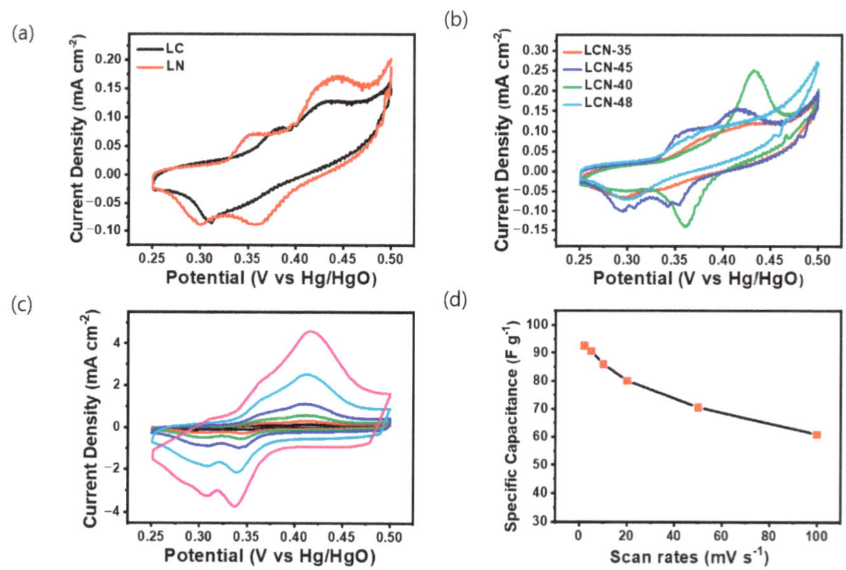

Figure 4. Cyclic voltammetry curves of (**a**) LC and LN, (**b**) LCN series at a scan rate of 2 mV s^{-1}. (**c**) Cyclic voltammetry curves of LCN-45 with different scan rates ranging from 2 to 100 mV s^{-1}. (**d**) Calculation of the specific capacitance of the LCN-45 sample based on (**c**).

The impedance spectroscopy curves of LN (Bare NiO) and LCN-45 (NiO-CNFs) nanocomposites are shown in Figure 5d. The Nyquist Plot of the LN electrode showed a highly suppressed semicircle in the high-frequency region and an oblique line in the low-frequency region, which may be due to the charge transfer process at the electrode/electrolyte interface and electrolyte ion diffusion process. At high frequencies, the arc in the curve of the LCN-45 electrode was difficult to discern, but the arc was clearly visible in the curve of the LN electrode. Through this, it can be seen that in the high-frequency range, the LN electrode has much faster charge transfer characteristics between the electrolyte and the lotus root-shaped active material, with a resistance of less than 0.2 Ω. The equivalent series resistance value of the LCN-45 electrode was found to be lower compared to the LN electrode, measuring 0.57 Ω and 0.78 Ω, respectively. This distinction arises due to the lotus root-shaped morphology and porous structure, which enhance the mobility of electrolyte ions across the electrode surface. Additionally, the LCN-45 electrode exhibits a steeper slope compared to the LN electrode, indicating a closer resemblance to an ideal electrode with a favorable electrolyte penetration pathway during the charge storage process. This suggests that the LN electrode has a lower charge transfer resistance than the LCN-45 electrode.

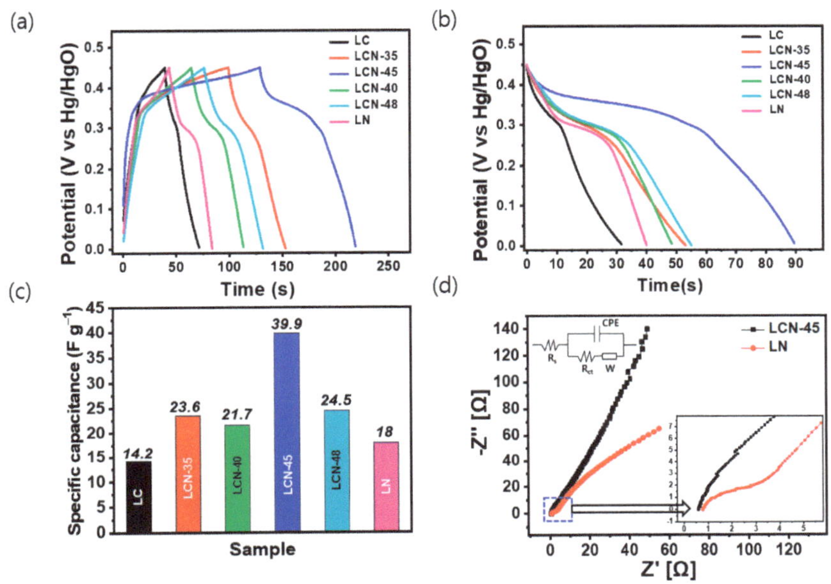

Figure 5. (**a**) Galvanostatic Charge–Discharge (GCD) curves of LC, LCN series, and LN at a current density of 0.2 A g^{-1}. (**b**) Discharge curve from (**a**). (**c**) Calculation of specific capacitances of LC, LCN series, and LN from (**b**). (**d**) Electrochemical Impedance Spectroscopy (EIS) Nyquist plot of LCN-45 and LN electrodes (Inset shows the enlarged region of high-frequency range).

Figure 6a shows the charge/discharge curve of LCN-45 at various current density values, which shows the highest capacitance among the electrodes analyzed previously. The measurements were taken at different current densities at a potential range of 0 to 0.45 V. The graph indicates that the charge time exceeds the discharge time due to ion diffusion resistance in the electrolyte during charging and discharging, which mitigates the ion migration during charging; the other factor is the polarization of the electrode at a slow rate leading to an increase in charging time. The graph in Figure 6b illustrates the variation in capacitance corresponding to each current density during discharge. The inserted graph in Figure S3 displays the capacitance calculated while considering the voltage drop, with measurements conducted at current densities of 0.2, 0.3, 0.5, 1, 2, and 3 A g^{-1}. It was observed that the capacitance decreased as the current density increased, measuring 142.8, 138.7, 124.5, 115.3, 87.6, and 75.1 F g^{-1} (32.13 C g^{-1}, 31.2 C g^{-1}, 28.01 C g^{-1}, 25.94 C g^{-1}, 19.71 C g^{-1}, and 16.89 C g^{-1}), respectively.

Figure 6c presents a Ragone plot depicting the energy and power density of the LCN-45 electrode. Cycling stability is demonstrated in Figure 6d, where continuous charge and discharge cycles were performed on the LCN-45 electrode for 1000 cycles at a constant current density of 1 A g^{-1} [31]. The capacitance and charge/discharge time exhibited stability, with the constant current charge/discharge curve of the final cycle closely resembling the initial one. The sample's capacitance retention rate was 82.6%. The constant current charge/discharge cycle of the LN electrode is depicted in Figure S4, demonstrating that the hybrid-type supercapacitor maintains a high lifespan and safety even after several charge/discharge cycles.

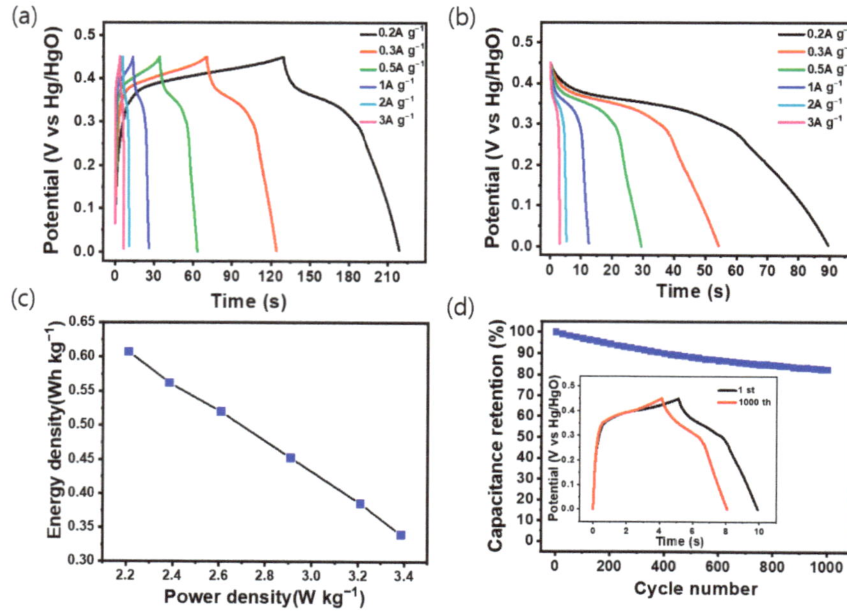

Figure 6. (**a**) Galvanostatic charge–discharge (GCD) curves of the LCN-45 electrode with various current densities from 0.2 to 3 A g^{-1}. (**b**) Discharge curve from (**a**). (**c**) Ragone plot of LCN-45 electrode. (**d**) Cycling stability for 1000 cycles at 1.0 A g^{-1} of the LCN-45 electrode (Inset shows the GCD curves for the 1st and 1000th cycles).

4. Conclusions

In summary, we generated polymer fibers through electrospinning, incorporating PAN, PS, and Ni sources. Subsequently, we fabricated highly porous lotus root-type hybrid supercapacitor materials comprising NiO and carbon using a straightforward carbonization process. Additionally, the TGA data revealed an optimized material ratio of 45% NiO causing a synergetic advanced electrochemical effect with carbon. Notably, the LCN-45 electrode, constructed with this ratio, demonstrated a higher specific capacitance (39.9 F g^{-1}) than materials with alternative ratios at a current density of 0.2 A g^{-1}. Regarding cycling durability, a capacitance retention rate of 82.6% was demonstrated after approximately 1000 cycles at a current density of 1.0 A g^{-1}. Likewise, an energy density as high as 2.83 Wh kg^{-1} was confirmed at a current density of 1.0 A g^{-1}. This demonstrates that the hybrid-type LCN-45 material outperforms EDLC, pseudo-materials, and LCN series materials of other ratios. It is believed that the discovery of the optimal ratio for hybrid materials combining metal oxides and carbon will pave the way for the development of high-performance supercapacitors.

Supplementary Materials: The following supporting information can be downloaded at: https://www.mdpi.com/article/10.3390/app14072977/s1. Figure S1: (a) Adsorption isotherms and (b) Pore size distribution graphs of LC, LCN-45, and LN; Figure S2: BET plots of LC, LCN series, and LN samples; Figure S3: Specific capacitances of LCN-45 electrode calculated from Figure 6b; Figure S4: Cycle stability of LN electrode at 1.0 A g^{-1} (Inset shows the GCD curves of the 1st and 1000th cycles); Table S1. Adsorption characteristics of LC, LCN series, and LN.

Author Contributions: Conceptualization, J.K. and J.L.; methodology, J.L., H.R. and T.C.; formal analysis, J.K.; investigation, J.K., H.R. and A.P.; data curation, Y.H.; writing—original draft preparation, J.K. and J.L.; writing—review and editing, Y.H. and S.N.; supervision, Y.S.; project administration, Y.S. All authors have read and agreed to the published version of the manuscript.

Funding: This work was supported by the Gwangju.Jeonnam local EnergyCluster Manpower training of the Korea Insitute of Energy Technology Evaluation and Planning (KETEP) grant funded by the Korea government Ministry of Knowledge Economy (No. 20214000000560). This research was funded by the Creative Convergence Research Project under the National Science and Technology Research Council: Development of Ultra-Light/Ultra-Thin Electronic Device Manufacturing Technology with Free-Form Multifunctional Structure, MSIT-No. CAP22082-101.

Institutional Review Board Statement: Not applicable.

Informed Consent Statement: Not applicable.

Data Availability Statement: The data that support the study are available from the corresponding author, Y.S., upon reasonable request. The data are not publicly available due to privacy.

Conflicts of Interest: The authors declare no conflict of interest.

References

1. Liu, Z.; Zhang, L.; Xu, G.; Zhang, L.; Jia, D.; Zhang, C. Mn_3O_4 hollow microcubes and solid nanospheres derived from a metal formate framework for electrochemical capacitor applications. *RSC Adv.* **2017**, *7*, 11129–11134. [CrossRef]
2. Simon, P.; Gogotsi, Y. Materials for electrochemical capacitors. *Nat. Mater.* **2008**, *7*, 845–854. [CrossRef] [PubMed]
3. Chmiola, J.; Yushin, G.; Dash, R.; Gogotsi, Y. Effect of pore size and surface area of carbide derived carbons on specific capacitance. *J. Power Sources* **2006**, *158*, 765–772. [CrossRef]
4. Xu, J.; Sun, Y.; Lu, M.; Wang, L.; Zhang, J.; Qian, J.; Liu, X. Fabrication of hierarchical $MnMoO_4 \cdot H_2O$@ MnO_2 core-shell nanosheet arrays on nickel foam as an advanced electrode for asymmetric supercapacitors. *Chem. Eng. J.* **2018**, *334*, 1466–1476. [CrossRef]
5. Kumar, R.; Singh, R.K.; Singh, D.P.; Joanni, E.; Yadav, R.M.; Moshkalev, S.A. Laser-assisted synthesis, reduction and micro-patterning of graphene: Recent progress and applications. *Coord. Chem. Rev.* **2017**, *342*, 34–79. [CrossRef]
6. Tian, D.; Wang, C.; Lu, X. Supercapacitors based on electrospun metal oxide nanofibers. In *Metal Oxide-Based Nanofibers and Their Applications*; Elsevier: Amsterdam, The Netherlands, 2022; pp. 361–393.
7. Liang, R.; Du, Y.; Xiao, P.; Cheng, J.; Yuan, S.; Chen, Y.; Yuan, J.; Chen, J. Transition Metal Oxide Electrode Materials for Supercapacitors: A Review of Recent Developments. *Nanomaterials* **2021**, *11*, 1248. [CrossRef]
8. Jadhav, A.L.; Jadhav, S.L.; Kadam, A.V. Effect of different metals doped in nickel oxide nanomaterials on electrochemical capacitive performance. In *Supercapacitors for the Next Generation*; IntechOpen: London, UK, 2019.
9. Chime, U.K.; Nkele, A.C.; Ezugwu, S.; Nwanya, A.C.; Shinde, N.M.; Kebede, M.; Ejikeme, P.M.; Maaza, M.; Ezema, F.I. Recent progress in nickel oxide-based electrodes for high-performance supercapacitors. *Curr. Opin. Electrochem.* **2020**, *21*, 175–181. [CrossRef]
10. Della Noce, R.; Eugénio, S.; Boudard, M.; Rapenne, L.; Silva, T.M.; Carmezim, M.J.; Donne, S.W.; Montemor, M.F. One-step process to form a nickel-based/carbon nanofoam composite supercapacitor electrode using Na_2SO_4 as an eco-friendly electrolyte. *RSC Adv.* **2016**, *6*, 15920–15928. [CrossRef]
11. Hussain, T.; Wang, Y.; Xiong, Z.; Yang, J.; Xie, Z.Z.; Liu, J. Fabrication of electrospun trace NiO-doped hierarchical porous carbon nanofiber electrode for capacitive deionization. *J. Colloid Interface Sci.* **2018**, *532*, 343–351. [CrossRef]
12. Shin, S.; Shin, M.W. Nickel metal–organic framework (Ni-MOF) derived NiO/C@ CNF composite for the application of high performance self-standing supercapacitor electrode. *Appl. Surf. Sci.* **2021**, *540*, 148295. [CrossRef]
13. Kumar, S.; Madhuri, R. Carbon-Based Electrodes for Flexible Supercapacitors Beyond Graphene. In *Flexible Supercapacitor Nanoarchitectonics*; Scrivener Publishing LLC: Beverly, MA, USA, 2021; pp. 177–210.
14. Minakshi, M.; Duraisamy, S.; Tirupathi, P.R.; Kandhasamy, S.; Nookala, M. Multi-component olivine for lithium-ion hybrid capacitor. *Int. J. Electrochem. Sci.* **2014**, *9*, 5974–5992. [CrossRef]
15. Zhai, Z.; Zhang, L.; Du, T.; Ren, B.; Xu, Y.; Wang, S.; Miao, J.; Liu, Z. A review of carbon materials for supercapacitors. *Mater. Des.* **2022**, *221*, 111017. [CrossRef]
16. Rajagopal, S.; Pulapparambil Vallikkattil, R.; Mohamed Ibrahim, M.; Velev, D.G. Electrode Materials for Supercapacitors in Hybrid Electric Vehicles: Challenges and Current Progress. *Condens. Matter* **2022**, *7*, 6. [CrossRef]
17. Conway, B. *Electrochemical Supercapacitors: Scientific Fundamentals and Technological Applications*; Springer Science & Business Media: Boston, MA, USA, 1999.
18. Szepcsik, B.; Pukánszky, B. The mechanism of thermal stabilization of polyacrylonitrile. *Thermochim. Acta* **2019**, *671*, 200–208. [CrossRef]
19. Park, G.S.; Lee, J.-S.; Kim, S.T.; Park, S.; Cho, J. Porous nitrogen doped carbon fiber with churros morphology derived from electrospun bicomponent polymer as highly efficient electrocatalyst for Zn–air batteries. *J. Power Sources* **2013**, *243*, 267–273. [CrossRef]
20. Lee, J.-J.; Kim, Y.-C. The Adsorption Characteristics by the Optimun Activation Process of PAN-based Carbon Fiber and SO_2 Adsorption Characteristics by the Impregnated Nanoparticles. *Appl. Chem. Eng.* **2006**, *17*, 532–538.
21. Newcomb, B.A. Processing, structure, and properties of carbon fibers. *Compos. Part A Appl. Sci. Manuf.* **2016**, *91*, 262–282. [CrossRef]

22. He, X.; Yang, M.; Ni, P.; Li, Y.; Liu, Z.-H. Rapid synthesis of hollow structured MnO_2 microspheres and their capacitance. *Colloids Surf. A Physicochem. Eng. Asp.* **2010**, *363*, 64–70. [CrossRef]
23. Chen, N.; Ren, Y.; Kong, P.; Tan, L.; Feng, H.; Luo, Y. In situ one-pot preparation of reduced graphene oxide/polyaniline composite for high-performance electrochemical capacitors. *Appl. Surf. Sci.* **2017**, *392*, 71–79. [CrossRef]
24. Wang, H.; Yi, H.; Chen, X.; Wang, X. Facile synthesis of a nano-structured nickel oxide electrode with outstanding pseudocapacitive properties. *Electrochim. Acta* **2013**, *105*, 353–361. [CrossRef]
25. Kim, S.-I.; Lee, J.-S.; Ahn, H.-J.; Song, H.-K.; Jang, J.-H. Facile route to an efficient NiO supercapacitor with a three-dimensional nanonetwork morphology. *ACS Appl. Mater. Interfaces* **2013**, *5*, 1596–1603. [CrossRef] [PubMed]
26. Wu, M.-S.; Zheng, Y.-R.; Lin, G.-W. Three-dimensional carbon nanotube networks with a supported nickel oxide nanonet for high-performance supercapacitors. *Chem. Commun.* **2014**, *50*, 8246–8248. [CrossRef] [PubMed]
27. Xu, X.; Liang, J.; Zhou, H.; Ding, S.; Yu, D. The preparation of hierarchical tubular structures comprised of NiO nanosheets with enhanced supercapacitive performance. *RSC Adv.* **2014**, *4*, 3181–3187. [CrossRef]
28. Zhang, M.; Li, Q.; Fang, D.; Ayhan, I.A.; Zhou, Y.; Dong, L.; Xiong, C.; Wang, Q. NiO hierarchical hollow nanofibers as high-performance supercapacitor electrodes. *RSC Adv.* **2015**, *5*, 96205–96212. [CrossRef]
29. Jadhav, S.A.; Dhavale, S.B.; Patil, A.H.; Patil, P.S. Brief overview of electrospun polyacrylonitrile carbon nanofibers: Preparation process with applications and recent trends. *Mater. Des. Process. Commun.* **2019**, *1*, e83. [CrossRef]
30. Yuan, Y.; Cai, F.; Yang, L. Pore structure characteristics and fractal structure evaluation of medium- and high-rank coal. *Energy Explor. Exploit.* **2022**, *40*, 328–342. [CrossRef]
31. Lo, A.Y.; Chang, C.C.; Lai, Y.W.; Chen, P.R.; Xu, B.C. Improving the supercapacitor performance by dispersing SiO_2 microspheres in electrodes. *ACS Omega* **2020**, *5*, 11522–11528. [CrossRef]

applied sciences

MDPI

Communication

$MoSe_2$ with Ultra-Fine Pt Decoration for Efficient Photodegradation

Yong Chen [1], Dawei Shao [1], Fupeng Xu [1], Zhongjia Huang [2] and Xinying Shi [1,*]

1 School of Physics and Electronic Engineering, Jiangsu Normal University, Xuzhou 221116, China
2 Anhui Key Laboratory of High-Performance Non-Ferrous Metal Materials, Anhui Polytechnic University, Wuhu 241000, China
* Correspondence: xshi@jsnu.edu.cn

Abstract: Transition metal dichalcogenides are widely studied for their photocatalytic ability due to the adjustable bandgap, high carrier mobility and possibility of foreign-element doping. In this work, multilayer molybdenum diselenide ($MoSe_2$) was decorated with ultra-fine Pt nanoparticles through the mild hydrothermal method. $MoSe_2$-Pt nanocomposites were synthesized and showed good structural and chemical stabilities. The incorporation of Pt nanoparticles provides plenty of active sites for $MoSe_2$. The dominant Pt particle sizes are 1.8 nm, 1.8 nm, and 1.9 nm for the three synthesized samples, respectively. The mean crystal sizes of Pt (111) were calculated from X-ray diffraction patterns and we found that they were in accordance with the particle sizes. Both the particle sizes and mean crystal sizes are related to the synthesis conditions. X-ray photoelectron spectroscopy (XPS) characterizations revealed the formation of Se–Pt bonding. The relative contents of Pt–Se bonding were also calculated from XPS results, and they show the same trends as the optical absorption properties. Combining the XPS and optical absorption results, the effects of Se–Pt bonding during the photo-related process could be further confirmed. By degrading methylene blue (MB) under visible light, the synthesized nanocomposites proved promising for application in real-case degradation of organic pollutants. The sample synthesized with a moderate content of $MoSe_2$ exhibited the best photodegradation efficiency, which could be explained by the maximum Pt-Se contents. Based on the experimental findings, we proposed a possible photodegradation mechanism.

Keywords: $MoSe_2$; photodegradation; photocatalyst; two-dimensional materials

Citation: Chen, Y.; Shao, D.; Xu, F.; Huang, Z.; Shi, X. $MoSe_2$ with Ultra-Fine Pt Decoration for Efficient Photodegradation. *Appl. Sci.* **2024**, *14*, 3592. https://doi.org/10.3390/app14093592

Academic Editors: Robert A. Varin and Geng Zhong

Received: 17 March 2024
Revised: 20 April 2024
Accepted: 21 April 2024
Published: 24 April 2024

1. Introduction

Photodegradation of organic pollutants in natural water has been studied for several decades [1,2]. By employing sunlight and semiconducting materials, redox reactions could be taking place to degrade and remove those dissolved pollutants, especially organic dyes [3] such as methylene blue, methyl orange, bisphenol A (BPA), etc. It provides a green and sustainable route for natural water purification. Among so many semiconducting candidates, transition metal dichalcogenides (TMDs) are known for their suitable bandgaps for visible light absorption and utilization [4,5]. As layered two-dimensional materials, TMDs are formed by covalence bonds in layers and van der Waals forces between layers [6]. Such a layered structure also brings an adjustable bandgap which can be easily manipulated by controlling the layer numbers [7]. They usually have active edge sites due to various types of defects but a relatively inert basal surface [8,9]. Thanks to their host ability for incorporating dopants, especially noble metal nanoparticles, the light–matter interaction as well as the photocatalytic performance could be improved. In this case, members of the TMD group are currently featured as photocatalysts for hydrogen evolution and organic-pollutant decomposition in water [10]. It is believed that the TMDs would hopefully provide a solution towards current fossil fuel deficiency and natural water pollution problems.

As a typical member of the TMD family, molybdenum diselenide ($MoSe_2$) has similar atomic arrangements to molybdenum disulfide (MoS_2) [11], and has been widely investigated for applications such as transistors, gas sensors, photodetectors and photocatalysts [12,13]. For example, n-$MoSe_2$ transistors perform well in detecting the streptavidin molecule, due to the controllable high reverse current [14]. As for photocatalytic ability, researchers proposed the hydrothermal synthesis method to prepare $MoSe_2$ for hydrogen evolution reactions and organic dye degradations [15]. $MoSe_2$ is stacked into layers of Se-Mo-Se sandwich planes through the van der Waals force. The typically tunable bandgap also exists in $MoSe_2$, making it suitable for light absorption among a wide range of wavelengths. It should be mentioned that $MoSe_2$ has relatively smaller band gap than MoS_2, which brings more efficient photoexcitation [16]. In the photocatalytic applications, researchers proposed several routes to improve its total activities. For instance, the typical core-shell structures are beneficial for strong chemical activities which could be achieved by incorporating carbon nanotubes, MoS_2 nanoflakes, and transition metals, especially noble metals [17–21]. The optical absorption of $MoSe_2$ has been successfully adjusted by phase engineering, providing another choice for obtaining efficient photocatalytic performance [22]. Moreover, it is also possible to form Z-scheme nanocomposites by introducing other semiconducting materials [23,24]. Among the above methods, noble-metal doping is an effective and controllable way. Noble metal nanoparticles could be easily anchored to $MoSe_2$ edges due to the commonly existing Se defects. However, it usually requires sophisticated synthesis to activate the inert $MoSe_2$ basal planes [8]. For instance, there should be a pre-synthesis to prepare $MoSe_2$@mesoporous hollow carbon spheres, and the following step to obtain the incorporation of Pt nanoparticles [25].

Herein, we report a feasible and straightforward synthetic method to introduce Pt nanoparticles onto $MoSe_2$. The Pt nanoparticles are well formed and their sizes are controlled within 2.3 nm, which could be called ultra-fine Pt nanoparticles. The synthesized $MoSe_2$-Pt nanocomposites exhibit great structural and chemical stabilities. The introduced Pt nanoparticles provide plenty of active sites for light absorption and photocatalysis. We performed the photodegradation of MB under visible light irradiation. It has been found that the synthesized materials could decompose MB efficiently compared to pristine $MoSe_2$. Such improved photocatalytic performance has been explained by the enhanced light absorption and photoelectron transportation. Using X-ray photoelectron spectroscopy (XPS), the chemical bonding between Se and Pt was confirmed. It explains not only the great stabilities of the compounds but also the efficient channels for charge transfer during photocatalytic reactions. The Pt–Se bonding also contributes to the enhanced visible light absorption. Considering the excellent photocatalytic properties of the $MoSe_2$-Pt nanocompounds, we believe that the reported synthetic method might be suitable for further investigation and mass production.

2. Materials and Methods

2.1. Materials and Sample Preparations

The materials were purchased from Sigma-Aldrich (St. Louis, MA, USA) with analytical grade, and deionized water was produced at our own lab. The $MoSe_2$-Pt samples were synthesized with the following steps: (a) multilayer $MoSe_2$ nanopowders were dispersed into deionized water with various ratios. The concentration of $MoSe_2$ suspension was 0.1 mg/mL, 0.2 mg/mL and 0.3 mg/mL, respectively. (b) In order to obtain Pt nanoparticles with an extremely small size, K_2PtCl_4 and trisodium citrate were used as precursors to obtain Pt nanoparticles [26,27]. K_2PtCl_4 and trisodium citrate were added into the $MoSe_2$ suspension by mechanically stirring. (3) The mixed solution was set under an ultrasonic bath at 50 °C for 150 min or 200 min. (4) Pt^0 can be reduced from $PtCl_4{}^{2-}$, and Pt nanoparticles would finally grow onto $MoSe_2$ nanoflakes. After synthesis, three samples with different mass ratios were obtained, as shown in Table 1. These samples are labelled according to their contents, which were M1Pt, M2Pt and M3Pt, respectively.

Table 1. The synthetic details of the samples.

Samples	$MoSe_2$	K_2PtCl_4	Conditions
M1Pt	0.1 mg/mL × 10 mL	41.5 mg	50 °C, 150 min
M2Pt	0.2 mg/mL × 10 mL	41.5 mg	50 °C, 150 min
M3Pt	0.3 mg/mL × 10 mL	41.5 mg	50 °C, 200 min

2.2. Characterizations

The sample morphologies were studied through a transmission electron microscope (TEM, JEOL 2200FS (JEOL, Tokyo, Japan)). The crystal structures of $MoSe_2$ and the synthesized complexes were measured through X-ray diffraction (XRD, Bruker (Billerica, MA, USA)). Cu Kα emission (λ = 1.5406 Å) was employed as the X-ray source. The optical properties were measured through a Lambda 950 UV-Vis spectrophotometer (Spectralab Scientific Inc., Markham, ON, Canada) with $BaSO_4$ as reference. The synthesized powder was dissolved in deionized water for UV–vis absorption tests. The chemical states were analyzed by Thermo Fisher (Waltham, MA, USA) K-Alpha X-ray photoelectron spectroscopy.

2.3. Photodegradation Measurements

The photocatalytic properties of the synthesized $MoSe_2$-Pt catalysts were studied by degrading Methylene blue (MB). During each photocatalytic reaction, 50 mg of the catalyst was put in 50 mL of MB solution (10 mg/L). Reactions were carried out under a 300 W Xe lamp. We used a cutoff filter (#84-689, TECHSPEC, Co., Ltd., Gauteng, South Africa) to block the light under the wavelength of 400 nm. Before degradation tests, MB solutions with catalysts inside were stirred under a dark environment for half an hour to dispose of the adsorption/desorption effects. During photocatalytic degradations, 2 mL of the solution was taken every 30 min and was centrifuged to remove the $MoSe_2$-Pt powder. Visible-light absorption spectra ranging from 400 nm to 800 nm were measured through a Lambda 950 UV-Vis spectrophotometer. During measurements, the absorption peak of MB was set at 664 nm and the concentration of residual MB could be determined by the integration of the MB absorption peaks. The photocatalytic reactions were therefore presented by time-course decrease in MB concertation.

3. Results and Discussion

3.1. $MoSe_2$-Pt Morphologies and Crystal Structures

After synthesis, a drop of the solution containing $MoSe_2$-Pt was placed onto a holey carbon grid and dried in air. Then the samples were observed under TEM. The morphologies of $MoSe_2$ flakes among the three samples were similar in shape and size. Therefore, we chose sample M1Pt as an example whose morphologies are shown in Figure 1. In Figure 1a, $MoSe_2$ nanoflakes are of a round shape with a typical size of 40 nm. There are tiny Pt dots on the edge of the $MoSe_2$. The Pt dots are clearer in the zoomed-in image in Figure 1b. Due to the mild synthetic conditions and low concentration of K_2PtCl_4 in solution 1, $PtCl_4^{2-}$ was reduced to ultrafine particles. There are more Pt nanoparticles formed on the edge sites compared to the surfaces of the $MoSe_2$ because the edge sites are more reactive than the basal planes [28]. Although very small in size, the Pt particles were well crystallized in (111) phase with a lattice spacing of 2.29 Å, as shown in Figure 1c. Such crystalline structure is compatible with the $MoSe_2$ (100) facet. It suggests the Pt particles are well grown on $MoSe_2$. Figure 1d–f show the distribution of Pt nanoparticles for the samples of M1Pt, M2Pt, and M3Pt, respectively. It can be seen that the samples of M1Pt and M2Pt have the same dominant Pt size of 1.8 nm, while M3Pt has a slightly larger Pt size of 1.9 nm.

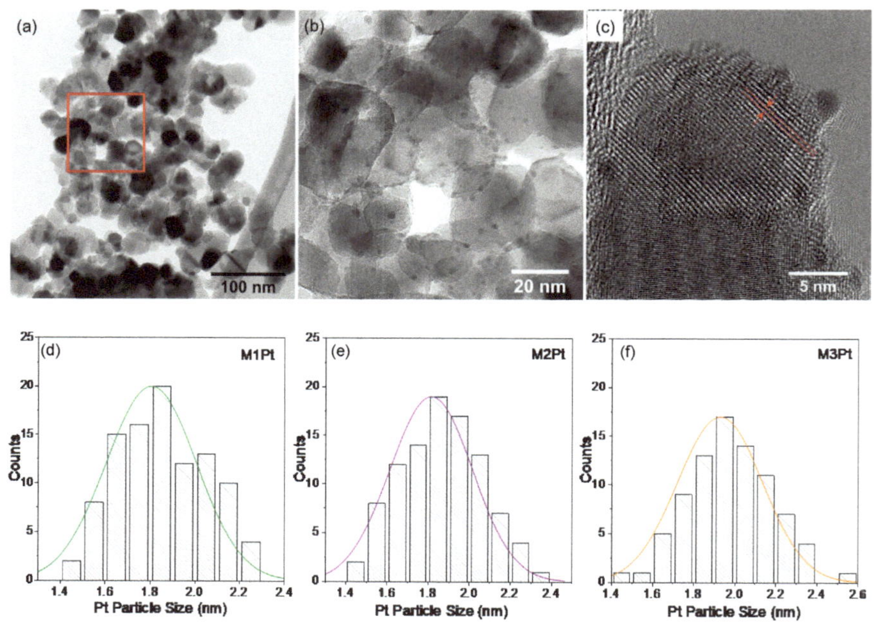

Figure 1. Morphologies of $MoSe_2$-Pt samples. (**a**) TEM image of $MoSe_2$-Pt; (**b**) zoomed-in image of the squared region in (**a**); (**c**) HR-TEM image shows the crystal structure of Pt; (**d–f**) Pt size distributions of the samples of M1Pt, M2Pt, and M3Pt, respectively.

In Figure 2, the XRD spectra show the crystal structure of the three samples of $MoSe_2$-Pt nanocomposites and pristine $MoSe_2$ nanopowder. The crystal planes of $MoSe_2$ and Pt were marked with diamonds and stars, respectively. The $MoSe_2$ (100), (103), (105), (110) and (200) planes were identified from both samples, which was in good agreement with previous reports [29]. Compared to the pristine $MoSe_2$, Figure 2 shows that $MoSe_2$ is stable and keeps its crystal structure after synthesis.

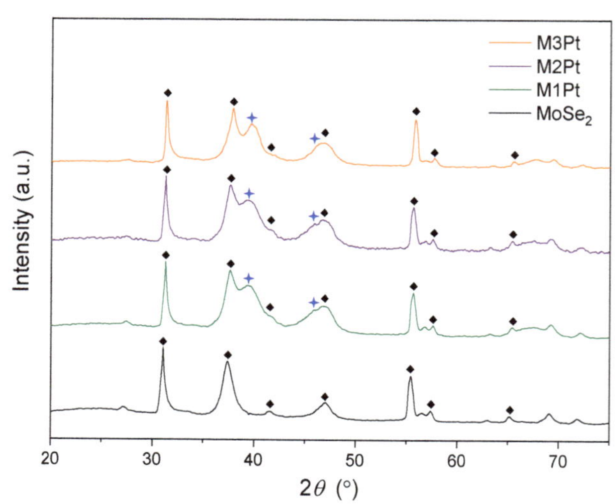

Figure 2. XRD spectra of the synthesized $MoSe_2$-Pt nanocomposite and $MoSe_2$ nanopowder.

Pt (111) and (200) planes were identified in the synthesized $MoSe_2$-Pt samples. The Pt (111) lattice spacings are shown in Table 2. We can see that Pt (111) peak positions were slightly varied among the three samples. The lattice spacings are calculated as 0.2285, 0.2291, and 0.2274 nm, respectively. Moreover, the mean crystal sizes (D (111) in Table 2) were also calculated through Scherrer's equation [30]. They are very close for the M1Pt and M2Pt samples, while slightly larger for the M2Pt sample. It should be noted that the calculated mean crystal sizes are not the particle sizes which could be directly measured under TEM; however, the D (111) values show the same trends with the particle sizes shown in Figure 1d–f.

Table 2. Pt (111) lattice spacings and mean crystal sizes.

Samples	2θ/°	FWHM/°	d (111)/nm	D (111)/nm
M1Pt	39.40	12.4217	0.2285	0.6793
M2Pt	39.30	12.4017	0.2291	0.6802
M3Pt	39.60	12.0752	0.2274	0.6992

3.2. Chemical States, Mechanism of Efficient Photocatalytic Reactions

The chemical electronic states of the $MoSe_2$-Pt samples were evaluated through XPS. The high-resolution XPS spectra of Pt 4f and Se 3d are shown in Figure 3. In Figure 3a–c, there are two Pt species which were identified by fitting the original Pt 4f spectra. Pt^0 peaks are the reduced Pt nanoparticles in metal form, and Pt-Se peaks are attributed to the Pt–Se bonding. The binding energies of these Pt peaks are slightly shifted among the three samples which may be caused by partial electron transfer between $MoSe_2$ and Pt [31]. In Figure 3d–f, two main contents are found in the Se 3d spectra, which are the Se-Mo peaks and Se-Pt peaks. Such results are typical for $MoSe_2$ and $PtSe_2$ characteristics [32], and further confirm the formation of Se–Pt bonding. Therefore, it is obvious that chemical bonding of Pt–Se formed after the synthesis. Such chemical bonding can provide an effective pathway for charge transportation during photocatalytic reactions [9]. The Se-Mo $3d_{5/2}$ peaks at 54.56 eV are barely shifted among the three samples, suggesting that $MoSe_2$ takes a dominant content compared to the Se-Pt species.

Based on the XPS results, it is noticed that the relative contents of Pt-Se varies among the samples, which may have influence on electron transportation. The relative contents of Pt-Se can be calculated from Pt 4f spectra, which are 34.78%, 40.02%, and 19.79%, respectively. It suggests that too much $MoSe_2$ involved in the synthesis (the M3Pt sample) may limit the formation of Pt–Se bonding.

In this case, there may be two ways that the photocatalytic abilities could be affected. Firstly, Pt–Se bonding provides an efficient pathway for charge transfer, so more contents of Pt-Se would be beneficial to the photocatalytic reactions. Secondly, the existence of Pt nanoparticles (Pt^0) can be confirmed by TEM results in Figure 1. Such ultra-fine Pt nanoparticles offer active sites for photocatalytic reactions where degradation could take place [33]. Introducing noble metal nanoparticles on edge sites of $MoSe_2$ is a typical method for enhancing the photocatalytic properties.

3.3. Optical Properties of the Catalysts

In order to evaluate the optical performance of the synthesized samples, they were prepared in aqua solution for UV–visible absorption measurements. Figure 4 shows the absorption spectra of $MoSe_2$ and the three samples. In the visible wavelength range, there are obvious absorption peaks at around 475 nm and 710 nm, respectively. Such optical behavior is typical for multilayer $MoSe_2$ on a nanometer scale [34]. It can be seen that the pristine $MoSe_2$ has a weak response to visible light, while the $MoSe_2$-Pt samples strongly absorb the short wavelength of visible light, mainly between 450 and 550 nm. Among the three synthesized samples, M2Pt shows the best absorption ability for visible light while M3Pt shows the worst. Moreover, there is slight red-shift between the pristine $MoSe_2$ and the catalysts. Such feature enhances the absorption of light in a longer wavelength, making it suitable for the utilization of the solar spectrum.

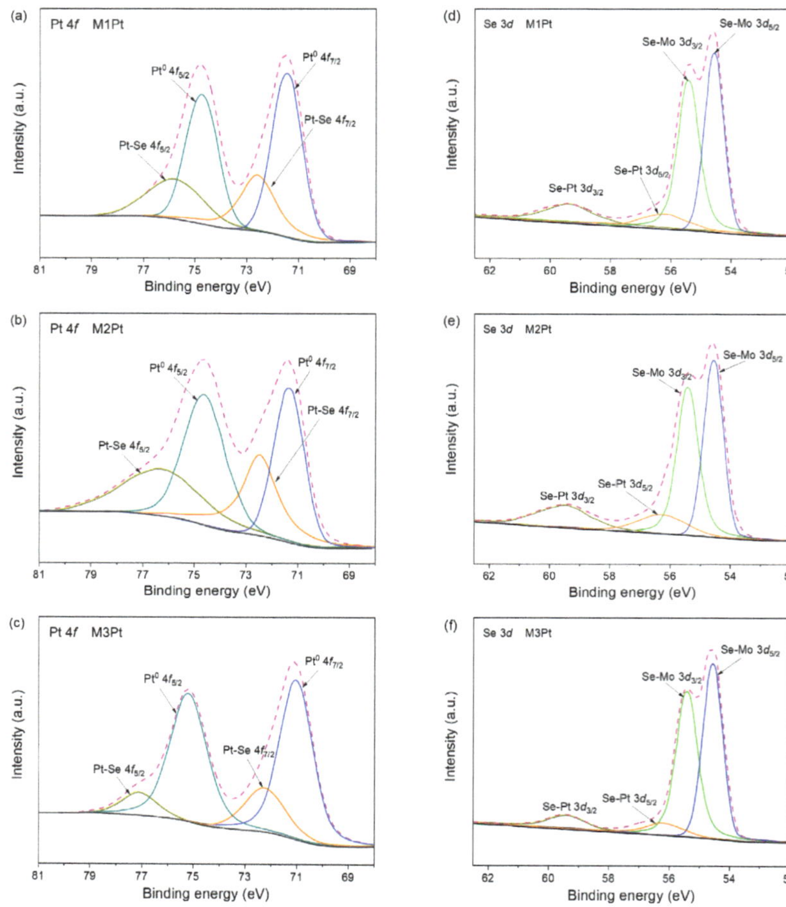

Figure 3. XPS spectra of the synthesized $MoSe_2$-Pt samples. (**a**) Pt 4*f* of M1Pt, (**b**) Pt 4*f* of M2Pt, (**c**) Pt 4*f* of M3Pt, (**d**) Se 3*d* of M1Pt, (**e**) Se 3*d* of M2Pt, (**f**) Se 3*d* of M3Pt.

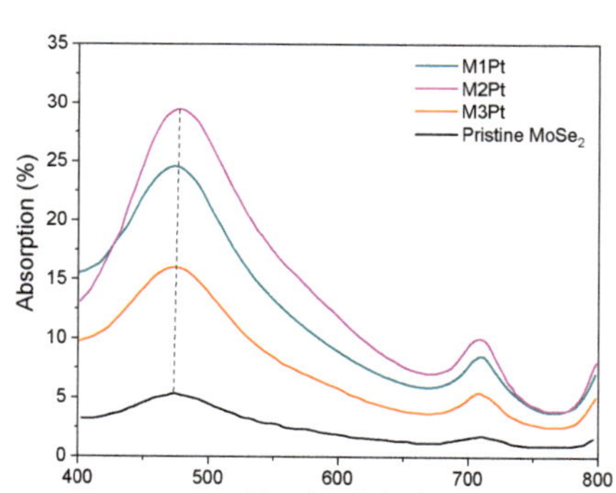

Figure 4. UV–vis absorption spectra of $MoSe_2$ and the synthesized samples.

3.4. Photodegradation

The aforementioned characterizations provide chemical and optical insights into the synthesized materials. Their photocatalytic performance is evaluated by MB degradation, as shown in Figure 5. The time-course photodegradation results are shown in Figure 5a. Here, C/C_0 is used to represent the decrease of MB concentration, where C_0 and C are the original MB concentration and the MB concentration at a certain time, respectively. The dark process suggests that the synthesized materials show obvious adsorption of MB, while the pristine $MoSe_2$ can hardly adsorb MB. Such adsorption abilities are usually related to the photocatalytic process. We can see that pristine $MoSe_2$ can hardly degrade MB and the samples are all reactive with MB. Within around 60 min, MB is nearly removed by the samples of M1Pt and M2Pt. However, the M3Pt sample performs slightly less efficiently, but it can still degrade more than 90% of MB after 120 min. Such photocatalytic results are in accordance with the optical absorption properties shown in Figure 4. Considering the catalyst amount used in the degradation process, the above photocatalytic performance is comparable to other popular photocatalysts, such as $Bi_2WO_6/BiOCl$ [35]. As a comparison, pristine $MoSe_2$ removes only 5.4% of the MB in 120 min with the same irradiation condition. It suggests that the introduction of Pt to $MoSe_2$ edges greatly improves the photocatalytic property.

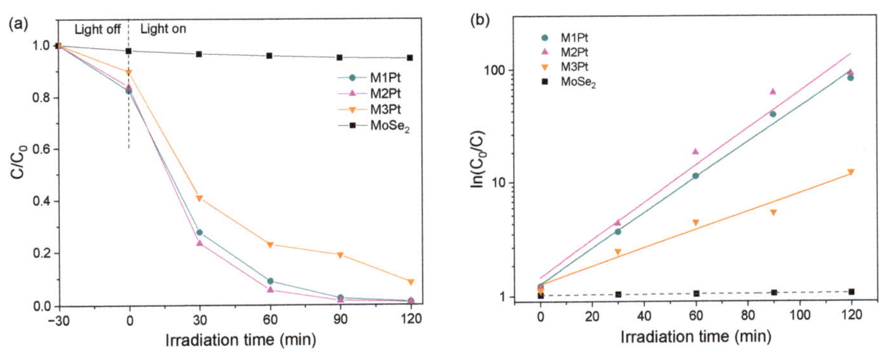

Figure 5. The time-course photocatalytic MB degradation under visible light irradiation. (**a**) Photodegradation of the synthesized samples and pristine $MoSe_2$. (**b**) Photodegradation kinetics of $\ln(C_0/C)$.

The photodegradation kinetics was evaluated by the pseudo-first-order model. With this model, the pseudo-first-order constant *k* could be expressed as

$$k = \frac{\ln(C_0/C)}{t} \quad (1)$$

The results of $\ln(C_0/C)$ are shown in Figure 5b. The constants were calculated as 0.0361 min^{-1}, 0.0379 min^{-1}, and 0.0185 min^{-1} for the three samples, respectively. It is obvious that the M2Pt sample has the best photocatalytic kinetics. The constant values have the same trends as the Pt-Se contents. This indicates the correlation between Se–Pt bonding and photocatalytic performance. On the other hand, *k* values are not lineally affected by either the Pt amount used in the synthesis nor the relative contents of Pt–Se bonding. This may indicate that the surface segregation occurred during synthesis [36].

The possible photocatalytic mechanism can therefore be discussed. During the degradation process, photo-generated electrons from valence bands of $MoSe_2$ reach the conduction bands, and finally migrate to the active sites provided by Pt nanoparticles. The electron-hole pairs participate in redox reactions, where the holes oxidize water to H^+ and ·OH. The hydroxyl radicals then react with MB [9]. Meanwhile, the electrons reduce H^+

to ·H. The Pt–Se bonding offers stable and efficient routes for charge transfer during the whole process.

4. Conclusions

In conclusion, we synthesized $MoSe_2$-Pt complexes with a facile method, where Pt is reduced to ultra-fine nanoparticles. The synthetic method is simple and straightforward, making it suitable for mass production. The synthesized samples show a greatly enhanced photocatalytic performance for degrading the organic pollutant of MB under visible-light irradiation. Such properties show the excellent ability of visible-light utilization. The photodegradation mechanism could be explained through XPS analysis. Effective Pt–Se bonding provides pathways for charge transportation and ultra-fine Pt nanoparticles offer active sites for photocatalytic reactions. This work may be helpful for further photocatalyst design and applications in real cases.

Author Contributions: Y.C., investigation, characterization, and writing—original draft; D.S., characterization; F.X., characterization; Z.H., investigation, revision; X.S., supervision and writing—review and editing. All authors have read and agreed to the published version of the manuscript.

Funding: This work was financially supported by the Natural Science Foundation of Jiangsu Higher Education Institutions of China (No. 19KJB140008), and the Anhui Provincial Natural Science Foundation (No. 2008085ME141).

Institutional Review Board Statement: Not applicable.

Informed Consent Statement: Not applicable.

Data Availability Statement: The original contributions presented in the study are included in the article, further inquiries can be directed to the corresponding author.

Conflicts of Interest: The authors declare no conflicts of interest.

References

1. Sellers, P.; Kelly, C.A.; Rudd, J.W.M.; MacHutchon, A.R. Photodegradation of methylmercury in lakes. *Nature* **1996**, *380*, 694–697. [CrossRef]
2. Saeed, M.; Muneer, M.; Haq, A.U.; Akram, N. Photocatalysis: An effective tool for photodegradation of dyes—A review. *Environ. Sci. Pollut. Res.* **2022**, *29*, 293–311. [CrossRef] [PubMed]
3. Jiang, Q.; Lu, Y.; Huang, Z.; Hu, J. Facile solvent-thermal synthesis of ultrathin $MoSe_2$ nanosheets for hydrogen evolution and organic dyes adsorption. *Appl. Surf. Sci.* **2017**, *402*, 277–285. [CrossRef]
4. Chen, C.; Ma, W.; Zhao, J. Semiconductor-mediated photodegradation of pollutants under visible-light irradiation. *Chem. Soc. Rev.* **2010**, *39*, 4206–4219. [CrossRef] [PubMed]
5. Krasnok, A.; Lepeshov, S.; Alú, A. Nanophotonics with 2D transition metal dichalcogenides. *Opt. Express* **2018**, *26*, 15972. [CrossRef]
6. Chhowalla, M.; Liu, Z.; Zhang, H. Two-dimensional transition metal dichalcogenide (TMD) nanosheets. *Chem. Soc. Rev.* **2015**, *44*, 2584. [CrossRef]
7. Tan, C.; Zhang, H. Two-dimensional transition metal dichalcogenide nanosheet-based composites. *Chem. Soc. Rev.* **2015**, *44*, 2713. [CrossRef]
8. Kibsgaard, J.; Chen, Z.; Reinecke, B.N.; Jaramillo, T.F. Engineering the surface structure of MoS_2 to preferentially expose active edge sites for electrocatalysis. *Nat. Mater.* **2012**, *11*, 963–969. [CrossRef]
9. Shi, X.; Posysaev, S.; Huttula, M.; Pankratov, V.; Hoszowska, J.; Dousse, J.C.; Zeeshan, F.; Niu, Y.; Zakharov, A.; Li, T.; et al. Metallic contact between MoS_2 and Ni via Au nanoglue. *Small* **2018**, *14*, 1704526. [CrossRef]
10. Shi, X.; Zhang, M.; Wang, X.; Kistanov, A.A.; Li, T.; Cao, W.; Huttula, M. Nickel nanoparticle-activated MoS_2 for efficient visible light photocatalytic hydrogen evolution. *Nanoscale* **2022**, *14*, 8601–8610. [CrossRef]
11. Shim, G.W.; Yoo, K.; Seo, S.B.; Shin, J.; Jung, D.Y.; Kang, I.S.; Ahn, C.W.; Cho, B.J.; Choi, S.Y. Large-area single-layer $MoSe_2$ and its van der Waals heterostructures. *ACS Nano* **2014**, *8*, 6655–6662. [PubMed]
12. Mathew, S.; Sai Chandu, K.K.; Halder, S.; Polumati, G.; Chakraborty, C.; Sahatiya, P.; Pal, S. Band alignment and charge transport study of Au nanoparticles decorated over $MoS_2/MoSe_2$ hybrid heterostructure for plasmon enhanced photodetection. *Mater. Sci. Semicond. Process.* **2023**, *156*, 107302. [CrossRef]
13. Kumar, S.; Mirzaei, A.; Kumar, A.; Lee, M.H.; Ghahremani, Z.; Kim, T.U.; Kim, J.Y.; Kwoka, M.; Kumar, M.; Kim, S.S.; et al. Nanoparticles anchored strategy to develop 2D MoS_2 and $MoSe_2$ based room temperature chemiresistive gas sensors. *Coord. Chem. Rev.* **2024**, *503*, 215657. [CrossRef]

14. Nisar, S.; Dastgeer, G.; Shahzadi, M.; Shahzad, Z.M.; Elahi, E.; Irfan, A.; Eom, J.; Kim, H.; Kim, D.K. Gate-assisted $MoSe_2$ transistor to detect the streptavidin via supporter molecule engineering. *Mater. Today Nano* **2023**, *24*, 100405. [CrossRef]
15. Shi, H.; Zhang, H.; Li, M.; Wang, Y.; Wang, D. Nanoflower-like 1T/2H mixed-phase $MoSe_2$ as an efficient electrocatalyst for hydrogen evolution. *J. Alloys Compd.* **2021**, *878*, 160381. [CrossRef]
16. Zhang, Y.; Chang, T.R.; Zhou, B.; Cui, Y.T.; Yan, H.; Liu, Z.; Schmitt, F.; Lee, J.; Moore, R.; Chen, Y.; et al. Direct observation of the transition from indirect to direct bandgap in atomically thin epitaxial $MoSe_2$. *Nat. Nanotech.* **2014**, *9*, 111–115. [CrossRef] [PubMed]
17. Maity, S.; Das, B.; Samanta, M.; Das, B.K.; Ghosh, S.; Chattopadhyay, K.K. $MoSe_2$-Amorphous CNT Hierarchical Hybrid Core–Shell Structure for Efficient Hydrogen Evolution Reaction. *ACS Appl. Energy Mater.* **2020**, *3*, 5067–5076. [CrossRef]
18. Hwang, Y.; Shin, N. Colloidal Synthesis of $MoSe_2/WSe_2$ Heterostructure Nanoflowers via Two-Step Growth. *Materials* **2021**, *14*, 7294. [CrossRef] [PubMed]
19. Su, J.; Nie, Z.; Feng, Y.; Hu, X.; Li, H.; Zhao, Z.; Zan, S.; Qi, S. Hollow core–shell structure Co/C@$MoSe_2$ composites for high-performance microwave absorption. *Compos. Part A Appl. Sci. Manuf.* **2022**, *162*, 107140. [CrossRef]
20. Rao, B.G.; Matte, H.S.S.R.; Rao, C.N.R. Decoration of few-layer graphene-like MoS_2 and $MoSe_2$ by noble metal nanoparticles. *J. Clust. Sci.* **2012**, *23*, 929–937.
21. Upadhyay, S.N.; Pakhira, S. Nanostructured Pt-doped 2D $MoSe_2$: An efficient bifunctional electrocatalyst for both hydrogen evolution and oxygen reduction reactions. *Phys. Chem. Chem. Phys.* **2022**, *37*, 22823–22844. [CrossRef] [PubMed]
22. Hanson, E.D.; Lilley, L.M.; Cain, J.D.; Hao, S.; Palacios, E.; Aydin, K.; Wolverton, C.; Meade, T.; Dravid, V.P. Phase engineering and optical properties of 2D $MoSe_2$: Promise and pitfalls. *Mater. Chem. Phys.* **2019**, *225*, 219–226. [CrossRef]
23. Wang, F.; Lu, T.; Cheng, X.; Zhang, Y.; Xiao, X. Fabrication of $MoSe_2/Bi_3O_4Br$ Z-scheme heterojunction with enhanced photocatalytic performance. *J. Environ. Chem. Eng.* **2024**, *12*, 111991. [CrossRef]
24. Wang, X.; Wang, X.; Huang, J.; Li, S.; Meng, A.; Li, Z. Interfacial chemical bond and internal electric field modulated Z-scheme S_v-$ZnIn_2S_4/MoSe_2$ photocatalyst for efficient hydrogen evolution. *Nat. Commun.* **2021**, *12*, 4112. [CrossRef] [PubMed]
25. Yang, F.; Qiao, W.; Yu, L.; Wang, S.; Feng, L. Support engineering modulated Pt/hierarchical $MoSe_2$@mesoporous hollow carbon spheres for efficient methanol-assisted water splitting. *Chem. Eng. J.* **2024**, *483*, 149055. [CrossRef]
26. Huang, X.; Zeng, Z.; Bao, S.; Wang, M.; Qi, X.; Fan, Z.; Zhang, H. Solution-phase epitaxial growth of noble metal nanostructures on dispersible single-layer molybdenum disulfide nanosheets. *Nat. Commun.* **2013**, *4*, 1444. [CrossRef]
27. Lin, C.S.; Khan, M.R.; Lin, S.D. Platinum states in citrate sols by EXAFS. *J. Colloid Interface Sci.* **2005**, *287*, 366–369. [CrossRef] [PubMed]
28. Tsai, C.; Chan, K.; Abild-Pedersenb, F.; Nørskov, J.K. Active edge sites in $MoSe_2$ and WSe_2 catalysts for the hydrogen evolution reaction: A density functional study. *Phys. Chem. Chem. Phys.* **2014**, *16*, 13156–13164. [CrossRef]
29. Shelke, N.T.; Late, D.J. Hydrothermal growth of $MoSe_2$ nanoflowers for photo- and humidity sensor applications. *Sens. Actuators A Phys.* **2019**, *295*, 160–168. [CrossRef]
30. Holzwarth, U.; Gibson, N. The Scherrer equation versus the 'Debye-Scherrer equation'. *Nat. Nanotechnol.* **2011**, *6*, 534. [CrossRef]
31. Qiao, W.; Yu, L.; Chang, J.; Yang, F.; Feng, L. Efficient bi-functional catalysis of coupled $MoSe_2$ nanosheet/Pt nanoparticles for methanol-assisted water splitting. *Chin. J. Catal.* **2023**, *51*, 113–123.
32. Avsar, A.; Ciarrocchi, A.; Pizzochero, M.; Unuchek, D.; Yazyev, O.V.; Kis, A. Defect induced, layer-modulated magnetism in ultrathin metallic $PtSe_2$. *Nat. Nanotech.* **2019**, *14*, 674–678. [CrossRef] [PubMed]
33. Li, Z.; Liao, Y.; Liu, Y.; Zeng, W.; Zhou, Q. Room temperature detection of nitrogen dioxide gas sensor based on Pt-modified $MoSe_2$ nanoflowers: Experimental and theoretical analysis. *Appl. Surf. Sci.* **2023**, *610*, 155527. [CrossRef]
34. Dong, N.; Li, Y.; Feng, Y.; Zhang, S.; Zhang, X.; Chang, C.; Fan, J.; Zhang, L.; Wang, J. Optical limiting and theoretical modelling of layered transition metal dichalcogenide nanosheets. *Sci. Rep.* **2015**, *5*, 14646. [CrossRef] [PubMed]
35. Zhu, S.; Yang, C.; Li, F.; Li, T.; Zhang, M.; Cao, W. Improved photocatalytic $Bi_2WO_6/BiOCl$ heterojunctions: One-step synthesis via an ionic-liquid assisted ultrasonic method and first-principles calculations. *Mol. Catal.* **2017**, *435*, 33–48.
36. Sidorowicz, A.; Yigit, N.; Wicht, T.; Stöger-Pollach, M.; Concas, A.; Orrù, R.; Cao, G.; Rupprechter, G. Microalgae-derived Co_3O_4 nanomaterials for catalytic CO oxidation. *RSC Adv.* **2024**, *14*, 4575–4586.

applied sciences

MDPI

Article

Removal of Bisphenol A from Water by Single-Walled Carbon Nanotubes Loaded with Iron Oxide Nanoparticles

Luying Chen [1,2], Jintao Jiang [3] and Leimei Sheng [1,2,*]

1 School of Physics, Shanghai University, Shanghai 200444, China; 3146617397@shu.edu.cn
2 Institute of Low-Dimensional Carbon and Device Physics, Shanghai University, Shanghai 200444, China
3 School of Materials Science and Engineering, Shanghai University, Shanghai 200444, China; jiangjintao1102@shu.edu.cn
* Correspondence: shenglm@staff.shu.edu.cn

Abstract: Single-walled carbon nanotubes (SWCNTs) loaded with magnetic iron oxide nanoparticles were prepared by the arc discharge method and air heat treatment. The nanocomposite was characterized by X-ray diffraction, scanning electron microscopy, Raman spectroscopy, vibrating sample magnetometry, etc. The results showed that the heat-treated nanocomposites (SWCNTs/Fe_xO_y) had iron oxide phases and superparamagnetic properties with a saturation magnetization of 33.32 emu/g. Compared with the non-heat-treated materials, SWCNTs/Fe_xO_y had a larger specific surface area and pore volume. Using SWCNTs/Fe_xO_y to remove the organic contaminant (bisphenol A, BPA), it was found that under the conditions of pH = 3 and adsorbent dosage of 0.2 g/L, the maximum adsorption capacity of the composite was 117 mg/g, and the adsorption could reach more than 90% in only 5 min when the BPA content was below 0.05 mmol/L. The fitting results of the Langmuir and D-R models are more consistent with the experimental data, indicating a relatively uniform distribution of the adsorption sites and that the adsorption process is more consistent with physical adsorption. The kinetic calculations showed that the SWCNTs/Fe_xO_y exhibits chemical effects on both the surface and the gap, and the adsorption process is controlled by the π-π bonds and the hydrophobicity of the SWCNTs/Fe_xO_y.

Keywords: SWCNT composites; Bisphenol A; magnetic iron oxide nanoparticles; adsorption; adsorption mechanism

Citation: Chen, L.; Jiang, J.; Sheng, L. Removal of Bisphenol A from Water by Single-Walled Carbon Nanotubes Loaded with Iron Oxide Nanoparticles. *Appl. Sci.* **2024**, *14*, 3943. https://doi.org/10.3390/app14093943

Academic Editor: Piotr Gas

Received: 24 March 2024
Revised: 23 April 2024
Accepted: 1 May 2024
Published: 6 May 2024

1. Introduction

In recent years, with the rapid development of industrialization, the problem of water pollution has become increasingly prominent. The pollutants of concern include bisphenol A (BPA), a ubiquitous chemical widely used in the production of plastics and resins. However, BPA is a quintessential endocrine disruptor in the environment [1] and can disrupt the normal physiological functions of the human body even in minute quantities. Numerous studies have revealed that BPA enters the human body through various common pathways, such as contaminating water, food, cooking fumes, and even through daily direct contact, posing threats to human health and affecting estrogen levels. The extensive production and use of BPA has led to its widespread presence in the environment, and it has even been detected in drinking water [2–4]. Currently, numerous technologies have been developed to eliminate BPA, including advanced oxidation processes, biodegradation, membrane separation, adsorption [5–8], and more. In fact, a large number of researchers believe that adsorption [9,10] is better than other methods, as it can quickly and effectively adsorb pollutants. Additionally, adsorption offers advantages such as cost-effectiveness, efficiency, simple design, and fewer harmful by-products.

With their exceptional adsorption efficacy, various carbon materials [11], such as graphene oxide (GO), activated carbon (AC), and carbon nanotubes (CNTs) [12–15], have been increasingly adopted as preferred adsorbents for the eradication of contaminants from aquatic environments. Among them, CNTs [16], as a new adsorbent material, have been used in many studies [17] to adsorb pollutants in water owing to their extensive surface area and plentiful reactive sites. Evidence confirms that CNTs possess a strong capacity to adsorb phenols, heavy metals, and natural organic compounds effectively [18]. Nevertheless, the isolation of CNTs from suspension in aqueous media poses a challenge because of their minute dimensions, and their introduction into water resources may result in secondary pollution [19]. On the other hand, magnetic nanoparticles such as Fe_3O_4, γ-Fe_2O_3, and Fe^0 have drawn considerable interest due to their excellent adsorption characteristics [20,21], and their inherent superparamagnetism enables them to be readily extricated from aqueous environments using magnetic attraction after application. The practical usage of magnetic nanoparticles often faces issues like pronounced clustering in suspension and a susceptibility to being lost via water runoff. Therefore, CNTs are usually used as a carrier [22,23] to support magnetic nanoparticles such as Fe_3O_4 and Fe^0, and their superparamagnetism [24] is used to enable them to effectively perform magnetic separation in water, which is not only conducive to improving the adsorption performance, but also to improving the dispersion and mechanical properties [25,26] of these magnetic nanoparticles in practical applications. For example, Li et al. [27] successfully fabricated a recyclable CNTs/Fe_3O_4 magnetic composite via a hydrothermal method, which exhibited robust adsorption behaviors towards BPA. Al-Musawi et al. [28] synthesized multi-walled carbon nanotubes coated with $CoFe_2O_4$ nanoparticles using solvothermal co-precipitation and electrophoretic precipitation methods, and applied it to the adsorption of BPA in water. The experimental findings indicated that the incorporation of a composite material significantly augmented the adsorption capabilities as well as enhanced the characterization attributes of multi-walled carbon nanotubes (MWCNTs). Guo et al. [29] synthesized MWCNTs modified with iron oxide and manganese dioxide (MWCNTs-FeO-MnO) via the co-precipitation method, and utilized them to determine the levels of residual BPA in bottled water supplies. The results demonstrated that the nanotubes possessed effective adsorptive qualities for BPA, along with commendable recyclability and concentration capabilities. Most studies have focused on MWCNTs, with fewer reports on SWCNTs [30]. Nevertheless, SWCNTs exhibit superior performance; they possess smaller diameters and a larger specific surface area compared to MWCNTs. Therefore, drawing on earlier investigations into phenol elimination from water, this study prepared magnetic SWCNT composites incorporating iron oxide nanoparticles via an electric arc process followed by thermal treatment. The adsorption behavior and underlying mechanisms of the composites towards BPA were investigated. These results offer foundational theoretical and empirical support for the application of magnetic SWCNT nanocomposites in the removal of organic contaminants during water purification processes.

2. Materials and Methods

2.1. Adsorbent Preparation

We employed the hydrogen arc discharge method to fabricate SWCNTs, utilizing a carbon rod doped with 4 wt% iron as the anode and the carbon rod's composition was pure toner and iron powder; a pure graphite carbon rod was used as the cathode. Both electrodes had a diameter of 10 mm. Within the arc reaction chamber, the anode and cathode were horizontally secured at the left and right flanks, respectively, maintaining a distance of 2–3 mm [31]. Under conditions of low vacuum, a hydrogen/argon (volume ratio of 2:3) mixture gas at 200 torr pressure was introduced into the reaction vacuum chamber. With a current of 100 A, after about 30 min of arc discharge, a substantial amount of black SWCNTs containing iron catalyst adhered to the interior chamber walls. The as-grown SWCNTs of iron-containing catalyst were then heated in dry air to 350 °C and kept for 1 h. Finally, SWCNT composites loaded with magnetic iron oxide nanoparticles (SWCNTs/Fe_xO_y) were obtained.

2.2. Adsorption Experiment

Pre-measured solid BPA particles were dissolved in deionized water to obtain BPA master solution (1 mmol/L). The master solution was further diluted with deionized water to create BPA solutions in different concentrations, adjusted the pH value of the solution by adding an appropriate amount of HCl or NaOH. Each experiment was conducted in a beaker containing 50 mL of BPA solution, with an adsorbent dosage of 10 mg (0.2 g/L). A magnetic stirrer was used to agitate the BPA solution and adsorbent at a rotational speed of 500 revolutions per minute, while the temperature for adsorption was kept constant at ambient conditions (25 °C). At predetermined time intervals, 2 mL of the suspension was extracted from the beaker and subsequently filtered utilizing a polyethersulfone (PES) membrane filter with a pore size of 0.22 μm. The residual organic compound concentration was analyzed by ultraviolet–visible–near-infrared absorption spectroscopy (UV-vis-NIR spectroscopy, UV-3600, Shimadzu, Japan). The rate of BPA adsorption α in the solution by SWCNTs/Fe_xO_y as the adsorbent and the adsorption capacity q_t as the unit mass of the adsorbent at a certain time are written as follows.

$$\alpha = \frac{C_0 - C_t}{C_t} \times 100\%, \quad (1)$$

$$q_t = \frac{(C_0 - C_t)V}{m}, \quad (2)$$

where C_0 is the concentration of BPA solution at the initial time (mmol/L); C_t is the concentration of BPA solution at time t (mmol/L); V is the volume (L) of the solution; m is the mass (g) of the adsorbent; α is the adsorption rate; and q_t is the adsorption capacity (mg/g).

2.3. Microstructure Analyses

The morphologies of the nanocomposites were observed by scanning electron microscopy on a Gemini 500 instrument (Zeiss, Oberkochen, Germany) at a magnification of 20,000–100,000.

The structural characterization of the nanocomposites was characterized using a X-ray diffractometer from Rigaku, JP, model D/max-2550. The X-ray source was Cu Kα radiation, with a wavelength of 1.54056 Å and a power of 18 KW. The scanning angle range was set from 15° to 85°.

The structure of the nanocomposites was studied using a Raman spectroscope from Renishaw, UK, model In Via-plus. The Raman spectroscope was equipped with a Leica microscope. The excitation wavelength used was 633 nm.

The surface elemental composition and chemical state of the nanocomposites were analyzed using an X-ray Photoelectron Spectrometer from Thermo Fisher Scientific, Winsford, UK, model ESCALAB 250Xi. To control the surface composition of the sample, survey spectra were collected at a step size of 1 eV over the range of 0–1100 eV.

The nitrogen adsorption–desorption isotherms of the nanocomposites were measured at 77 K using a surface area and pore size distribution analyzer (ASAP2020, Micromeritics Instrument Co., Norcross, GA, USA). Before measurement, the nanocomposite samples were degassed at 105 °C for over 16 h. The specific surface area of the nanocomposites was calculated using the Brunauer–Emmett–Teller (BET) method, while the pore diameter and volume were determined from the desorption isotherm using the Barrett–Joyner–Halenda (BJH) method.

The measurement of magnetic properties was conducted using the Magnetic Property Measurement System (MPMS3) from Quantum Design, San Diego, CA, USA. In this study, measurements were carried out in VSM (Vibrating Sample Magnetometer) mode.

3. Results and Discussion

3.1. Characterization

Figure 1a shows the SEM micrograph of the as-grown SWCNTs. The image reveals that the SWCNTs, synthesized with the aid of an iron catalyst, possess a loose arrangement. The bundles of CNTs are observed to be entangled with each other, and they are coated with large amounts of amorphous carbon. The SEM image at a high magnification (Figure 1b) more clearly reveals the prepared SWCNT bundle loaded with iron nanoparticles. From the energy dispersive spectra (EDS) finding (Figure 1c), it can be seen that the region comprised 51.79 wt% C element, 6.41 wt% O element, and 41.62 wt% Fe element, indicating that the iron nanoparticles were successfully attached to the surface of the CNTs.

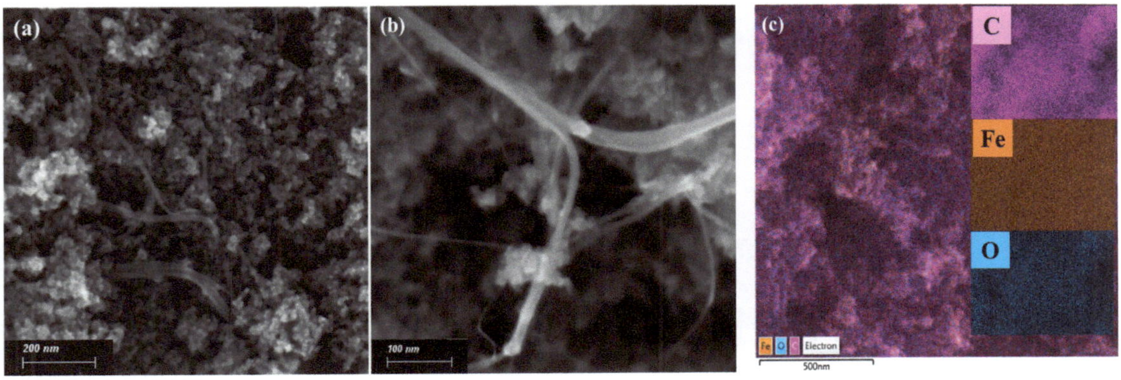

Figure 1. Scanning electron micrographs of (**a**) low-magnification as-grown SWCNTs and (**b**) high-magnification as-grown SWCNTs; (**c**) EDS image.

Figure 2a displays the XRD spectra of the as-grown SWCNTs and the air-heat-treated SWCNTs. It shows that the diffraction peaks of C in the as-grown SWCNTs diminish after the air heat treatment, indicating the effective removal of amorphous carbon from the composite, thereby enhancing its adsorption capacity. Simultaneously, the Fe phase is transformed into the Fe_2O_3 and Fe_3O_4 phases after heat treatment. Figure 2b depicts the Raman spectra obtained under 633 nm laser excitation for the as-grown SWCNTs and SWCNTs/Fe_xO_y. Within the range of 100–400 cm^{-1}, resonant breathing mode (RBM) peaks of the SWCNTs are evident, allowing for the calculation of the diameter of the SWCNTs using the frequency (ω_{RBM}) of the RBM peak. For the SWCNTs, the calculation formula is ω_{RBM} (cm^{-1}) = 234/d (nm) + 10, and the diameter of the SWCNTs in the original sample ranges between 0.89 and 1.78 nm. The weak Raman peak at 1350 cm^{-1} is the D peak, which arises from defects like pentagon and heptagon rings within the CNTs, reflecting the defect density directly. The prominent Raman peak near 1590 cm^{-1} is the G peak, which is associated with in-plane tangential vibrations of carbon atoms. The intensity ratio of the D peak to the G peak (I_D/I_G) is typically utilized to characterize the defect content in SWCNTs; a lower ratio signals fewer defects and higher crystallinity. The I_D/I_G ratio of the SWCNTs/Fe_xO_y and as-grown SWCNTs was 0.06 and 0.2196, respectively. The findings demonstrate that the crystallinity of the sample was enhanced subsequent to the heat treatment and the D peak was significantly decreased after the air heat treatment.

Figure 3 displays the XPS spectra of SWCNTs/Fe_xO_y (Figure 3a–c) and as-grown SWCNTs (Figure 3d–f), including the regions for Fe 2p, O 1s, and C 1s. The analysis shows that the main valence state of iron in the as-grown SWCNTs prepared by the arc discharge method exists in the form of Fe^0, although a small amount of Fe_3C phase is formed during the arc evaporation process. After air oxidation treatment, the iron valence states in SWCNTs/Fe_xO_y are mainly Fe^{2+} and Fe^{3+}, while a small amount of Fe^0 phase still exists, which may be due to the incomplete oxidation of Fe nanoparticles in the internal region. It is worth noting that although the XRD analysis failed to detect the Fe^0 signal, this may be due to its low content, which is below the detection limit of X-rays.

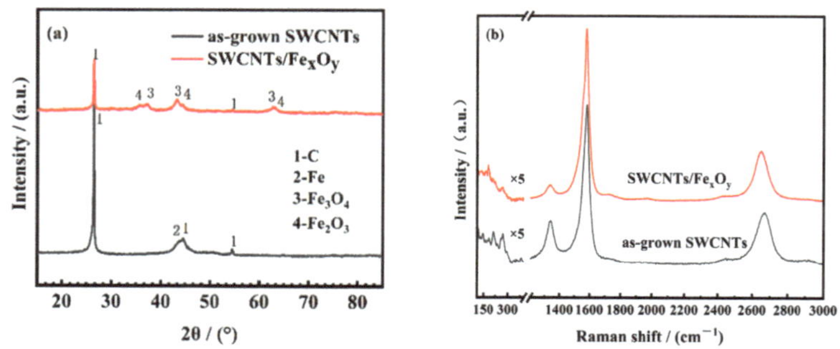

Figure 2. (**a**) XRD images and (**b**) Raman spectra of as-grown SWCNTs and SWCNTs/Fe_xO_y.

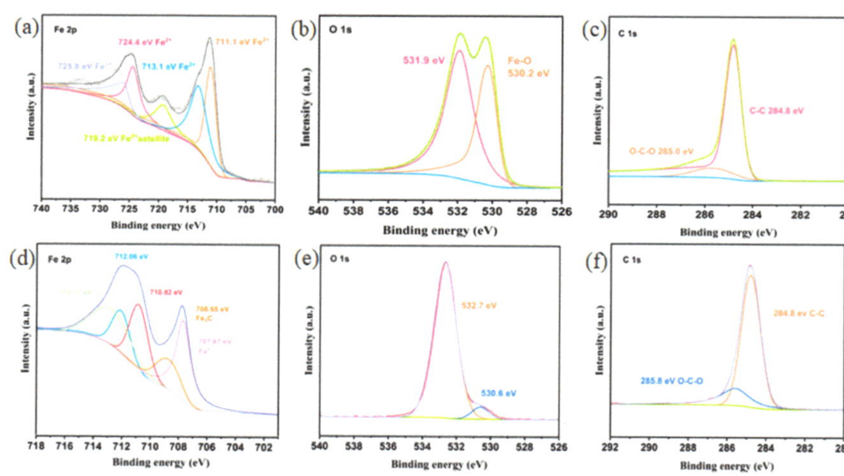

Figure 3. XPS spectra of SWCNTs/Fe_xO_y and as-grown SWCNTs.

Figure 4 presents the N_2 adsorption–desorption curves and pore size distribution for the as-grown SWCNTs and SWCNTs/Fe_xO_y, and the specific surface area, pore volume, and average pore diameter of the two nanocomposites are given in Table 1. As demonstrated in Figure 4 and Table 1, it appears that the adsorption–desorption isotherm exhibits an H_3 hysteresis loop, and compared to the as-grown SWCNTs, the SWCNTs/Fe_xO_y has a larger specific surface area and pore volume and a larger pore diameter. The specific surface area of a material is determined by the synergistic contribution of micropores, mesopores, and macropores. Generally speaking, a larger capacity of micropores correlates with a higher specific surface area of the material, and when mesopores and macropores are present in significant quantities, the average pore size of the material tends to decrease. It is evident that thermal treatment effectively removes amorphous carbon, which is favorable for the formation of micropores. This not only increases the number of adsorption channels but also prolongs the adsorbate residence time within these channels due to the abundance of micropores. As a result, adsorbate molecules are more easily adsorbed onto the SWCNTs/Fe_xO_y.

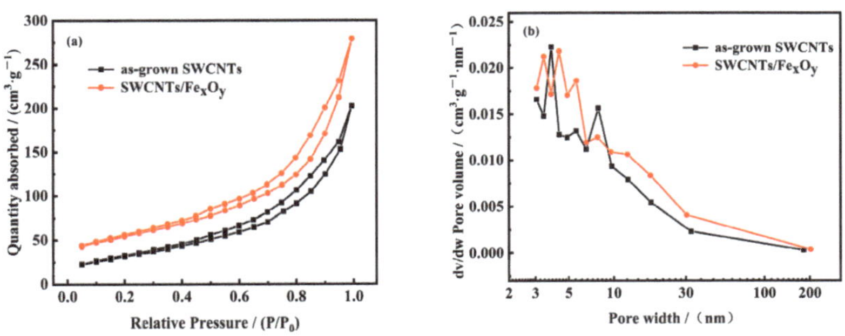

Figure 4. (**a**) N_2 adsorption–desorption curves and (**b**) pore size distribution of as-grown SWCNTs and SWCNTs/Fe_xO_y.

Table 1. Pore structure analysis of SWCNTs/Fe_xO_y and as-grown SWCNTs.

Sample	BET $m^2{\cdot}g^{-1}$	Pore Volume $cm^3{\cdot}g^{-1}$	Pore Diameter nm
SWCNTs/Fe_xO_y	191	0.389	4.314
As-grown SWCNTs	115	0.299	3.828

Figure 5a is the hysteresis loop of the SWCNTs/Fe_xO_y. The coercivity and remanence of the SWCNTs/Fe_xO_y are both zero, indicating the paramagnetic nature of this nanocomposite material. The saturation magnetization strength at room temperature is measured at 33.32 emu/g, suggesting that this material is capable of achieving solid–liquid separation. Figure 5b is a photo of a small experiment in which magnetic SWCNTs/Fe_xO_y can be rapidly collected in a BPA solution with a concentration of 0.25 mmol/L using magnets.

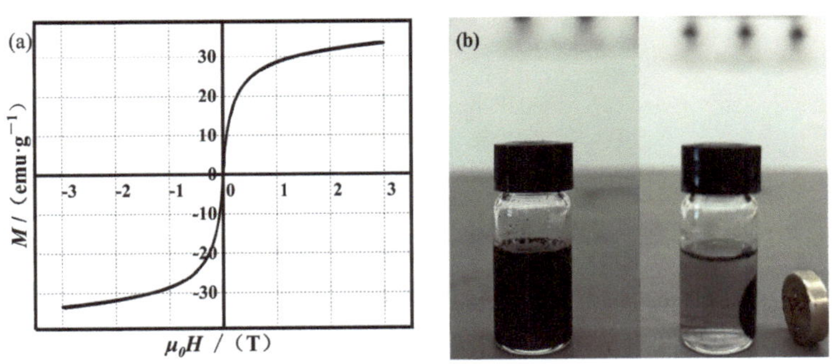

Figure 5. (**a**) Hysteresis loop of SWCNTs/Fe_xO_y and (**b**) a photo of a small experiment.

3.2. Adsorption Studies

Under the conditions of T = 25 °C, pH = 6, the absorption rate of the as-grown SWCNTs and SWCNTs/Fe_xO_y on BPA was studied. As depicted in Figure 6, the adsorption rate of BPA by both nanocomposites increased with time. Both adsorbents reached equilibrium within 15 min, with the SWCNTs/Fe_xO_y exhibiting an equilibrium adsorption rate of 97%, which represents a 50% improvement over the as-grown SWCNTs, indicating that the heat-treated composites exhibited a superior adsorption performance. Consequently, further experiments will utilize SWCNTs/Fe_xO_y as the adsorbent for research purposes.

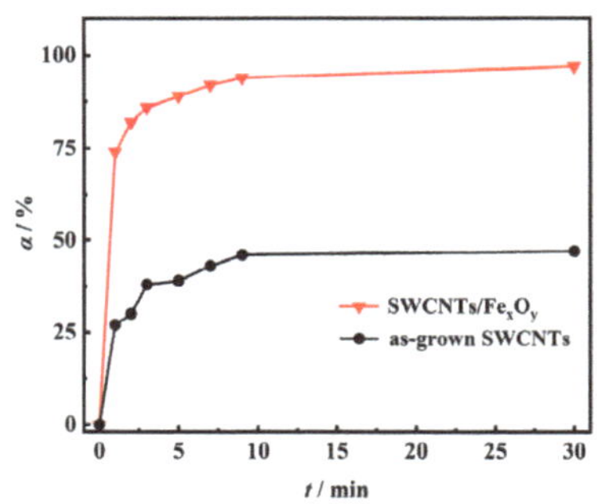

Figure 6. As-grown SWCNTs and SWCNTs/Fe_xO_y adsorption curves (T = 25 °C, pH = 6, adsorbent dosage = 0.2 g/L, C_{BPA} = 0.025 mmol/L).

Figure 7 illustrates the relationship between the rate of adsorption of BPA and the absorption time under different starting concentrations of BPA. Solutions of varying concentrations reached adsorption equilibrium within 15 min, and the starting concentration significantly influenced the BPA equilibrium adsorption rate. The adsorption rate decreased as the initial BPA concentration increased. When the concentration of the BPA solution increased from 0.025 to 0.15 mmol/L, the equilibrium adsorption rate decreased from 98% to 51%. This reduction is primarily due to the limited adsorption capacity of the SWCNTs/Fe_xO_y towards BPA. Exceeding the adsorption capacity of the SWCNTs/Fe_xO_y with a high initial concentration of BPA solution results in the adsorption of merely a fraction of the BPA molecules. Within the initial 0–10 min, the adsorption rate of the SWCNTs/Fe_xO_y increased rapidly, followed by a slower progression. This phenomenon can be attributed to the initial abundance of adsorption sites on the adsorbent, which allowed for the rapid adsorption of BPA from the solution. As adsorption continued, these sites became gradually saturated with BPA molecules, reducing the pore transport capacity and leading to a gradual decrease in the adsorption efficiency until equilibrium was reached. Furthermore, it was observed that for BPA at low concentrations, such as concentrations of 0.025 and 0.05 mmol/L, an adsorption rate of 90% can be achieved after 5 min. Hence, the SWCNT composite loaded with iron oxide nanoparticles can be considered a rapid and effective adsorbent for BPA, especially at lower concentrations.

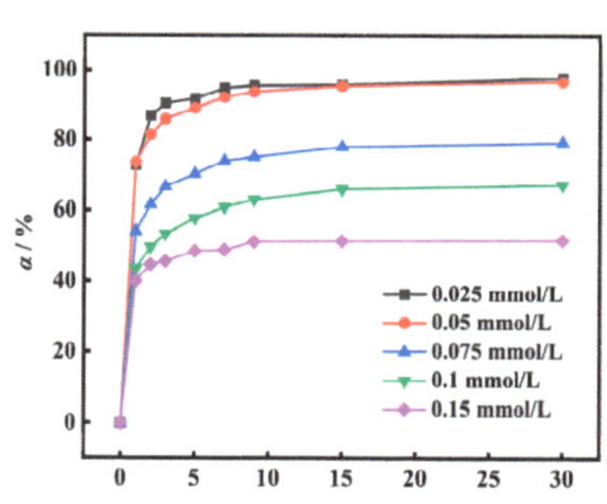

Figure 7. Effect of different initial concentrations (T = 25 °C, pH = 6, adsorbent dosage = 0.2 g/L).

The relationship between the removal rate of BPA and the dosage of the SWCNTs/Fe_xO_y is depicted in Figure 8, with the dosage of ranging from 0.05 to 0.35 g/L. An adsorbent dosage of 0.05 g/L resulted in a BPA adsorption rate of 34%. As the adsorbent dosage increased to 0.3 g/L, the adsorption rate reached 96.8% and, upon further increment to 0.35 g/L, exhibited little change, remaining at approximately 97%. Nonetheless, an increment in the adsorbent dosage resulted in a diminution of the adsorption capacity. This phenomenon can be attributed to the fact that while the adsorbent dosage increased, the amount of adsorbate remained unchanged, resulting in an excess of the adsorbent competing with the BPA molecules. At the same time, excessive adsorbents caused the overlap of effective adsorption sites on SWCNTs/Fe_xO_y, consequently reducing the BPA content per unit mass of the adsorbent.

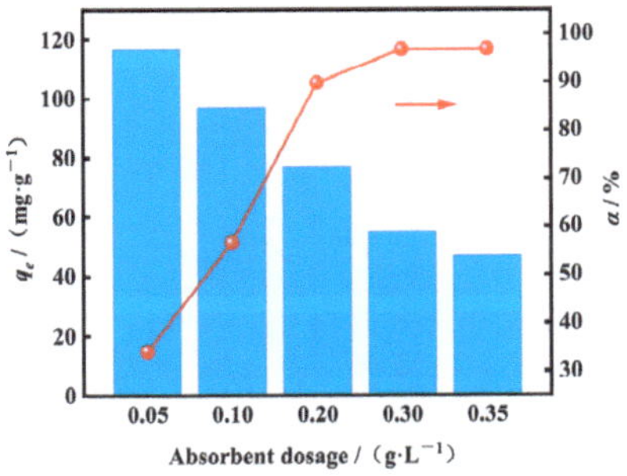

Figure 8. Effect of adsorbent dosage (T = 25 °C, pH = 6, C_{BPA} = 0.075 mmol/L).

3.3. Effect of pH Value on Adsorption

The acid or base nature of a solution exerts influence not merely on the solubility of the adsorbent but also affects the speciation of ions within the solution. To investigate the effect of pH on adsorption, the pH of a 0.15 mmol/L BPA solution was adjusted using HCl or NaOH. This was conducted at an adsorption temperature of 25 °C and an adsorbent dosage of 0.2 g/L. As Figure 9a illustrates, the adsorption rate of BPA decreased with increasing pH. At acidic pH (pH = 3), the nanocomposite exhibited an adsorption rate of 68% towards a BPA solution with C_0 = 0.15 mmol/L. Under neutral and alkaline conditions, the adsorption rates were 56% and 50%, respectively. The maximum adsorption rate was observed at pH = 3, and the adsorption capacity of the SWCNTs/Fe_xO_y for BPA gradually diminished as the pH value increased. Therefore, for practical applications, wastewater acidification is often used as a pre-treatment step for adsorption methods. This process can break down complex polymeric organic matter into smaller molecular organic compounds, but the acidification of wastewater will also increase the cost and affect the biological treatment.

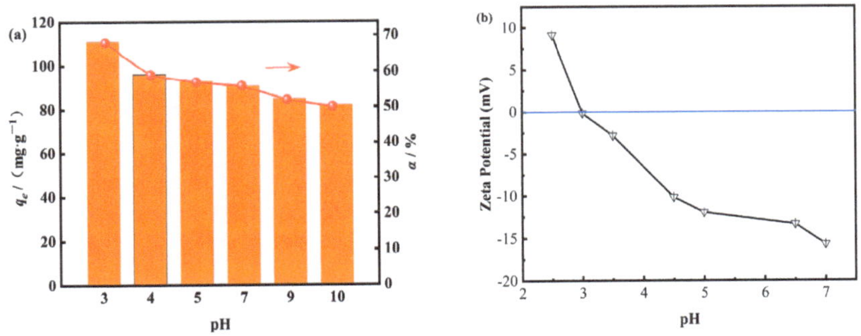

Figure 9. (**a**) Effect of pH value and (**b**) the point of zero charge of SWCNTs/Fe_xO_y.

In general, BPA is present in its molecular form in the solution. However, the deprotonation of the BPA single anion occurs at pH 8, and further ionization occurs when the pH exceeds 9.6. Additionally, we also identified the point zero charge (pH_{pzc}) of the SWCNTs/Fe_xO_y in the study, which is about 2.9, so the nanocomposite is positively charged at a pH below 2.9 and negatively charged at a pH above 2.9. Therefore, BPA^{2-} and negatively charged SWCNTs experienced electrostatic repulsion [32], which resulted in a decrease in the adsorption rate of BPA onto the SWCNTs/Fe_xO_y as pH increased. Moreover, it is evident that the electrostatic interaction between the BPA and SWCNTs/Fe_xO_y is not the primary factor controlling adsorption.

3.4. Adsorption Isotherm

An adsorption isotherm describes the relationship between the equilibrium concentration of a solution and the corresponding equilibrium adsorption capacity, which is attained when the adsorption process reaches equilibrium at a given temperature. To model this equilibrium, two widely used isotherm equations, the Langmuir [33] and Freundlich [34] models, are often employed to fit experimental data. The mathematical expressions for these models are as follows:

$$q_e = \frac{K_L q_{max} C_e}{1 + K_L C_e}, \tag{3}$$

$$q_e = K_F C_e^{\frac{1}{n}}, \tag{4}$$

where q_e is the adsorption capacity of the unit adsorbent to BPA at the adsorption equilibrium (mg/g); C_e is the concentration of BPA in solution at equilibrium (mmol/L); q_{max} is the theoretical maximum adsorption capacity (mg/g); and K_L, K_F, and n are all model constants. The value of n serves as an indicator of the ease or difficulty associated with the adsorption process. Adsorption is more likely to occur when the value of n is within the range of 1 to 10.

The data on the adsorption of BPA by the SWCNTs/Fe_xO_y were also analyzed using the Dubinin–Radushkevich (D-R) adsorption isotherm model. The mathematical expressions for the model are as follows:

$$q_e = q_{max} exp\left[-\beta\left\{RTln\left(1 + \frac{1}{C_e}\right)\right\}^2\right], \tag{5}$$

$$\varepsilon = RTln\left(1 + \frac{1}{C_e}\right), \tag{6}$$

$$E_a = \frac{1}{\sqrt{2\beta}}, \tag{7}$$

where q_{max} is the maximum adsorption capacity calculated according to the D-R model, $mg \cdot g^{-1}$; β is a constant related to the adsorption energy E_a ($kJ \cdot mol^{-1}$), $mol^2 \cdot kJ^{-2}$; ε is the Polanyi adsorption potential. When the E_a value is in the range of 1–8 $kJ \cdot mol^{-1}$, it indicates physical adsorption. When the E_a value exceeds 8 $kJ \cdot mol^{-1}$, chemisorption is predominant.

Figure 10 illustrates the Langmuir, Freundlich, and D-R adsorption isotherms associated with the removal of BPA from solution using SWCNTs/Fe_xO_y. The corresponding fitting parameters were presented in Table 2. The data demonstrate a better fit with the Langmuir and D-R isotherm models, as indicated by the correlation fitting coefficient (R^2) values of 0.903 and 0.913, respectively, compared to the R^2 value of 0.874 for the Freundlich model. This implies that the BPA molecules are likely adsorbed onto the SWCNTs/Fe_xO_y in a uniform monolayer. The maximum adsorption capacity as determined by the Langmuir model is 78.89 mg/g. The BPA adsorption energy calculated using the D-R model is 1.543 kJ/mol, indicating that the adsorption process is predominantly physical adsorption. As indicated in the table, the value of n is 5.559, suggesting that the adsorption process occurs easily.

Table 2. Fitting parameters of isothermal adsorption models.

Langmuir			Freundlich			Dubinin–Radushkevich (D-R)			
q_{max} ($mg \cdot g^{-1}$)	K_L ($L \cdot mg^{-1}$)	R^2	K_F	n	R^2	q_{max} ($mg \cdot g^{-1}$)	β ($mol^2 \cdot kJ^{-2}$)	E_a ($kJ \cdot mol^{-1}$)	R^2
78.89	4.714	0.903	52.68	5.559	0.874	76.22	0.20989	1.543	0.913

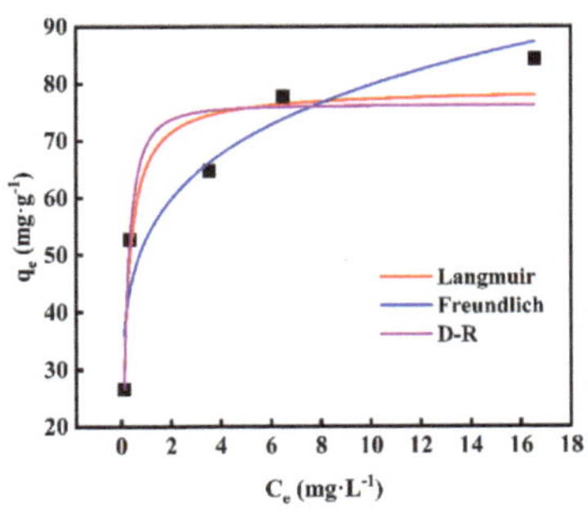

Figure 10. Langmuir, Freundlich, and D-R adsorption isotherms (T = 25 °C, pH = 6, adsorbent dosage = 0.2 g/L).

3.5. Adsorption Kinetics

In this study, the adsorption rate and mechanism of SWCNTs/Fe_xO_y adsorbing BPA were examined by fitting the data to pseudo first-order and pseudo second-order kinetic models. The equations for these two adsorption kinetic models are as follows [35]:

$$q_t = q_e\left(1 - e^{-k_1 t}\right), \tag{8}$$

$$\frac{t}{q_t} = \frac{1}{k_2 {q_e}^2} + \frac{1}{q_e}t, \tag{9}$$

where q_t is the adsorption capacity of the unit adsorbent to the BPA at time t (mg/g); q_e is the adsorption capacity of the unit adsorbent to BPA at adsorption equilibrium (mg/g); k_1 and k_2 are both adsorption rate constants.

Figure 11a,b illustrate the kinetic fittings of the SWCNTs/Fe_xO_y for the BPA solutions with varying initial concentrations. The corresponding kinetic fitting parameters are listed in Table 3. The data demonstrate that the second-order kinetics model fits the experimental data of BPA better than the first-order kinetic model, with the correlation coefficients (R^2) exceeding 0.99 in both cases. The equilibrium adsorption capacity, as calculated from the pseudo second-order kinetic curves, is closer to the experimental values, suggesting that chemisorption was the primary mechanism driving the adsorption process. The adsorption mechanism was likely influenced by various forces, such as π-π electron donor–acceptor interactions, hydrogen bonding, etc.

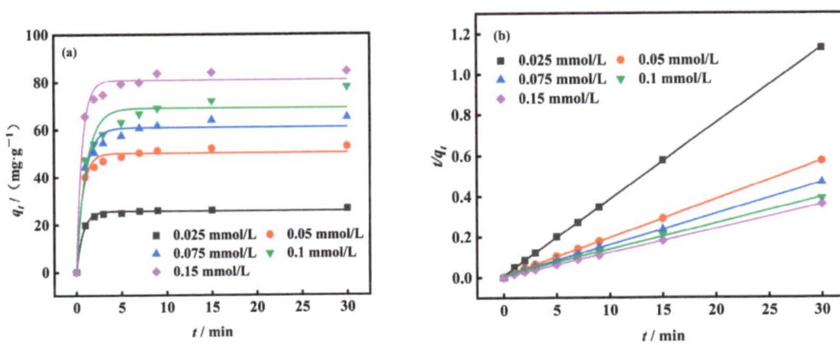

Figure 11. Kinetic fitting curves: (**a**) pseudo first-order kinetics; (**b**) pseudo second-order kinetic (T = 25 °C, pH = 6, adsorbent dosage = 0.2 g/L).

Table 3. Fitting parameters of kinetics.

C_0 mmol·L^{-1}	Pseudo First-Order			Pseudo Second-Order		
	k_1 min^{-1}	q_e mg·g^{-1}	R^2	k_2 g·mg^{-1}·min^{-1}	q_e mg·g^{-1}	R^2
0.025	1.40	25.69	0.9950	0.14	26.75	0.9998
0.05	1.43	50.08	0.9830	0.06	53.16	0.9998
0.075	1.08	60.69	0.9724	0.03	65.62	0.9994
0.1	0.90	68.77	0.9422	0.01	78.86	0.9969
0.15	1.50	80.71	0.9834	0.04	85.03	0.9997

To further elucidate the principal mechanism governing the adsorption process, the Weber–Morris model [36] was employed to fit the data depicted in Figure 12.

$$q_t = k_{int}t^{1/2} + I, \quad (10)$$

where k_{int} is the intraparticle diffusion rate constant. The value I represents the boundary layer thickness, i.e., the larger the intercept, the greater the boundary layer effect will be [37]. According to this model, if the fitting line of q_t and $t^{1/2}$ passes through the origin, then the internal diffusion of the particle determines the adsorption rate.

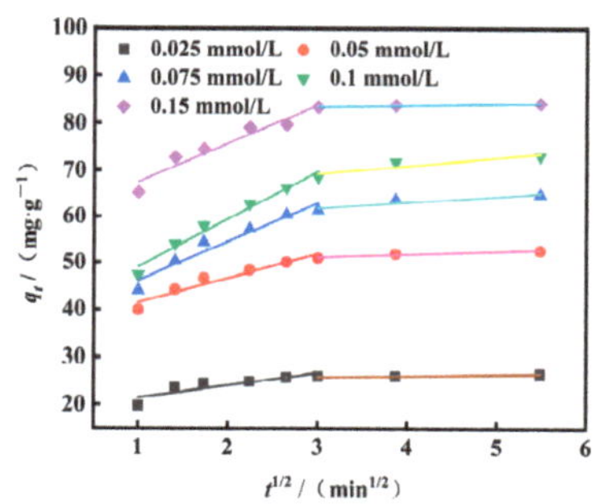

Figure 12. Weber–Morris curves (T = 25 °C, pH = 6, adsorbent dosage = 0.2 g/L).

The relationship between q_t and $t^{1/2}$ was observed to be nonlinear, indicating a multi-line relationship. The adsorption process can be divided into two stages: Firstly, there is the rapid adsorption stage, which is considered to be the swift diffusion of the adsorbate to the surface of the adsorbent; the SWCNTs/Fe_xO_y contain abundant π electrons, which form π-π conjugations [38] with π electrons in the benzene ring in the structure of BPA. In the early stages of adsorption, the SWCNTs/Fe_xO_y had a larger specific surface area and numerous available adsorption sites, allowing for more π electrons to be exposed for binding with BPA. Consequently, the rate of adsorption of BPA increased rapidly with the adsorption time. The second stage was a slower adsorption phase, where the BPA molecules diffused into the mesopore and macropore of the SWCNTs/Fe_xO_y. As the BPA concentration decreased, the number of effective adsorption sites on the SWCNTs/Fe_xO_y also diminished, and the quantity of BPA molecules within the adsorbent gradually increased. Adsorption was influenced by the concentration and molecular size of the adsorbate. As adsorption approached equilibrium, the diffusion resistance rose. The fact that $I \neq 0$ under the test conditions suggested that intra-particle diffusion was not the sole rate-limiting step, and that several other controlling factors significantly influenced the adsorption of BPA by the CNTs.

3.6. Comparison of Adsorption Quantity

The adsorption capacity of the SWCNTs/Fe_xO_y composites for the removal of BPA, as illustrated in this study, was comparatively assessed with data from other adsorbents reported in the literature, with the findings being summarized in Table 4. Figure 13 shows the adsorption mechanism of the SWCNTs/Fe_xO_y for BPA. Combined with Table 4 and Figure 13, it can be seen that the modified SWCNTs/Fe_xO_y had an excellent adsorption capacity for BPA. Previous studies have suggested that the two phenyl rings of the BPA molecule can be adsorbed onto the surface of CNTs in various configurations: namely, aligned parallel to the longitudinal axis of the tube, encircling the circumference, or positioned diagonally. This adsorption is purported to occur via a π-π electron donor–acceptor interaction, culminating in the establishment of a π-π conjugation with the π electron system of the CNTs. Furthermore, it has been suggested that the reliance of these π-π interactions on pH is minimal, indicating a robustness in the adsorption mechanism across varying pH conditions. In this research, the modification of the SWCNTs results in a larger surface area and the presence of numerous micropores and mesopores, which allows for more π electrons to be exposed for binding with BPA and to form π-π conjugations, thus accelerating the adsorption process. Furthermore, the hydrophobic nature of the carbon nanotube surfaces [39,40] has been found to be effective in BPA adsorption, exhibiting a pH independence similar to that of π-π interactions. Additionally, the water solubility of BPA should be considered; in alkaline conditions, BPA exists in a salt form, which can also affect adsorption. Finally, the unique magnetic properties of the SWCNTs/Fe_xO_y are expected to facilitate the recovery and reuse of the adsorbent, minimizing the potential harm to humans or aquatic organisms.

Table 4. The adsorption capacity of BPA by SWCNTs observed in this work is compared with that reported for other adsorbents in the literature.

Target Adsorbate	Adsorbent	pH Value	Maximum Capacity ($mg \cdot g^{-1}$)	References
BPA	N-NiO@N-Fe_3O_4@N-ZnO	5	100	[41]
BPA	activated carbon	7	44.1	[42]
BPA	MWCNTs/SiO_2/CS	6.22	46.2	[43]
BPA	HMWCNTs/Fe_3O_4	-	113	[44]
BPA	MWCNTs/Fe_3O_4	6.2	45.31	[27]
BPA	Fe_3O_4/GO	-	72.8	[45]
BPA	SWCNTs/Fe_xO_y	3	117	This work
		4	96	This work
		5	93	This work
		7	91	This work

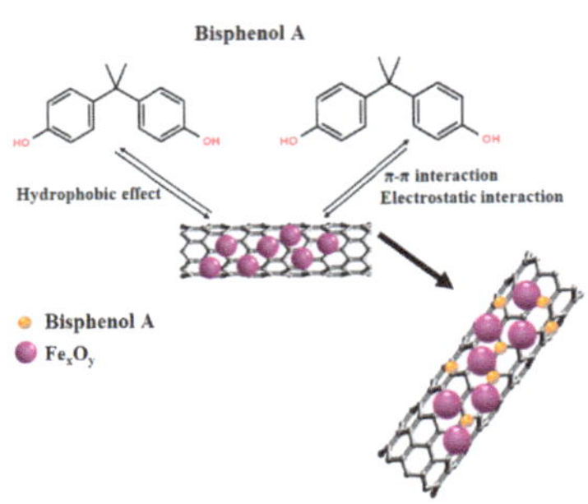

Figure 13. The main mechanism of adsorption of SWCNTs/Fe_xO_y on BPA.

4. Conclusions

In this work, high-quality SWCNTs containing an iron catalyst were prepared by a simple arc discharge method in a hydrogen–argon mixed atmosphere, and the iron nanoparticles were oxidized in situ into magnetic iron oxide nanoparticles by air heat treatment. Based on the unique structure of this nanocomposite material, the adsorption performance of BPA in water as an adsorbent was studied. The main findings are as follows:

(1) At the same pH value, compared to other CNTs or carbon materials coated with different oxides, this composite material demonstrates a higher efficiency in adsorbing BPA. After air heat treatment, the SWCNT composite material possesses an increased surface area, which enhances its adsorption efficiency.
(2) At low initial BPA concentrations, especially below 0.05 mmol/L, the adsorption efficiency quickly reaches 90% and quickly reaches equilibrium. Acidic conditions were more conducive to the adsorption of BPA by SWCNTs/Fe_xO_y.
(3) The kinetic calculations of adsorption indicate that the process follows a pseudo second-order kinetic model. The experimental data fit well with the Langmuir and D-R adsorption isotherm models. Therefore, the adsorption of BPA on the SWCNTs/Fe_xO_y is likely to be in the form of a uniform monolayer, and the adsorption process is more consistent with physical adsorption.

The nanocomposite not only exhibits excellent adsorption capacity for hydrophobic phenols (BPA) but also exhibits effective magnetic separation performance, and is easy to recover from water, so it is a promising material for removing BPA from water. Its adsorption capacity is higher under acidic conditions, so the next step will be to study its adsorption performance in actual wastewater. We also need to pay attention to the effect of the Fe doping ratio on adsorption.

Author Contributions: Conceptualization, L.C.; methodology, L.C., J.J. and L.S.; software, L.C.; validation, L.C. and J.J.; formal analysis, L.C.; investigation, L.C. and J.J.; resources, L.S.; data curation, L.C. and J.J.; writing—original draft preparation, L.C.; writing—review and editing, L.S. and J.J.; visualization, L.C.; supervision, L.S.; project administration, L.S.; funding acquisition, L.S. All authors have read and agreed to the published version of the manuscript.

Funding: This work was financially supported by the Foundation Strengthening Program Project Fund (No. 2020-JCJQ-JJ-417).

Institutional Review Board Statement: Not applicable.

Informed Consent Statement: Not applicable.

Data Availability Statement: The data presented in this study are available on request from the corresponding author.

Conflicts of Interest: The authors declare no conflicts of interest.

References

1. Jin, H.; Zhu, L. Occurrence and partitioning of bisphenol analogues in water and sediment from Liaohe River Basin and Taihu Lake, China. *Water Res.* **2016**, *103*, 343–351. [CrossRef] [PubMed]
2. Wu, D.; He, L.; Sun, R.; Tong, M.; Kim, H. Influence of Bisphenol A on the transport and deposition behaviors of bacteria in quartz sand. *Water Res.* **2017**, *121*, 1–10. [CrossRef] [PubMed]
3. Yuan, M.; Chen, S.; Zeng, C.; Fan, Y.; Ge, W.; Chen, W. Estrogenic and non-estrogenic effects of bisphenol A and its action mechanism in the zebrafish model: An overview of the past two decades of work. *Environ. Int.* **2023**, *176*, 107976. [CrossRef] [PubMed]
4. Cimmino, I.; Fiory, F.; Perruolo, G.; Miele, C.; Beguinot, F.; Formisano, P.; Oriente, F. Potential Mechanisms of Bisphenol A (BPA) Contributing to Human Disease. *Int. J. Mol. Sci.* **2020**, *21*, 5761. [CrossRef] [PubMed]
5. Adebowale, K.O.; Egbedina, A.O. Facile green synthesis of bio-carbon material from eggshells and its application for the removal of Bisphenol A and 2,4,6-trichlorophenol from water. *Environ. Nanotechnol. Monit. Manag.* **2022**, *17*, 100622. [CrossRef]
6. Modi, A.; Bellare, J. Copper sulfide nanoparticles/carboxylated graphene oxide nanosheets blended polyethersulfone hollow fiber membranes: Development and characterization for efficient separation of oxybenzone and bisphenol A from water. *Polymer* **2019**, *163*, 57–67. [CrossRef]
7. Huang, Y.; Yang, T.; Liang, M.; Wang, Y.; Xu, Z.; Zhang, D.; Li, L. Ni-Fe layered double hydroxides catalized ozonation of synthetic wastewater containing Bisphenol A and municipal secondary effluent. *Chemosphere* **2019**, *235*, 143–152. [CrossRef]
8. Yue, W.; Yin, C.-F.; Sun, L.; Zhang, J.; Xu, Y.; Zhou, N.-Y. Biodegradation of bisphenol-A polycarbonate plastic by Pseudoxanthomonas sp. strain NyZ600. *J. Hazard. Mater.* **2021**, *416*, 125775. [CrossRef]
9. Dehghani, M.H.; Ghadermazi, M.; Bhatnagar, A.; Sadighara, P.; Jahed-Khaniki, G.; Heibati, B.; McKay, G. Adsorptive removal of endocrine disrupting bisphenol A from aqueous solution using chitosan. *J. Environ. Chem. Eng.* **2016**, *4*, 2647–2655. [CrossRef]

10. Li, Y.; Zhu, H.; Zhang, C.; Cheng, M.; He, H. PEI-grafted magnetic cellulose for Cr(VI) removal from aqueous solution. *Cellulose* **2018**, *25*, 4757–4769. [CrossRef]
11. Kah, M.; Sigmund, G.; Xiao, F.; Hofmann, T. Sorption of ionizable and ionic organic compounds to biochar, activated carbon and other carbonaceous materials. *Water Res.* **2017**, *124*, 673–692. [CrossRef] [PubMed]
12. Jabeen, A.; Kamran, U.; Noreen, S.; Park, S.-J.; Bhatti, H.N. Mango seed-derived hybrid composites and sodium alginate beads for the efficient uptake of 2,4,6-trichlorophenol from simulated wastewater. *Catalysts* **2022**, *12*, 972. [CrossRef]
13. Kamran, U.; Bhatti, H.N.; Noreen, S.; Tahir, M.A.; Park, S.-J. Chemically modified sugarcane bagasse-based biocomposites for efficient removal of acid red 1 dye: Kinetics, isotherms, thermodynamics, and desorption studies. *Chemosphere* **2022**, *291*, 132796. [CrossRef] [PubMed]
14. Mariyam, A.; Mittal, J.; Sakina, F.; Baker, R.T.; Sharma, A.K.; Mittal, A. Efficient batch and Fixed-Bed sequestration of a basic dye using a novel variant of ordered mesoporous carbon as adsorbent. *Arab. J. Chem.* **2021**, *14*, 103186. [CrossRef]
15. Mishakov, I.V.; Bauman, Y.I.; Brzhezinskaya, M.; Netskina, O.V.; Shubin, Y.V.; Kibis, L.S.; Stoyanovskii, V.O.; Larionov, K.B.; Serkova, A.N.; Vedyagin, A.A. Water purification from chlorobenzenes using heteroatom-functionalized carbon nanofibers produced on self-organizing Ni-Pd catalyst. *J. Environ. Chem. Eng.* **2022**, *10*, 107873. [CrossRef]
16. Rahman, G.; Najaf, Z.; Mehmood, A.; Bilal, S.; Shah, A.U.H.A.; Mian, S.A.; Ali, G. An overview of the recent progress in the synthesis and applications of carbon nanotubes. *C* **2019**, *5*, 3. [CrossRef]
17. Krishna, R.H.; Chandraprabha, M.; Samrat, K.; Murthy, T.K.; Manjunatha, C.; Kumar, S.G. Carbon nanotubes and graphene-based materials for adsorptive removal of metal ions–a review on surface functionalization and related adsorption mechanism. *Appl. Surf. Sci. Adv.* **2023**, *16*, 100431. [CrossRef]
18. Pan, B.; Xing, B. Adsorption mechanisms of organic chemicals on carbon nanotubes. *Environ. Sci. Technol.* **2008**, *42*, 9005–9013. [CrossRef] [PubMed]
19. Kang, S.; Herzberg, M.; Rodrigues, D.F.; Elimelech, M. Antibacterial Effects of Carbon Nanotubes: Size Does Matter! *Langmuir* **2008**, *24*, 6409–6413. [CrossRef]
20. Donadelli, J.A.; Carlos, L.; Arques, A.; Einschlag, F.S.G. Kinetic and mechanistic analysis of azo dyes decolorization by ZVI-assisted Fenton systems: pH-dependent shift in the contributions of reductive and oxidative transformation pathways. *Appl. Catal. B* **2018**, *231*, 51–61. [CrossRef]
21. Wang, H.; Jia, S.; Wang, H.; Li, B.; Liu, W.; Li, N.; Qiao, J.; Li, C.-Z. A novel-green adsorbent based on betaine-modified magnetic nanoparticles for removal of methyl blue. *Sci. Bull.* **2017**, *62*, 319–325. [CrossRef] [PubMed]
22. Yadav, A.; Bagotia, N.; Yadav, S.; Sharma, A.K.; Kumar, S. In-situ fabrication of surfactant modified CNT-based novel bio-composite and its performance evaluation for simultaneous removal of anionic dyes: Optimization by Box-Behnken design. *Sep. Purif. Technol.* **2022**, *284*, 120262. [CrossRef]
23. Sarkar, B.; Mandal, S.; Tsang, Y.F.; Kumar, P.; Kim, K.-H.; Ok, Y.S. Designer carbon nanotubes for contaminant removal in water and wastewater: A critical review. *Sci. Total Environ.* **2018**, *612*, 561–581. [CrossRef] [PubMed]
24. Mou, F.; Guan, J.; Xiao, Z.; Sun, Z.; Shi, W.; Fan, X.-A. Solvent-mediated synthesis of magnetic Fe_2O_3 chestnut-like amorphous-core/γ-phase-shell hierarchical nanostructures with strong As (v) removal capability. *J. Mater. Chem.* **2011**, *21*, 5414–5421. [CrossRef]
25. Cuzzola, A.; Bernini, M.; Salvadori, P. A preliminary study on iron species as heterogeneous catalysts for the degradation of linear alkylbenzene sulphonic acids by H_2O_2. *Appl. Catal. B* **2002**, *36*, 231–237. [CrossRef]
26. Munoz, M.; de Pedro, Z.M.; Casas, J.A.; Rodriguez, J.J. Preparation of magnetite-based catalysts and their application in heterogeneous Fenton oxidation—A review. *Appl. Catal. B* **2015**, *176*, 249–265. [CrossRef]
27. Li, S.; Gong, Y.; Yang, Y.; He, C.; Hu, L.; Zhu, L.; Sun, L.; Shu, D. Recyclable CNTs/Fe_3O_4 magnetic nanocomposites as adsorbents to remove bisphenol A from water and their regeneration. *Chem. Eng. J.* **2015**, *260*, 231–239. [CrossRef]
28. Al-Musawi, T.J.; Mengelizadeh, N.; Ganji, F.; Wang, C.; Balarak, D. Preparation of multi-walled carbon nanotubes coated with $CoFe_2O_4$ nanoparticles and their adsorption performance for Bisphenol A compound. *Adv. Powder Technol.* **2022**, *33*, 103438. [CrossRef]
29. Guo, X.; Huang, Y.; Yu, W.; Yu, X.; Han, X.; Zhai, H. Multi-walled carbon nanotubes modified with iron oxide and manganese dioxide (MWCNTs-Fe_3O_4- MnO_2) as a novel adsorbent for the determination of BPA. *Microchem. J.* **2020**, *157*, 104867. [CrossRef]
30. Xu, G.; Zhang, B.; Wang, X.; Li, N.; Liu, L.; Lin, J.-M.; Zhao, R.-S. Nitrogen-doped flower-like porous carbon nanostructures for fast removal of sulfamethazine from water. *Environ. Pollut.* **2019**, *255*, 113229. [CrossRef]
31. Zhao, X.; Inoue, S.; Jinno, M.; Suzuki, T.; Ando, Y. Macroscopic oriented web of single-wall carbon nanotubes. *Chem. Phys. Lett.* **2003**, *373*, 266–271. [CrossRef]
32. Yu, F.; Sun, S.; Han, S.; Zheng, J.; Ma, J. Adsorption removal of ciprofloxacin by multi-walled carbon nanotubes with different oxygen contents from aqueous solutions. *Chem. Eng. J.* **2016**, *285*, 588–595. [CrossRef]
33. Athalathil, S.; Stüber, F.; Bengoa, C.; Font, J.; Fortuny, A.; Fabregat, A. Characterization and performance of carbonaceous materials obtained from exhausted sludges for the anaerobic biodecolorization of the azo dye Acid Orange II. *J. Hazard. Mater.* **2014**, *267*, 21–30. [CrossRef] [PubMed]
34. Joseph, L.; Boateng, L.K.; Flora, J.R.; Park, Y.-G.; Son, A.; Badawy, M.; Yoon, Y. Removal of bisphenol A and 17α-ethinyl estradiol by combined coagulation and adsorption using carbon nanomaterials and powdered activated carbon. *Sep. Purif. Technol.* **2013**, *107*, 37–47. [CrossRef]

35. Zhang, S.; Shao, T.; Kose, H.S.; Karanfil, T. Adsorption kinetics of aromatic compounds on carbon nanotubes and activated carbons. *Environ. Toxicol. Chem.* **2012**, *31*, 79–85. [CrossRef] [PubMed]
36. Faria, M.C.; Rosemberg, R.S.; Bomfeti, C.A.; Monteiro, D.S.; Barbosa, F.; Oliveira, L.C.; Rodriguez, M.; Pereira, M.C.; Rodrigues, J.L. Arsenic removal from contaminated water by ultrafine δ-FeOOH adsorbents. *Chem. Eng. J.* **2014**, *237*, 47–54. [CrossRef]
37. Mane, V.S.; Mall, I.D.; Srivastava, V.C. Kinetic and equilibrium isotherm studies for the adsorptive removal of Brilliant Green dye from aqueous solution by rice husk ash. *J. Environ. Manag.* **2007**, *84*, 390–400. [CrossRef] [PubMed]
38. Sun, Z.; Zhao, L.; Liu, C.; Zhen, Y.; Ma, J. Fast adsorption of BPA with high capacity based on π-π electron donor-acceptor and hydrophobicity mechanism using an in-situ sp^2 C dominant N-doped carbon. *Chem. Eng. J.* **2020**, *381*, 122510. [CrossRef]
39. Xiao, P.; Wang, P.; Li, H.; Li, Q.; Shi, Y.; Wu, X.-L.; Lin, H.; Chen, J.; Wang, X. New insights into bisphenols removal by nitrogen-rich nanocarbons: Synergistic effect between adsorption and oxidative degradation. *J. Hazard. Mater.* **2018**, *345*, 123–130. [CrossRef]
40. Oyetade, O.A.; Martincigh, B.S.; Skelton, A.A. Interplay between electrostatic and hydrophobic interactions in the pH-dependent adsorption of ibuprofen onto acid-functionalized multiwalled carbon nanotubes. *J. Phys. Chem. C* **2018**, *122*, 22556–22568. [CrossRef]
41. Allam, E.A.; Ali, A.S.M.; Elsharkawy, R.M.; Mahmoud, M.E. Framework of nano metal oxides N-NiO@N-Fe_3O_4@N-ZnO for adsorptive removal of atrazine and bisphenol-A from wastewater: Kinetic and adsorption studies. *Environ. Nanotechnol. Monit. Manag.* **2021**, *16*, 100481. [CrossRef]
42. Bautista-Toledo, I.; Ferro-García, M.; Rivera-Utrilla, J.; Moreno-Castilla, C.; Vegas Fernández, F. Bisphenol A removal from water by activated carbon. Effects of carbon characteristics and solution chemistry. *Environ. Sci. Technol.* **2005**, *39*, 6246–6250. [CrossRef] [PubMed]
43. Mohammadi, A.A.; Dehghani, M.H.; Mesdaghinia, A.; Yaghmaian, K.; Es'haghi, Z. Adsorptive removal of endocrine disrupting compounds from aqueous solutions using magnetic multi-wall carbon nanotubes modified with chitosan biopolymer based on response surface methodology: Functionalization, kinetics, and isotherms studies. *Int. J. Biol. Macromol.* **2020**, *155*, 1019–1029. [CrossRef] [PubMed]
44. Gao, H.; Han, X.; Wang, R.; Zhu, K.; Han, R. Adsorption and catalytic degradation of bisphenol A and p-chlorophenol by magnetic carbon nanotubes. *Environ. Res.* **2023**, *231*, 116314. [CrossRef]
45. Ouyang, K.; Zhu, C.; Zhao, Y.; Wang, L.; Xie, S.; Wang, Q. Adsorption mechanism of magnetically separable Fe_3O_4/graphene oxide hybrids. *Appl. Surf. Sci.* **2015**, *355*, 562–569. [CrossRef]

materials

MDPI

Article

Optimized Adsorption–Catalytic Conversion for Lithium Polysulfides by Constructing Bimetallic Compounds for Lithium–Sulfur Batteries

Liping Chen, Runhua Wang, Nan Li, Yang Bai, Yimo Zhou and Juan Wang *

Shaanxi Key Laboratory of Nanomaterials and Nanotechnology, Xi'an University of Architecture and Technology, Xi'an 710055, China; chenlp202203@xauat.edu.cn (L.C.)

* Correspondence: juanwang618@126.com

Abstract: Although lithium–sulfur batteries possess the advantage of high theoretical specific capacity, the inevitable shuttle effect of lithium polysulfides is still a difficult problem restricting its application. The design of highly active catalysts to promote the redox reaction during charge–discharge and thus reduce the existence time of lithium polysulfides in the electrolyte is the mainstream solution at present. In particular, bimetallic compounds can provide more active sites and exhibit better catalytic properties than single-component metal compounds by regulating the electronic structure of the catalysts. In this work, bimetallic compounds-nitrogen-doped carbon nanotubes (NiCo)Se_2-NCNT and (CuCo)Se_2-NCNT are designed by introducing Ni and Cu into $CoSe_2$, respectively. The (CuCo)Se_2-NCNT delivers an optimized adsorption–catalytic conversion for lithium polysulfide, benefitting from adjusted electron structure with downshifted *d*-band center and increased electron fill number of Co in (CuCo)Se_2 compared with that of (NiCo)Se_2. This endows (CuCo)Se_2 moderate adsorption strength for lithium polysulfides and better catalytic properties for their conversion. As a result, the lithium–sulfur batteries with (CuCo)Se_2-NCNT achieve a high specific capacity of 1051.06 mAh g^{-1} at 1C and an enhanced rate property with a specific capacity of 838.27 mAh g^{-1} at 4C. The work provides meaningful insights into the design of bimetallic compounds as catalysts for lithium–sulfur batteries.

Keywords: catalysts; metallic compounds; electron structure; lithium sulfur batteries; electrochemical performance

Citation: Chen, L.; Wang, R.; Li, N.; Bai, Y.; Zhou, Y.; Wang, J. Optimized Adsorption–Catalytic Conversion for Lithium Polysulfides by Constructing Bimetallic Compounds for Lithium–Sulfur Batteries. *Materials* **2024**, *17*, 3075. https://doi.org/10.3390/ma17133075

Academic Editor: Giovanni Battista Appetecchi

Received: 17 May 2024
Revised: 12 June 2024
Accepted: 19 June 2024
Published: 22 June 2024

1. Introduction

Lithium–sulfur (Li-S) batteries have gotten a lot of attention since 2009 benefitting from their high theoretical specific capacity (1675 mAh g^{-1}), energy density (2600 Wh kg^{-1}), and application possibilities [1,2]. However, lithium polysulfide (LiPSs, Li_2S_x, $2 \leq x \leq 8$) intermediates are generate during the discharge process, which is easily dissolved and suffers sluggish reaction kinetics. The accumulated LiPSs in the electrolyte will diffuse through the separator to the lithium anode and return to the cathode during the charging process, resulting in the so-called infamous shuttle effect [3]. This causes a loss of sulfur-active material and a decrease in capacity, slowing down the commercialization of batteries, which remains a major challenge [4]. In recent years, the solution strategies have focused on the design of catalysts for sulfur cathode or separator modification to adsorb the LiPSs and catalyze their conversion, reducing the shuttle effect and improving the charge–discharge property of Li-S batteries [5–7].

Currently, catalyst materials for Li-S batteries mainly include metal oxides, metal sulfides, metal nitrides, and metal phosphates with single metal components [8,9]. Considering the complex and multistep conversion of LiPSs in the process of charge–discharge and the difficulty of adjusting the electron structure of catalysts by single, it is extremely necessary to modify metal components for increasing active site and adjusting its catalytic activity with introduced metal ions [10–12]. The methods involve doping and the construction of bimetallic compounds [8,13–15]. Among them, doping modifications are deeply studied to boost the catalytic activity of catalysts for Li-S batteries, while bimetallic compounds should be further developed even though some reports. For example, $NiCo_2S_4$, NiCoP, $Li_4Ti_5O_{12}$, Co_3Mo_3N, $CoSn(OH)_6$, $Ni_2Co_4P_3$, et al. had been designed, which delivered improved catalytic performance than corresponding to one component metallic compounds [16–21]. This could be ascribed to optimized chemical interaction between bimetallic compounds and LiPSs as well as catalytic activity, contributing from adjusted electron structure, such as *d* orbital electron filling and *d* band center [22–26]. This is because the LiPSs on catalysts include adsorption–conversion–desorption processes, weak adsorption cannot effectively trigger the reaction, while too strong adsorption passivates the active sites to hinder the subsequent reaction and moderate adsorption capacity will obtain the best catalytic performance [27]. Among them, *d* band center is an important parameter related to adsorption energy for LiPSs: the higher *d* band center, the stronger the chemical interaction, and vice versa [23]. For the *d* orbital fill number, the more it is filled, the weaker the binding capacity with LiPSs, and vice versa [28,29]. Introducing metal ions can optimize the adsorption–conversion–desorption process of LiPSs by adjusting the *d*-band center as well as electron filling of metal-based catalysts [21]. Ni element is widely selected for modifying Co-based metallic compounds due to their similar element characteristics, such as Ni-doped WS_2, Ni-doped MoS_2, and $Ni_{0.2}Mo_{0.8}N$ [30–32]. Whether other elements can achieve better modification effects and the reason for this are worth exploring to promote the research of catalysts for Li-S batteries.

In this work, Ni and Cu are introduced into $CoSe_2$ and construct bimetallic compounds-nitrogen-doped carbon nanotubes, $(NiCo)Se_2$-NCNT and $(CuCo)Se_2$-NCNT, respectively. Even though the $(NiCo)Se_2$-NCNT presents a better adsorption effect, the $(CuCo)Se_2$-NCNT delivers an improved catalytic performance compared to $(NiCo)Se_2$-NCNT. Consequently, the $(CuCo)Se_2$-NCNT modified separator (Figure S1) endows the corresponding Li-S batteries with a higher specific capacity and enhanced rate performance. The electron structure analysis reveals the better catalytic properties of $(CuCo)Se_2$ resulting from a downshifted *d* band center and increased electron fill number.

2. Experimental

2.1. Preparation of Bimetallic Selenides

A total of 8 g of melamine and 0.1 g of anhydrous glucose were added into 60 mL of deionized water, after being stirred and mixed thoroughly. Ten mL of $Ni(NO_3)_2{\cdot}6H_2O$ and $Co(NO_3)_2{\cdot}6H_2O$ aqueous solution was then added drop by drop, in which the total mass of the two metal salt ions was 0.8 g and the molar ratio was 1:1. The mixture was then stirred at 60 °C until a dry powder was obtained. The powder was then heated at 800 °C for 3 h in N_2 to obtain NiCo-NCNT. Finally, NiCo-NCNT was mixed with selenium powder at a mass ratio of 1:2 and heated at 400 °C for 2 h under nitrogen at a heating rate of 2 °C min^{-1} to obtain $(NiCo)Se_2$-NCNT. $(CuCo)Se_2$-NCNT was prepared by replacing $Ni(NO_3)_2{\cdot}6H_2O$ with equimolar $Cu(NO_3)_2{\cdot}3H_2O$.

2.2. Preparation of Modified Separators and Sulfur Cathodes

$(NiCo)Se_2$-NCNT/$(CuCo)Se_2$-NCNT, acetylene black, and polyvinylidene fluoride (PVDF) were ground evenly at the mass ratio of 4:5:1, which was then stirred for 12 h with N-methylpyrrolidone (NMP). The prepared paste was coated on Celgard 2500 separator, and the modified separator was cut after vacuum drying at 60 °C for 12 h.

Sublimed sulfur powder and acetylene black were ground evenly at the mass ratio of 7:3 and heat treated at 155 °C under a nitrogen atmosphere for 12 h to obtain sulfur composite. Sulfur composite, acetylene black, and PVDF with a mass ratio of 7:2:1 were mixed in NMP and coated on carbon-coated aluminum foil. Sulfur cathode was obtained with a sulfur load of about 1.1 mg cm^{-2} after being dried at 60 °C for 12 h in a vacuum.

2.3. Characterization of Bimetallic Selenide Materials

X-ray diffraction (XRD) using Cu Kα radiation source, transmission electron microscopy (TEM), selected area electron diffraction (SAED), and energy dispersive spectroscopy were carried out to analyze the morphology, crystal structure, and composition of the prepared materials. X-ray photoelectron spectroscopy (XPS) was conducted to study the valence states of the materials. N_2 adsorption/desorption analyzer was conducted to measure the pore size distribution.

2.4. Electrochemical Performance Test of Li-S Batteries

Sulfur cathode, modified separator, and lithium sheet were used to assemble 2032 coin batteries, using 1 M lithium bis(trifluoromethanesulfonyl) imide (LiTFSI) in DOL/DME with 2 wt% $LiNO_3$ as electrolyte. Constant current charge–discharge tests were conducted on the NEWARE system at 1.7~2.8 V. Electrochemical impedance spectroscopy (EIS) was carried out at 10^2~10^5 Hz and $\pm$5 mV. Cyclic voltammetry (CV) was performed on a CHI760E electrochemical workstation with a scanning speed of 0.1 mV s^{-1} at 1.7~2.8 V. Galvanostatic intermittent titration technique (GITT) was conducted to obtain open circuit voltage (OCV) with the current pulse of 30 min, and the quasi-open circuit voltage (QOCV) was obtained by standing for 1 h at 0.1C.

2.5. Analysis of Adsorption and Catalytic Properties of Bimetallic Selenides

2.5.1. Adsorption Effect Evaluation

First, S and Li_2S with a molar ratio of 5:1 were dissolved in 1,3-dioxolane (DOL) and dimethoxymethane (DME) (v:v = 1:1) to prepare Li_2S_6 solution. Then, an equal amount of bimetallic selenides was added to 2 mL Li_2S_6, respectively. The color change of the solution was recorded. And, after adsorption, the supernatant was removed and the sample was dried for the XPS test to analyze the change of valence state.

2.5.2. CV test of Li_2S_6 Symmetric Batteries

The bimetallic selenides and PVDF were mixed at a mass ratio of 9:1 in NMP and coated on aluminum foil to make electrodes for symmetric batteries. A total of 15 µL of 0.5 M Li_2S_6 solution was used to assemble Li_2S_6 symmetric batteries. CV test was conducted in the range of -1.0~1.0 V at 10 mV s^{-1}.

2.5.3. Li_2S Deposition and Decomposition Test

S and Li_2S in a molar ratio of 7:1 were dissolved in tetraethylene glycol dimethyl ether at 60 °C to prepare 0.1 M Li_2S_8 solution. CR2032 coin batteries were assembled with the cathode sheets used for symmetric batteries, and lithium sheets as the anodes. The electrolyte on the cathode side was 20 µL of 0.1 M Li_2S_8, and that on the anode side was the same with Li-S batteries. The batteries were discharged at 0.113 mA to 2.06 V and then held at 2.05 V until the current was below 10^{-2} mA for Li_2S deposition. The batteries were constantly discharged to 1.8 V, then potentiostatically discharged to a current lower than 0.01 mA, and then potentiostatically charged at 2.4 V for 10 h to evaluate Li_2S decomposition ability.

3. Results and Discussion

3.1. Materials Characterization

XRD patterns of $CoSe_2$-NCNT, (NiCo)Se_2-NCNT, and (CuCo)Se_2-NCNT are shown in Figure 1a, which correspond to PDF#88-1712, PDF#29-1417, and PDF#25-0309, respectively, proving the successful synthesis of bimetallic selenides. The three materials are cubic crystals with similar crystal structures, whose cell parameters are a = b = c = 5.859, a = b = c = 5.891, and a = b = c = 6.056, respectively. Moreover, the diffraction peaks of (NiCo)Se_2-NCNT and (CuCo)Se_2-NCNT are left-shifted compared to that of $CoSe_2$-NCNT by 2° and 3°, respectively, indicating that the crystal plane spacing increases. To study the chemical state of different $CoSe_2$-based catalysts, XPS tests were conducted. As shown in

Figure 1b, the peaks of 932.1 and 952.1 eV are Cu^+ $2p_{3/2}$ and Cu^+ $2p_{1/2}$, and the peaks of 934.3 eV and 954.4 eV are Cu^{2+} $2p_{3/2}$ and Cu^{2+} $2p_{1/2}$ [33]. As for Ni 2p XPS spectrum of $(NiCo)Se_2$-NCNT (Figure 1c), peaks at 854.40 eV and 856.00 eV are Ni^{2+} $2p_{3/2}$ and Ni^{3+} $2p_{3/2}$, peaks of 871.90 eV and 876.18 eV relate to Ni^{2+} $2p_{1/2}$ and Ni^{3+} $2p_{1/2}$ [34,35]. The effects of the introduction of Ni and Cu on the electronic structure is also compared, as displayed in Figure 1d,e. In comparison to $CoSe_2$, the peaks of Co^{2+} $2p_{1/2}$ and Co^{3+} $2p_{1/2}$ of $(NiCo)Se_2$-NCNT and $(CuCo)Se_2$-NCNT are shifted to lower binding energies. The peaks of Se 3d of $(CuCo)Se_2$-NCNT are upshifted, while that of $(NiCo)Se_2$-NCNT is downshifted. This means the change of electron density $CoSe_2$ resulting from the electron interaction of metal ions after the incorporation of Ni or Cu [19]. Figure 1f presents the N 1s spectrum, in which the Pyridinic N, Pyrrolic N, and Graphitic N are regarded as active sites to anchor LiPSs [36]. The microstructure and crystal structure of $(CuCo)Se_2$-NCNT TEM are further studied by TEM. The material displays a one-dimensional tubular morphology with a diameter ranging from 10 to 60 nm, as shown in Figure 1g. In addition, most $(CuCo)Se_2$ nanoparticles are located inside the carbon nanotubes. Moreover, the high-resolution TEM (HRTEM) (Figure 1h) shows that the nanoparticles are about 10–150 nm in diameter and wrapped in a carbon layer. The carbon nanotubes are in-situ generated during the high-temperature stage resulting from the catalytic action of metal particles with glucose as a carbon source. Furthermore, the nanoparticles are generally encapsulated in nanotubes and the diameters of the tubes are consistent with the nanoparticles. The size of nanoparticles has a great effect on the electrochemical performances, due to the fact that the smaller nanoparticles, the larger surface areas, meaning active sites to anchor LiPSs and catalyze their conversion [37]. The fast Fourier transform (FFT) corresponding to the nanoparticle in Figure 1h is presented in Figure 1i. The lattice spacing is 2.08 Å assigned to the (220) crystal plane of $(CuCo)Se_2$. The SAED pattern (Figure S2) displays the diffraction ring of (002) of amorphous carbon and (220) lattice planes of $(CuCo)Se_2$. According to the EDS result of $(CuCo)Se_2$ (Table S1), the doping amount of Cu is 36.56 at.%. The pore structure of $CoSe_2$-NCNT was characterized by N_2 adsorption–desorption experiments. It can be seen from Figure S3 that the material mainly contains mesoporous of 4 nm.

3.2. Electrochemical Performance of Li-S Batteries

The cycle performance at 0.1C was first tested to compare the electrochemical performance of Li-S batteries with various separators (Figure 2a). The initial discharge capacities of Li-S batteries with $(CuCo)Se_2$-NCNT, $(NiCo)Se_2$-NCNT, and $CoSe_2$-NCNT reach 1670.08, 1410.41, and 814.81 mAh g^{-1}, respectively. After 140 cycles, the capacities are maintained at 634.74, 682.81, and 578.23 mAh g^{-1}, respectively. In addition, the rate performance of the battery equipped with $(CuCo)Se_2$-NCNT modified separator is much higher than that of the battery with $(NiCo)Se_2$-NCNT and $CoSe_2$-NCNT (Figure 2b). The discharge capacities at 0.1C, 0.2C, 0.5C, 1C, 2C, and 4C reach 1504.40, 1388.32, 1206.94, 1083.17, 959.77, and 838.27 mAh g^{-1}, respectively, and the battery with $(CuCo)Se_2$-NCNT achieves a discharge capacity of 1294.91 mAh g^{-1} when returns to 0.1C again, showing excellent rate performance. The initial charge and discharge curves at 0.1C are displayed in Figure 2c. Two platforms are presented during discharging, which correspond to the conversion of S_8 into LiPSs and further reduction into Li_2S, and their capacities are marked as Q_H and Q_L, respectively. Li-S battery containing $(CuCo)Se_2$-NCNT possess larger Q_H and Q_L, meaning more S_8 is converted to LiPSs and more Li_2S are formed [38]. To compare the reaction kinetics of Li-S battery, the voltage difference is marked as ΔE, and the ΔE under different rates is compared, as presented in Figure 2d,e and Figure S4. Compared with Li-S batteries with $(NiCo)Se_2$-NCNT and $CoSe_2$-NCNT, that of $(CuCo)Se_2$-NCNT exhibits the least polarization at various current densities (Figure 2f), meaning improved reaction kinetics [38]. The battery with $(CuCo)Se_2$-NCNT achieves an initial capacity of 1051.06 mAh g^{-1} at 1C, which is much higher than that of $(NiCo)Se_2$-NCNT (838.63 mAh g^{-1}) and $CoSe_2$-NCNT (657.44 mAh g^{-1}). All in all, the Li-S batteries with bimetallic compounds exhibit superior electrochemical performance, including higher specific capacity, more stable cycle performance, and optimal reaction kinetics, especially $(CuCo)Se_2$, as displayed in Table S2. This means that the $(CuCo)Se_2$ is more conducive to the application of Li-S batteries for higher capacity and longer cycle life.

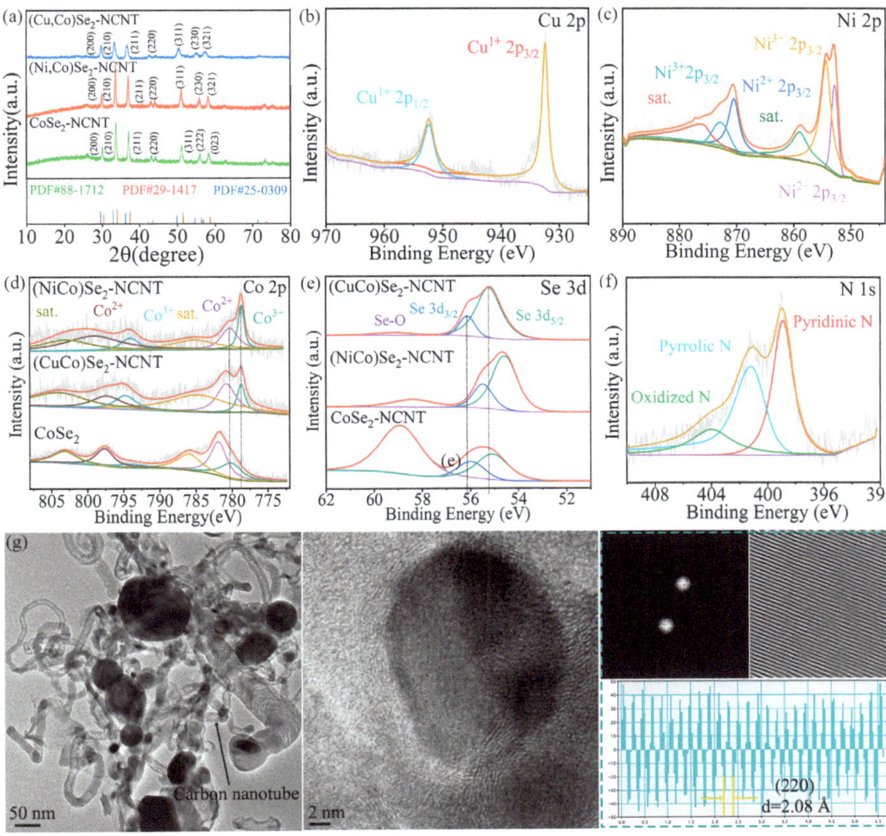

Figure 1. Physical and chemical property analyses of $CoSe_2$-NCNT, $(NiCo)Se_2$-NCNT, and $(CuCo)Se_2$-NCNT: (**a**) XRD patterns, XPS: (**b**) Cu 2p of $(CuCo)Se_2$-NCNT, (**c**) Ni 2p of $(NiCo)Se_2$-NCNT, (**d**) comparison of Co 2p of three samples, (**e**) comparison of Se 3d, (**f**) N 1s of $(CuCo)Se_2$-NCNT, (**g**) TEM, (**h**) HRTEM, (**i**)corresponding Fast Fourier transform patterns of $(CuCo)Se_2$-NCNT.

3.3. *Comparison of Adsorption–Catalytic Properties of Bimetallic Selenides*

To compare the adsorption capacity of the three materials for LiPSs, equal amounts of $(NiCo)Se_2$-NCNT, $(CuCo)Se_2$-NCNT, and $CoSe_2$-NCNTy were added to Li_2S_6 solution, and the color change was recorded. As shown in Figure 3a, the solution containing $(NiCo)Se_2$-NCNT began to fade first, suggesting the best adsorption effect, followed by $(CuCo)Se_2$-NCNT. The $(NiCo)Se_2$-NCNT after adsorption was analyzed by XPS. Figure 3b shows that a Li-N bond is formed between the catalyst and Li_2S_6 besides the Li-S bond of Li_2S_6. Furthermore, as displayed in Figure 3c,d, after adsorbing Li_2S_6, the peak of Co 2p moves towards lower binding energy, while the peak of Se 3d moves in the direction of higher binding energy compared to the pristine sample, indicating the occurred chemical interaction of $(NiCo)Se_2$-NCNT and LiPSs as the electron transfer between them [39]. In addition, the S 2p XPS (Figure 3e) displays the existence of polythionate and thiosulfate, suggesting the oxidation of polysulfides, which is conductive to anchoring LiPSs.

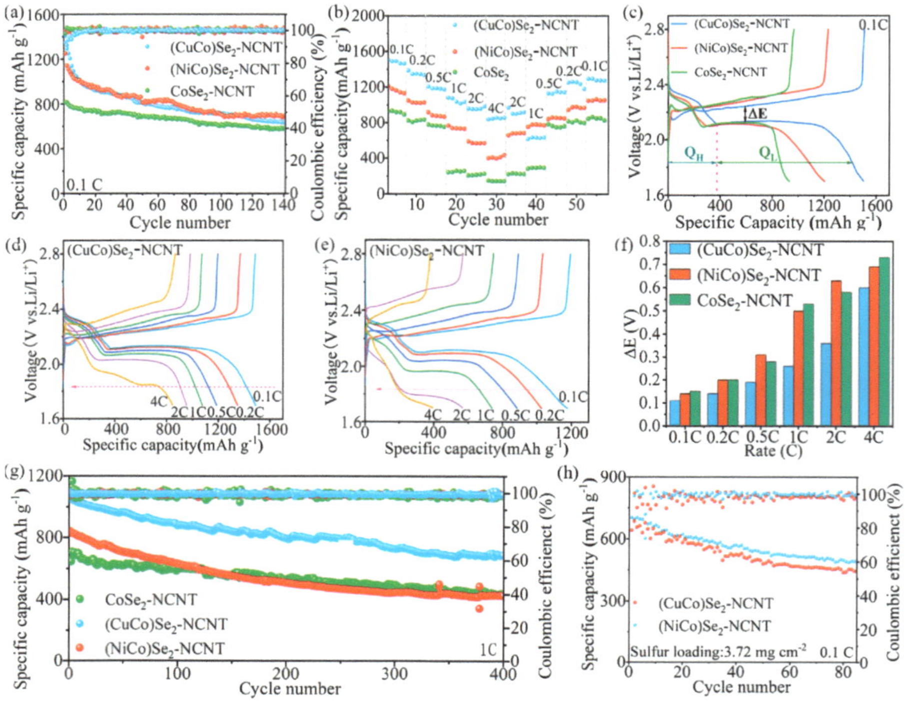

Figure 2. Electrochemical properties of various Li-S batteries: (**a**) cycling performance at 0.1C, (**b**) rate capability, (**c**) initial charge–discharge curves at 0.1C, charge–discharge curves of battery at different rate with (**d**) (CuCo)Se_2-NCNT, (**e**) (NiCo)Se_2-NCNT, (**f**) comparison of voltage difference (ΔE), (**g**) cycle performance at 1C, and (**h**) cycle stability of batteries with higher sulfur loading at 0.1C.

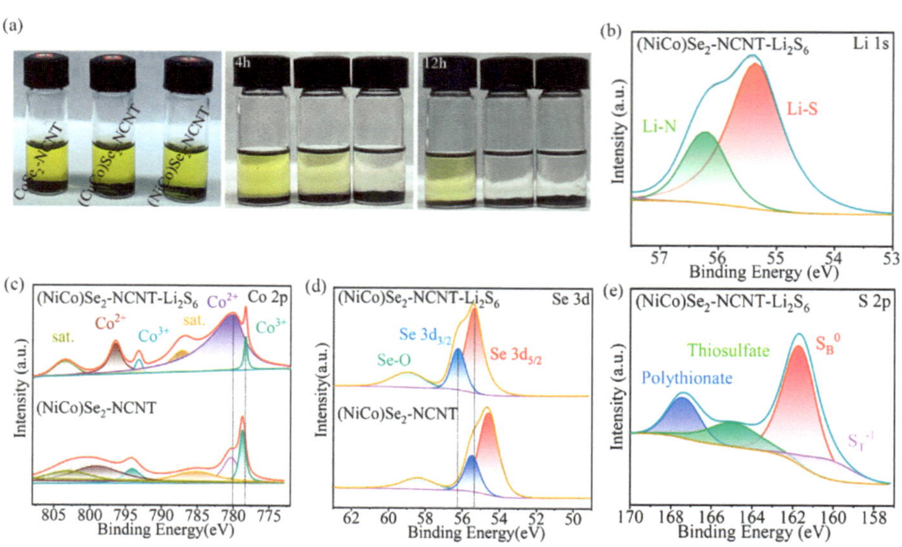

Figure 3. (**a**) Comparison of visible adsorption effect; XPS spectra of (NiCo)Se_2-NCNT after adsorption test: (**b**) Li 1s, (**c**) Co 2p, (**d**) Se 3d, (**e**) S 2p.

The EIS spectra of different Li-S batteries present semicircular in high-frequency regions and oblique in low-frequency regions (Figure 4a). The value of the semicircle radius depends on the resistance generated by the charge transfer between the electrolyte and the cathode, and the slope of the diagonal depends on the rate of ion diffusion [35]. It can be seen that the battery of (CuCo)Se_2-NCNT delivers the smallest resistance and the largest Li^+ diffusion rate in comparison to the other samples. Furthermore, the reduction peaks (Peak A) around 2.3 V and 2.0 V (Peak B) in CV curves are attributed to S_8 reduced to Li_2S_x ($4 \leq x \leq 8$) and then Li_2S/Li_2S_2 (Figure 4b) [5]. The oxidation peak (Peak C) is related to the conversion from Li_2S to LiPSs/S_8 [40]. (CuCo)Se_2-NCNT endows the corresponding battery with the optimized potential difference and highest peak current compared with $CoSe_2$-NCNT and (NiCo)Se_2-NCNT, meaning an improvement in sulfur utilization and reaction kinetics [41]. Tafel slopes were further analyzed according to CV curves to assess the catalytic capacity (Figure 4c–e). The Tafel slopes of the battery with (CuCo)Se_2-NCNT are 82 and 66 mV dec^{-1} for Peak A and Peak B, and 73 mV dec^{-1} for Peak C. It is much smaller than (NiCo)Se_2-NCNT (108, 73, 67 mV dec^{-1}) and $CoSe_2$-NCNT (280, 293, 180 mV dec^{-1}). This means that (CuCo)Se_2-NCNT can effectively promote the conversion of sulfur species [42]. As for the CV of Li_2S_6 symmetric batteries (Figure 4f), the peaks are related to the redox of Li_2S_6. The battery with (CuCo)Se_2-NCNT delivers the largest peak current, indicating better catalytic activity than other samples, thus enhancing LiPSs conversion kinetics [19,43]. GITT was also applied to analyze the ohmic impedance during charge–discharge, and the calculation basises of ohmic impedance is shown in Figure 4g,h and Figure S5, and Equation (S1) [44]. It can be seen from Figure 4i that the battery with (CuCo)Se_2-NCNT presents the lowest impedance in comparison to (NiCo)Se_2-NCNT and $CoSe_2$-NCNT, meaning facilitated reaction kinetics [45].

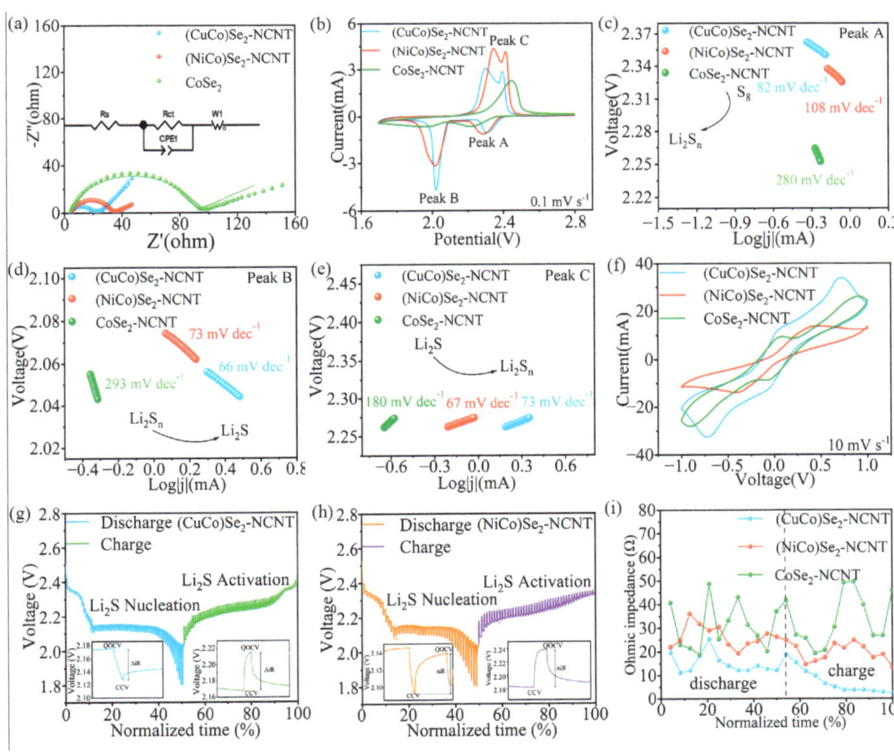

Figure 4. Catalytic performance analyses: (**a**) EIS, (**b**) CV at 0.1 mV s^{-1}, Tafel curves of (**c**) Peak A, (**d**) Peak B, (**e**) Peak C, (**f**) CV of Li_2S_6 symmetrical cell at scanning rate of 10 mV s^{-1}, GITT plots of Li-S batteries with (**g**) (CuCo)Se_2-NCNT, (**h**) (NiCo)Se_2-NCNT, and (**i**) comparison of impedance during charging–discharging.

To analyze the catalytic effect of various bimetallic selenides on the liquid–solid conversion from LiPSs to Li_2S, nucleation experiments of Li_2S were carried out at 2.05 V. The (CuCo)Se_2-NCNT achieves a larger Li_2S deposition capacity of 180.59 mAh g^{-1} than (NiCo)Se_2-NCNT (126.78 mAh g^{-1}) and $CoSe_2$-NCNT (147.23 mAh g^{-1}) (Figure 5a–c). In the meanwhile, Li_2S has a faster nucleation response (2361 s) on (CuCo)Se_2-NCNT than on (NiCo)Se_2-NCNT and $CoSe_2$-NCNT. This suggests that (CuCo)Se_2-NCNT can significantly promote the nucleation and growth of Li_2S [21]. Similarly, the Li_2S decomposition experiments were carried out at 2.4 V to evaluate the decomposition ability of Li_2S under different catalysts. The Li_2S decomposition capacity on the surface of (NiCo)Se_2-NCNT (436.84 mAh g^{-1}) is superior to (CuCo)Se_2-NCNT (364.72.8 mAh g^{-1}) and $CoSe_2$-NCNT (264.37 mAh g^{-1}) (Figure 5d–f), presenting a good catalytic effect on the decomposition of Li_2S [46].

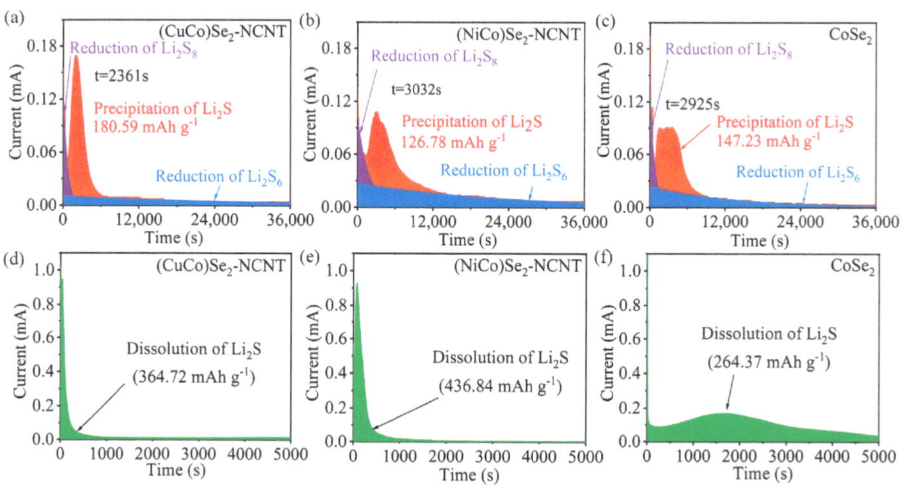

Figure 5. Potentiostatic discharging curves of Li_2S_8 solution on different catalysts at 2.05 V: (**a**) (CuCo)Se_2-NCNT; (**b**) (NiCo)Se_2-NCNT; (**c**) $CoSe_2$-NCNT, potentiostatic charging curves of Li_2S_8 solution at 2.4 V on different catalysts: (**d**) (CuCo)Se_2-NCNT and (**e**) (NiCo)Se_2-NCNT, (**f**) $CoSe_2$-NCNT.

To reveal the fundamental reason for the superior catalytic performance of (CuCo)Se_2, the electronic structure was analyzed by first-principles calculation. The theoretical calculation structure models of the three samples and the model structure parameters are displayed in Figure S6 and Table S3. (CuCo)Se_2 presents larger lattice parameters than the other two, which is in line with the XRD analysis result. Moreover, the density of states of different materials was also calculated. The (CuCo)Se_2 presents a higher density of states around the Fermi level in comparison to $CoSe_2$ and (NiCo)Se_2 (Figure 6a), meaning a better electron conductivity, which can help to accelerate the electrochemical reaction [8,47,48]. Charge density difference analyses are further displayed in Figure 6b,c, the yellow region stands for charge accumulation and the cyan region means charge depletion. As a result, electron consumption occurs around Ni and Cu, suggesting charge transfer from Ni/Cu to other ions and causing charge redistribution. Moreover, the *d* orbital electronic structure of Co and incorporated metal ions is presented in Figure 6d. Compared to $CoSe_2$, the *d* band center of Co in (NiCo)Se_2 is upshifted, while that of Co in (CuCo)Se_2 is downshifted. Meanwhile, the *d* band center of Ni is also higher than that of Cu. According to *d* band center theory, the upshifted *d* band center, the higher the adsorption energy [28,49]. However, the catalyzing LiPSs conversion involves the adsorption–catalysis–desorption process, and too strong adsorption could inhibit the desorption process and occupy active sites, which is not conductive to further reaction [23,27,50]. In addition, Cu increases the *d* orbital occupation number of Co while the Ni decreases it (Figure 6d and Table S4), thus improving the desorption and catalytic property of $(CuCo)Se_2$ for LiPSs [21,26]. Consequently, the introduction of Cu into $CoSe_2$ endows the $(CuCo)Se_2$ with a relatively low d band center and d orbital occupation number of Co, which can render it to achieve proper adsorption energy for LiPSs and catalyze their conversion, promoting the electrochemical properties of Li-S batteries.

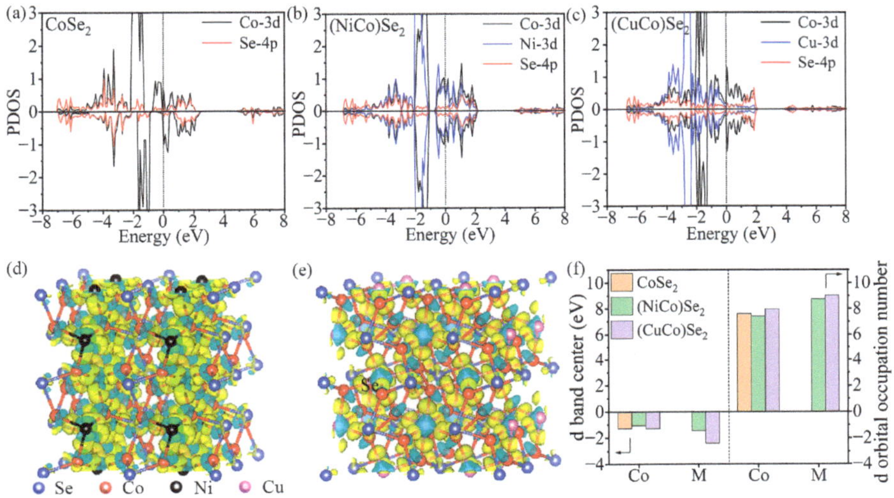

Figure 6. Theoretical analyses of d orbital electronic structure. DOS of: (**a**) $CoSe_2$, (**b**) $(NiCo)Se_2$, (**c**) $(CuCo)Se_2$, Electron density difference of (**d**) $(NiCo)Se_2$, (**e**) $(CuCo)Se_2$, (**f**) Comparison of d orbital electronic structure parameters.

In addition to being used in Li-S batteries, bimetallic selenides can also be used in other types of batteries (lithium-ion batteries, sodium-ion batteries, magnesium-ion batteries, et al.), various supercapacitors, electrocatalysts for hydrogen evolution reaction, oxygen evolution reaction, etc. [51–57]. Therefore, the designed bimetallic selenides in this paper can also be extended to other fields.

4. Conclusions

In summary, Ni and Cu were successfully introduced to $CoSe_2$ and constructting bimetallic compounds $(NiCo)Se_2$ and $(CuCo)Se_2$. The electron structure of $(NiCo)Se_2$ and $(CuCo)Se_2$ were adjusted to optimize the adsorption–conversion of LiPSs. To be specific, the Ni rendered the d band center of Co in $(NiCo)Se_2$ upshifted, while Cu downshifted it. In the meanwhile, d orbital occupation number of Co in $(CuCo)Se_2$ was increased but that of $(NiCo)Se_2$ was decreased. The downshifted d band center and increase of d orbital occupation number of Co in $(CuCo)Se_2$ could endow the LiPSs with moderate adsorption energy, optimizing their adsorption–conversion process. In contrast, too strong adsorption of $(NiCo)Se_2$ would inhibit the desorption of LiPSs and was not conductive to further catalyze the LiPSs conversion. Consequently, the $(CuCo)Se_2$ delivered better catalytic properties for LiPSs and endowed the corresponding Li-S batteries with improved electrochemical performance.

Supplementary Materials: The following supporting information can be downloaded at: https://www.mdpi.com/article/10.3390/ma17133075/s1, Figure S1. Scheme of Li-S battery with modified separator. Figure S2. Selected area electron diffraction (SAED) pattern of $(CuCo)Se_2$. Figure S3. Pore size distribution of $CoSe_2$-NCNT. Figure S4. Charge-discharge curves of battery with $CoSe_2$-NCNT at different rate. Figure S5. GITT plots of Li-S batteries with $CoSe_2$-NCNT. Figure S6. Theoretical calculation structure model of (a) $CoSe_2$, (b) $(NiCo)Se_2$, (c) $(CuCo)Se_2$. Table S1. The content of each element in $(CuCo)Se_2$. Table S2. Comparison of electrochemical performance of Li-S batteries with different catalysts. Table S3. Theoretical model structural parameters of different materials. Table S4. *d* orbital electronic structure parameters of different $(MCo)Se_2$.

Author Contributions: Formal analysis, N.L.; Investigation, Y.Z.; Data curation, R.W. and Y.B.; Writing—original draft, L.C.; Supervision, J.W. All authors have read and agreed to the published version of the manuscript.

Funding: The authors acknowledge funding from the National Natural Science Foundation of China (52302307), Shaanxi Province (2023-ZDLGY-24, 2023-JC-QN-0473), a project funded by China Postdoctoral Science Foundation (2023MD734210).

Institutional Review Board Statement: Not applicable.

Informed Consent Statement: Not applicable.

Data Availability Statement: The data that support the findings of this study are available on request from the corresponding author.

Conflicts of Interest: The authors declare no conflicts of interest.

References

1. Liang, Z.; Shen, J.; Xu, X.; Li, F.; Liu, J.; Yuan, B.; Yu, Y.; Zhu, M. Advances in the development of single-atom catalysts for high-energy-density lithium-sulfur batteries. *Adv. Mater.* **2022**, *34*, 2200102. [CrossRef] [PubMed]
2. Liu, T.; Hu, H.; Ding, X.; Yuan, H.; Jin, C.; Nai, J.; Liu, Y.; Wang, Y.; Wan, Y.; Tao, X. 12 years roadmap of the sulfur cathode for lithium sulfur batteries. *Energy Storage Mater.* **2020**, *30*, 346–366. [CrossRef]
3. Long, J.J.; Yu, H.; Liu, W.B. Structure engineering of cathode host materials for Li-S batteries. *Rare Metals* **2024**, *43*, 1370–1389. [CrossRef]
4. Lv, Z.C.; Wang, P.F.; Wang, J.C.; Tian, S.H.; Yi, T.F. Key challenges, recent advances and future perspectives of rechargeable lithium-sulfur batteries. *J. Ind. Eng. Chem.* **2023**, *124*, 68–88. [CrossRef]
5. Cao, G.; Duan, R.; Li, X. Controllable catalysis behavior for high performance lithium sulfur batteries: From kinetics to strategies. *EnergyChem* **2023**, *5*, 100096. [CrossRef]
6. Zhou, L.; Danilov, D.L.; Eichel, R.A.; Notten, P.H.L. Host materials anchoring polysulfides in Li-S batteries reviewed. *Adv. Energy Mater.* **2021**, *11*, 2001304. [CrossRef]
7. Han, J.; Wang, P.; Zhang, H.; Song, N.; An, X.; Xi, B.; Xiong, S. Performance optimization of chalcogenide catalytic materials in lithium-sulfur batteries: Structural and electronic engineering. *Chin. Chem. Lett.* **2024**, *35*, 109543. [CrossRef]
8. Chen, L.; Cao, G.; Li, Y.; Zu, G.; Duan, R.; Bai, Y.; Xue, K.; Fu, Y.; Xu, Y.; Wang, J.; et al. review on engineering transition metal compound catalysts to accelerate the redox kinetics of sulfur cathodes for lithium-sulfur batteries. *Nano-Micro Lett.* **2024**, *16*, 97. [CrossRef] [PubMed]
9. Wang, J.; Du, R.; Yu, C.; Xu, C.; Shi, Z. Application of transition metal compounds in cathode materials for lithium-sulfur battery. *Ionics* **2022**, *28*, 5275–5288. [CrossRef]
10. Shi, F.; Yu, J.; Chen, C.; Lau, S.P.; Lv, W.; Xu, Z.-L. Advances in understanding and regulation of sulfur conversion processes in metal-sulfur batteries. *J. Mater. Chem. A* **2022**, *10*, 19412–19443. [CrossRef]
11. Du, S.; Yu, Y.; Liu, X.; Lu, D.; Yue, X.; Liu, T.; Yin, Y.; Wu, Z. Catalytic engineering for polysulfide conversion in high-performance lithium-sulfur batteries. *J. Mater. Sci. Technol.* **2024**, *186*, 110–131. [CrossRef]
12. Yao, Y.; Chen, J.; Niu, R.; Zhao, Z.; Wang, X. High-entropy materials: Features for lithium-sulfur battery applications. *Metals* **2023**, *13*, 833. [CrossRef]
13. Jiang, Y.C.; Arshad, H.M.U.; Li, H.J.; Liu, S.; Li, G.R.; Gao, X.P. Crystalline multi-metallic compounds as host materials in cathode for lithium-sulfur batteries. *Small* **2021**, *17*, 2005332. [CrossRef] [PubMed]
14. Huang, S.; Wang, Z.; Von Lim, Y.; Wang, Y.; Li, Y.; Zhang, D.; Yang, H.Y. Recent advances in heterostructure engineering for lithium-sulfur batteries. *Adv. Energy Mater.* **2021**, *11*, 2003689. [CrossRef]
15. Wang, J.; Han, W.-Q. A review of heteroatom doped materials for advanced lithium-sulfur batteries. *Adv. Funct. Mater.* **2022**, *32*, 2107166. [CrossRef]
16. Li, S.; Xu, P.; Aslam, M.K.; Chen, C.; Rashid, A.; Wang, G.; Zhang, L.; Mao, B. Propelling polysulfide conversion for high-loading lithium-sulfur batteries through highly sulfiphilic $NiCo_2S_4$ nanotubes. *Energy Storage Mater.* **2020**, *27*, 51–60. [CrossRef]
17. Duan, J.; Zou, Y.; Li, Z.; Long, B.; Du, Y. Hollow quasi-polyhedron structure of NiCoP with strong constraint sulfur effect for lithium sulfur battery. *Electroanal. Chem.* **2019**, *847*, 113187. [CrossRef]
18. Zhao, S.; Li, Y.; Zhang, F.; Guo, J. $Li_4Ti_5O_{12}$ nanowire array as a sulfur host for high performance lithium sulfur battery. *J. Alloys Compd.* **2019**, *805*, 873–879. [CrossRef]
19. Yan, Y.; Li, H.; Cheng, C.; Yan, T.; Gao, W.; Mao, J.; Dai, K.; Zhang, L. Boosting polysulfide redox conversion of Li-S batteries by one-step-synthesized Co-Mo bimetallic nitride. *J. Energy Chem.* **2021**, *61*, 336–346. [CrossRef]

20. Wang, X.; Han, J.; Luo, C.; Zhang, B.; Ma, J.; Li, Z.; He, Y.-B.; Yang, Q.-H.; Kang, F.; Lv, W. Coordinated adsorption and catalytic conversion of polysulfides enabled by perovskite bimetallic hydroxide nanocages for lithium-sulfur batteries. *Small* **2021**, *17*, 2101538. [CrossRef]
21. Shen, Z.; Cao, M.; Zhang, Z.; Pu, J.; Zhong, C.; Li, J.; Ma, H.; Li, F.; Zhu, J.; Pan, F.; et al. Efficient $Ni_2Co_4P_3$ nanowires catalysts enhance ultrahigh-loading lithium-sulfur conversion in a microreactor-like battery. *Adv. Funct. Mater.* **2020**, *30*, 1906661. [CrossRef]
22. Li, Y.; Zhang, J.; Chen, Q.; Xia, X.; Chen, M. Emerging of heterostructure materials in energy storage: A review. *Adv. Mater.* **2021**, *33*, 2100855. [CrossRef] [PubMed]
23. Wang, B.; Wang, L.; Ding, D.; Zhai, Y.; Wang, F.; Jing, Z.; Yang, X.; Kong, Y.; Qian, Y.; Xu, L. Zinc-assisted cobalt ditelluride polyhedra inducing lattice strain to endow efficient adsorption-catalysis for high-energy lithium-sulfur batteries. *Adv. Mater.* **2022**, *34*, 2204403. [CrossRef]
24. Shang, C.; Li, G.; Wei, B.; Wang, J.; Gao, R.; Tian, Y.; Chen, Q.; Zhang, Y.; Shui, L.; Zhou, G.; et al. Dissolving vanadium into titanium nitride lattice framework for rational polysulfide regulation in Li-S batteries. *Adv. Energy Mater.* **2021**, *11*, 2003020. [CrossRef]
25. Wang, L.; Hu, Z.; Wan, X.; Hua, W.; Li, H.; Yang, Q.H.; Wang, W. Li_2S_4 anchoring governs the catalytic sulfur reduction on defective $SmMn_2O_5$ in lithium-sulfur battery. *Adv. Energy Mater.* **2022**, *12*, 2200340. [CrossRef]
26. Cheng, Z.; Wang, Y.; Zhang, W.; Xu, M. Boosting polysulfide conversion in lithium-sulfur batteries by cobalt-doped vanadium nitride microflowers. *ACS Appl. Energy Mater.* **2020**, *3*, 4523–4530. [CrossRef]
27. Shen, Z.; Jin, X.; Tian, J.; Li, M.; Yuan, Y.; Zhang, S.; Fang, S.; Fan, X.; Xu, W.; Lu, H.; et al. Cation-doped ZnS catalysts for polysulfide conversion in lithium-sulfur batteries. *Nat. Catal.* **2022**, *5*, 555–563. [CrossRef]
28. Wang, M.; Fan, L.; Sun, X.; Guan, B.; Jiang, B.; Wu, X.; Tian, D.; Sun, K.; Qiu, Y.; Yin, X.; et al. Nitrogen-doped $CoSe_2$ as a bifunctional catalyst for high areal capacity and lean electrolyte of Li-S battery. *ACS Energy Lett.* **2020**, *5*, 3041–3050. [CrossRef]
29. Sun, D.; Zhou, J.; Rao, D.; Zhu, L.; Niu, S.; Cai, J.; Fang, Y.; Liu, Y.; Liu, X.; Zang, Y.; et al. Regulating the electron filling state of d orbitals in Ta-based compounds for tunable lithium-sulfur chemistry. *Sustain. Mater. Techno.* **2021**, *28*, e00271. [CrossRef]
30. Al-Tahan, M.A.; Dong, Y.; Shrshr, A.E.; Liu, X.; Zhang, R.; Guan, H.; Zhang, J. Enormous-sulfur-content cathode and excellent electrochemical performance of Li-S battery accouched by surface engineering of Ni-doped WS_2@rGO nanohybrid as a modified separator. *J. Colloid Interf. Sci.* **2022**, *609*, 235–248. [CrossRef]
31. Zhang, R.; Dong, Y.; Al-Tahan, M.A.; Zhang, Y.; Wei, R.; Ma, Y.; Yang, C.; Zhang, J. Insights into the sandwich-like ultrathin Ni-doped MoS_2/rGO hybrid as effective sulfur hosts with excellent adsorption and electrocatalysis effects for lithium-sulfur batteries. *J. Energy Chem.* **2021**, *60*, 85–94. [CrossRef]
32. Zhang, H.; Dai, R.; Zhu, S.; Zhou, L.; Xu, Q.; Min, Y. Bimetallic nitride modified separator constructs internal electric field for high-performance lithium-sulfur battery. *Chem. Eng. J.* **2022**, *429*, 132454. [CrossRef]
33. Jiang, Z.; Yuan, Y.; Tan, L.; Li, M.; Peng, K. Self-reconstruction of (CoNiFeCuCr)Se high-entropy selenide for efficient oxygen evolution reaction. *Appl. Surf. Sci.* **2023**, *627*, 157282. [CrossRef]
34. Tian, L.; Zhang, Z.; Liu, S.; Li, G.; Gao, X. High-entropy spinel oxide nanofibers as catalytic sulfur hosts promise the high gravimetric and volumetric capacities for lithium-sulfur batteries. *Energy Environ. Mater.* **2022**, *5*, 645–654. [CrossRef]
35. Yao, W.; Tian, C.; Yang, C.; Xu, J.; Meng, Y.; Manke, I.; Chen, N.; Wu, Z.; Zhan, L.; Wang, Y.; et al. P-doped $NiTe_2$ with Te-vacancies in lithium-sulfur batteries prevents shuttling and promotes polysulfide conversion. *Adv. Mater.* **2022**, *34*, 2106370. [CrossRef] [PubMed]
36. Ao, J.; Xie, Y.; Lai, Y.; Yang, M.; Xu, J.; Wu, F.; Cheng, S.; Wang, X. $CoSe_2$ nanoparticles-decorated carbon nanofibers as a hierarchical self-supported sulfur host for high-energy lithium-sulfur batteries. *Sci. China Mater.* **2023**, *66*, 3075–3083. [CrossRef]
37. Zhang, X.; Zhang, Y.; Wei, X.; Wei, C.; Song, Y. A review of size engineering-enabled electrocatalysts for Li-S chemistry. *Nanoscale Adv.* **2021**, *3*, 5777–5784. [CrossRef] [PubMed]
38. Gu, Z.; Cheng, C.; Yan, T.; Liu, G.; Jiang, J.; Mao, J.; Dai, K.; Li, J.; Wu, J.; Zhang, L. Synergistic effect of Co_3Fe_7 alloy and N-doped hollow carbon spheres with high activity and stability for high-performance lithium-sulfur batteries. *Nano Energy* **2021**, *86*, 106111. [CrossRef]
39. Jiang, B.; Qiu, Y.; Tian, D.; Zhang, Y.; Song, X.; Zhao, C.; Wang, M.; Sun, X.; Huang, H.; Zhao, C.; et al. Crystal facet engineering induced active tin dioxide nanocatalysts for highly stable lithium-sulfur batteries. *Adv. Energy Mater.* **2021**, *11*, 2102995. [CrossRef]
40. Yao, W.; Zheng, W.; Xu, J.; Tian, C.; Han, K.; Sun, W.; Xiao, S. ZnS-SnS@NC heterostructure as robust lithiophilicity and sulfiphilicity mediator toward high-rate and long-life lithium-sulfur batteries. *ACS Nano.* **2021**, *15*, 7114–7130. [CrossRef]
41. Zhang, Y.; Qiu, Y.; Fan, L.; Sun, X.; Jiang, B.; Wang, M.; Wu, X.; Tian, D.; Song, X.; Yin, X.; et al. Dual-atoms iron sites boost the kinetics of reversible conversion of polysulfide for high-performance lithium-sulfur batteries. *Energy Storage Mater.* **2023**, *63*, 103026. [CrossRef]
42. Song, X.; Tian, D.; Qiu, Y.; Sun, X.; Jiang, B.; Zhao, C.; Zhang, Y.; Fan, L.; Zhang, N. Accelerating sulfur redox reactions by topological insulator Bi_2Te_3 for high-performance Li-S batteries. *Adv. Funct. Mater.* **2022**, *32*, 2109413. [CrossRef]
43. Cai, D.; Liu, B.; Zhu, D.; Chen, D.; Lu, M.; Cao, J.; Han, W. Ultrafine Co_3Se_4 nanoparticles in nitrogen-doped 3D carbon matrix for high-stable and long-cycle-life lithium sulfur batteries. *Adv. Energy Mater.* **2020**, *10*, 1904273. [CrossRef]

44. Wen, K.; Huang, L.; Qu, L.; Deng, T.; Men, X.; Chen, L.; Wang, J. g-C_3N_4/MoO_3 composite with optimized crystal face: A synergistic adsorption-catalysis for boosting cathode performance of lithium-sulfur batteries. *J. Colloid Interf. Sci.* **2023**, *649*, 890–899. [CrossRef] [PubMed]
45. Qu, L.; Wang, J.; Chen, L.; Men, X.; Deng, T.; Wen, K.; Huang, L. Accelerated polysulfide conversion using nitrogen-doped carbon nanofibers embedded with V_2O_3 as interlayers for lithium-sulfur batteries. *Ionics* **2023**, *29*, 2299–2310. [CrossRef]
46. Guo, Y.; Li, J.; Yuan, G.; Guo, J.; Zheng, Y.; Huang, Y.; Shao, H. Elucidating the volcanic-type catalytic behavior in lithium-sulfur batteries via defect engineering. *ACS Nano.* **2023**, *17*, 18253–18265. [CrossRef] [PubMed]
47. Song, X.; Qu, Y.; Zhao, L.; Zhao, M. Monolayer Fe_3GeX_2 (X = S, Se, and Te) as highly efficient electrocatalysts for lithium-sulfur batteries. *ACS Appl. Mater. Interfaces* **2021**, *13*, 11845–11851. [CrossRef] [PubMed]
48. Shi, L.; Fang, H.; Yang, X.; Xue, J.; Li, C.; Hou, S.; Hu, C. Fe-cation doping in $NiSe_2$ as an effective method of electronic structure modulation towards high-performance lithium-sulfur batteries. *ChemSusChem* **2021**, *14*, 1710–1719. [CrossRef] [PubMed]
49. Xu, H.; Hu, R.; Zhang, Y.; Yan, H.; Zhu, Q.; Shang, J.; Yang, S.; Li, B. Nano high-entropy alloy with strong affinity driving fast polysulfide conversion towards stable lithium sulfur batteries. *Energy Storage Mater.* **2021**, *43*, 212–220. [CrossRef]
50. Zhang, W.; Pan, H.; Han, N.; Feng, S.; Zhang, X.; Guo, W.; Tan, P.; Xie, S.; Zhou, Z.; Ma, Q.; et al. Balancing adsorption, catalysis, and desorption in cathode catalyst for Li-S batteries. *Adv. Energy Mater.* **2023**, *13*, 2301551. [CrossRef]
51. Seo, H.Y.; Choi, J.H.; Kim, Y.B.; Cho, J.S.; Kang, Y.C.; Park, G.D. Tailoring the shell thickness of yolk-shell structured carbon microspheres: Applications in metal selenide and carbon composite microspheres for enhanced sodium ion storage properties. *J. Mater. Chem. A* **2023**, *11*, 24738–24753. [CrossRef]
52. Wang, Z.; Zhang, F.; Chen, S.; Gao, S.; Wang, L.; Liu, X.; Li, M.; Shangguan, E. Facile preparation of bimetallic cobalt-copper selenide nanosheets as stable cathode materials for rechargeable magnesium batteries. *J. Alloys Compd.* **2024**, *976*, 173223. [CrossRef]
53. Meshesha, M.M.; Chanda, D.; Jang, S.G.; Yang, B.L. Enhancing cobalt-based bimetallic selenide performance for urea and water electrolysis through interface engineering. *Chem. Eng. J.* **2023**, *474*, 145708. [CrossRef]
54. Chae, S.H.; Muthurasu, A.; Kim, T.; Kim, J.S.; Khil, M.S.; Lee, M.; Kim, H.; Lee, J.Y.; Kim, H.Y. Templated fabrication of perfectly aligned metal-organic framework-supported iron-doped copper-cobalt selenide nanostructure on hollow carbon nanofibers for an efficient trifunctional electrode material. *Appl. Catal. B Environ.* **2021**, *293*, 120209. [CrossRef]
55. Lee, S.J.; Lee, S.H.; Ha, J.S.; Kim, I.T.; Park, S.H.; Park, H.K.; Park, H.J.; Kang, B.K.; Park, Y.S. Highly Active cobalt-copper-selenide electrocatalysts for solar-driven oxygen evolution reaction: An electrochemical activation energy aspect. *ACS Sustain. Chem. Eng.* **2024**, *12*, 275–282. [CrossRef]
56. Hui, X.; Zhao, J.; Mao, J.; Zhao, H. Reduced graphene oxide-wrapped copper cobalt selenide composites as anode materials for high-performance lithium-ion batteries. *Colloids Surf. A Physicochem. Eng. Asp.* **2023**, *663*, 130979. [CrossRef]
57. Sun, P.; Zhang, J.; Huang, J.; Wang, L.; Wang, P.; Cai, C.; Lu, M.; Yao, Z.; Yang, Y. Bimetallic MOF-derived (CuCo)Se nanoparticles embedded in nitrogen-doped carbon framework with boosted electrochemical performance for hybrid supercapacitor. *Mater. Res. Bull.* **2021**, *137*, 111196. [CrossRef]

nanoenergy advances

Article

Spray-Coated Transition Metal Dichalcogenides as Hole Transport Layers in Inverted NFA-Based Organic Photovoltaics with Enhanced Stability under Solar and Artificial Light

Marinos Tountas [1,*,†], Katerina Anagnostou [1,†], Evangelos Sotiropoulos [1], Christos Polyzoidis [1] and Emmanuel Kymakis [1,2,*]

1 Department of Electrical & Computer Engineering, Hellenic Mediterranean University (HMU), 71410 Heraklion, Greece; katerinanag@hmu.gr (K.A.); evangelossot@hmu.gr (E.S.); polyzoidis@hmu.gr (C.P.)
2 Institute of Emerging Technologies (i-EMERGE), HMU Research Center, 71410 Heraklion, Greece
* Correspondence: tountasm@hmu.gr (M.T.); kymakis@hmu.gr (E.K.)
† These authors contributed equally to this work.

Abstract: In this study, we explored the potential of exfoliated transition metal dichalcogenides (TMDs) as innovative spray-coated hole transport layers (HTLs) in organic photovoltaics (OPVs), addressing the need for efficient and stable materials in solar cell technology. This research was motivated by the need for alternative HTLs that can offer enhanced performance under varying lighting conditions, particularly in indoor environments. Employing UV-visible absorption and Raman spectroscopy, we characterized the optical properties of MoS_2, $MoSe_2$, WS_2, and WSe_2, confirming their distinct excitonic transitions and direct bandgap features. The nanocrystalline nature of these TMDs, revealed through XRD patterns and crystallite size estimation using the Scherrer method, significantly contributes to their enhanced physical properties and operational efficiency as HTLs in OPVs. These TMDs were then integrated into OPV devices and evaluated under standard solar and indoor lighting conditions, to assess their effectiveness as HTLs. The results demonstrated that MoS_2, in particular, displayed remarkable performance, rivalling traditional HTL materials like MoO_3. It maintained high power conversion efficiency across a spectrum of light intensities, illustrating its versatility for both outdoor and indoor applications. Additionally, MoS_2 showed superior stability over extended periods, suggesting its potential for long-term usage in OPVs. This study contributes significantly to the field of photovoltaic materials, presenting TMDs, especially MoS_2, as promising candidates for efficient and stable OPVs in diverse lighting conditions, thereby broadening the scope of solar cell applications.

Keywords: organic photovoltaics (OPVs); indoor organic photovoltaics (iOPVs); transition metal dichalcogenide (TMD); liquid phase exfoliation (LPE); spray-coating

Citation: Tountas, M.; Anagnostou, K.; Sotiropoulos, E.; Polyzoidis, C.; Kymakis, E. Spray-Coated Transition Metal Dichalcogenides as Hole Transport Layers in Inverted NFA-Based Organic Photovoltaics with Enhanced Stability under Solar and Artificial Light. *Nanoenergy Adv.* **2024**, *4*, 221–234. https://doi.org/10.3390/nanoenergyadv4030014

Academic Editors: Robert A. Varin, Geng Zhong and Ya Yang

Received: 26 February 2024
Revised: 4 July 2024
Accepted: 5 July 2024
Published: 10 July 2024

1. Introduction

Organic photovoltaics (OPVs) have evolved substantially since the first devices were developed in the 1980s [1], with their reported power conversion efficiency (PCE) displaying an overall increase of >19% as of 2023 [2]. However, it has become increasingly clear that the objective of this class of emerging photovoltaics is not to replace conventional silicon (Si)-based solar panels but to complement them. Organic solar cells (OSCs) offer a lightweight and cost-effective alternative for solar power generation, and while recent advancements have enabled PCE values to approach those of commercial Si solar panels, their role is seen as supplementary, particularly in applications where flexibility and lower weight are advantageous [2]. There are persisting drawbacks that generally hinder their commercial viability and market adoption. OPVs exhibit low stability, as exposure to moisture, oxygen, and UV light cause degradation of the cell, especially to the polymer active layer (AL) [3,4].

Therefore, providing protection of the AL is a pragmatic approach for increasing the overall stability of OPVs.

To address these challenges effectively and enhance the performance and stability of OPVs, it is crucial to not only focus on the photoactive materials but also on other critical components of the cell where performance losses are likely to occur [5]. The carrier-selective charge transport interlayers, essential for enhancing photogenerated carrier extraction from the bulk heterojunction (BHJ) [6], are of particular interest. The ideal interlayer, whether it is an n-or a p-type semiconductor, should align energetically with the BHJ layer and offer processing flexibility to prevent issues like de-wetting and pinhole formation [7]. While there are several successful examples of electron-transporting layers (ETLs), like ZnO, LiF, and PFN-Br in organic photovoltaics (OPVs) [8,9], progress in hole-transporting layers (HTLs) has been more constrained, with fewer materials available [10–12]. Currently, the conjugated polymer Poly(3,4-ethylenedioxythiophene) (PEDOT:PSS) stands out as the most prevalent HTL in standard OPV architectures, while the vacuum-deposited MoO_3 is typically employed in inverted structures [13]. However, the hygroscopic and acidic nature of PEDOT:PSS can degrade the commonly used indium tin oxide (ITO) electrode material. To overcome these challenges, alternative HTL materials, including various transition metal oxides, have been suggested [14,15]. Yet, these alternatives often require vacuum processing or complicated chemistry, involving high-temperature annealing, which add complexity and energy intensity to the cell manufacturing process.

The advent of Internet of Things (IoT) technology has catalyzed the development of low-energy electronics, such as wireless sensors and router nodes, that can be powered by the minimal energy harvested from indoor lighting. This advancement has carved a specialized market for OPVs. Compared to their inorganic counterparts [16,17], OPVs have demonstrated superior performance in indoor environments, attributed to their highly tunable light absorption properties and lower current leakage. Studies indicate that OPVs can achieve indoor PCE values exceeding 30% [18,19], underscoring their potential in powering low-energy-consuming electronics for indoor use. A key factor distinguishing indoor from outdoor power generation is the light source. Indoor artificial lighting, predominantly within the visible spectrum, has a varied emission spectrum and significantly lower intensity—often less than one-thousandth—compared to outdoor sunlight [20].

TMDs are a class of layered semiconductor that consist of one transition metal atom (M) and two chalcogen atoms (X) in the general chemical formula of MX_2 (Figure 1a). In their bulk or three-dimensional (3D) form, TMDs appear as stacked layers of X–M–X held together by weak van der Waals forces, wherein the metal and chalcogen atoms are connected with strong covalent bonds, creating a honeycomb-shaped lattice. Bulk TMDs are semiconductors with a tunable indirect energy bandgap that widens and shifts toward a direct bandgap with the transition toward a monolayer structure [21–23]. Due to their optical and electrical properties [24,25], TMDs such as MoS_2, WS_2, $MoSe_2$, and WSe_2 have been utilized in electronic and optoelectronic technologies [25,26], photovoltaic devices [27,28], and energy storage [29].

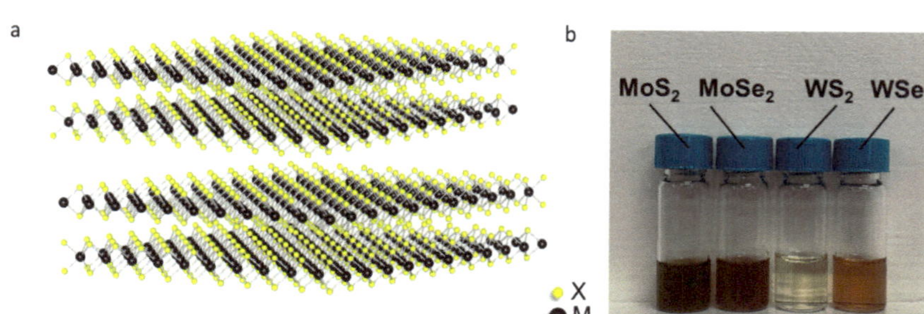

Figure 1. (**a**) Three-dimensional structure of TMDs, where M = transition metal and X = chalcogen. (**b**) Digital photograph of TMD dispersions prepared in IPA via liquid phase exfoliation (LPE).

Monolayer or few-layer TMDs can be produced with exfoliation, which involves the isolation of individual flakes from the bulk crystal by applying an external force which is sufficient to overcome the van der Waals forces holding the individual flakes together. Exfoliation is a top-down production method that can be achieved by mechanical cleavage, also known as "the Scotch^{TM} tape method" [30] and chemical exfoliation [31,32]. In this study, liquid phase exfoliation (LPE) was employed to produce exfoliated TMD flakes directly from pristine TMD powders.

LPE is a simple, high-yield, scalable, and cost-effective method of producing exfoliated mono- and few-layer nanomaterials such as graphene, graphene-related materials, and TMDs [33–36]. The driving force behind LPE is cavitation [37,38]. Cavitation involves the formation and growth of bubbles within the liquid medium due to the applied high ultrasonic frequencies, the intercalation of these microbubbles between the individual sheets of the bulk nanomaterial, and subsequent collapse of said bubbles. Upon implosion, the energy that is released from the generated shock wave is sufficient to overcome the van der Waals forces between the materials layers resulting in the separation of individual nanosheets. Another important benefit of LPE as a production method for exfoliated nanomaterials is that the produced nanosheets are dispersed directly into a liquid medium from the bulk crystal and are ready to use for further modification through chemical reactions or for thin film formulation using several deposition methods, such as spin-coating [27] or inkjet printing [39].

For the deposition of the LPE-produced TMD flakes onto the AL, spray-coating was employed. This deposition method was selected as an alternative to spin-coating, as it ensures the distribution of TMD flakes all along the AL surface and allows for thickness control by monitoring the total volume of the dispersion that is sprayed, with results ranging from TMD flake decoration to thick, compact TMD films. The spray-coating method involves spraying a solution or dispersion of a material onto the desired substrate, which is mildly heated to aid the evaporation of the solvent, thus leaving behind a solid thin film of the material.

In this study, we embarked on a detailed investigation into the properties and applications of exfoliated transition metal dichalcogenides (TMDs), focusing on their potential as alternative HTLs in OPVs. Our approach involved a thorough characterization of the optical properties of MoS_2, $MoSe_2$, WS_2, and WSe_2 using UV-visible absorption and Raman spectroscopy. These analyses confirmed the unique excitonic transitions and direct bandgap features of the TMDs, aligning well with the existing literature. A significant aspect of our research was the integration of these TMDs into OPV systems, where they were tested under both standard and indoor lighting conditions. This dual-environment testing was crucial in evaluating the real-world applicability of the TMDs in OPVs, especially considering the varying light conditions in practical scenarios. The results were particularly promising for MoS_2, which emerged as a strong contender among the TMDs for its effective performance as an HTL. This material demonstrated comparable, and in some cases superior, efficacy to traditional HTL materials like MoO_3, both under standard solar illumination and indoor lighting conditions. The adaptability of MoS_2 across different light intensities, essential for indoor applications, was a standout finding, highlighting its potential versatility in photovoltaic applications. Furthermore, the long-term stability of MoS_2, surpassing that of MoO_3 in extended aging tests, pointed to its robustness and suitability for sustained use in OPV applications. This comprehensive study not only advances our understanding of TMDs in photovoltaic applications but also opens new avenues for the development of more efficient and versatile OPVs, suitable for a wide range of environmental conditions.

2. Materials and Methods

For the preparation of the exfoliated TMDs, molybdenum(IV) sulfide (98%), molybdenum(IV) selenide (99.9%, metals basis), tungsten(IV) sulfide, (99.8%, metals basis) and tungsten(IV) selenide (99.8%, metals basis) were purchased from Thermo Fisher Scientific Inc. (Waltham, MA, USA), while isopropanol (IPA, $\geq$99.8%) was purchased from Honeywell (Charlotte, NC, USA). LPE was performed using a Hielscher (Teltow, Germany) UP200Ht (200 W, 26 KHz) ultrasonic probe. The resulting TMD dispersions were isolated via centrifugation using a Hettich (Tuttlingen, Germany) UNIVERSAL 320 centrifuge. The UV-Vis absorption spectra of the TMD dispersions and their respective sprayed film were recorded using a Shimadzu (Kyoto, Japan) UV-2401 PC Recording Spectrophotometer and UV Probe Software v2.34. Raman spectra were obtained at room temperature using a modified LabRAM HR Raman Spectrometer (HORIBA Scientific, Kyoto, Japan). Deposition of TMD dispersions via spray-coating was carried out using a commercial spray gun and air compressor.

To prepare the TMD dispersions, 135 mg of each TMD powder was added to 45 mL IPA for an initial concentration of 3 mg/mL. Each mixture was ultrasonicated using an ultrasonic probe (200 W, 26 KHz, 100% Amplitude) for 2 h to achieve LPE. An ice bath was used throughout LPE to maintain low temperatures and avoid overheating. The dispersions were then centrifugated for 20 min to remove unexfoliated particles and aggregates. MoS_2(IPA) was centrifugated at 5000 rpm (3075 g), while $MoSe_2$(IPA), WS_2(IPA), and WSe_2(IPA) were centrifugated at 4000 rpm (1968 g). The resulting dispersions (Figure 1b) were diluted to C/2 and their UV-Vis absorption spectra were recorded (Figure 2). Solid samples for Raman spectroscopy measurements (Figure S1) were prepared by drop-casting 10 μL of each LPE-prepared TMD dispersion onto O_2-plasma-treated (50 W, 5 min) silicon (Si) substrates.

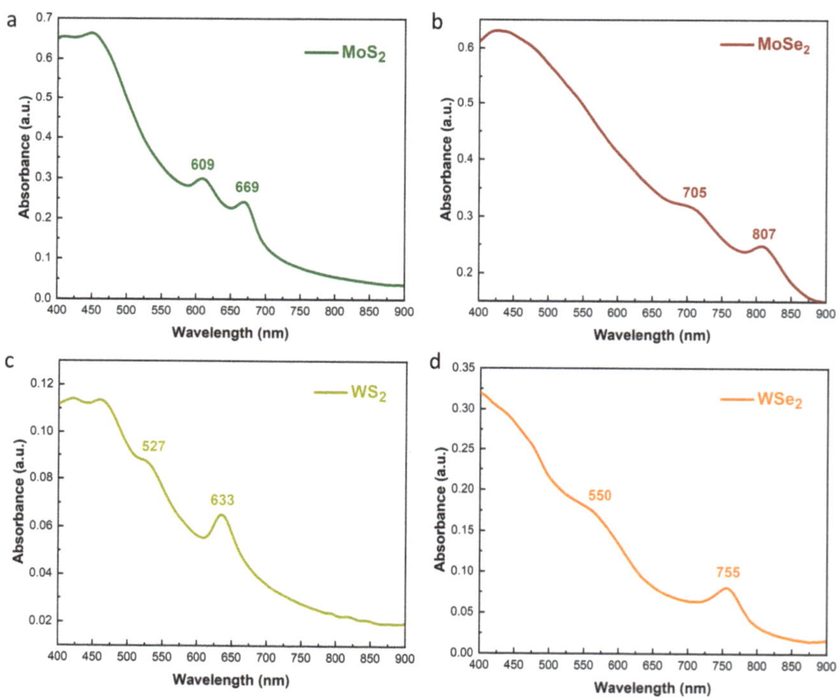

Figure 2. UV-visible absorption spectra of LPE-prepared (**a**) MoS_2; (**b**) $MoSe_2$; (**c**) WS_2; (**d**) WSe_2 dispersions in IPA.

For the device fabrication, the glass substrates with the indium tin oxide (ITO) pattern from Naranjo first underwent a thorough cleaning process. Initially, they were subjected to ultrasonic cleaning (37 KHz) in a sequence of deionized water, acetone, and isopropanol, each for a duration of 10 min. This was followed by drying using a flow of nitrogen gas and then an oxygen plasma treatment (50 W) for 5 min. Subsequently, a zinc oxide (ZnO) solution was prepared by mixing zinc acetate dihydrate (0.5 g) and ethanolamine (0.14 g) in 10 mL of 2-methoxyethanol. This mixture was stirred vigorously for 12 h under ambient conditions. The ZnO ETL layer was deposited at 4000 rpm for 40 s and was annealed at 160 °C for 30 min. For the active layer (AL), a solution comprising Poly[(2,6-(4,8-bis(5-(2-ethylhexyl)thiophen-2-yl)-benzo[1,2-b:4,5-b′]dithiophene))-alt-(5,5-(1′,3′-di-2-thienyl-5′,7′-bis(2-ethylhexyl)benzo[1′,2′-c:4′,5′-c′]dithiophene-4,8-dione)] (PBDB-T) and 3,9-bis(2-methylene-(3-(1,1-dicyanomethylene)-indanone))-5,5,11,11-tetrakis (4-hexylphenyl)-dithieno[2,3-d:2′,3′-d′]-s-indaceno [1,2-b:5,6-b′]dithiophene (ITIC) in a 1:1 ratio (10:10 mg) was dissolved in 992.5 μL of chlorobenzene (CB) and 7.5 μL of 1,8-diiodooctane (DIO). This solution was then spin-coated onto the prepared substrates at a speed of 2500 rpm for 60 s. To complete the device, molybdenum oxide (MoO_x) and silver (Ag), measuring 10 nm and 100 nm, respectively, were thermally evaporated onto the substrates. This was realized in a high-vacuum thermal evaporator operating at less than 3×10^{-6} Torr, creating an active area of 0.4 cm^2. To incorporate the TMD flakes as the HTL into the OPV devices, 0.5 mL of each TMD (IPA) dispersion were spray-coated onto the previously deposited AL. The spray-coating deposition was performed using a commercial airbrush and the devices were placed on a hotplate (40 °C) to aid solvent removal. The carrier gas pressure was set at 2 bar, the deposition rate was 5 spray cycles per ml, the deposition height was set at 15 cm, and the length was set at ~7 cm. The dispersions were sprayed in 30 s intervals to allow time for the solvent to evaporate before beginning the next spray cycle.

The electrical performance of the solar cell devices was evaluated through current density-voltage (J-V) testing conducted in a nitrogen-filled glovebox. This process utilized an Oriel solar simulator equipped with a 450 W Xenon lamp and an AM1.5G filter. To ensure accuracy, the lamp's intensity was carefully calibrated to 100 mW/cm^2 using a KG5 silicon reference cell before any measurements were taken. Regarding the indoor light measurements, an ultra-low-cost, commercial, and remotely-controlled smart LED lamp was used that has the ability to change its emission spectrum and intensity per user's taste. The X-ray diffraction (XRD) measurements were conducted using a RIGAKU (Tokyo, Japan) D/MAX-2500 powder diffractometer equipped with a monochromated Cu Kα radiation [(λ = 1.5418 Å)]. Steady-state and time-resolved photoluminescence spectra were measured using a fluorescence spectrophotometer (FS5, Edinburgh Instruments, Livingston, UK) at room temperature. The samples were excited using a 478.4 nm laser as a source.

3. Results

3.1. Characterization of Exfoliated TMDs

When observing the recorded UV-Vis absorption spectra of all four TMD dispersions in IPA (Figure 2), characteristic peaks of each sample are visible in the region of 500–900 nm and are in agreement with previous reports on the optical absorbance of TMD nanoflakes prepared via LPE with ultrasonication [40–42]. The absorption peaks appearing at 609 and 669 nm for MoS_2, 705 and 807 nm for $MoSe_2$, 527 and 633 nm for WS_2, and ~550 and 755 nm for WSe_2 are attributed to excitonic transitions that arise at the K point of the Brillouin zone and are a characteristic of the 2H-phase [40,43–45]. Absorbance at $\leq$500 nm is owed to direct transitions from the valence band to the conduction band [40,43]. The Raman spectra obtained for the prepared TMD samples are available in the Supporting Information (Figure S1). Raman spectra were obtained at room temperature using a modified LabRAM HR Raman Spectrometer (HORIBA Scientific, Kyoto, Japan). Raman excitation was achieved with a 532 nm central wavelength solid-state laser module with a maximum laser output power of 90 mW. The microscope was coupled with a 50× microscopic objective lens with a 0.5 numerical aperture and 10.6 mm working distance (LMPlanFLN 50×/0.5, Olympus, Tokyo, Japan) that delivered the excitation light and collected the Raman signals. The laser spot size was approximately 1.7 μm laterally and about 2 μm axially. A 600 grooves/mm grating was used, resulting in a Raman spectral resolution of ~2 cm^{-1}. Figure S1 shows the recorded Raman spectra of the four LPE-prepared TMD flakes.

For MoS_2, two prominent peaks are observed at 376 cm^{-1} and 401 cm^{-1}, and are attributed to first-order Raman modes, which are associated with the in-plane (E^1_{2g}) and out-of-plane (A_{1g}) vibrational modes, respectively [46–49]. For $MoSe_2$, the E^1_{2g} peak appears at 281 cm^{-1} and is lower in intensity than the A_{1g}, which is observed at 236 cm^{-1} [49–51]. An additional low-intensity peak is observed at 346 cm^{-1}, which is assigned to the B^1_{2g} vibrational mode. This mode is not active in bulk $MoSe_2$; instead, it becomes Raman active only in few-layer $MoSe_2$ samples. In the WS_2 spectrum, two signals are detected: a strong peak at 338 cm^{-1}, which is attributed to the second-order Raman peak, associated with the 2LA(M) phonon mode, and the A_{2g} peak at 418 cm^{-1} [40,52]. In WS_2, the E^1_{2g} peak, which appears at ~355 cm^{-1}, overlaps with the high-intensity 2LA peak [52]. The resonance of the 2LA(M) phonon mode is activated in monolayer WS_2, thus providing ample evidence of the successful exfoliation of the bulk material [53]. Two additional low-intensity peaks appear at 171 and 227 cm^{-1}. Lastly, the Raman spectrum recorded for WSe_2 exhibits a strong peak at 247 cm^{-1}, which is interpreted as the combination signal of the A_{1g} and E^1_{2g} Raman modes [49,50,54]. This can be an indicator of successful exfoliation, as distinct A_{1g} and E^1_{2g} peaks are observed in the bulk WSe_2 samples. The appearance of the B^1_{2g} peak, which is inactive in the bulk WSe_2, further supports the generation of few-layer flakes upon ultrasonication.

X-ray diffraction (XRD) was utilized to investigate the crystalline phase, the degree of crystallinity, and the flake size of the TMD samples [55,56]. The TMD powders, which were measured as reference samples, exhibit multiple narrow, sharp peaks with high intensity, indicating the existence of multiple layers within well-ordered, highly crystalline materials, a long-range order, and well-defined scattering planes (Figure 3) [57]. The XRD patterns of the TMD powders are in agreement with previous reports for MoS_2, $MoSe_2$, WS_2, and WSe_2 that match 2H polytype, which has a hexagonal (D6h) symmetry (space group: P63/mmc) [58–66]. The LPE-prepared TMD flakes have fewer and broader diffraction peaks compared to their respective powders, indicating that the crystallites are of smaller size and/or have a less ordered crystalline structure. Overall, the obtained results for our LPE-prepared samples are in accordance with previously reported XRD patterns for the same exfoliated nanomaterials [67,68]. The absence of any other diffraction peaks is evidence that the exfoliated TMD flakes remain pristine and unoxidized during the LPE process [65,67,69]. A low-intensity peak appears at 44° for the $MoSe_2$ flakes, which are attributed to the aluminum (Al) holder [70]. A broad peak appears for the spray-coated TMD samples, which is attributed to the signal derived from the glass substrates that the TMD flakes were deposited onto [71,72]. The (002) peak is present in both the bulk and exfoliated samples and corresponds to a typical Bragg's plane for a nanocrystalline structure [73]. The (002) peak of all the LPE-prepared flakes is much lower than that of the respective powder, indicating successful exfoliation of the bulk material and the generation of few-layer TMDs. Apart from the crystal structure analysis, information on the particle size can be derived from the XRD peaks through the Scherrer equation [74–76]. The average crystallite size D was calculated for the four TMDs at 14.44 nm for MoS_2, 35 nm for $MoSe_2$, 18.27 nm for WS_2, and 31.11 nm for WSe_2, according to the Scherrer equation, where $K = 0.94$, $\lambda = 0.154$ nm, β = FWHM, and θ = peak position. The nanocrystalline size of TMD flakes affects their physical, optoelectronic, mechanical, and structural properties [25,57,77–80]. The reduced crystallite sizes, estimated at 14.44 nm for MoS_2, 35 nm for $MoSe_2$, 18.27 nm for WS_2, and 31.11 nm for WSe_2, contribute to a significant increase in surface area and active sites [81]. This nanocrystalline feature is crucial for improving the charge transfer efficiency and stability of the TMDs when applied as HTLs in OPVs. The enhanced performance of MoS_2, notably its high PCE and superior stability under various lighting conditions, can be attributed to its nanocrystalline nature, which facilitates efficient charge transfer properties and effective charge carrier dynamics [82,83]. The size-dependent optical and electrical properties of these nanocrystalline TMDs underscore their suitability for integration into OPVs, wherein their ability to enhance light absorption and charge separation can significantly boost device performance.

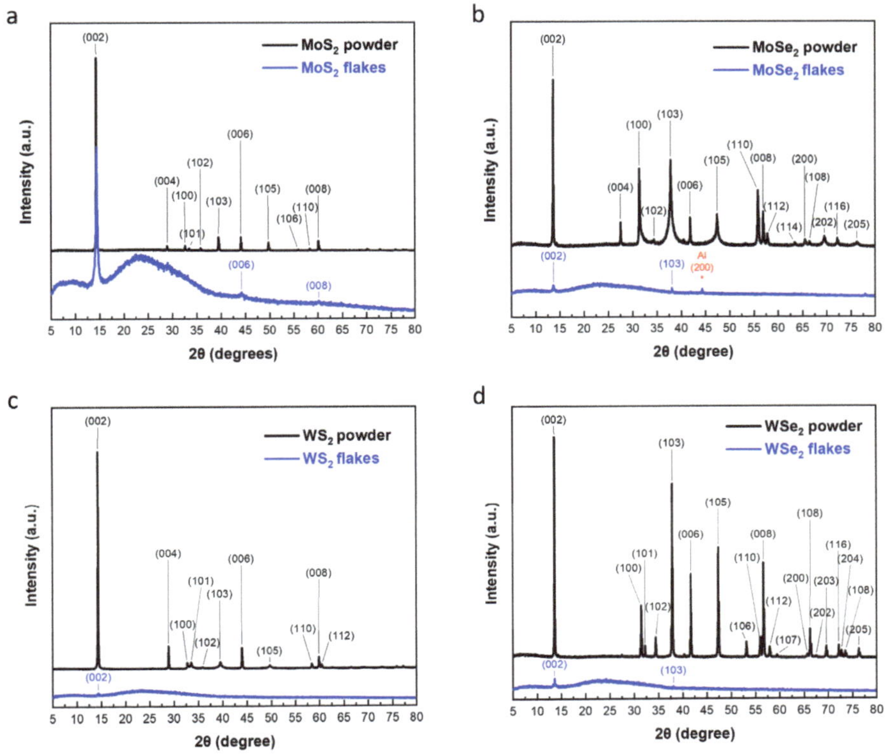

Figure 3. XRD patterns of the four TMDs (**a**) MoS_2; (**b**) $MoSe_2$; (**c**) WS_2; (**d**) WSe_2. The black line corresponds to the bulk material powder and the blue line corresponds to the sprayed TMD flakes.

3.2. Device Fabrication and Characterization

The inverted architecture followed for the development of the OPVs in our study is displayed in Figure 4a. A thin layer of the four TMDs was inserted via spray-coating between the AL and the metal anode, in order to fabricate the OPVs. The active layer that was used is PBDB-T:ITIC. The J-V curves in Figure 4b compare the devices fabricated without HTL, with MoO_3, and with the four TMDs MoS_2, $MoSe_2$, WS_2, and WSe_2 taken under 100 mW/cm^2 AM 1.5G illumination. Table 1 and Figure 4c provide a summary of the average values for short-circuit current density (Jsc), open-circuit voltage (Voc), fill factor (FF), and power conversion efficiency (PCE) for each device set. The absence of an HTL results in a lower PCE of 5.2 $\pm$ 0.14%, attributed to a diminished built-in electric field. This is evident from the moderate open-circuit voltage (Voc) and short-circuit current density (Jsc) values. The devices incorporating the commonly used MoO_3 as an HTL show a significant improvement, elevating the PCE to 9.68 $\pm$ 0.12%. This improvement underscores the pivotal role of an efficient HTL in photovoltaic devices, as evidenced by the increased Voc and Jsc values, along with a substantial FF. Shifting focus to the four TMDs, MoS_2, $MoSe_2$, WS_2, and WSe_2, as spray-coated replacement HTLs, each MX_2 demonstrates unique performance characteristics. MoS_2 emerges as a strong contender, closely rivaling the performance of MoO_3 with a PCE of 9.78 $\pm$ 0.10%. This suggests that MoS_2 is an effective alternative to traditional HTL materials. $MoSe_2$, while showing a higher Voc, yielded a PCE of 7.21 $\pm$ 0.41%, indicating a compromise in other performance metrics like Jsc and FF. WS_2 and WSe_2, although exhibiting comparable Voc values to MoS_2 and MoO_3, fall short in their overall performance, with PCEs of 7.76 $\pm$ 0.64% and 6.51 $\pm$ 0.62%, respectively. Their lower Jsc and FF values suggest that while they are viable HTL materials, their efficiency in converting solar energy is less optimal compared to MoS_2 and MoO_3.

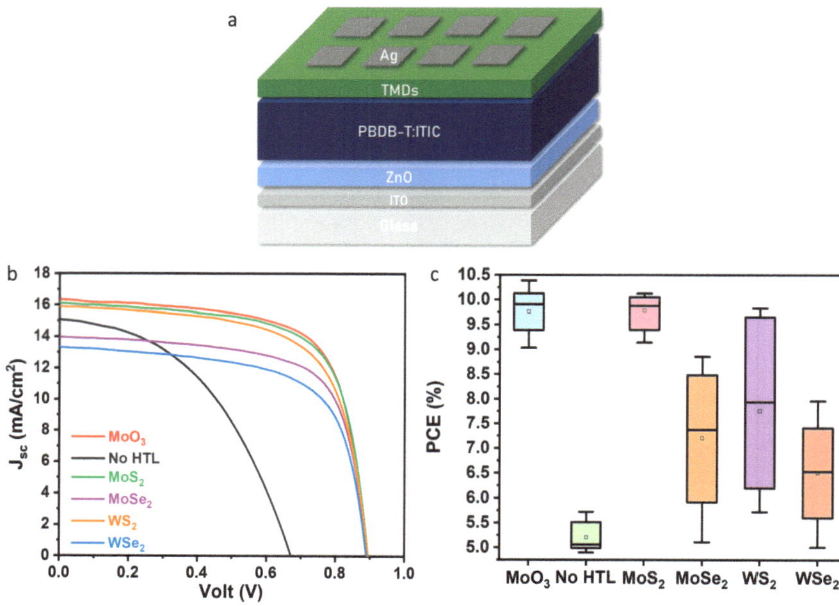

Figure 4. (**a**) The inverted OPV device architecture. (**b**) J-V curves of the best performing devices for each parameter. (**c**) Statistical distribution of PCE from 20 cells for each parameter.

Table 1. Average photovoltaic parameters from J-V characterization of 20 devices, with and without TMDs as the HTL. The boldface values denote the parameters of the best performing devices.

HTL	Voc (V)	Jsc (mA/cm^2)	FF (%)	PCE (%)
No HTL	0.73 ± 0.02 **(0.74)**	14.35 ± 0.31 **(14.63)**	48.87 ± 0.32 **(49.15)**	5.2 ± 0.14 **(5.34)**
MoO_3	0.88 ± 0.01 **(0.89)**	15.71 ± 0.1 **(16.36)**	69.27 ± 0.52 **(71.13)**	9.68 ± 0.12 **(10.38)**
MoS_2	0.89 ± 0.01 **(0.90)**	15.88 ± 0.09 **(16.07)**	69.08 ± 0.45 **(70.31)**	9.78 ± 0.10 **(10.12)**
$MoSe_2$	0.77 ± 0.05 **(0.89)**	14.25 ± 0.44 **(13.94)**	62.73 ± 3.34 **(71.04)**	7.21 ± 0.41 **(8.86)**
WS_2	0.86 ± 0.02 **(0.89)**	14.84 ± 0.55 **(16.06)**	60.42 ± 3.43 **(68.46)**	7.76 ± 0.64 **(9.82)**
WSe_2	0.8 ± 0.05 **(0.89)**	13.81 ± 0.56 **(13.29)**	59.03 ± 4.09 **(67.28)**	6.51 ± 0.62 **(7.96)**

To study the charge carrier dynamics and understand the efficiency of the various HTLs in the OPVs, we performed photoluminescence (PL) measurements (Figure S2). In the provided PL graph, MoO_3 is denoted with the lowest PL intensity among the tested HTL materials, suggesting a highly efficient charge separation. The TMDs also demonstrate quenched PL intensities in comparison to the device without an HTL, indicating enhanced charge transfer processes and potential photovoltaic performance benefits over the non-HTL baseline. This quenching effect is a positive attribute for the application of these materials in solar cell technologies.

3.3. Indoor Devices Fabrication and Characterization

Next, we tested the OPV cells under indoor lighting conditions. This step was crucial to further understand the real-world applicability of these cells in environments where artificial lighting is the primary light source. Specifically, we focused on the discrete illuminance steps of 200, 500, and 1000 lux, which are representative of conditions typical for indoor environments, like homes, offices, and commercial spaces. The corresponding light power intensities measured at these lux levels were 61, 151, and 303 μW/cm^2, respectively. The data in Figure 5 and Table 2 offer compelling insight into the performance of the devices under different indoor lighting conditions, particularly highlighting the efficacy of MoS_2 as an HTL, similar to the results seen before under AM 1.5G illumination. The efficacy of MoS_2 as a viable HTL, comparable to the commonly used MoO_3, is notably highlighted. At the outset, under the 200 lux condition, which simulates dim indoor environments, OPV cells without any HTL demonstrate a baseline PCE of 8.55%. When MoS_2 is employed as an HTL, there is a significant leap in performance, with the PCE reaching 17.29%, a figure that closely rivals the 17.97% achieved by cells using MoO_3. This parity in efficiency at a lower light intensity underscores the potential of MoS_2 as a strong alternative to MoO_3 in less illuminated settings. As the light intensity increases to 500 lux, representing moderately lit indoor spaces, the trend affirming the competitiveness of MoS_2 continues, as seen in Figure 5a. In this scenario, cells without an HTL show a modest improvement in PCE to 9.68%, but the introduction of MoS_2 as an HTL results in a substantial boost in efficiency to 19.16%, a value that nearly matches the 20.02% PCE of cells with MoO_3. This consistency in performance under medium light intensity further cements MoS_2's position as an effective and viable HTL option. At the highest light intensity of 1000 lux, akin to well-lit indoor areas, the efficiency of cells without an HTL reaches 10.83%. However, OPV cells incorporating MoS_2 as an HTL exhibit an impressive PCE of 22.26%, maintaining a close gap with the 22.34% PCE observed in cells using MoO_3. This sustained high performance of MoS_2 at elevated light levels highlights its capability to operate effectively across a broad spectrum of lighting conditions, making it a compelling alternative to the traditional MoO_3 HTL.

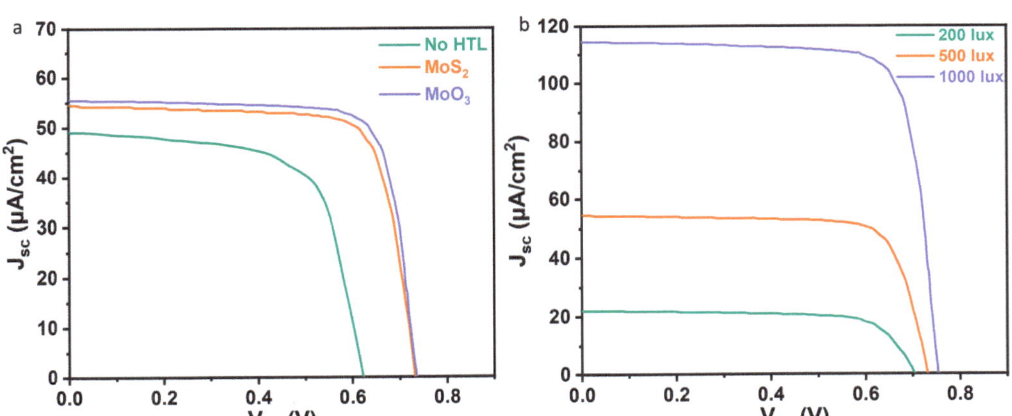

Figure 5. J-V curves of the best performing devices (**a**) comparing the different HTLs under 500 lux indoor lighting intensity and (**b**) comparing MoS_2 under different indoor light intensities.

Table 2. Parameters of the OPV cells with various HTLs under different indoor light intensities.

HTL	Light Source (lux)	P_{in} (μW/cm^2)	Voc (V)	Jsc (μA/cm^2)	FF (%)	PCE (%)
No HTL	200	61	0.58	20.04	44.51	8.55
	500	151	0.61	49.59	48.55	9.68
	1000	303	0.62	104.49	50.31	10.83
MoO_3	200	61	0.71	22.41	68.90	17.97
	500	151	0.73	55.45	74.74	20.02
	1000	303	0.75	116.84	77.28	22.34

Table 2. *Cont.*

HTL	Light Source (lux)	P_{in} (μW/cm^2)	Voc (V)	Jsc (μA/cm^2)	FF (%)	PCE (%)
	200	61	0.70	22.01	67.99	17.29
MoS_2	500	151	0.72	54.47	73.77	19.16
	1000	303	0.75	114.79	78.09	22.26

3.4. Device Stability

Finally, we evaluated the stability of MoS_2 by comparing it to the devices without HTL and MoO_3 by implementing the ISOS D1 protocol in unencapsulated devices. Figure 6 illustrates the comparative analysis of the normalized PCE over time, under aging conditions, for the OPVs with different HTLs. This analysis is valuable for understanding the longevity and stability of OPV cells under prolonged use. Initially, all HTL conditions start with a normalized PCE set at unity for ease of comparison. As the aging time progresses, there is a discernible decline in the normalized PCE for all conditions, which is indicative of the degradation phenomena typical in OPV cells. However, the rate and degree of degradation vary notably between the different HTLs. Cells without an HTL suffer the most pronounced degradation, with their efficiency dropping precipitously to about 30% of the initial value within 100 h, before entering a more gradual decline. Conversely, the OPV cells incorporating MoS_2 as an HTL exhibit remarkable stability. The normalized PCE of MoS_2-equipped cells shows a slight initial reduction but remains notably high, above 80%, even after several hundred hours of aging. This gentle slope indicates that MoS_2 significantly decelerates the degradation process, suggesting an inherent stability that is beneficial for the longevity of OPV cells. Intriguingly, while MoO_3 is known for its stability-enhancing properties, the graph indicates that over extended aging times, its stability is surpassed by that of MoS_2. MoO_3 begins with a strong retention of PCE but experiences a more pronounced degradation past the 300 h mark. The decline in the stability of MoO_3 could be attributable to several factors. MoO_3 is known to be sensitive to moisture and could undergo morphological and structural changes over time when exposed to environmental factors. Additionally, the interaction between MoO_3 and the photoactive layers may lead to interfacial reactions that compromise stability. In contrast, MoS_2's superior stability might be ascribed to its robustness against environmental factors and a more stable interfacial contact with the photoactive layer. Owing to its hydrophobic nature [84] and nonacidic and chemically stable properties [85] the layered structure of MoS_2 could provide a barrier from ambient moisture penetration, which is a critical factor in the aging of OPV cells. Moreover, the chemical stability of MoS_2 might contribute to fewer interfacial reactions, thereby preserving the integrity of the photoactive layer over time.

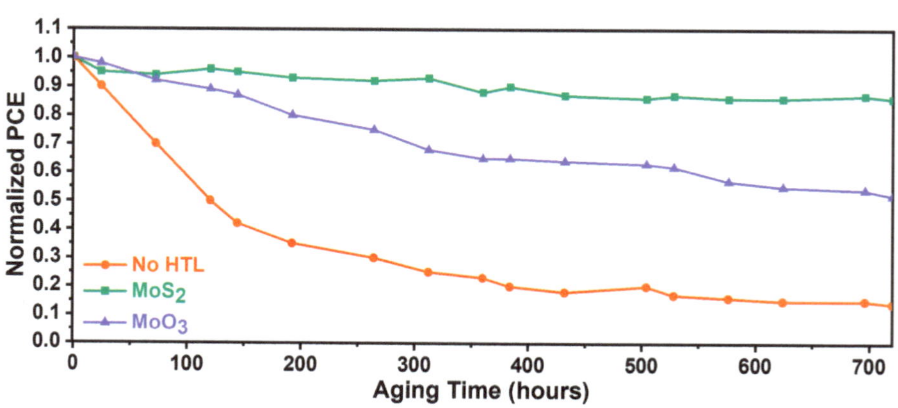

Figure 6. Normalized PCE measured for the different HTLs as a function of the aging time.

4. Conclusions

In this study, we studied exfoliated transition metal dichalcogenides (TMDs) and explored their applications as innovative hole transport layers (HTLs) in organic photovoltaics (OPVs), both under standard and indoor lighting conditions. Our analysis, supported by UV-visible absorption and Raman spectroscopy, confirmed the unique optical properties of MoS_2, $MoSe_2$, WS_2, and WSe_2, in line with the existing literature on their excitonic transitions and direct bandgap features. The integration of these TMDs into OPVs not only enhanced power conversion efficiency (PCE) compared to the devices without HTL in standard light conditions but also showed remarkable performance under indoor lighting, an aspect critical for real-world applications. Particularly noteworthy was the performance of MoS_2, which demonstrated its effectiveness as an HTL, rivaling traditional materials like MoO_3 in both standard and indoor lighting conditions. This adaptability of MoS_2, maintaining high PCE across a range of light intensities typical of indoor environments, underscores its potential for diverse photovoltaic applications. Moreover, this study revealed the superior stability of MoS_2 compared to MoO_3 over extended aging periods, suggesting its robustness against environmental factors and affirming its suitability for long-term usage in OPVs. These findings not only contribute significantly to the field of photovoltaic materials science but also pave the way for the development of more efficient and stable OPVs suitable for a variety of lighting conditions.

Supplementary Materials: The following supporting information can be downloaded at: https://www.mdpi.com/article/10.3390/nanoenergyadv4030014/s1, Figure S1: Raman spectra of LPE-prepared (a) MoS_2; (b) $MoSe_2$; (c) WS_2; (d) WSe_2 flakes; Figure S2: Photoluminescence (PL) spectra of the four TMDs deposited on BHJ on glass.

Author Contributions: Conceptualization, M.T., K.A. and E.K; methodology, M.T. and K.A.; validation, M.T., K.A., E.S. and C.P.; formal analysis, M.T., K.A. and C.P.; investigation, M.T., K.A., E.S. and C.P.; resources, E.K.; data curation, M.T., K.A. and C.P.; writing—original draft preparation, M.T., K.A.; writing—review and editing, M.T., K.A. and E.K.; visualization, M.T., K.A., E.S., C.P. and E.K.; supervision, E.K.; project administration, E.K.; funding acquisition, E.K. All authors have read and agreed to the published version of the manuscript.

Funding: This research project was supported by the Hellenic Foundation for Research and Innovation (H.F.R.I.) under the "2nd Call for H.F.R.I. Research Projects to support Faculty Members & Researchers" (Project Number: 02712).

Data Availability Statement: The original contributions presented in the study are included in the article/Supplementary Material, further inquiries can be directed to the corresponding author/s.

Conflicts of Interest: The authors declare no conflicts of interest.

References

1. Chamberlain, G.A. Organic Solar Cells: A Review. *Sol. Cells* **1983**, *8*, 47–83. [CrossRef]
2. Best Research-Cell Efficiency Chart. Available online: https://www.nrel.gov/pv/cell-efficiency.html (accessed on 3 December 2023).
3. Chander, N.; Singh, S.; Iyer, S.S.K. Stability and Reliability of P3HT:PC61BM Inverted Organic Solar Cells. *Sol. Energy Mater. Sol. Cells* **2017**, *161*, 407–415. [CrossRef]
4. Solak, E.K.; Irmak, E. Advances in Organic Photovoltaic Cells: A Comprehensive Review of Materials, Technologies, and Performance. *RSC Adv.* **2023**, *13*, 12244–12269. [CrossRef] [PubMed]
5. Firdaus, Y.; Le Corre, V.M.; Khan, J.I.; Kan, Z.; Laquai, F.; Beaujuge, P.M.; Anthopoulos, T.D. Key Parameters Requirements for Non-Fullerene-Based Organic Solar Cells with Power Conversion Efficiency >20%. *Adv. Sci.* **2019**, *6*, 1802028. [CrossRef]
6. Zheng, Z.; Hu, Q.; Zhang, S.; Zhang, D.; Wang, J.; Xie, S.; Wang, R.; Qin, Y.; Li, W.; Hong, L.; et al. A Highly Efficient Non-Fullerene Organic Solar Cell with a Fill Factor over 0.80 Enabled by a Fine-Tuned Hole-Transporting Layer. *Adv. Mater.* **2018**, *30*, 1801801. [CrossRef]
7. Zhou, H.; Zhang, Y.; Mai, C.-K.; Collins, S.D.; Nguyen, T.-Q.; Bazan, G.C.; Heeger, A.J. Conductive Conjugated Polyelectrolyte as Hole-Transporting Layer for Organic Bulk Heterojunction Solar Cells. *Adv. Mater.* **2014**, *26*, 780–785. [CrossRef]
8. He, Z.; Zhong, C.; Su, S.; Xu, M.; Wu, H.; Cao, Y. Enhanced Power-Conversion Efficiency in Polymer Solar Cells Using an Inverted Device Structure. *Nat. Photon* **2012**, *6*, 591–595. [CrossRef]
9. Sun, C.; Wu, Z.; Hu, Z.; Xiao, J.; Zhao, W.; Li, H.-W.; Li, Q.-Y.; Tsang, S.-W.; Xu, Y.-X.; Zhang, K.; et al. Interface Design for High-Efficiency Non-Fullerene Polymer Solar Cells. *Energy Environ. Sci.* **2017**, *10*, 1784–1791. [CrossRef]

10. Sun, Y.; Takacs, C.J.; Cowan, S.R.; Seo, J.H.; Gong, X.; Roy, A.; Heeger, A.J. Efficient, Air-Stable Bulk Heterojunction Polymer Solar Cells Using MoO_x as the Anode Interfacial Layer. *Adv. Mater.* **2011**, *23*, 2226–2230. [CrossRef]
11. de Jong, M.P.; van IJzendoorn, L.J.; de Voigt, M.J.A. Stability of the Interface between Indium-Tin-Oxide and Poly(3,4-Ethylenedioxythiophene)/Poly(Styrenesulfonate) in Polymer Light-Emitting Diodes. *Appl. Phys. Lett.* **2000**, *77*, 2255–2257. [CrossRef]
12. Wijeyasinghe, N.; Regoutz, A.; Eisner, F.; Du, T.; Tsetseris, L.; Lin, Y.-H.; Faber, H.; Pattanasattayavong, P.; Li, J.; Yan, F.; et al. Copper(I) Thiocyanate (CuSCN) Hole-Transport Layers Processed from Aqueous Precursor Solutions and Their Application in Thin-Film Transistors and Highly Efficient Organic and Organometal Halide Perovskite Solar Cells. *Adv. Funct. Mater.* **2017**, *27*, 1701818. [CrossRef]
13. Kyaw, A.K.K.; Sun, X.W.; Jiang, C.Y.; Lo, G.Q.; Zhao, D.W.; Kwong, D.L. An Inverted Organic Solar Cell Employing a Sol-Gel Derived ZnO Electron Selective Layer and Thermal Evaporated MoO_3 Hole Selective Layer. *Appl. Phys. Lett.* **2008**, *93*, 221107. [CrossRef]
14. Manders, J.R.; Tsang, S.-W.; Hartel, M.J.; Lai, T.-H.; Chen, S.; Amb, C.M.; Reynolds, J.R.; So, F. Solution-Processed Nickel Oxide Hole Transport Layers in High Efficiency Polymer Photovoltaic Cells. *Adv. Funct. Mater.* **2013**, *23*, 2993–3001. [CrossRef]
15. Yu, X.; Marks, T.J.; Facchetti, A. Metal Oxides for Optoelectronic Applications. *Nat. Mater* **2016**, *15*, 383–396. [CrossRef] [PubMed]
16. Cutting, C.L.; Bag, M.; Venkataraman, D. Indoor Light Recycling: A New Home for Organic Photovoltaics. *J. Mater. Chem. C* **2016**, *4*, 10367–10370. [CrossRef]
17. Arai, R.; Furukawa, S.; Hidaka, Y.; Komiyama, H.; Yasuda, T. High-Performance Organic Energy-Harvesting Devices and Modules for Self-Sustainable Power Generation under Ambient Indoor Lighting Environments. *ACS Appl. Mater. Interfaces* **2019**, *11*, 9259–9264. [CrossRef] [PubMed]
18. Lee, C.; Lee, J.-H.; Lee, H.H.; Nam, M.; Ko, D.-H. Over 30% Efficient Indoor Organic Photovoltaics Enabled by Morphological Modification Using Two Compatible Non-Fullerene Acceptors. *Adv. Energy Mater.* **2022**, *12*, 2200275. [CrossRef]
19. Bi, P.; An, C.; Zhang, T.; Chen, Z.; Xu, Y.; Cui, Y.; Wang, J.; Li, J.; Wang, Y.; Ren, J.; et al. Achieving 31% Efficiency in Organic Photovoltaic Cells under Indoor Light Using a Low Energetic Disorder Polymer Donor. *J. Mater. Chem. A* **2023**, *11*, 983–991. [CrossRef]
20. Polyzoidis, C.; Rogdakis, K.; Kymakis, E. Indoor Perovskite Photovoltaics for the Internet of Things—Challenges and Opportunities toward Market Uptake. *Adv. Energy Mater.* **2021**, *11*, 2101854. [CrossRef]
21. Chaves, A.; Azadani, J.G.; Alsalman, H.; da Costa, D.R.; Frisenda, R.; Chaves, A.J.; Song, S.H.; Kim, Y.D.; He, D.; Zhou, J.; et al. Bandgap Engineering of Two-Dimensional Semiconductor Materials. *Npj 2D Mater. Appl.* **2020**, *4*, 29. [CrossRef]
22. Tongay, S.; Zhou, J.; Ataca, C.; Lo, K.; Matthews, T.S.; Li, J.; Grossman, J.C.; Wu, J. Thermally Driven Crossover from Indirect toward Direct Bandgap in 2D Semiconductors: $MoSe_2$ versus MoS_2. *Nano Lett.* **2012**, *12*, 5576–5580. [CrossRef] [PubMed]
23. Wang, G.; Chernikov, A.; Glazov, M.M.; Heinz, T.F.; Marie, X.; Amand, T.; Urbaszek, B. Colloquium: Excitons in Atomically Thin Transition Metal Dichalcogenides. *Rev. Mod. Phys.* **2018**, *90*, 021001. [CrossRef]
24. Mak, K.F.; Shan, J. Photonics and Optoelectronics of 2D Semiconductor Transition Metal Dichalcogenides. *Nat. Photon* **2016**, *10*, 216–226. [CrossRef]
25. Wang, Q.H.; Kalantar-Zadeh, K.; Kis, A.; Coleman, J.N.; Strano, M.S. Electronics and Optoelectronics of Two-Dimensional Transition Metal Dichalcogenides. *Nat. Nanotechnol.* **2012**, *7*, 699–712. [CrossRef] [PubMed]
26. Pospischil, A.; Mueller, T. Optoelectronic Devices Based on Atomically Thin Transition Metal Dichalcogenides. *Appl. Sci.* **2016**, *6*, 78. [CrossRef]
27. Lin, Y.; Adilbekova, B.; Firdaus, Y.; Yengel, E.; Faber, H.; Sajjad, M.; Zheng, X.; Yarali, E.; Seitkhan, A.; Bakr, O.M.; et al. 17% Efficient Organic Solar Cells Based on Liquid Exfoliated WS_2 as a Replacement for PEDOT:PSS. *Adv. Mater.* **2019**, *31*, 1902965. [CrossRef] [PubMed]
28. Barrera, D.; Jawaid, A.; Daunis, T.B.; Cheng, L.; Wang, Q.; Lee, Y.-J.; Kim, M.J.; Kim, J.; Vaia, R.A.; Hsu, J.W.P. Inverted OPVs with MoS_2 Hole Transport Layer Deposited by Spray Coating. *Mater. Today Energy* **2017**, *5*, 107–111. [CrossRef]
29. Hoang Huy, V.P.; Ahn, Y.N.; Hur, J. Recent Advances in Transition Metal Dichalcogenide Cathode Materials for Aqueous Rechargeable Multivalent Metal-Ion Batteries. *Nanomaterials* **2021**, *11*, 1517. [CrossRef] [PubMed]
30. Frindt, R.F. Single Crystals of MoS_2 Several Molecular Layers Thick. *J. Appl. Phys.* **2004**, *37*, 1928–1929. [CrossRef]
31. Dines, M.B. Lithium Intercalation via N-Butyllithium of the Layered Transition Metal Dichalcogenides. *Mater. Res. Bull.* **1975**, *10*, 287–291. [CrossRef]
32. Yang, R.; Fan, Y.; Mei, L.; Shin, H.S.; Voiry, D.; Lu, Q.; Li, J.; Zeng, Z. Synthesis of Atomically Thin Sheets by the Intercalation-Based Exfoliation of Layered Materials. *Nat. Synth.* **2023**, *2*, 101–118. [CrossRef]
33. Hernandez, Y.; Nicolosi, V.; Lotya, M.; Blighe, F.M.; Sun, Z.; De, S.; McGovern, I.T.; Holland, B.; Byrne, M.; Gun'Ko, Y.K.; et al. High-Yield Production of Graphene by Liquid-Phase Exfoliation of Graphite. *Nat. Nanotechnol.* **2008**, *3*, 563–568. [CrossRef] [PubMed]
34. Paton, K.R.; Varrla, E.; Backes, C.; Smith, R.J.; Khan, U.; O'Neill, A.; Boland, C.; Lotya, M.; Istrate, O.M.; King, P.; et al. Scalable Production of Large Quantities of Defect-Free Few-Layer Graphene by Shear Exfoliation in Liquids. *Nat. Mater.* **2014**, *13*, 624–630. [CrossRef] [PubMed]
35. Ciesielski, A.; Samorì, P. Graphene via Sonication Assisted Liquid-Phase Exfoliation. *Chem. Soc. Rev.* **2013**, *43*, 381–398. [CrossRef] [PubMed]

36. Bonaccorso, F.; Bartolotta, A.; Coleman, J.N.; Backes, C. 2D-Crystal-Based Functional Inks. *Adv. Mater.* **2016**, *28*, 6136–6166. [CrossRef] [PubMed]
37. Suslick, K.S. Sonochemistry. *Science* **1990**, *247*, 1439–1445. [CrossRef]
38. Lauterborn, W.; Kurz, T.; Geisler, R.; Schanz, D.; Lindau, O. Acoustic Cavitation, Bubble Dynamics and Sonoluminescence. *Ultrason. Sonochem.* **2007**, *14*, 484–491. [CrossRef] [PubMed]
39. Finn, D.J.; Lotya, M.; Cunningham, G.; Smith, R.J.; McCloskey, D.; Donegan, J.F.; Coleman, J.N. Inkjet Deposition of Liquid-Exfoliated Graphene and MoS_2 Nanosheets for Printed Device Applications. *J. Mater. Chem. C* **2014**, *2*, 925–932. [CrossRef]
40. Mishra, A.K.; Lakshmi, K.V.; Huang, L. Eco-Friendly Synthesis of Metal Dichalcogenides Nanosheets and Their Environmental Remediation Potential Driven by Visible Light. *Sci. Rep.* **2015**, *5*, 15718. [CrossRef]
41. Dong, N.; Li, Y.; Feng, Y.; Zhang, S.; Zhang, X.; Chang, C.; Fan, J.; Zhang, L.; Wang, J. Optical Limiting and Theoretical Modelling of Layered Transition Metal Dichalcogenide Nanosheets. *Sci. Rep.* **2015**, *5*, 14646. [CrossRef]
42. Li, H.; Zhang, Q.; Yap, C.C.R.; Tay, B.K.; Edwin, T.H.T.; Olivier, A.; Baillargeat, D. From Bulk to Monolayer MoS_2: Evolution of Raman Scattering. *Adv. Funct. Mater.* **2012**, *22*, 1385–1390. [CrossRef]
43. Muscuso, L.; Cravanzola, S.; Cesano, F.; Scarano, D.; Zecchina, A. Optical, Vibrational, and Structural Properties of MoS_2 Nanoparticles Obtained by Exfoliation and Fragmentation via Ultrasound Cavitation in Isopropyl Alcohol. *J. Phys. Chem. C* **2015**, *119*, 3791–3801. [CrossRef]
44. Zhao, W.; Ghorannevis, Z.; Chu, L.; Toh, M.; Kloc, C.; Tan, P.-H.; Eda, G. Evolution of Electronic Structure in Atomically Thin Sheets of WS_2 and WSe_2. *ACS Nano* **2013**, *7*, 791–797. [CrossRef] [PubMed]
45. Wang, K.; Feng, Y.; Chang, C.; Zhan, J.; Wang, C.; Zhao, Q.; Coleman, J.N.; Zhang, L.; Blau, W.J.; Wang, J. Broadband Ultrafast Nonlinear Absorption and Nonlinear Refraction of Layered Molybdenum Dichalcogenide Semiconductors. *Nanoscale* **2014**, *6*, 10530–10535. [CrossRef] [PubMed]
46. Lee, C.; Yan, H.; Brus, L.E.; Heinz, T.F.; Hone, J.; Ryu, S. Anomalous Lattice Vibrations of Single- and Few-Layer MoS_2. *ACS Nano* **2010**, *4*, 2695–2700. [CrossRef] [PubMed]
47. Mignuzzi, S.; Pollard, A.J.; Bonini, N.; Brennan, B.; Gilmore, I.S.; Pimenta, M.A.; Richards, D.; Roy, D. Effect of Disorder on Raman Scattering of Single-Layer MoS_2. *Phys. Rev. B* **2015**, *91*, 195411. [CrossRef]
48. Korn, T.; Heydrich, S.; Hirmer, M.; Schmutzler, J.; Schüller, C. Low-Temperature Photocarrier Dynamics in Monolayer MoS_2. *Appl. Phys. Lett.* **2011**, *99*, 102109. [CrossRef]
49. Niu, Y.; Gonzalez-Abad, S.; Frisenda, R.; Marauhn, P.; Drüppel, M.; Gant, P.; Schmidt, R.; Taghavi, N.S.; Barcons, D.; Molina-Mendoza, A.J.; et al. Thickness-Dependent Differential Reflectance Spectra of Monolayer and Few-Layer MoS_2, $MoSe_2$, WS_2 and WSe_2. *Nanomaterials* **2018**, *8*, 725. [CrossRef] [PubMed]
50. Tonndorf, P.; Schmidt, R.; Böttger, P.; Zhang, X.; Börner, J.; Liebig, A.; Albrecht, M.; Kloc, C.; Gordan, O.; Zahn, D.R.T.; et al. Photoluminescence Emission and Raman Response of Monolayer MoS_2, $MoSe_2$, and WSe_2. *Opt. Express* **2013**, *21*, 4908–4916. [CrossRef]
51. Nam, D.; Lee, J.-U.; Cheong, H. Excitation Energy Dependent Raman Spectrum of $MoSe_2$. *Sci. Rep.* **2015**, *5*, 17113. [CrossRef]
52. Berkdemir, A.; Gutiérrez, H.R.; Botello-Méndez, A.R.; Perea-López, N.; Elías, A.L.; Chia, C.-I.; Wang, B.; Crespi, V.H.; López-Urías, F.; Charlier, J.-C.; et al. Identification of Individual and Few Layers of WS_2 Using Raman Spectroscopy. *Sci Rep* **2013**, *3*, 1755. [CrossRef]
53. McCreary, A.; Berkdemir, A.; Wang, J.; Nguyen, M.A.; Elías, A.L.; Perea-López, N.; Fujisawa, K.; Kabius, B.; Carozo, V.; Cullen, D.A.; et al. Distinct Photoluminescence and Raman Spectroscopy Signatures for Identifying Highly Crystalline WS_2 Monolayers Produced by Different Growth Methods. *J. Mater. Res.* **2016**, *31*, 931–944. [CrossRef]
54. Sahin, H.; Tongay, S.; Horzum, S.; Fan, W.; Zhou, J.; Li, J.; Wu, J.; Peeters, F.M. Anomalous Raman Spectra and Thickness-Dependent Electronic Properties of WSe_2. *Phys. Rev. B* **2013**, *87*, 165409. [CrossRef]
55. Harrington, G.F.; Santiso, J. Back-to-Basics Tutorial: X-Ray Diffraction of Thin Films. *J. Electroceram.* **2021**, *47*, 141–163. [CrossRef]
56. Khan, H.; Yerramilli, A.S.; D'Oliveira, A.; Alford, T.L.; Boffito, D.C.; Patience, G.S. Experimental Methods in Chemical Engineering: X-Ray Diffraction Spectroscopy—XRD. *Can. J. Chem. Eng.* **2020**, *98*, 1255–1266. [CrossRef]
57. Nguyen, T.P.; Sohn, W.; Oh, J.H.; Jang, H.W.; Kim, S.Y. Size-Dependent Properties of Two-Dimensional MoS_2 and WS_2. *J. Phys. Chem. C* **2016**, *120*, 10078–10085. [CrossRef]
58. Wu, Z.; Wang, D.; Wang, Y.; Sun, A. Preparation and Tribological Properties of MoS_2 Nanosheets. *Adv. Eng. Mater.* **2010**, *12*, 534–538. [CrossRef]
59. Ratha, S.; Rout, C.S. Supercapacitor Electrodes Based on Layered Tungsten Disulfide-Reduced Graphene Oxide Hybrids Synthesized by a Facile Hydrothermal Method. *ACS Appl. Mater. Interfaces* **2013**, *5*, 11427–11433. [CrossRef]
60. Luo, X.; Huang, J.; Li, J.; Cao, L.; Wang, Y.; Xu, Z.; Guo, L.; Cheng, Y.; Kajiyoshi, K.; Chen, S. Controlled WS_2 Crystallinity Effectively Dominating Sodium Storage Performance. *J. Energy Chem.* **2020**, *51*, 143–153. [CrossRef]
61. Gupta, D.; Chauhan, V.; Upadhyay, S.; Koratkar, N.; Singh, F.; Kumar, S.; Mahajan, A.; Chandra, R.; Kumar, R. Defects Engineering and Enhancement in Optical and Structural Properties of 2D-MoS_2 Thin Films by High Energy Ion Beam Irradiation. *Mater. Chem. Phys.* **2022**, *276*, 125422. [CrossRef]
62. Krishnamoorthy, K.; Pazhamalai, P.; Veerasubramani, G.K.; Kim, S.J. Mechanically Delaminated Few Layered MoS_2 Nanosheets Based High Performance Wire Type Solid-State Symmetric Supercapacitors. *J. Power Sources* **2016**, *321*, 112–119. [CrossRef]

63. Liu, Y.; Zhu, M.; Chen, D. Sheet-like $MoSe_2/C$ Composites with Enhanced Li-Ion Storage Properties. *J. Mater. Chem. A* **2015**, *3*, 11857–11862. [CrossRef]
64. Zhang, X.; Wang, J.; Xu, H.; Tan, H.; Ye, X. Preparation and Tribological Properties of WS_2 Hexagonal Nanoplates and Nanoflowers. *Nanomaterials* **2019**, *9*, 840. [CrossRef] [PubMed]
65. Wang, X.; Chen, Y.; Zheng, B.; Qi, F.; He, J.; Li, Q.; Li, P.; Zhang, W. Graphene-like WSe_2 Nanosheets for Efficient and Stable Hydrogen Evolution. *J. Alloys Compd.* **2017**, *691*, 698–704. [CrossRef]
66. Rahul; Arora, S.K. Quantitative Evaluation of Nonlinear Temperature Dependence of Raman Shift in Exfoliated WSe_2 Nanosheets. *J. Electron. Mater.* **2021**, *50*, 7126–7132. [CrossRef]
67. Ni, P.; Dieng, M.; Vanel, J.-C.; Florea, I.; Bouanis, F.Z.; Yassar, A. Liquid Shear Exfoliation of MoS_2: Preparation, Characterization, and NO2-Sensing Properties. *Nanomaterials* **2023**, *13*, 2502. [CrossRef]
68. Chen, I.-W.P.; Lai, Y.-M.; Liao, W.-S. One-Pot Synthesis of Chlorophyll-Assisted Exfoliated MoS_2/WS_2 Heterostructures via Liquid-Phase Exfoliation Method for Photocatalytic Hydrogen Production. *Nanomaterials* **2021**, *11*, 2436. [CrossRef] [PubMed]
69. Zhang, X.; Xue, M.; Yang, X.; Luo, G.; Yang, F. Hydrothermal Synthesis and Tribological Properties of $MoSe_2$ Nanoflowers. *Micro Nano Lett.* **2015**, *10*, 339–342. [CrossRef]
70. Ayieko, C.O.; Musembi, R.J.; Ogacho, A.A.; Aduda, B.O.; Muthoka, B.M.; Jain, P.K. Controlled Texturing of Aluminum Sheet for Solar Energy Applications. *Adv. Mater. Phys. Chem.* **2015**, *5*, 458–466. [CrossRef]
71. Mikhalitsyna, E.A.; Kataev, V.A.; Larrañaga, A.; Lepalovskij, V.N.; Kurlyandskaya, G.V. Nanocrystallization in FINEMET-Type $Fe_{73.5}Nb_3Cu_1Si_{13.5}B_9$ and $Fe_{72.5}Nb_{1.5}Mo_2Cu_{1.1}Si_{14.2}B_{8.7}$ Thin Films. *Materials* **2020**, *13*, 348. [CrossRef]
72. Amjad, R.J.; Sahar, M.R.; Ghoshal, S.K.; Dousti, M.R.; Riaz, S.; Samavati, A.R.; Jamaludin, M.N.A.; Naseem, S. Plasmon-Enhanced Upconversion Fluorescence in Er^{3+}:Ag Phosphate Glass: The Effect of Heat Treatment. *Chin. Phys. Lett.* **2013**, *30*, 027301. [CrossRef]
73. Wang, S.; Wang, Z.; Qin, J.; Wang, W.; Li, W.; He, D. Nanocrystalline MoS_2 through Directional Growth along the (002) Crystal Plane under High Pressure. *Mater. Chem. Phys.* **2011**, *130*, 170–174. [CrossRef]
74. Zsigmondy, R.; Scherrer, P. *Kolloidchemie: Ein Lehrbuch—Primary Source Edition*; NABU Press: Berlin, Germany, 2014; ISBN 978-1-294-87694-6.
75. Muniz, F.T.L.; Miranda, M.A.R.; Morilla dos Santos, C.; Sasaki, J.M. The Scherrer Equation and the Dynamical Theory of X-ray Diffraction. *Acta Cryst. A* **2016**, *72*, 385–390. [CrossRef] [PubMed]
76. Monshi, A.; Foroughi, M.R.; Monshi, M.R. Modified Scherrer Equation to Estimate More Accurately Nano-Crystallite Size Using XRD. *World J. Nano Sci. Eng.* **2012**, *2*, 154–160. [CrossRef]
77. Huang, Y.; Liu, L.; Sha, J.; Chen, Y. Size-Dependent Piezoelectricity of Molybdenum Disulfide (MoS_2) Films Obtained by Atomic Layer Deposition (ALD). *Appl. Phys. Lett.* **2017**, *111*, 063902. [CrossRef]
78. Saisopa, T.; Jitapunkul, K.; Bunpheng, A.; Nakajima, H.; Supruangnet, R.; Busayaporn, W.; Sukprom, T.; Hirunpinyopas, W.; Seubsai, A.; Songsiriritthigul, P.; et al. The Structure Analysis and Chemical Properties Probing during Recycling Processes of Transition Metal Dichalcogenides Exfoliation. *Electrochim. Acta* **2023**, *449*, 142171. [CrossRef]
79. Chowdhury, T.; Kim, J.; Sadler, E.C.; Li, C.; Lee, S.W.; Jo, K.; Xu, W.; Gracias, D.H.; Drichko, N.V.; Jariwala, D.; et al. Substrate-Directed Synthesis of MoS_2 Nanocrystals with Tunable Dimensionality and Optical Properties. *Nat. Nanotechnol.* **2020**, *15*, 29–34. [CrossRef] [PubMed]
80. Voiry, D.; Yang, J.; Chhowalla, M. Recent Strategies for Improving the Catalytic Activity of 2D TMD Nanosheets Toward the Hydrogen Evolution Reaction. *Adv. Mater.* **2016**, *28*, 6197–6206. [CrossRef]
81. Kadiev, K.M.; Maximov, A.L.; Kadieva, M.K. The Effect of MoS_2 Active Site Dispersion on Suppression of Polycondensation Reactions during Heavy Oil Hydroconversion. *Catalysts* **2021**, *11*, 676. [CrossRef]
82. Mukherjee, S.; Biswas, S.; Ghorai, A.; Midya, A.; Das, S.; Ray, S.K. Tunable Optical and Electrical Transport Properties of Size- and Temperature-Controlled Polymorph MoS_2 Nanocrystals. *J. Phys. Chem. C* **2018**, *122*, 12502–12511. [CrossRef]
83. Mukherjee, S.; Maiti, R.; Midya, A.; Das, S.; Ray, S.K. Tunable Direct Bandgap Optical Transitions in MoS_2 Nanocrystals for Photonic Devices. *ACS Photonics* **2015**, *2*, 760–768. [CrossRef]
84. Fang, L.J.; Chen, J.H.; Wang, J.M.; Lin, W.W.; Lin, X.G.; Lin, Q.J.; He, Y. Hydrophobic Two-Dimensional MoS_2 Nanosheets Embedded in a Polyether Copolymer Block Amide (PEBA) Membrane for Recovering Pyridine from a Dilute Solution. *ACS Omega* **2021**, *6*, 2675–2685. [CrossRef] [PubMed]
85. Kwon, K.C.; Kim, C.; Le, Q.V.; Gim, S.; Jeon, J.-M.; Ham, J.Y.; Lee, J.-L.; Jang, H.W.; Kim, S.Y. Synthesis of Atomically Thin Transition Metal Disulfides for Charge Transport Layers in Optoelectronic Devices. *ACS Nano* **2015**, *9*, 4146–4155. [CrossRef] [PubMed]

energies

Article

Assessment of Cytotoxicity and Genotoxicity Induced by Diesel Exhaust Particles (DEPs) on Cell Line A549 and the Potential Role of Amide-Functionalized Carbon Nanotubes as Fuel Additive

Juan Sebastian Pino [1], Pedro Nel Alvarado [2], Winston Rojas [1], Karen Cacua [2] and Natalia Gomez-Lopera [3,*]

1 Grupo de Genética Molecular, Facultad de Ciencias Exactas y Naturales, Universidad de Antioquia, Medellín 050010, Colombia; juan.pino@udea.edu.co (J.S.P.)
2 Grupo de Investigación Materiales Avanzados y Energía, Facultad de Ingenierías, Instituto Tecnológico Metropolitano, Medellín 050010, Colombia
3 Grupo de Genética Médica, Facultad de Medicina, Universidad de Antioquia, Medellín 050010, Colombia
* Correspondence: natalia.gomezl@udea.edu.co

Abstract: Epidemiological studies have consistently linked air pollution to severe health risks. One strategy to reduce the impact of combustion products from engines is adding additives to the fuel. Potential benefits have been observed in terms of performance and emissions, as well as in decreasing fuel consumption. However, the associated emission of particulate matter into the environment may have unforeseen health effects. This study examines the effects of diesel exhaust particles (DEPs) from diesel fuel mixed with amide-functionalized carbon nanotubes (CNTF). The aim is to analyze the properties of DEPs and determine their toxic effects on lung cells. The DEPs were characterized using scanning and transmission electron microscopy, while the polycyclic aromatic hydrocarbons (PAHs) were analyzed through gas chromatography. Various assays were conducted to assess cell viability, apoptosis, oxidative stress, and DNA damage. The addition of CNTF to diesel fuel altered the morphology and size of the particles, as well as the quantity and composition of PAHs. At the cellular level, diesel DEPs induce higher levels of reactive oxygen species (ROS) production, DNA damage, apoptosis, and cytotoxicity compared to both CNTF and diesel–CNTF DEPs. These findings suggest that the nano-additives enhance energy efficiency by reducing pollutants without significantly increasing cell toxicity.

Keywords: functionalized carbon nanotubes; diesel exhaust particles; human epithelial cells; cytotoxicity; apoptosis; DNA damage

Citation: Pino, J.S.; Alvarado, P.N.; Rojas, W.; Cacua, K.; Gomez-Lopera, N. Assessment of Cytotoxicity and Genotoxicity Induced by Diesel Exhaust Particles (DEPs) on Cell Line A549 and the Potential Role of Amide-Functionalized Carbon Nanotubes as Fuel Additive. *Energies* **2024**, *17*, 4646. https://doi.org/10.3390/en17184646

Academic Editors: Robert A. Varin and Geng Zhong

Received: 8 June 2024
Revised: 26 June 2024
Accepted: 27 June 2024
Published: 18 September 2024

1. Introduction

Epidemiological studies have shown that exposure to air pollution is associated with serious health risks. Cardiovascular diseases, respiratory diseases, and various types of cancer are among the health issues empirically linked to air pollution [1]. The 2015 Global Burden of Disease Study revealed that outdoor exposure to fine particles (PM2.5) is the fifth leading risk factor for death worldwide, causing 4.2 million deaths that year. Moreover, the World Health Organization (WHO) reports that 92% of the world's population breathes air with dangerously high levels of pollutants [2,3]. Prolonged exposure to these fine particles can lead to lung inflammation, reduced lung function, and the exacerbation of pre-existing respiratory and cardiovascular diseases [4,5].

The transportation sector is responsible for a significant amount of particulate matter (PM) emissions, especially from compression-ignition (CI) engines that predominantly use diesel fuel [6,7]. One potential strategy to improve air quality worldwide and reduce PM emissions is incorporating nanomaterials into diesel fuel. Nanomaterials have exceptional properties, such as a large surface area, excellent stability, high catalytic performance, rapid oxidation rate, and high combustion heat [8]. When mixed with biofuels or liquid diesel, these characteristics result in better engine performance, optimized combustion, and reduced PM emissions [9,10]. A recent comprehensive review [11] shows that CuO, Al_2O_3, multi-walled carbon nanotubes (MWCNT), CeO_2, GO, carbon nanotubes (CNT), and TiO_2 are the most utilized nano-additives in diesel–biodiesel fuel blends. These additives have demonstrated significant improvements in combustion processes. It has been reported that adding CeO_2 nanoparticles to diesel fuel, whose use is increasing in the United States and Europe, reduced fuel consumption by 5% to 8% and the release of PM and unburned hydrocarbons by up to 15% [12].

Nevertheless, the adverse health effects of releasing CeO_2 and other metallic nanoparticles into the environment have been thoroughly documented. Both in vivo and in vitro studies have demonstrated that these nanoparticles primarily impact the respiratory, nervous, endocrine, immune, and reproductive systems [13]. Extensive research has been conducted to understand nanoparticle toxicity mechanisms and identify the contributing factors to their adverse effects. These factors include the generation of reactive oxygen species (ROS), mitochondrial impairment, inflammation, apoptosis, DNA damage, alterations in cell cycle regulation, and changes in epigenetic control [14,15]. For example, research indicates that exposure to TiO_2 nanoparticles leads to the activation of microglia, the production of ROS, and the activation of signaling pathways involved in inflammation and cell death [16,17]. CeO_2 nanoparticles are associated with various risks, such as the induction of cytotoxicity, oxidative stress, and pulmonary inflammation [18]. Therefore, their use should be limited and controlled. Evidence also suggests that exposure to aluminum nanoparticles can lead to increased oxidative stress, inflammatory events, and the degradation of the blood–brain barrier [19].

A promising avenue to mitigate health risks is using nano-additives that do not contain metals in their chemical structure, such as carbonaceous nano-additives. These additives offer comparable benefits in fuel efficiency and pollutant reduction to metallic additives [20]. CNTs are among the most commonly used nanomaterials as additives to diesel and biodiesel fuel. CNTs are commonly produced by the chemical vapor deposition method (CVD) [21] and have remarkable mechanical, electrical, optical, and thermal properties. However, CNTs often agglomerate and exhibit poor dispersion in various fluids and matrices. To address this issue, covalent and non-covalent functionalization methods have been employed to attach functional groups to the surface of CNTs. Covalent functionalization involves creating a covalent bond between functional entities and the carbon framework of the nanotubes, while non-covalent functionalization relies on supramolecular complexation utilizing forces such as van der Waals forces, hydrogen bonds, electrostatic forces, and π-stacking interactions [22]. In the case of covalent functionalization with amide groups, these can be conducted using oxidized CNT and amines via the solvent method [23] or using an amine treatment in the gas phase at a vacuum pressure. This approach demonstrates rapid and efficient results [24]. Acevedo et al. assessed the impact of functionalizing pristine carbon nanotubes with amide groups and evaluated their stability in commercial diesel. They found that the functional groups on the surface of the carbon nanotubes allowed for the stable dispersion to last up to 70 days [20]. This nanomaterial was additionally evaluated in a stationary compression engine. The results showed that the use of CNTF as an additive to diesel decreased ignition delay and PM emissions [25]. Table 1 shows the effects of adding functionalized carbon nanotubes on the properties of the fuel and engine performance, compared to the properties without the addition of nanotubes.

Table 1. Comparison of the properties of the fuel and engine performance with different CNTF additives.

Functionalized Carbon Nanotubes	Concentration in Diesel Fuel	Effect on Diesel Fuel Properties	Effect on Diesel Engine	Ref.
N-doped MWCNTs	20, 50, 75, and 100 ppm	Kinematic viscosity decreased by 10.6% on average There is a slight decrease in density.	Increased cylinder pressure (except at 25 ppm) and heat release rate (13.87% for 100 ppm); decreased ignition delay and reduced NOx, CO, UHC, PM, and smoke.	[26]
Amide functionalized MWCNTs	50 and 100 mg/L (in a blend with 10% vol. biodiesel)	There is no significant effect on kinematic viscosity and heating value at 100 mg/L. Increase in cetane index by 3.57% at 100 mg/L.	No impact on thermal efficiency. Increase in CO emissions and decrease in NO and PM by up to 29.4% and 39.8%, respectively.	[25]
MWCNT-OH	30, 60, and 90 ppm (in a blend of 90% diesel, 5% vol. biodiesel, and 4% or 8% vol. ethanol)	Increase in viscosity and density.	Increases in torque and power and decreases in specific fuel consumption. Decreases in CO, UHC (unburned hydrocarbons), and increases in NOx emissions.	[27]
MWCNT-COOH	30, 60, and 90 ppm	For all MWCNT-COOH concentrations (with 5% biodiesel), kinematic viscosity and density slightly decreased.	Increases in power by 13.01% and decrease in specific fuel consumption by 9.48%. CO and UHC decreased by up to 10.26% and 16.33%, respectively.	[28]
NH2-MWCNTs	25, 50, 75, and 100 ppm	Kinematic viscosity decreased (an average drop of 8.95%), while density was only slightly reduced.	-There was a 13.76% increase in the HHR (100 ppm). -At 50% load, there was a 5.19% decrease in brake-specific fuel consumption and a 5.47% increase in brake thermal efficiency (100 ppm). -At 75% load, CO emissions have decreased by 18.28% (75 ppm) and NOx by 28.97% (25 ppm); there has also been a 14.12% decrease in soot (75 ppm).	[29]
MWCNTs-COOH	30, 60, and 90 ppm	Kinematic viscosity and density were only slightly reduced.	Average power increased by 13.01%; Average torque increased by 17.61%; CO decreased by 10.26%; and UHC by 16.33%.	[28]

Beyond the advantages of engine combustion, surface functionalization, both covalent and non-covalent, has enhanced the biocompatibility of carbon nanotubes [30]. CNTF typically exhibits superior hydrophobicity and dispersion in biological media, lower immunogenicity, a reduced tendency to agglomerate, and greater biological compatibility compared to non-functionalized CNT [31,32]. However, available in vitro studies on the toxicity of CNTF yield controversial results. Several in vitro studies have shown that cytotoxic effects are lower in CNTF than in non-functionalized ones. For instance, CNTF with carboxyl, phosphorylcholine, and polystyrene groups induces lower production of cytokines, lung inflammation, and fibrosis compared to non-functionalized MWCNT [33]. However, it is important to note that some studies indicate that the functional group, such as acid functionalization, can significantly affect cellular toxicity, causing oxidative stress, damage to lysosomes, mitochondrial dysfunction, inflammation, and genotoxicity in terms of DNA damage [34–36]. The conflicting results mentioned above show that the toxicity of MWCNT depends on its physical and chemical properties, such as purity, chemical composition, size, surface area, and functionalization type.

Despite the numerous reported advantages of using nano-additives in diesel engine combustion, few studies have explored the direct correlation between the emissions of these particles and their effects on health. Currently, there are no reports on the biological effects of amide-CNTF after the combustion process. Therefore, despite the recognized benefits of adding amide-CNTF to reduce pollutants, as shown in Table 1, and improve the properties of the fuels for long-term dispersion, it is essential to evaluate their potential toxic effects and identify the physicochemical properties responsible for their toxicity. This study aims to fill this knowledge gap by examining the cytotoxic and genotoxic effects of diesel exhaust particles (DEP) resulting from using stable mixtures of amide and CNTF in diesel fuel.

2. Materials and Methods

The pristine carbon nanotubes (CNTs) obtained from SkySpring Nanomateriales, Inc. (ref 0554CA) were oxidized using a 1:1 volume ratio mixture of H_2SO_4 (96%, Merck, Rahway, NJ, USA) and HNO_3 (65%, Merck). Acylation was carried out using thionyl chloride (≥99%, Merck) and *N,N*-dimethylformamide (99.8%, Fischer Scientific, Hampton, NH, USA), and amide functionalization was conducted using technical-grade oleylamine. The amide-functionalized carbon nanotubes (CNTF) were obtained following the experimental methodology established by [23] and adapted by [20,25].

The CNTFs were introduced into Colombian commercial diesel. Prior to the experiments, the fuels underwent filtration using a 6 μm filter paper, as no fuel filter was installed before the engine inlet. The concentration of CNTF in the mixtures was 100 mg/L. The CNTF, in the form of a dry powder, was dispersed in diesel. Ultrasound probe dispersion (model EP225-DR, Hielscher, Ringwood, NJ, USA) was utilized to disperse the CNTF.

2.1. DEPs Collection

The DEPs collection utilized a stationary compression ignition engine connected to a synchronous generator (Figure 1). The engine tests were carried out under an engine load of 50% (6 kW) at a constant speed of 1800 rpm in a three-cylinder diesel engine (Yanmar, 3TNE84, Yanmar Holdings Co., Ltd., Tokio, Japan). An electrical resistance test rig was employed to simulate this load.

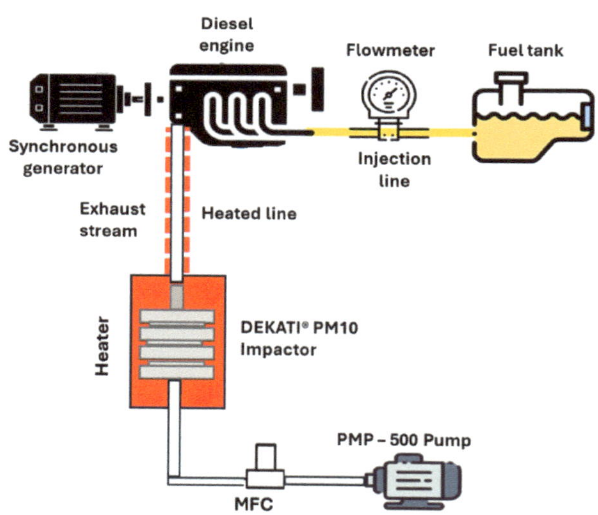

Figure 1. Diesel exhaust particles sampling in a non-road engine.

DEPs were collected using a DEKATI cascade impactor sampler (PM10 impactor) attached to the engine's exhaust gasses. The DEKATI impactor is based on the inertial classification of aerosol particles. The sample collection process involved a heated sampling probe, a particle impactor, an impactor heating zone, a pump for extracting combustion products, and a volumetric flow controller. The sampling line was heated meticulously to a temperature of 180 °C to avoid vapor condensation. The volumetric flow was maintained at a constant rate of 10 L/min. Figure 1 demonstrates the sampling process. Before collecting DEPs, the engine was warmed up for 10 min to ensure that the operating conditions were in a steady state and to avoid high DEP emissions associated with a cold engine start [37,38]. The DEPs were sampled to obtain particles with an aerodynamic size of less than 2.5 μm.

This size of particle has been classified as a class I carcinogen by the International Agency for Research on Cancer due to its negative impact on health, attributed to its small size and chemical composition [39,40]. This justification supports the selection of this particle size for the present study.

2.2. Physicochemical Characterization of DEPs

The structure of DEPs collected from the test engines underwent examination through transmission electron microscopy (TEM) and scanning electron microscopy (SEM) to characterize their carbon composition and particle size distribution, following the procedures reported by [41,42]. A small volume (3 μL) of each DEP solution was deposited onto a copper grid coated with a polymer support film (Structure Probe, Inc., West Chester, PA, USA) for TEM analysis. Following solvent evaporation in a vacuum environment, the particles' dimensions and morphology were evaluated using a JEOL 1200EX transmission electron microscope (JEOL, Toyko, Japan) at an acceleration voltage of 120 kV. Images were recorded using an 11-megapixel SIS Morada CCD camera and processed using AnalySIS software V2. The histograms were generated by measuring primary particle diameters derived from TEM images. These images were subjected to analysis using the public-domain image processing software ImageJ V. 1.54j. Primary particles with clearly defined boundaries within the PM agglomerates or non-agglomerated particles were identified from the TEM micrographs, and their respective diameters were determined. Specifically, 200 particles were utilized for the diesel mode, while 160 primary particles were employed for the diesel–CNTF. The frequency histograms were generated through the measurement of primary particle diameters derived from TEM images. These images were subjected to analysis using the public-domain image processing software ImageJ. Primary particles with clearly defined boundaries within the PM agglomerates or non-agglomerated particles were identified from the TEM micrographs, and their respective diameters were determined. Specifically, 200 particles were utilized for the diesel mode, while 160 primary particles were employed for the diesel–CNTF.

For SEM analysis, a 6 mm diameter segment of the filters containing the DEP samples was affixed to a dedicated holder using Quick Drying Silver Paint conductive adhesive (Agar, London, UK). Subsequently, the samples underwent a thin gold coating (approximately 10 nm) using a vacuum evaporator (JEE-4X, JEOL, Japan) to ensure conductivity and safeguard against heat-induced damage. Coated samples were inspected using an HRSEM-AURIGA scanning electron microscope (Zeiss, Oberkochen, Germany) operating in secondary electron (SE) mode. The microscope's settings were configured with a high voltage (HV) of 10 kV and a working distance (WD) of 20 mm.

The chemical analysis of DEPs was made to determine the PAH composition of diesel and diesel–CNTF fuels. Filters from diesel and diesel–CNTF and a control filter were prepared for PAH analysis. Each filter received an internal standard (Pentachloronitrobenzene) and 5 mL of dichloromethane, then stood for 24 h after sealing and shaking. Following 10 min of ultrasound treatment, the filters were removed, and dichloromethane was evaporated to 0.1 mL using nitrogen. Acetone (1 mL) was added, and samples were transferred to vials for PAH analysis. A calibration curve was created for 16 PAH compounds: naphthalene (Naf), acenaphthylene (Acyl), acenaphthene (Ac), fluorene (Fl), pentachloronitrobenzene (PCNB), phenanthrene (Fen), anthracene (Ant), fluoranthene (Flu), pyrene (Pir), benz(a)anthracene (BaA), chrysene (Ch), benzo(k)fluoranthene (Bkf), benzo(a)pyrene (BaP), benzo(b)fluoranthene (BbF), indeno(1,2,3-cd)pyrene (IP), dibenz(a,h)anthracene (DahA), and benzo(g,h,i)perylene (Bper).

2.3. Preparation of Functionalized Nanoparticle and DEP Samples

To prepare a stock solution of 0.5 mg/mL, 0.5 mg of CNTF was weighed and resuspended in 5 mL of PBS, and 1 μL of DMSO was added as a surfactant to facilitate dispersion. This CNTF suspension was sonicated using a Bransonic CPX-952-338R sonicator (Branson Ultrasonics, Brookfield, CT, USA) for 30 min in three cycles of 10 min each to prevent overheating of the sonication bath. Once the dispersion was obtained, dilutions with concentrations of 2.5, 5.0, and 10 μg/mL were prepared as a treatment for the cell line. This study involved collecting diesel and diesel–CNTF DEPs using Emfab 47 mm filters wrapped in aluminum foil and stored at −20 °C until extraction. Before extraction, the filters were placed in amber bottles, moistened, and sonicated in distilled water for 3 h, with intermittent intervals of 15 min. Following sonication, the filters were air-dried for 3 days in the bottles and then weighed to calculate the concentration of DEPs (mg/mL) based on the difference in weight before and after extraction, determined through 3–5 measurements. The extracted DEPs were then stored at −80 °C for future use.

2.4. Cell Culture Maintenance

Human lung carcinoma cells A549 (ATCC # CCL-185) were cultured in DMEM (Dulbecco's modified Eagle's medium, Sigma-Aldrich, St. Louis, MO, USA) with 5% fetal bovine serum (GIBCO, Thermo Fisher Scientific, Waltham, MA, USA) and 1% penicillin/streptomycin. A549 cells were exposed to CNTF and DEPS (diesel and diesel–CNTF) at different concentrations (2.5, 5.0, and 10 μg/mL) for 48 h. Extracts stored at −80 °C were used for exposure. Negative and positive controls were included, with cells subsequently incubated for an additional 24 h at 37 °C with 5% CO_2 after treatments.

2.5. Cellular Uptake of CNTF andDEPs Using Transmission Electron Microscopy

Transmission electron microscopy (TEM) confirmed the cellular uptake of CNTF and DEPs. Cells cultured in T25 flasks were treated with 10 μg/mL of DEPs (diesel or diesel–CNTF) for 48 h. Following trypsinization, the A549 cell suspension was fixed with glutaraldehyde. Sample preparation for TEM included chemical fixation, washing, dehydration, embedding in resin, and polymerization. Ultra-thin sections were prepared using an ultramicrotome and stained with heavy metals for visualization in a Tecnai F20 Super Twin TMP Transmission Electron Microscope from FEI Technologies Inc. (Hillsboro, OR, USA).

2.6. Cell Viability Assay

The MTT assay evaluated the cell viability of A549 cells exposed to CNTF or DEPs (diesel and diesel–CNTF). A549 cells were plated in 96-well plates at a density of 2×10^4 cells/well and incubated for 24 h before exposure to different concentrations of nanoparticles and DEPs (diesel and diesel + CNTF) for 48 h. Following exposure, MTT solution (200 μg/mL) was added, and after a 4 h incubation, formazan crystals were dissolved in 200 μL DMSO. The optical density (OD) was measured at 492 nm using a plate reader (BioTek, Synergy H4, Shoreline, WA, USA) to determine the cell viability percentage compared to control cells.

2.7. Cell Apoptosis Assay

A549 cells were exposed to different concentrations of CNTF and DEPs (2.5, 5.0, and 10 μg/mL) for 48 h in six-well plates. After treatment, the cells were harvested and analyzed for apoptosis using the FITC Annexin V/PI apoptosis detection kit (BD Biosciences, San Jose, CA, USA) and flow cytometry in a cytometer FACScantoII (BD Biosciences, Franklin Lakes, NJ, USA). The analysis, performed with FlowJo™ v10 software (BD Biosciences), identified three cell populations: early apoptotic cells (Annexin V+/PI−), late apoptotic/necrotic cells (Annexin V+/PI+), and necrotic cells (Annexin V−/PI+). The flow cytometry also excluded small debris and collected ten thousand events from the cell population for analysis.

2.8. ROS Assay

A549 cells were cultured in six-well plates for 24 h, then exposed to CNTF and DEPs (diesel and diesel–CNTF) at specified concentrations (2.5, 5.0, and 10 μg/mL) for 48 h. Following exposure, the cells were harvested and treated with 10 μM DCFH-DA (2′,7′-Dichlorofluorescin Diacetate, Merck KGaA, Darmstadt, Alemania) and incubated at 37 °C for 30 min. The fluorescently labeled cells were analyzed using flow cytometry on a FACSCantoII cytometer (BD Biosciences) to measure the signal in the FITC channel (excitation wavelength = 491 nm, emission wavelength = 525 nm). Data analysis was conducted using the FlowJo™ v10 software (BD Biosciences).

2.9. DNA Damage Assay

The comet assay (pH 13) was utilized to evaluate DNA damage in A549 cells exposed to CNTF and DEPs (diesel and diesel–CNTF) at different concentrations for 48 h. Cells were harvested at a concentration of 1×10^6 cells/mL, embedded on slides (in 0.5% low melting-point agarose on slides coated with 1.5% normal melting-point agarose), lysed at 4 °C for 2 h, and subjected to alkaline electrophoresis followed by neutralization (0.4 mmol/L Tris-HCl, pH 7.5) at 4 °C for 10 min. Ethidium bromide staining and fluorescence microscopy imaging (Olympus BX51) were performed to evaluate DNA damage through the Olive Tail Moment (OTM). Data analysis involved the examination of 100 randomly selected cells per sample using the CometScore image analysis system (TriTek Corp., Williamsburg, OH, USA).

2.10. Statistical Analysis

In analyzing the responses of A549 cells to CNTF and DEP exposure, statistical analysis was conducted using GraphPad Prism 8 software for Windows (GraphPad Software, Boston, MA, USA). The statistical comparisons between the experimental and control groups involved using the homogeneity of variance test and Dunnett's test for independent samples. Additionally, an ANOVA was applied to compare the responses of the test cells with those of the control cells. Statistical significance levels were defined as *** $p < 0.001$, ** $p < 0.01$, and * $p < 0.05$.

3. Results and Discussion

DEPs were collected from an internal combustion engine using either diesel fuel or diesel–CNTF. In general, the formation of particulate material involves several stages, including pyrolysis, nucleation, surface growth, agglomeration, and oxidation. Pyrolysis reactions occur in an oxygen-free atmosphere and cause the breakdown of fuel molecules into precursor molecules such as acetylene, benzene, or gas-phase PAHs. Subsequently, the nucleation process occurs, where the first soot nuclei are formed by aggregating larger PAH molecules. In the surface growth stage, the mass of the nascent soot particles increases by adding species, leading to an increase in the C/H ratio. The particles undergo agglomeration due to particle–particle collisions, leading to an increase in size and a decrease in number. Finally, the oxygen oxidation stage of PAHs and soot occurs, forming CO and CO_2. This stage competes with its formation during combustion [41,42].

Figure 2A,B show that when diesel is used as fuel, chain-like structures and irregularly shaped particle agglomerates are observed, along with larger agglomerates close to 1 μm in size. Conversely, when the diesel–CNTF mixture is employed as fuel, spherules-like structures are formed (as shown in Figure 3A,B), and agglomerates of particles display fractal-like shapes. Figure 4A shows the presence of individual particles with diameters ranging from 10 to 40 nm when diesel is used as fuel. On the other hand, when the diesel–CNTF mixture is employed as fuel, spherules with diameters between 20 and 150 nm are formed. As a result, using a diesel–CNTF mixture yields larger primary particle sizes than diesel. In both fuel cases, these primary particles serve as the basis for forming particles in the accumulation mode.

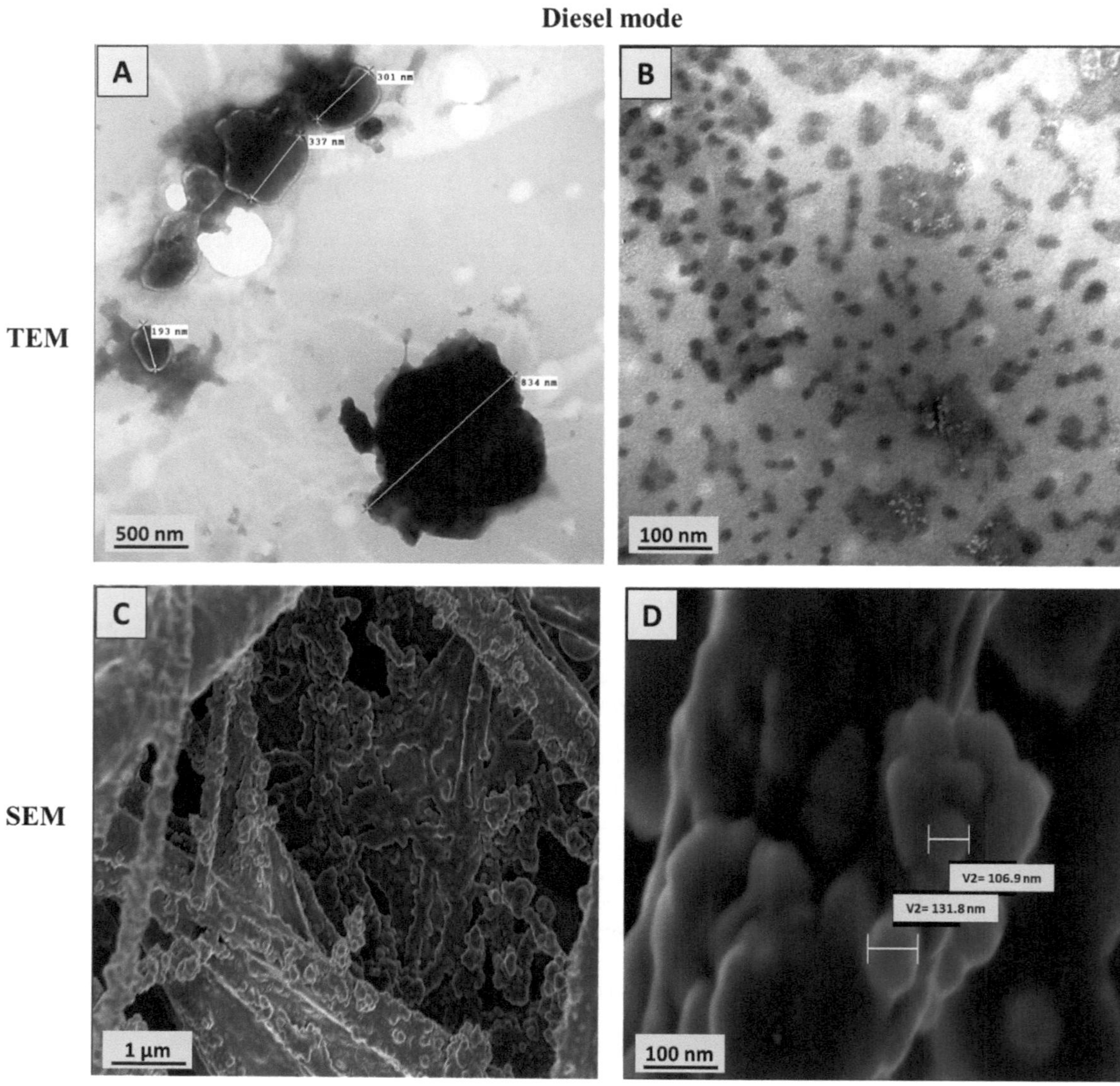

Figure 2. Images of DEPs in diesel mode. (**A**,**B**) show TEM microscopy. (**C**,**D**) show SEM microscopy. Different magnifications are used to show details of DEPs.

In the SEM micrographs (Figure 2C,D), it can be seen that many particles are deposited in the filter structure, with a high degree of overlap. Additionally, larger sizes are observed for the diesel–CNTF case compared to when diesel is used as fuel (Figure 3C,D). This result agrees with the one obtained through the TEM technique, which offers higher resolution. The reason for this is that when primary particles collide, they agglomerate and then form larger particles. If these primary particles have a larger diameter in the nucleation mode, it is expected that the higher diameter observed by SEM would correspond to this phenomenon.

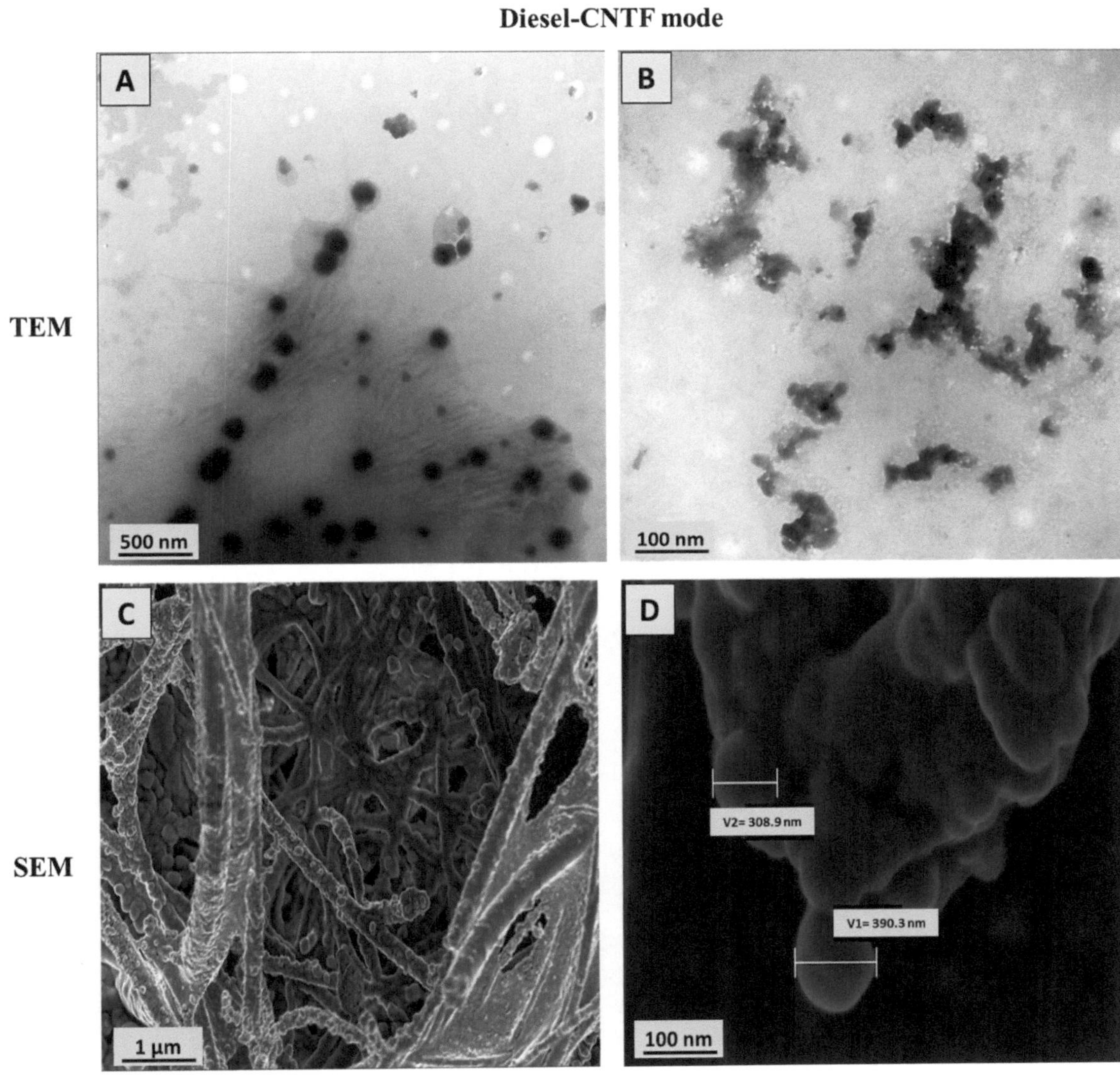

Figure 3. Images of DEPs in diesel–CNTF mode. (**A**,**B**) show TEM microscopy. (**C**,**D**) show SEM microscopy. Different magnifications are used to show details of DEPs.

Important factors affecting the formation of particulate matter include the temperature and pressure in the cylinder, as well as the effectiveness of the mixing process between fuel and air [42,43]. On the other hand, in this study, the addition of carbon nanotubes plays a significant role by introducing aromatic-type structures into the fuel. The aromatic structures in the fuel are important for soot emissions and affect the particle size distribution, which is related to variations in the carbon/hydrogen ratio, fuel evaporation, and the formation of intermediate radicals caused by the aromatic structures. These radicals can impact the initial stages of soot formation (e.g., pyrolysis and nucleation) as well as the final oxidation stage [44]. Although the C/H ratio is slightly modified with the concentration of CNTF used in this study (100 mg/L), the addition of these additives could mainly affect the initial stages of soot formation by reducing the ignition delay time [25] and providing precursor molecules for soot formation and surface growth.

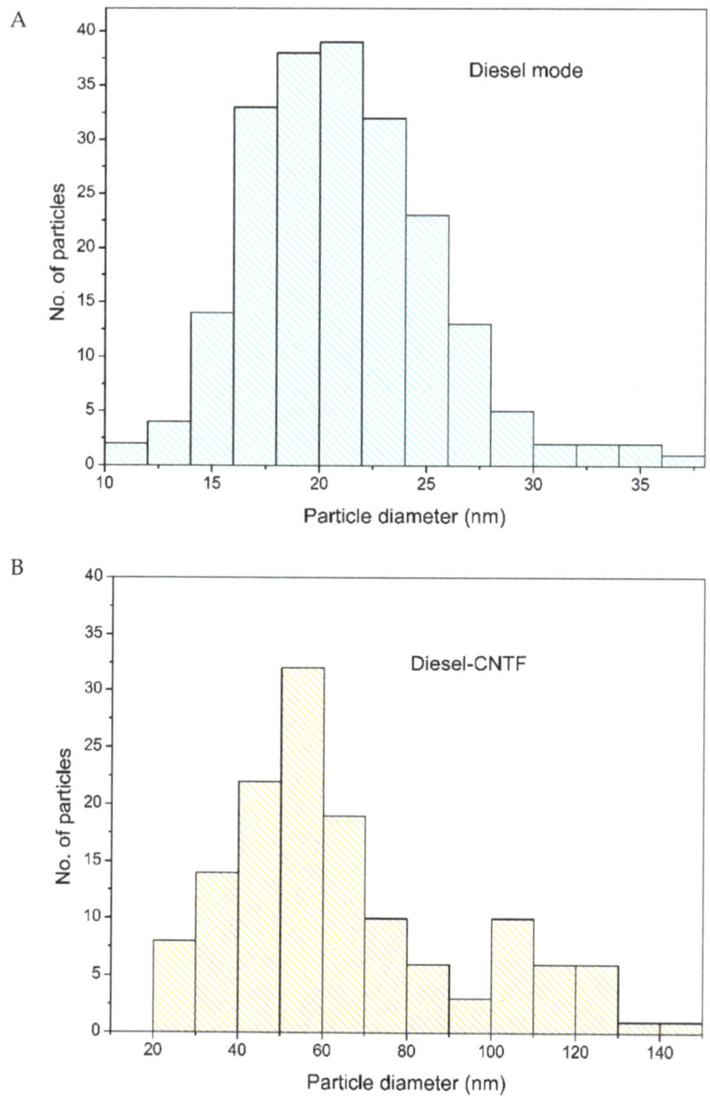

Figure 4. Particle size distribution in diesel and diesel–CNTF DEPs. (**A**), in light blue, shows the particle diameter in diesel mode. (**B**), in light orange, shows the particle diameter in the diesel-CNTF mode.

Aromatics are important for polycyclic aromatic hydrocarbon (PAH) formation and are a significant soot precursor during diesel fuel combustion due to their high unsaturation. Studies have been conducted both on a laboratory scale using flames or with diesel engines to assess how different fuels affect the production of particulate matter [45,46]. These studies have used mixtures of heptane or diesel with aromatic liquids such as benzene, xylene, or tetralin. They have observed an impact on the morphological characteristics and chemical composition of the resulting particulate material. For instance, it was noted that the particle diameter of the diesel/aromatic soot was larger than when pure diesel was used as fuel, which aligns with the findings of the present study [44]. It is important to note that, unlike those studies, the present study involves the use of a solid nano-additive, significantly increasing the complexity of our research.

The data presented in Figure 5 illustrate the concentrations of PAHs measured when using diesel or diesel–CNTF as fuels. Out of the 16 PAHs examined, 8 depicted in the figure were determined to be above the detection limits of the employed method. The literature has documented the presence of these PAHs in studies analyzing heavy-duty vehicles that utilize diesel [47,48] or in areas characterized by high vehicular volume, such as tunnels [49]. These compounds are primarily attributed to incomplete combustion processes. When comparing the use of diesel fuel with and without functionalized carbon nanotubes (CNTF), it was observed that the general concentration of polycyclic aromatic hydrocarbons (PAHs) was higher in the case of diesel alone. Specifically, the relative concentration of PAHs with five aromatic rings (BkF, BbF, BaP, and DahA) accounted for 86.5% of the total PAHs in diesel mode, followed by PAHs with six aromatic rings (BPer) at 7.9%, and the remaining percentage was attributed to PAHs with four rings (Flu, BaA, and Ch). In contrast, in the diesel–CNTF case, PAHs with five aromatic rings still constituted the largest fraction, but at a notably lower value of 66.4% compared to diesel. The semi-volatile fraction of PAHs with four rings accounted for 31.8%. This indicates that adding functionalized carbon nanotubes to the fuel affects the amount and composition of the PAHs, particularly by favoring the semi-volatile fraction.

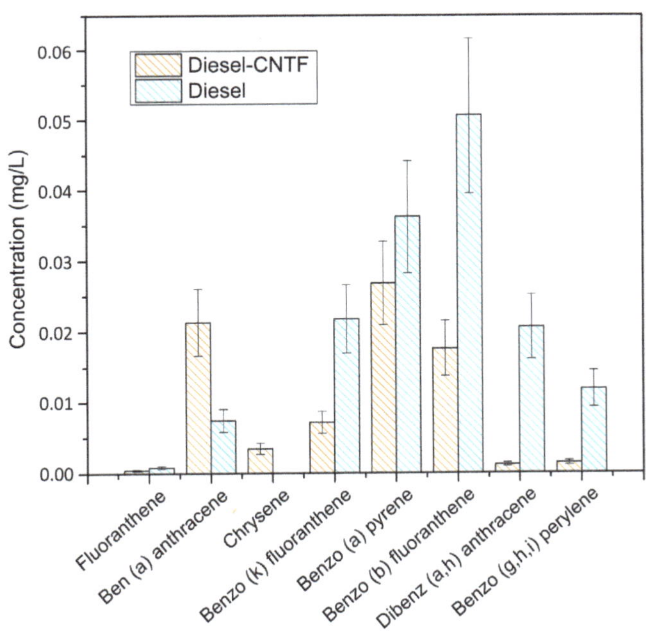

Figure 5. PAH concentration in diesel and diesel–CNTF.

The combined findings, along with those obtained through SEM and TEM, could indicate that the introduction of CNTF to diesel may facilitate the surface growth of particles by incorporating PAH species. This process results in increased carbon content and particle size. Furthermore, this addition may also decrease the available quantity of PAHs. However, further investigation is warranted due to the intricate mechanisms underlying particle formation from PAHs and the pyrolysis and oxidation reactions inherent in diesel combustion when CNTF is introduced.

Regarding toxicity, Flu and BPer are categorized by the International Agency for Research on Cancer (IARC) as belonging to group 3, which means they are not classifiable as carcinogenic for humans. Conversely, BaA, Ch, BkF, and BbF fall into Group 2B, signifying that they are possibly carcinogenic to humans. DahA is categorized under Group 2A, indicating that it is probably carcinogenic to humans, while BaP is classified in Group 1 as being carcinogenic to humans [50]. In considering the differential toxicity of PAHs, the potency equivalence factor (PEF) was utilized to assess the overall health impact of PAHs in both diesel and diesel–CNTF modes. The methodology outlined by [51] was adopted to determine the equivalent BaP (benzo[a]pyrene) toxicity of the identified PAHs. This analysis revealed a BaPeq (benzo(a)pyrene equivalent) value of 0.065 mg/L for standard diesel and 0.033 mg/L for diesel–CNTF, indicating that the incorporation of CNTF in diesel yields significant health benefits by reducing toxicity, as measured by BaPeq.

3.1. A549 Cellular Uptake of CNTF and DEPs

TEM enabled the ultrastructural study of A549 cells exposed to CNTF and DEPs (diesel and diesel–CNTF) to determine their cellular entry. It was demonstrated that both CNTF and DEPs enter the cells. In Figure 6A, the control cells showed no structural changes, with no vesicle formation, and organelles such as mitochondria maintained their normal shape. On the other hand, CNTF and both types of DEPs were found within phagosomes in the cytoplasmic region (Figure 6B). There were no free DEP or CNTF agglomerates observed in the cytoplasm or nucleus. While no signs of cell death, such as apoptosis, were observed at the ultrastructural level, treatment with diesel DEPs resulted in increased lysosomes and changes in mitochondrial morphology (Figure 6D). Treatment with diesel–CNTF DEPs led to fewer phagosomes but a higher number of lysosomes compared to treatment with diesel DEPs (Figure 6D).

This study demonstrated that CNTF can enter cells, but this alone does not result in significant cytotoxic effects. It was observed that treating cells with CNTF induces structural changes, notably the formation of phagosomes. This finding aligns with other research showing that endocytosis encapsulates CNT within membrane invaginations, which then bud and fold to form endocytic vesicles. These vesicles are transported to specialized intracellular sorting and trafficking compartments [52]. The cellular uptake mechanism of CNT may vary based on its functionalization and size. However, the specific pathways of entry, subsequent trafficking, and intracellular distribution of CNT remain unclear. Ghosh et al. state that carbon nanotubes trigger a stress response upon entering cells, activating autophagy and/or apoptosis-related pathways. The study suggests that while various mechanisms may contribute to CNT-induced pathogenesis, a comprehensive understanding of the relationship between different types of CNT exposures (multi-walled CNT vs. single-walled CNT) and the processes of autophagy, apoptosis, and necrosis in human lung epithelial cells is still lacking [53].

Our results suggest that the entry of CNTF may occur via autophagy, potentially leading to its elimination through the lysosomal pathway. Autophagy is a highly conserved lysosomal degradation pathway that operates at basal levels in all cells [54,55]. Upon internalization, CNT has been detected in lysosomes, cytoplasmic vacuoles, and phagosomes, free in the cytoplasm, near the nucleus, and occasionally piercing the nuclear membrane [56–59]. However, further studies are needed to evaluate these pathways and better understand the mechanisms involved.

Unlike CNTF, we observed that exposure of A549 cells to DEPs (diesel and diesel–CNTF) significantly impacts their structure and potential functioning, possibly related to mitochondria and lysosomes. Pierdominici et al. characterized the cytotoxicity of DEPs by examining changes in mitochondrial structure and function. They found that, in addition to morphological alterations, DEPs caused a significant loss of membrane potential ($\Delta\Psi m$) detectable after just 24 h of treatment. However, this loss of $\Delta\Psi m$ did not correlate with an increase in the percentage of apoptotic or necrotic cells, which remained unchanged in treated cells compared to untreated cells [60]. Like mitochondria, lysosomes are crucial for assessing the effects of DEPs because they process and digest materials through endocytosis of small molecules and cell surface proteins, phagocytosis of large particles such as apoptotic cell remnants and pathogenic bacteria, or autophagy of cytoplasmic contents, including damaged mitochondria, ER, and lysosomes [61–64]. In a review by Nie et al., it is mentioned that lysosomes play a role in cellular responses to airborne particulate matter

(APM) and DEP-induced stimuli, which can lead to the production of inflammatory factors, interfere with autophagy, alter iron homeostasis, and affect various other mechanisms. Consequently, lysosomal dysfunction or increased lysosomal activity may occur [65].

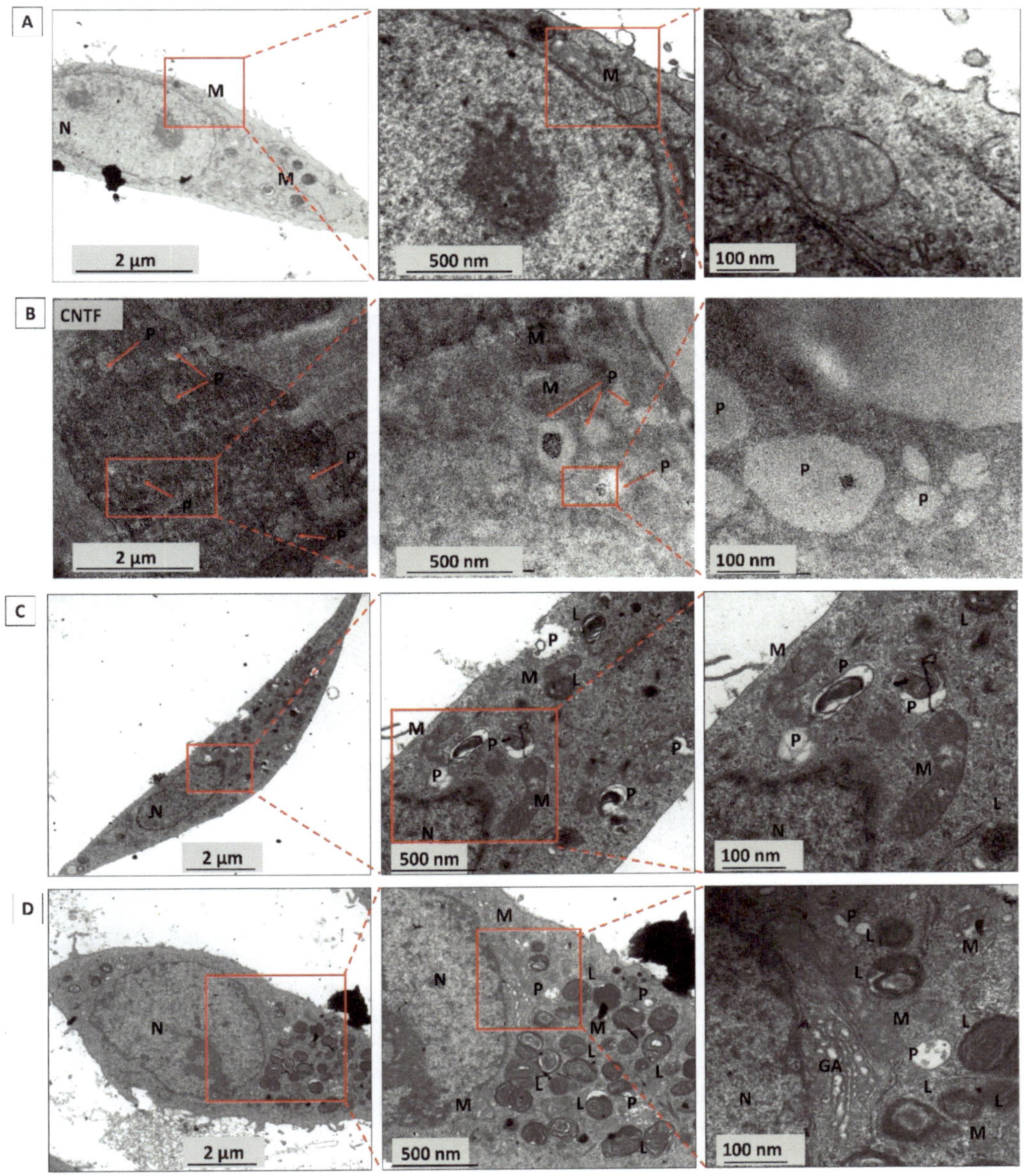

Figure 6. TEM cross-sectional images of A549 cells after 24 h of CNTF and DEPs incubation. (**A**) The control cells and some magnified images show details of the cellular structures. (**B**) Images of cells treated with CNTF zoom in on the intracellular localization of nanoparticles in the cytoplasm and phagosomes. (**C**) Cells treated with diesel DEPs show phagosomes and cellular structures. (**D**) Cells treated with diesel–CNTF DEPs show fewer phagosomes and more lysosomes. The cellular structures are labeled as follows: nuclei (N), mitochondria (M), Golgi apparatus (GA), and lysosome (L). Phagosomes are indicated by P. Arrows indicate nanoparticles in the phagosomes.

We focused on the role of diesel–CNTF DEP particles to determine their effect, as this has not yet been reported in the existing literature due to their unique physicochemical characteristics. It should be noted that although adding CNTF to diesel alters the physicochemical properties of the crude soot particles formed by combustion [25], both diesel and diesel–CNTF DEPs enter the cells and potentially cause damage. However, an increase in lysosomes in cells treated with diesel–CNTF DEPs might indicate that this elimination mechanism is activated, thereby reducing potential cytotoxic effects. It is advisable to evaluate autophagosomal markers to understand the mechanism and how this pathway is affected by diesel–CNTF DEPs. Additionally, assessing mitochondrial involvement and lysosomal function is recommended.

3.2. *Effect of CNTF and DEPs on Cell Viability*

Cell viability percentages (%) were measured using the MTT assay after cells were treated with CNTF, DEPs, and diesel–CNTF DEPs and incubated for 48 h with the corresponding treatment at doses of 0, 2.5, 5, and 10 μg/mL. Each concentration was analyzed in two replicates with three repetitions. Figure 7A illustrates the independent reduction in cell viability induced by each treatment. The black bars indicate that CNTF at concentrations of 5 and 10 μg/mL decrease cell viability by 17.2% and 12.8%, respectively, with statistical significance ($p < 0.0001$ and $p = 0.0001$) concerning control cells. Notably, the observed behavior is not dose-dependent, as the 5 μg/mL concentration demonstrates the highest induction of cell death. In the orange bars, it is evident that as the concentration of diesel DEPs increases, cell viability decreases, with effects of 13.8%, 19.6%, and 26.4%, respectively, all significant at $p < 0.0001$. This damage exhibits dose-dependent behavior. Moving to the blue bars, treatments with diesel–CNTF reduce viability by 15.2% and 18.8% for concentrations of 5 and 10 μg/mL, respectively, with a p-value < 0.0001. Here, it is determined that the behavior may be dose-dependent, particularly at higher concentrations. Additionally, adding CNTF to diesel appears to marginally reverse the harmful effects of the emissions (diesel decreases viability by 26.4%, and diesel–CNTF decreases viability by 18.8%).

A comparison between the CNTF treatments and the DEPs, considering the concentrations used (Figure 7B), reveals significant differences. At concentrations of 2.5 and 5.0 μg/mL, diesel demonstrates a more significant decrease in viability compared to CNTF and diesel–CNTF, with p-values of 0.0087 and 0.0046, respectively. Similarly, at 10 μg/mL, a similar trend is observed, where diesel vs. CNTF shows a significant decrease (13.6%) in viability ($p < 0.0001$). In contrast, diesel vs. diesel–CNTF indicates a significance level of $p = 0.0162$ (7.6% decrease in viability). No differences are observed at the 5 μg/mL concentration, as this is the point where CNTF begins to affect viability.

These assays demonstrated that CNTF significantly decreased cell viability at the highest concentrations compared to the control. Notably, a slight impairment of cell viability was observed at an intermediate concentration of 5.0 μg/mL. Toxicological studies on CNTs have shown their impact on cell viability across various cell lines. For instance, they reduce viability in the human epithelial cell line A549 [66] and induce chromatin condensation and DNA fragmentation in mesenchymal stem cells [67]. Additionally, their fiber-like structure and persistence in lung tissues raise concerns about potential negative health effects similar to those caused by asbestos fibers [68]. However, Kharlamova and Kramberger suggest that these adverse effects can be mitigated through the chemical functionalization of CNTFs [69]. Similarly, Coccini et al. reported that amide-functionalized carbon nanotubes exhibit reduced cytotoxic effects compared to non-functionalized nanotubes [70]. Chowdhry et al. also found that functionalized CNTs exhibited lower cytotoxicity compared to non-functionalized ones in both in vitro analyses with HEK 293 cells and in vivo assays with zebrafish [71].

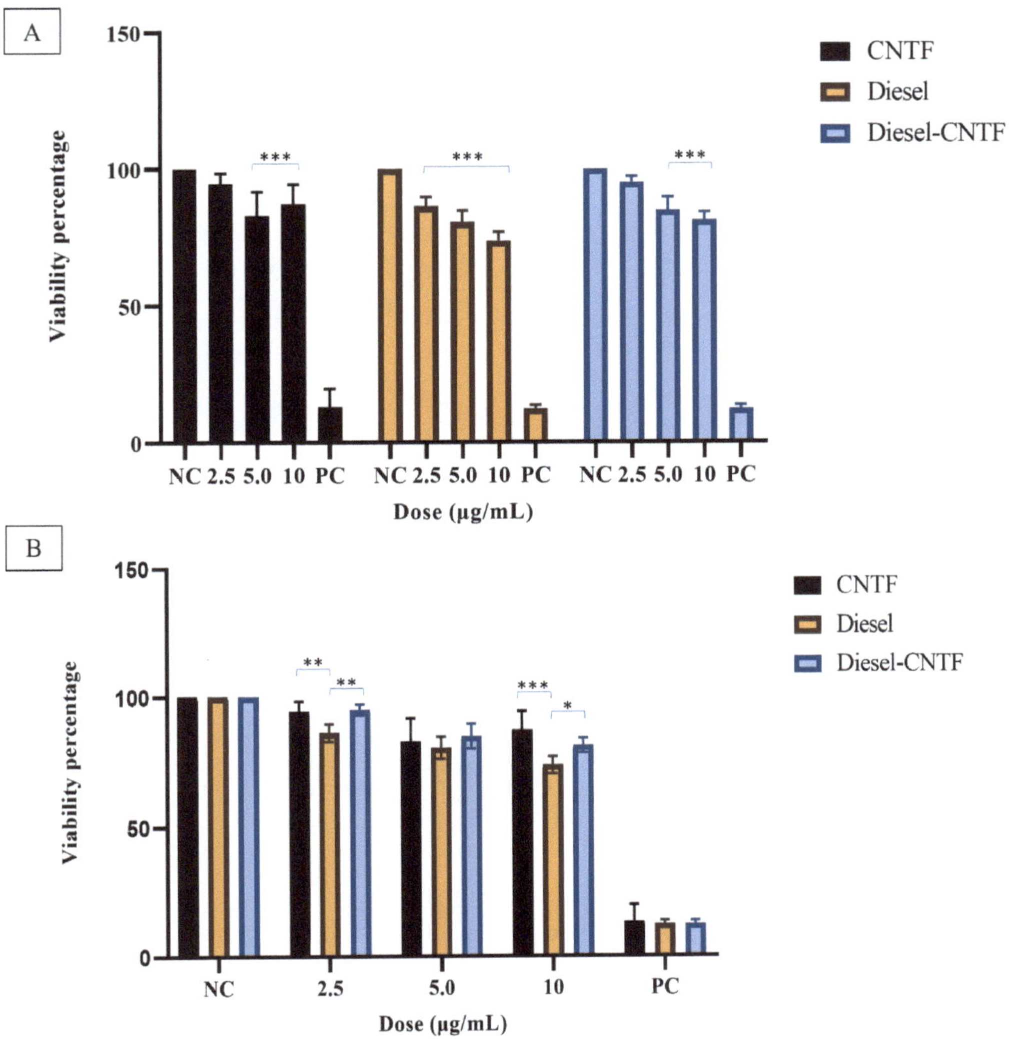

Figure 7. Effect of CNTF and DEPs on Cell Viability. (**A**) illustrates the percentage decrease in viability for each treatment independently. (**B**) compares different concentrations of the treatments. Black bars represent CNTF, orange bars represent diesel DEPs, and blue bars represent diesel + CNTF DEPs. Statistical significance is indicated by $p < 0.05$ *, $p < 0.01$ **, and $p < 0.001$ ***.

Concerning diesel DEPs, it has been found that, independent of the treatment concentration, there is a decrease in cell viability. This may be due to different factors that can interact and contribute to cell death through various mechanisms, including apoptosis, necrosis, autophagy, and other forms of cell death, such as physical damage. When this effect is compared to the toxicity of DEPs reported in other in vitro studies, the findings of our study are supported by several of these, in which it is shown that diesel engine particles induce cell death in murine RAW 264.7, macrophages, and human A549 lung cells [72], murine endothelial cells SVEC4-10 [73], mesothelial cell line (MeT-5A) and mesothelioma cell line (CRL-5820) [74], co-culture of A549 cells and THP-1 monocyte cells [75], human iPSC-derived microglia [76], and as found here in A549 cells [67]. In the above studies, this effect of decreased cell viability is due to one of the death mechanisms.

This study found differences in PAH content, particle distribution, and morphology between diesel DEPs and diesel–CNTF DEPs. Particle size distribution is a crucial parameter for assessing the cytotoxic effects of DEPs, as it provides insights into the distribution of particles in nucleation and accumulation modes, as well as their potential mechanisms of cellular entry [43]. Likewise, the morphology and microstructure of DEPs are essential for a comprehensive understanding of the potential mechanisms of cytotoxicity [77]. Diesel particles form branched aggregates at a microscopic scale, each consisting of tens to hundreds of spherical or nearly spherical primary particles [78]. On the other hand, the chemical composition of DEPs is also very important when measuring their cytotoxic effects. Different studies mention that PAHs can damage cells and that their content varies according to the type of fuel used [79,80].

Based on the above findings, it could be hypothesized that the effect on cell viability induced by diesel–CNTF DEPs remains marginal or similar to that of diesel DEPs. This may be due to the lower PAH content and the more compact morphology compared to diesel; these factors do not result in a highly significant change in cytotoxic effects. Studies on metallic nanoparticles have shown that their biological effects are diminished when mixed with diesel. However, this reduction is not so significant that it becomes almost imperceptible or negligible [81–84]. These results are similar to those presented here.

3.3. Induction of Apoptosis by Carbonaceous CNTF in the A549 Cell Line

To evaluate the effect of CNTFs and DEPs (diesel and diesel–CNTF) on A549 cells, an apoptosis and membrane integrity assay was performed using annexin and propidium iodide (PI) double staining. This assay was measured via flow cytometry. Figure 8A shows that different concentrations (2.5, 5.0, and 10 μg/mL) of CNTF and DEP treatments did not compromise membrane integrity compared to the negative control. However, some statistically significant differences were observed when comparing treatments across different concentrations. At 2.5 μg/mL, diesel DEPs induce greater membrane damage compared to CNTFs, with a p-value of 0.0480. At 5 μg/mL, diesel DEPs caused significantly more damage than CNTFs ($p = 0.0011$) and diesel–CNTF DEPs ($p = 0.0005$). Finally, at a concentration of 10 μg/mL, a similar pattern to the 5 μg/mL concentration was observed. Diesel DEPs induced more damage than CNTFs ($p = 0.0030$) and diesel–CNTF DEPs ($p = 0.0170$).

Regarding the induction of apoptosis, Figure 8C shows that diesel DEPs significantly induced apoptosis at concentrations of 5 μg/mL ($p = 0.0092$) and 10 μg/mL ($p < 0.0001$). CNTF and diesel–CNTF DEPs did not show observable effects. When comparing treatments based on the concentrations used, some differences became evident. At 2.5 μg/mL, treatment with diesel DEPs induced more apoptotic cells than CNTF ($p = 0.0003$), while there were no differences with diesel–CNTF DEPs. At 5 μg/mL, the induction of apoptosis by diesel DEPs was significantly higher than both CNTF ($p < 0.0001$) and diesel–CNTF DEPs ($p = 0.0021$). Additionally, diesel DEPs induced more apoptosis than diesel–CNTF DEPs ($p = 0.0233$). A similar pattern was observed at the highest concentration (10 μg/mL). Both diesel and diesel–CNTF DEPs induced a higher number of apoptotic cells compared to CNTF, with p-values of 0.0068 and 0.0017, respectively. Moreover, diesel DEPs caused greater apoptosis compared to diesel–CNTF DEPs at this concentration ($p = 0.0384$).

Overall, concentration-dependent diesel DEPs induce greater cell membrane damage and apoptosis induction than CNTFs and diesel–CNTF DEPs. CNTFs alone do not affect cells in terms of membrane damage and apoptosis induction. While diesel–CNTF DEPs induce damage, the use of CNTFs in combination with diesel appears to reverse or reduce the effects seen when using diesel DEPs alone.

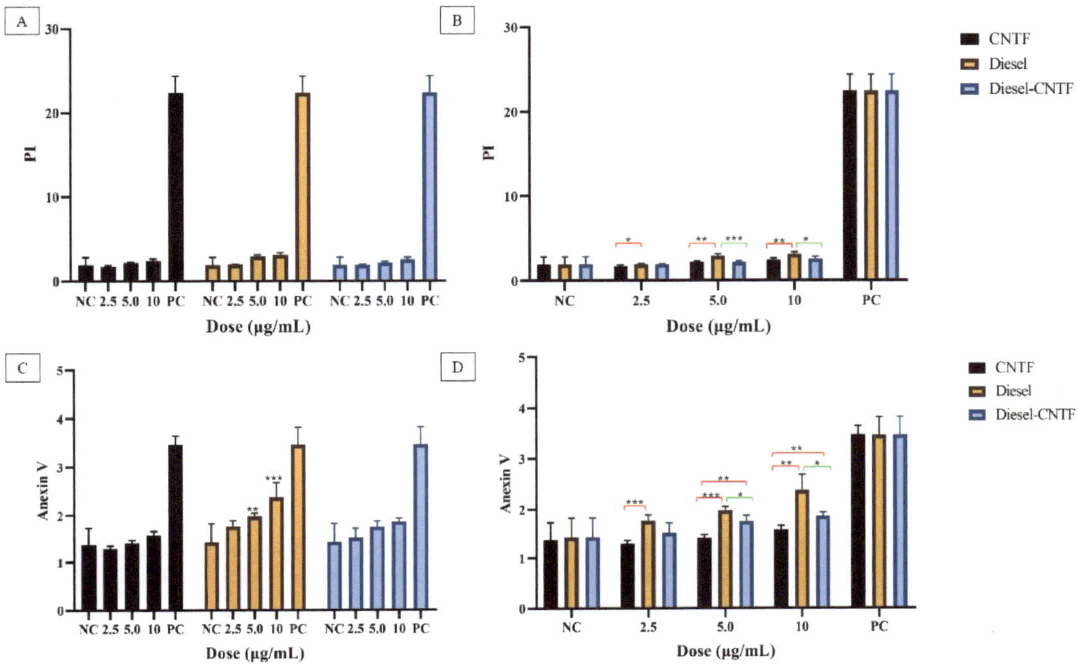

Figure 8. Apoptosis induction by CNTF and DEPS (diesel and diesel–CNTF) in A549 cells. (**A**,**B**) show the cell membrane damage measured by PI. (**A**) shows the comparison according to the type of treatment, where there are no differences. (**B**) shows the differences between treatments according to concentration. (**C**,**D**) show the induction of apoptosis measured with Annexin V. (**C**) shows that only treatment with diesel DEPs at each dose induces apoptotic cells. (**D**) shows that depending on the dose of treatment used, there are differences in the induction of apoptosis. Red brackets indicate comparisons between CNTF versus DEPs, and green brackets show comparisons between diesel DEPs and diesel–CNTF DEPs. Black bars show CNTF in orange diesel and blue diesel–CNTF. Statistical significance is indicated by $p < 0.05$ *, $p < 0.01$ **, and $p < 0.001$ ***.

This study evaluated cell membrane damage induced by CNTFs and DEPs by double staining with Annexin V and PI. In general, it was observed that membrane integrity was maintained in each of the treatments. This coincides with that reported by Ursini et al., who mention that exposure to CNTFs does not damage the cell membrane of A549 cells [85]. With respect to DEPs, it has been established that, generally, the effect of DEPs did not induce a compromise of the cell membrane [79], supporting the results obtained in this study. On the other hand, apoptosis induction by CNTF was found to have a dose-dependent behavior but without statistically significant differences. This relative induction of apoptosis by CNTF aligns with the study by Patlolla et al., which demonstrated a dose-dependent increase in the proportion of apoptotic cells after 48 h of exposure in human dermal fibroblasts [86]. However, another study indicates that CNTF exposure increased the proportion of apoptotic cells compared to untreated cells, starting with a concentration-dependent trend of 10 μg/mL [87]. In this context, Song et al. state that the cytotoxic effects largely depend on the type of functionalization applied to CNTs [87].

The results observed with DEPs align with numerous publications indicating that, compared to untreated cells, exposure to diesel particles leads to a statistically significant, concentration-dependent increase in the number of apoptotic cells [63–65,67]. DEPs can cause inflammation and produce ROS, both of which are associated with mitochondrial dysfunction. Cattani-Cavalieri et al., in their study performed in BEAS-2B cells, found that DEPs alter mitochondrial morphology and reduce mitochondrial bioenergetics, affecting respiration, ATP production, and reserve capacity, all leading to THE induction of apoptosis [88]. Alternatively, Lawal discusses how the organic components of DEPs are accountable for these effects in cells, emphasizing the involvement of DEP-induced stress-activated protein kinases in prompting apoptosis in macrophages [89].

Considering the separate results of CNTF and diesel DEPs and the lack of information regarding their potential toxicological effects when used together, the results obtained with diesel–CNTF DEPs suggest that CNTF could help reduce damage related to cell death and membrane integrity, as no statistically significant induction of apoptosis was observed. Taylor-Just and colleagues suggest that functionalization alone is not the important driver of toxicity but rather the purification method used prior to functionalization [90]. Furthermore, Gamboa et al. have shown that incorporating amide-functionalized CNT can effectively reduce pollutant emissions, such as diesel particulates (DEPs), and enhance the thermal efficiency of compression ignition engines (CIE) [25]. The non-induction of apoptosis by diesel–CNTF DEPs is significant, as CNTFs enhance engine performance and reduce harmful emissions when used with diesel. This beneficial effect is evident in our results. Modifying the physicochemical properties of CNT can impact their toxicity [91,92].

3.4. CNTF and DEPs Induced Oxidative Stress Changes

We compared the intracellular ROS production dynamics after exposing the A549 cell line to different concentrations of CNTF and DEPs (diesel and diesel–CNTF) for 48 h using flow cytometry with the ROS-sensitive fluorescent indicator DCFDA. Figure 9A shows that exposure to CNTF did not increase the DCF signal at any of the concentrations evaluated compared to the negative control, indicating no increase in ROS production. However, the DEPs (diesel and diesel–CNTF) induced ROS production at each of the different concentrations used as treatments, with a p-value < 0.0001.

Figure 9B presents the results of the comparison between treatments based on the dose used. Like the previous findings, diesel DEPs caused higher ROS production than CNTFs and diesel–CNTF DEPs, with a p-value < 0.0001. Combining CNTF with diesel significantly reduced ROS production; however, ROS production was still higher than the negative controls. A dose-dependent behavior was observed throughout the study.

The generation of ROS and the resulting oxidative stress can trigger various cellular processes, including DNA damage and apoptosis [93,94]. Nel et al. identified oxidative stress as the primary mechanism of toxicity associated with nanoparticles, attributing it to their small size and large surface area, which promotes ROS generation [95]. Our research found that CNTF did not induce intracellular ROS production, which was confirmed using the cell-permeable dye DCFH-DA. Functionalization of MWCNTs has been suggested as a strategy to potentially reduce ROS production [68,96,97]. In our study, no significant changes in intracellular ROS levels were observed after exposure to various concentrations of CNTF with amides, contrary to findings with metal- or acid-functionalized CNT, which have shown ROS-mediated cytotoxic damage [32,98,99]. These results align with Vijayalakshmi et al. (2023), who demonstrated that functionalized MWCNTs reduced cytotoxic effects by lowering impacts on cell viability and oxidative stress [100].

Another possible reason why the CNTF used in our study did not generate ROS could be related to their cellular uptake characteristics. Studies indicate that CNTF internalization varies among cell types; for example, some cells internalize CNTF through phagocytosis, while others, such as fibroblasts, show deficient phagocytosis or lack the machinery for endocytosis. Additionally, CNTF may enter cells via energy-dependent mechanisms, which do not induce ROS production [101–104]. Furthermore, the physicochemical properties of CNT, including their manufacture, application, shape, and durability, significantly influence their biopersistence and potential for ROS-dependent toxicity or cytotoxicity [105,106].

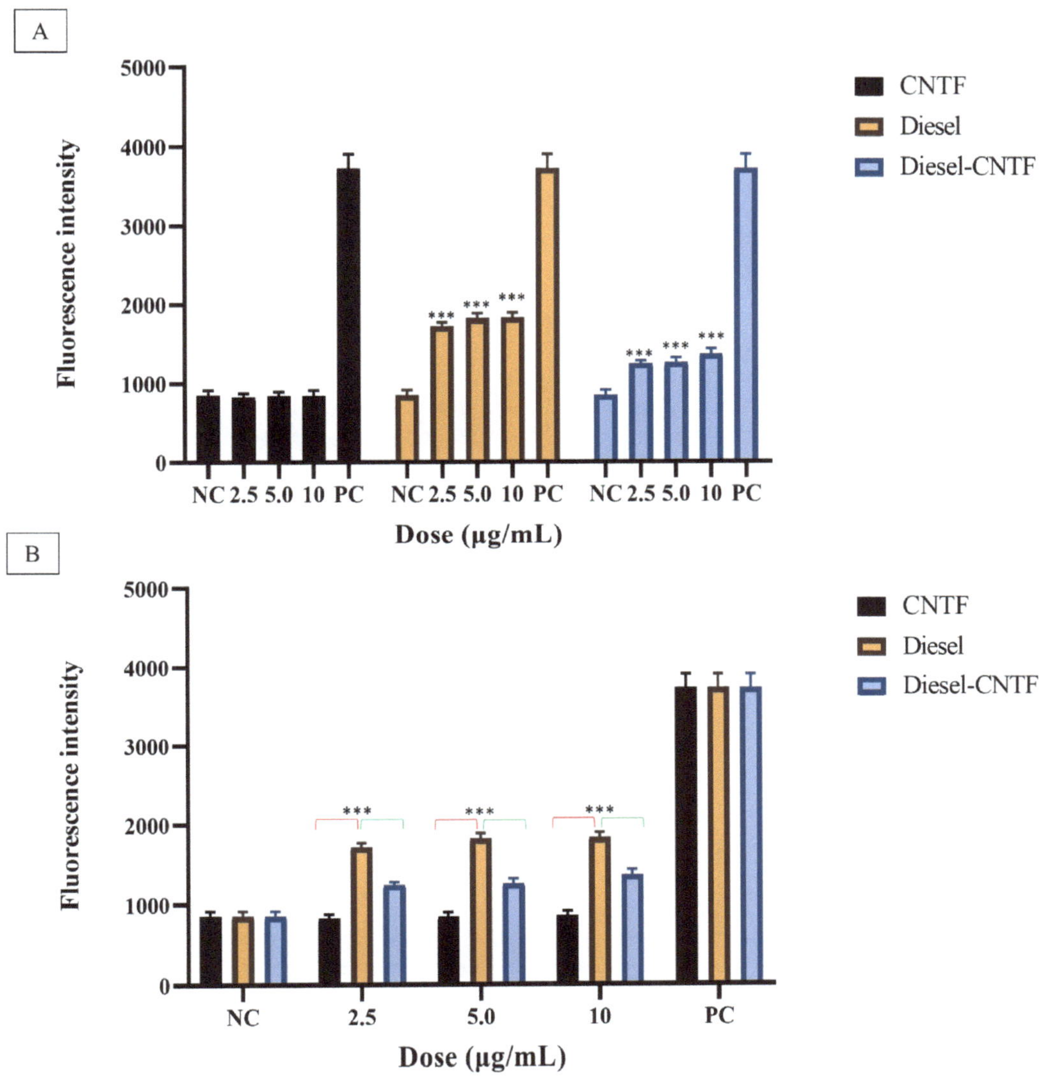

Figure 9. Analysis of Intracellular ROS in Cells Treated with CNTF and DEPs (diesel and diesel–CNTF). In (**A**), an increase in fluorescence intensity (arbitrary units) is observed in the treatments with DEPs, indicating an increase in intracellular ROS production. (**B**) shows that each of the concentrations increases ROS production, with higher levels in the case of diesel DEPs. Red brackets indicate comparisons between CNTF versus DEPs, and green brackets indicate comparisons between diesel DEPs and diesel–CNTF DEPs. Black bars show CNTF in orange diesel and in blue diesel–CNTF. Statistical significance is indicated by $p < 0.001$ ***.

ROS has a dual role in regulating biological processes: at low concentrations, it acts as a second messenger in signal transduction, while at high concentrations, it leads to biomolecule oxidation [107]. In DEP-treated A549 cells, we observed a considerable increase in ROS production. This is observable in both diesel and diesel–CNTF DEPs. However, the addition of CNTF considerably reduces this ROS concentration. It is widely recognized that excessive intracellular oxidative stress induced by DEPs can harm cells by oxidizing lipids, proteins, and DNA [108,109]. In this study, we found a correlation between ROS production and DNA damage induced by DEPs. This will be discussed in the test results below. This higher induction of ROS could be due to the PAH content and structure of diesel DEPs, which have been associated as the major inducers of damage at the cellular level in multiple studies [110–112]. Our results show that diesel DEPs have a higher content of PAHs and that their elongated and less condensed structure may be the damage inducers. It is important to mention that although diesel–CNTF DEPs also induce ROS production, at the tested conditions, we were able to detect intracellular ROS generation with a 20% reduction when compared to diesel DEPs. In this sense, the PAH content and the shape of diesel–CNTF DEPs may be involved. Here, we found that, compared with diesel DEPs, diesel–CNTF DEPs have a lower PAH content and a less condensed form, which may be associated with the mechanisms of cellular entry and possible impairment in function and structure. Zerboni et al. found that combining metal oxide nanoparticles with diesel reduced the adverse effects of diesel exhaust particles (DEPs) produced after combustion. Their study noted that no intracellular ROS generation was detected, and only DEPs induced a slight, but not significant, increase in ROS levels. Furthermore, the autophagic pathway could be activated in the presence of nanoparticles compartmentalizing harmful compounds in lysosomes for degradation [83]. The use of nanoparticles can induce a reduction in the effects, and if we extrapolate it to carbon nanotubes, we could be observing a similar behavior.

To gain a more comprehensive understanding of how ROS production induced by DEPs impacts cells, additional assays could be conducted to assess the signaling pathways involved, in addition to evaluating damage. Mitochondria are recognized as the primary generators of ROS, yet they are also highly susceptible to its effects [107]. The harmful effects induced by NOx should also be evaluated. Although it has been reported that amide-functionalized CNTF, when used in a mixture with diesel, significantly reduces NOx emissions by up to 29.4% [25], NOx can affect cell metabolism like ROS [103].

3.5. DNA Damage on A549 Cells Treated with CNTF and DEPs

Finally, we examined DNA damage in A549 cells treated with CNTF and DEPs (diesel and diesel–CNTF) using a comet assay to assess their genotoxic impacts. Figure 10A–E shows images of the tested cells and the comets formed in some of them after treatment. Figure 10F presents the results of the comparisons made according to the type of treatment. Treatment with CNTFs did not induce DNA damage at any of the concentrations used. Conversely, diesel DEPs and diesel–CNTF DEPs induce DNA damage, which can be visualized in the tails of the comets formed (D-E) with a significance of $p < 0.0001$. The comet tails are longer in the diesel DEP treatment at each concentration.

Figure 10G shows the results of the comparisons made according to the different concentrations and types of treatment. In general, it is observed that DEPs induce greater DNA damage than diesel–CNTF particles and CNTF alone. At concentrations of 2.5 μg/mL, 5.0 μg/mL, and 10 μg/mL, DEPs caused the most damage, followed by diesel–CNTF particles, compared to CNTF. Additionally, when comparing the genotoxic effects between diesel DEPs and diesel–CNTF DEPs, it was found that the former induced greater DNA damage, with larger comet tails and statistically significant differences between these two treatments ($p = 0.0044$). These findings suggest that diesel DEPs exhibit the highest DNA-damaging potential. Furthermore, exposure of the A549 cell line to the functionalized nanoparticles did not increase DNA damage. Although diesel–CNTF DEPs also induced DNA damage, the size of the comets formed was much smaller than that produced by diesel DEPs. As with previous assays, using CNTFs appears to mitigate the damage induced by diesel. The observed comet formation behavior is dose-dependent.

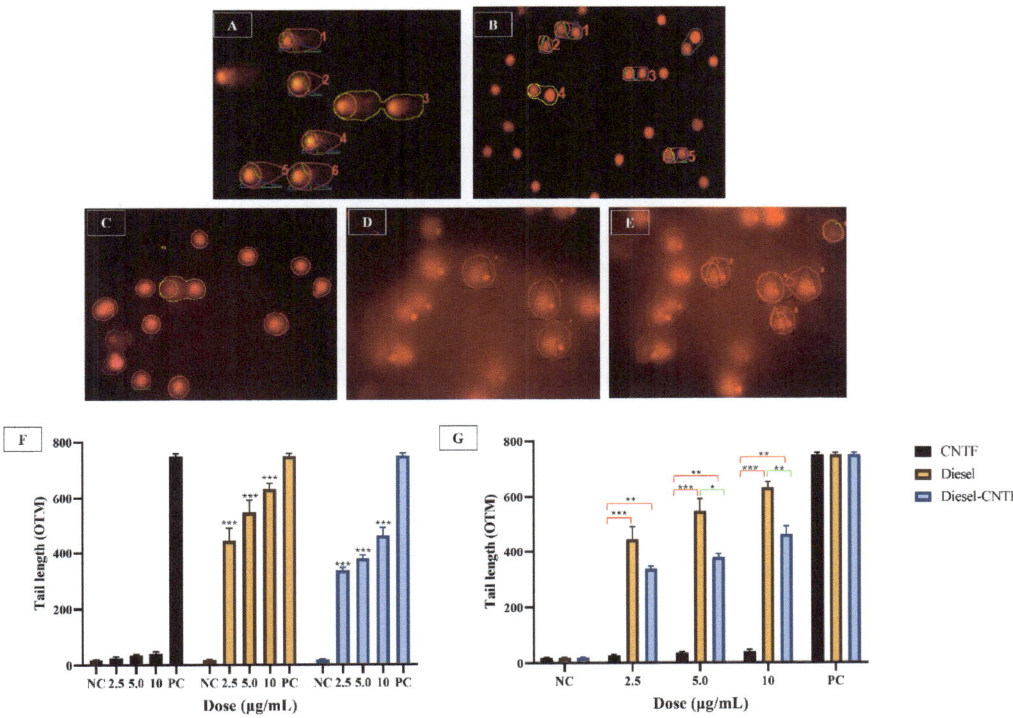

Figure 10. Comet assay of A549 cells exposed to CNTF and DEPs (diesel and diesel–CNTF). The upper part shows the positive control (**A**), negative control (**B**), treatment with CNTF (**C**), treatment with diesel DEPs (**D**), and treatment with diesel–CNTF DEPs (**E**). The lower part shows the statistical tests applied according to the treatment (**F**) and the comparison between treatments according to the doses applied (**G**). Red brackets indicate comparisons between CNTF and DEPs, and green brackets indicate comparisons between diesel DEPs and diesel–CNTF DEPs. Black bars show CNTF in orange diesel and in blue diesel–CNTF. Statistical significance is indicated by $p < 0.05$ *, $p < 0.01$ **, and $p < 0.001$ ***.

Our study revealed no genotoxic effect in cells exposed to CNTF. These results suggest that CNTFs possibly did not enter the nucleus and interact with DNA. Perhaps after a longer exposure time or an increase in concentration, this effect may be more pronounced. Studies investigating the toxicological effects of functionalized CNTs have primarily utilized carboxylic acid and amide-functionalized CNTs [85–87,113]. In their study, Jos and colleagues discovered that exposing differentiated and non-differentiated Caco-2 cells to COOH-CNTs at concentrations ranging from 5 to 1000 μg/mL for 24 h resulted in cytotoxic effects, particularly at concentrations exceeding 100 μg/mL [114]. Similarly, Patlolla et al. observed that the maximum increase in tail DNA was observed in COOH-functionalized (400 μg/mL) at 48 h post-treatment compared to the control [86]. Using low concentrations of CNTF could be a viable alternative to avoid significant biological effects. However, it is recommended to conduct measurements over extended exposure periods.

The effect of the type of functionalization on genotoxicity has been widely evaluated. It has been determined that COOH-CNTs have the highest level of toxicity and that NH2-CNTs have a lower cytotoxic and genotoxic effect [87,115,116]. In addition, other studies have reported that the physiological morphology and structure of CNTs and target cells were one of the main factors in inducing toxicity and genotoxicity [87,117]. This could explain what was found in our study, since the amide functionalization of CNTs and the type of cells used could be reducing their entry and subsequent damage. It has been suggested that early DNA damage highlights the carcinogenic potential of CNTs in lung cells; however, repair mechanisms can restore DNA integrity in most cases [118–122]. This is consistent with the lack of an increase in DNA damage frequency observed in this study in cells treated with CNTF, leading to the hypothesis that DNA breakage induced by these functionalized CNTs is poorly related to potential carcinogenesis.

Our results also indicate that DEPs at each of the concentrations used cause genotoxic damage. However, all concentrations of diesel DEPs differed statistically significantly from diesel–CNTF DEPs. Different studies have shown that DEPs can generate genotoxic damage that has been associated with oxidative damage and DNA strand breaks [65,67,97,99,113]. We found that there is a relationship between increased ROS and DNA damage that is not associated with direct damage to cell structure and function. Among the mechanisms that can generate DNA damage induced by increased ROS produced by DEPs are single and double-strand breaks, abasic sites, base and sugar lesions, and numerous oxidative lesions derived from pyrimidine and purine, such as the formation of 8-oxo-7,8-dihydro-2′-deoxyguanosine [123–129].

Exhaust emissions from the two fuel combinations (diesel or diesel–CNTF) showed different toxicological effects. The general mechanisms of DEP-induced toxicity are complex because of the chemical complexity of DEPs, the biological complexity of the cells involved, and the interactions between them [75]. The chemical composition of DEPs, including the concentration of metals and PAHs, has been shown to impact health due to their ability to induce oxidative stress and disrupt normal homeostasis that can lead to toxicological endpoints—cytotoxicity, genotoxicity, and ultimately cell death [130–134]. In addition, the ability of DEPs to absorb and transport compounds to deeper regions of tissues and cells is also associated with their morphological characteristics, with their size and shape being important aspects to take into consideration [135–138].

The differences in biological effects between diesel and diesel–CNTF DEPs can be attributed to variations in chemical composition and morphology. When CNTFs are used with diesel, both the composition of PAHs and the morphology change, resulting in a lower hydrocarbon content and a more compact and larger particle morphology. Zhang and Balasubramanian note that incorporating metal nanoparticles into fuels can effectively reduce particle mass concentrations, alter chemical composition, and modify particle shape [84]. Similar findings have been reported by other researchers, indicating that the inclusion of carbon nanotubes or metallic particles can significantly alter the physicochemical characteristics of DEPs, or particulate matter [139–141]. Given these observations, future research should focus on investigating the toxicological responses to DEP-CNTF at higher concentrations and extended exposure times. Additionally, evaluating signaling pathways related to DNA damage repair, cell cycle regulation, and cell death induction will be crucial to understanding the broader implications of these modifications.

4. Conclusions

This article presents original research findings pertaining to the impact of DEPs derived from an internal combustion engine utilizing either diesel fuel or a combination of diesel and CNTF. Specifically, this study assesses the effects of these DEPs on cytotoxicity and genotoxicity in lung cells while also examining the physicochemical properties of the DEPs and their correlation with cellular damage. Significantly, the research addresses a gap in the existing literature by focusing on the cellular damage caused by DEPs resulting from the presence of carbonaceous additives (CNTF), an area that has received limited attention compared to studies involving additives containing metallic components.

In the physicochemical characterization, it was observed that the addition of CNTF to diesel fuel altered the morphology and size of particles, as well as the quantity and composition of PAHs. This resulted in the formation of spheroid structures and aggregates of fractal-type particles, with an increase in aerodynamic diameters compared to diesel alone. The concentration of PAHs decreased overall, and the fraction of four aromatic rings (Flu, BaA, and Ch) increased by 31.8% in comparison to the total PAHs in diesel, which only accounted for 5.5% of this fraction. In terms of toxicity, this study identified a BaPeq (benzo(a)pyrene equivalent) of 0.033 mg/L, lower than the 0.065 mg/L found in diesel, suggesting that the incorporation of CNTF in diesel can effectively reduce the toxicity associated with these compounds.

Our research found that diesel DEPs induced higher levels of ROS production compared to both CNTF and diesel–CNTF DEPs. The combination of CNTF with diesel significantly reduced ROS production. We also observed that exposure to DEPs caused genotoxic damage, whereas exposure to functionalized nanoparticles did not increase DNA damage. In addition, the use of CNTF appeared to mitigate diesel-induced damage. We can conclude that the functionalization of carbon nanotubes can potentially reduce the toxic effects of diesel exhaust particles and improve their safety for energy applications.

In conclusion, this investigation provides a deeper understanding of the effects of non-metallic nano-additives on cell toxicity and their potential to improve fuel efficiency. This adds a new perspective to research on fuel nanomaterials, suggesting that amide-functionalized nano-additives could offer a safer alternative to traditional metallic additives with a lower risk of adverse health effects. This approach may influence the development and adoption of new, cleaner, and less toxic fuel technologies.

Author Contributions: Conceptualization, N.G.-L. and P.N.A. methodology, K.C., P.N.A. and N.G.-L.; software, J.S.P.; validation, J.S.P., P.N.A. and N.G.-L.; formal analysis, J.S.P., P.N.A., K.C. and N.G.-L.; investigation, J.S.P., P.N.A. and N.G.-L.; resources, N.G.-L.; data curation, J.S.P., P.N.A. and N.G.-L.; writing—original draft preparation, J.S.P., P.N.A., N.G.-L.; writing—review and editing, J.S.P., P.N.A., K.C., W.R. and N.G.-L.; visualization, J.S.P., P.N.A. and N.G.-L.; supervision, N.G.-L. and W.R.; project administration, W.R.; funding acquisition, N.G.-L. and W.R. All authors have read and agreed to the published version of the manuscript.

Funding: The APC was funded by Grupo genética molecular (GenMol), Universidad de Antioquia. This study received financial support from the Ministerio de Ciencia y Tecnología de Colombia–Minciencias for the research program "Use of carbon nanomaterials as additives in diesel for internal combustion engines working in mode dual with natural gas and their effects on yield, pollutant emissions, and cell damage" (Grant number: No. 508-2020). W.R. was funded by Universidad de Antioquia (CODI sostenibilidad de grupos 2023-2024).

Data Availability Statement: The data that support the findings of this study are available from the corresponding author upon reasonable request.

Acknowledgments: We are grateful to the Group of Chemistry of Energy Resources and Environment, QUIREMA, from the Universidad de Antioquia (Medellín, Colombia) for providing the functionalized CNTs. This study received financial support from the Ministerio de Ciencia y Tecnología de Colombia–Minciencias for the research program "Use of carbon nanomaterials as additives in diesel for internal combustion engines working in mode dual with natural gas and their effects on yield, pollutant emissions, and cell damage" (Grant number: No. 508-2020).

Conflicts of Interest: The authors declare no conflicts of interest.

References

1. Burnett, R.; Chen, H.; Szyszkowicz, M.; Fann, N.; Hubbell, B.; Pope, C.A.; Apte, J.S.; Brauer, M.; Cohen, A.; Weichenthal, S.; et al. Global Estimates of Mortality Associated with Long-Term Exposure to Outdoor Fine Particulate Matter. *Proc. Natl. Acad. Sci. USA* **2018**, *115*, 9592–9597. [CrossRef] [PubMed]
2. McDuffie, E.E.; Martin, R.V.; Spadaro, J.V.; Burnett, R.; Smith, S.J.; O'Rourke, P.; Hammer, M.S.; van Donkelaar, A.; Bindle, L.; Shah, V.; et al. Source Sector and Fuel Contributions to Ambient PM2.5 and Attributable Mortality across Multiple Spatial Scales. *Nat. Commun.* **2021**, *12*, 3594. [CrossRef] [PubMed]
3. McDuffie, E.; Martin, R.; Yin, H.; Brauer, M. Global Burden of Disease from Major Air Pollution Sources (GBD MAPS): A Global Approach. *Res. Rep. Health Eff. Inst.* **2021**, *2021*, 1–45.
4. Leikauf, G.D.; Kim, S.H.; Jang, A.S. Mechanisms of Ultrafine Particle-Induced Respiratory Health Effects. *Exp. Mol. Med.* **2020**, *52*, 329–337. [CrossRef] [PubMed]
5. Thangavel, P.; Park, D.; Lee, Y.C. Recent Insights into Particulate Matter (PM2.5)-Mediated Toxicity in Humans: An Overview. *Int. J. Environ. Res. Public Health* **2022**, *19*, 7511. [CrossRef] [PubMed]

6. Kegl, T.; Kovač Kralj, A.; Kegl, B.; Kegl, M. Nanomaterials as Fuel Additives in Diesel Engines: A Review of Current State, Opportunities, and Challenges. *Prog. Energy Combust. Sci.* **2021**, *83*, 100897. [CrossRef]
7. Cung, K.D.; Wallace, J.; Kalaskar, V.; Smith, E.M.; Briggs, T.; Bitsis, D.C. Experimental Study on Engine and Emissions Performance of Renewable Diesel Methanol Dual Fuel (RMDF) Combustion. *Fuel* **2024**, *357*, 129664. [CrossRef]
8. Lv, J.; Wang, S.; Meng, B. The Effects of Nano-Additives Added to Diesel-Biodiesel Fuel Blends on Combustion and Emission Characteristics of Diesel Engine: A Review. *Energies* **2022**, *15*, 1032. [CrossRef]
9. El-Seesy, A.I.; Abdel-Rahman, A.K.; Bady, M.; Ookawara, S. Performance, Combustion, and Emission Characteristics of a Diesel Engine Fueled by Biodiesel-Diesel Mixtures with Multi-Walled Carbon Nanotubes Additives. *Energy Convers. Manag.* **2017**, *135*, 373–393. [CrossRef]
10. Jin, C.; Wei, J.; Chen, B.; Li, X.; Ying, D.; Gong, L.; Fang, W. Effect of Nanoparticles on Diesel Engines Driven by Biodiesel and Its Blends: A Review of 10 Years of Research. *Energy Convers. Manag.* **2023**, *291*, 117276. [CrossRef]
11. Ampah, J.D.; Yusuf, A.A.; Agyekum, E.B.; Afrane, S.; Jin, C.; Liu, H.; Fattah, I.M.R.; Show, P.L.; Shouran, M.; Habil, M.; et al. Progress and Recent Trends in the Application of Nanoparticles as Low Carbon Fuel Additives—A State of the Art Review. *Nanomaterials* **2022**, *12*, 1515. [CrossRef] [PubMed]
12. Nemmar, A.; Yuvaraju, P.; Beegam, S.; Fahim, M.A.; Ali, B.H. Cerium Oxide Nanoparticles in Lung Acutely Induce Oxidative Stress, Inflammation, and DNA Damage in Various Organs of Mice. *Oxid. Med. Cell. Longev.* **2017**, *2017*, 9639035. [CrossRef] [PubMed]
13. Medici, S.; Peana, M.; Pelucelli, A.; Zoroddu, M.A. An Updated Overview on Metal Nanoparticles Toxicity. *Semin. Cancer Biol.* **2021**, *76*, 17–26. [CrossRef] [PubMed]
14. Ajdary, M.; Moosavi, M.A.; Rahmati, M.; Falahati, M.; Mahboubi, M.; Mandegary, A.; Jangjoo, S.; Mohammadinejad, R.; Varma, R.S. Health Concerns of Various Nanoparticles: A Review of Their in Vitro and in Vivo Toxicity. *Nanomaterials* **2018**, *8*, 634. [CrossRef] [PubMed]
15. Solorio-Rodriguez, S.A.; Wu, D.; Boyadzhiev, A.; Christ, C.; Williams, A.; Halappanavar, S. A Systematic Genotoxicity Assessment of a Suite of Metal Oxide Nanoparticles Reveals Their DNA Damaging and Clastogenic Potential. *Nanomaterials* **2024**, *14*, 743. [CrossRef] [PubMed]
16. Czajka, M.; Sawicki, K.; Sikorska, K.; Popek, S.; Kruszewski, M.; Kapka-Skrzypczak, L. Toxicity of Titanium Dioxide Nanoparticles in Central Nervous System. *Toxicol. In Vitro* **2015**, *29*, 1042–1052. [CrossRef] [PubMed]
17. Rizk, M.Z.; Ali, S.A.; Hamed, M.A.; El-Rigal, N.S.; Aly, H.F.; Salah, H.H. Toxicity of Titanium Dioxide Nanoparticles: Effect of Dose and Time on Biochemical Disturbance, Oxidative Stress and Genotoxicity in Mice. *Biomed. Pharmacother.* **2017**, *90*, 466–472. [CrossRef] [PubMed]
18. Guo, C.; Robertson, S.; Weber, R.J.M.; Buckley, A.; Warren, J.; Hodgson, A.; Rappoport, J.Z.; Ignatyev, K.; Meldrum, K.; Römer, I.; et al. Pulmonary Toxicity of Inhaled Nano-Sized Cerium Oxide Aerosols in Sprague-Dawley Rats. *Nanotoxicology* **2019**, *13*, 733–750. [CrossRef] [PubMed]
19. Chen, L.; Yokel, R.A.; Hennig, B.; Toborek, M. Manufactured Aluminum Oxide Nanoparticles Decrease Expression of Tight Junction Proteins in Brain Vasculature. *J. Neuroimmune Pharmacol.* **2008**, *3*, 286–295. [CrossRef]
20. Acevedo, J.; Escobar, G.; López, D. Functionalized Multiwalled Carbon Nanotubes for the Improvement of Dispersion Stability in Diesel Fuel-Type for Potential Application as Nano Additives. *ChemistrySelect* **2022**, *7*, e202202626. [CrossRef]
21. Sivamaran, V.; Balasubramanian, V.; Gopalakrishnan, M.; Viswabaskaran, V.; Gourav Rao, A.; Selvamani, S.T. Carbon Nanotubes, Nanorings, and Nanospheres: Synthesis and Fabrication via Chemical Vapor Deposition—A Review. *Nanomater. Nanotechnol.* **2022**, *12*. [CrossRef]
22. Meng, L.; Fu, C.; Lu, Q. Advanced Technology for Functionalization of Carbon Nanotubes. *Prog. Nat. Sci.* **2009**, *19*, 801–810. [CrossRef]
23. Kish, S.S.; Rashidi, A.; Aghabozorg, H.R.; Moradi, L. Increasing the Octane Number of Gasoline Using Functionalized Carbon Nanotubes. *Appl. Surf. Sci.* **2010**, *256*, 3472–3477. [CrossRef]
24. Basiuk, E.V.; Ramírez-Calera, I.J.; Meza-Laguna, V.; Abarca-Morales, E.; Pérez-Rey, L.A.; Re, M.; Prete, P.; Lovergine, N.; Álvarez-Zauco, E.; Basiuk, V.A. Solvent-Free Functionalization of Carbon Nanotube Buckypaper with Amines. *Appl. Surf. Sci.* **2015**, *357*, 1355–1368. [CrossRef]
25. Gamboa, D.; Herrera, B.; Acevedo, J.; López, D.; Cacua, K. Experimental Evaluation of a Diesel Engine Using Amide-Functionalized Carbon Nanotubes as Additives in Commercial Diesel and Palm-Oil Biodiesel. *Int. J. Thermofluids* **2024**, *22*, 100669. [CrossRef]
26. EL-Seesy, A.I.; Waly, M.S.; El-Batsh, H.M.; El-Zoheiry, R.M. Enhancement of the Diesel Fuel Characteristics by Using Nitrogen-Doped Multi-Walled Carbon Nanotube Additives. *Process Saf. Environ. Prot.* **2023**, *171*, 561–577. [CrossRef]
27. Shekofteh, M.; Gundoshmian, T.M.; Jahanbakhshi, A.; Heidari-Maleni, A. Performance and Emission Characteristics of a Diesel Engine Fueled with Functionalized Multi-Wall Carbon Nanotubes (MWCNTs-OH) and Diesel–Biodiesel–Bioethanol Blends. *Energy Rep.* **2020**, *6*, 1438–1447. [CrossRef]
28. Mesri Gundoshmian, T.; Heidari-Maleni, A.; Jahanbakhshi, A. Evaluation of Performance and Emission Characteristics of a CI Engine Using Functional Multi-Walled Carbon Nanotubes (MWCNTs-COOH) Additives in Biodiesel-Diesel Blends. *Fuel* **2021**, *287*, 119525. [CrossRef]

29. Waly, M.S.; EL-Seesy, A.I.; El-Batsh, H.M.; El-Zoheiry, R.M. Combustion and Emissions Characteristics of a Diesel Engine Fuelled with Diesel Fuel and Different Concentrations of Amino-Functionalized Multi-Walled Carbon Nanotube. *Atmos. Pollut. Res.* **2023**, *14*, 101831. [CrossRef]
30. Raphey, V.R.; Henna, T.K.; Nivitha, K.P.; Mufeedha, P.; Sabu, C.; Pramod, K. Advanced Biomedical Applications of Carbon Nanotube. *Mater. Sci. Eng. C* **2019**, *100*, 616–630. [CrossRef]
31. Jiang, Y.; Zhang, H.; Wang, Y.; Chen, M.; Ye, S.; Hou, Z.; Ren, L. Modulation of Apoptotic Pathways of Macrophages by Surface-Functionalized Multi-Walled Carbon Nanotubes. *PLoS ONE* **2013**, *8*, e65756. [CrossRef] [PubMed]
32. Yuan, X.; Zhang, X.; Sun, L.; Wei, Y.; Wei, X. Cellular Toxicity and Immunological Effects of Carbon-Based Nanomaterials. *Part. Fibre Toxicol.* **2019**, *16*, 18. [CrossRef] [PubMed]
33. Zhou, L.; Forman, H.J.; Ge, Y.; Lunec, J. Multi-Walled Carbon Nanotubes: A Cytotoxicity Study in Relation to Functionalization, Dose and Dispersion. *Toxicol. In Vitro* **2017**, *42*, 292. [CrossRef] [PubMed]
34. Tong, H.; McGee, J.K.; Saxena, R.K.; Kodavanti, U.P.; Devlin, R.B.; Gilmour, M.I. Influence of Acid Functionalization on the Cardiopulmonary Toxicity of Carbon Nanotubes and Carbon Black Particles in Mice. *Toxicol. Appl. Pharmacol.* **2009**, *239*, 224–232. [CrossRef] [PubMed]
35. Hirano, S.; Fujitani, Y.; Furuyama, A.; Kanno, S. Uptake and Cytotoxic Effects of Multi-Walled Carbon Nanotubes in Human Bronchial Epithelial Cells. *Toxicol. Appl. Pharmacol.* **2010**, *249*, 8–15. [CrossRef] [PubMed]
36. Roda, E.; Coccini, T.; Acerbi, D.; Barni, S.; Vaccarone, R.; Manzo, L. Comparative Pulmonary Toxicity Assessment of Pristine and Functionalized Multi-Walled Carbon Nanotubes Intratracheally Instilled in Rats: Morphohistochemical Evaluations. *Histol. Histopathol.* **2011**, *26*, 357–367. [CrossRef]
37. Reiter, M.S.; Kockelman, K.M. The Problem of Cold Starts: A Closer Look at Mobile Source Emissions Levels. *Transp. Res. Part D Transp. Environ.* **2016**, *43*, 123–132. [CrossRef]
38. Zare, A.; Bodisco, T.A.; Verma, P.; Jafari, M.; Babaie, M.; Yang, L.; Rahman, M.M.; Banks, A.P.W.; Ristovski, Z.D.; Brown, R.J.; et al. Particulate Number Emissions during Cold-Start with Diesel and Biofuels: A Special Focus on Particle Size Distribution. *Sustain. Energy Technol. Assess.* **2022**, *51*, 101953. [CrossRef]
39. Loomis, D.; Grosse, Y.; Lauby-Secretan, B.; El Ghissassi, F.; Bouvard, V.; Benbrahim-Tallaa, L.; Guha, N.; Baan, R.; Mattock, H.; Straif, K. The Carcinogenicity of Outdoor Air Pollution. *Lancet Oncol.* **2013**, *14*, 1262–1263. [CrossRef]
40. IARC Working Group on the Evaluation of Carcinogenic Risks to Humans. In Proceedings of the International Agency for Research on Cancer Diesel and Gasoline Engine Exhausts and Some Nitroarenes, Lyon, France, 5–12 June 2014; Volume 703.
41. Richter, H.; Howard, J.B. Formation of Polycyclic Aromatic Hydrocarbons and Their Growth to Soot—A Review of Chemical Reaction Pathways. *Prog. Energy Combust. Sci.* **2000**, *26*, 565–608. [CrossRef]
42. Khobragade, R.; Singh, S.K.; Shukla, P.C.; Gupta, T.; Al-Fatesh, A.S.; Agarwal, A.K.; Labhasetwar, N.K. Chemical Composition of Diesel Particulate Matter and Its Control. *Catal. Rev.* **2019**, *61*, 447–515. [CrossRef]
43. Wang, X.; Wang, Y.; Bai, Y.; Wang, P.; Zhao, Y. An Overview of Physical and Chemical Features of Diesel Exhaust Particles. *J. Energy Inst.* **2019**, *92*, 1864–1888. [CrossRef]
44. Wang, X.; Wang, Y.; Guo, F.; Wang, D.; Bai, Y. Physicochemical Characteristics of Particulate Matter Emitted by Diesel Blending with Various Aromatics. *Fuel* **2020**, *275*, 117928. [CrossRef]
45. Xiao, Z.; Ladommatos, N.; Zhao, H. The Effect of Aromatic Hydrocarbons and Oxygenates on Diesel Engine Emissions. *Proc. Inst. Mech. Eng. Part D J. Automob. Eng.* **2000**, *214*, 307–332. [CrossRef]
46. Qian, Y.; Qiu, Y.; Zhang, Y.; Lu, X. Effects of Different Aromatics Blended with Diesel on Combustion and Emission Characteristics with a Common Rail Diesel Engine. *Appl. Therm. Eng.* **2017**, *125*, 1530–1538. [CrossRef]
47. Jin, T.; Qu, L.; Liu, S.; Gao, J.; Wang, J.; Wang, F.; Zhang, P.; Bai, Z.; Xu, X. Chemical Characteristics of Particulate Matter Emitted from a Heavy Duty Diesel Engine and Correlation among Inorganic and PAH Components. *Fuel* **2014**, *116*, 655–661. [CrossRef]
48. Keyte, I.J.; Albinet, A.; Harrison, R.M. On-Road Traffic Emissions of Polycyclic Aromatic Hydrocarbons and Their Oxy- and Nitro- Derivative Compounds Measured in Road Tunnel Environments. *Sci. Total Environ.* **2016**, *566–567*, 1131–1142. [CrossRef] [PubMed]
49. Ho, K.F.; Ho, S.S.H.; Lee, S.C.; Cheng, Y.; Chow, J.C.; Watson, J.G.; Louie, P.K.K.; Tian, L. Emissions of Gas- and Particle-Phase Polycyclic Aromatic Hydrocarbons (PAHs) in the Shing Mun Tunnel, Hong Kong. *Atmos. Environ.* **2009**, *43*, 6343–6351. [CrossRef]
50. Patel, A.B.; Shaikh, S.; Jain, K.R.; Desai, C.; Madamwar, D. Polycyclic Aromatic Hydrocarbons: Sources, Toxicity, and Remediation Approaches. *Front. Microbiol.* **2020**, *11*, 562813. [CrossRef]
51. Collins, J.F.; Brown, J.P.; Alexeeff, G.V.; Salmon, A.G. Potency Equivalency Factors for Some Polycyclic Aromatic Hydrocarbons and Polycyclic Aromatic Hydrocarbon Derivatives. *Regul. Toxicol. Pharmacol.* **1998**, *28*, 45–54. [CrossRef]
52. Behzadi, S.; Serpooshan, V.; Tao, W.; Hamaly, M.A.; Alkawareek, M.Y.; Dreaden, E.C.; Brown, D.; Alkilany, A.M.; Farokhzad, O.C.; Mahmoudi, M. Cellular Uptake of Nanoparticles: Journey inside the Cell. *Chem. Soc. Rev.* **2017**, *46*, 4218–4244. [CrossRef]
53. Ghosh, M.; Murugadoss, S.; Janssen, L.; Cokic, S.; Mathyssen, C.; Van Landuyt, K.; Janssens, W.; Carpentier, S.; Godderis, L.; Hoet, P. Distinct Autophagy-Apoptosis Related Pathways Activated by Multi-Walled (NM 400) and Single-Walled Carbon Nanotubes (NIST-SRM2483) in Human Bronchial Epithelial (16HBE14o-) Cells. *J. Hazard. Mater.* **2020**, *387*, 121691. [CrossRef]
54. Chun, Y.; Kim, J. Autophagy: An Essential Degradation Program for Cellular Homeostasis and Life. *Cells* **2018**, *7*, 278. [CrossRef] [PubMed]

55. Gómez-Virgilio, L.; Silva-Lucero, M.D.C.; Flores-Morelos, D.S.; Gallardo-Nieto, J.; Lopez-Toledo, G.; Abarca-Fernandez, A.M.; Zacapala-Gómez, A.E.; Luna-Muñoz, J.; Montiel-Sosa, F.; Soto-Rojas, L.O.; et al. Autophagy: A Key Regulator of Homeostasis and Disease: An Overview of Molecular Mechanisms and Modulators. *Cells* **2022**, *11*, 2262. [CrossRef] [PubMed]
56. Haniu, H.; Saito, N.; Matsuda, Y.; Tsukahara, T.; Maruyama, K.; Usui, Y.; Aoki, K.; Takanashi, S.; Kobayashi, S.; Nomura, H.; et al. Culture Medium Type Affects Endocytosis of Multi-Walled Carbon Nanotubes in BEAS-2B Cells and Subsequent Biological Response. *Toxicol. In Vitro* **2013**, *27*, 1679–1685. [CrossRef] [PubMed]
57. Tsukahara, T.; Matsuda, Y.; Haniu, H. The Role of Autophagy as a Mechanism of Toxicity Induced by Multi-Walled Carbon Nanotubes in Human Lung Cells. *Int. J. Mol. Sci.* **2015**, *16*, 40. [CrossRef] [PubMed]
58. Ursini, C.L.; Maiello, R.; Ciervo, A.; Fresegna, A.M.; Buresti, G.; Superti, F.; Marchetti, M.; Iavicoli, S.; Cavallo, D. Evaluation of Uptake, Cytotoxicity and Inflammatory Effects in Respiratory Cells Exposed to Pristine and -OH and -COOH Functionalized Multi-Wall Carbon Nanotubes. *J. Appl. Toxicol.* **2016**, *36*, 394–403. [CrossRef] [PubMed]
59. Li, Y.; Cao, J. The Impact of Multi-Walled Carbon Nanotubes (MWCNTs) on Macrophages: Contribution of MWCNT Characteristics. *Sci. China Life Sci.* **2018**, *61*, 1333–1351. [CrossRef] [PubMed]
60. Pierdominici, M.; Maselli, A.; Cecchetti, S.; Tinari, A.; Mastrofrancesco, A.; Alfè, M.; Gargiulo, V.; Beatrice, C.; Di Blasio, G.; Carpinelli, G.; et al. Diesel Exhaust Particle Exposure in Vitro Impacts T Lymphocyte Phenotype and Function. *Part. Fibre Toxicol.* **2014**, *11*, 74. [CrossRef]
61. Cinque, L.; De Leonibus, C.; Iavazzo, M.; Krahmer, N.; Intartaglia, D.; Salierno, F.G.; De Cegli, R.; Di Malta, C.; Svelto, M.; Lanzara, C.; et al. MiT/TFE Factors Control ER-Phagy via Transcriptional Regulation of FAM134B. *EMBO J.* **2020**, *39*, 5696. [CrossRef]
62. Xu, Y.; Zhou, P.; Cheng, S.; Lu, Q.; Nowak, K.; Hopp, A.K.; Li, L.; Shi, X.; Zhou, Z.; Gao, W.; et al. A Bacterial Effector Reveals the V-ATPase-ATG16L1 Axis That Initiates Xenophagy. *Cell* **2019**, *178*, 552–566. [CrossRef]
63. Pickles, S.; Vigié, P.; Youle, R.J. Mitophagy and Quality Control Mechanisms in Mitochondrial Maintenance. *Curr. Biol.* **2018**, *28*, R170–R185. [CrossRef]
64. Yang, C.; Wang, X. Lysosome Biogenesis: Regulation and Functions. *J. Cell Biol.* **2021**, *220*, 2001. [CrossRef]
65. Nie, B.; Liu, X.; Lei, C.; Liang, X.; Zhang, D.; Zhang, J. The Role of Lysosomes in Airborne Particulate Matter-Induced Pulmonary Toxicity. *Sci. Total Environ.* **2024**, *919*, 170893. [CrossRef] [PubMed]
66. Tabet, L.; Bussy, C.; Amara, N.; Setyan, A.; Grodet, A.; Rossi, M.J.; Pairon, J.C.; Boczkowski, J.; Lanone, S. Adverse Effects of Industrial Multiwalled Carbon Nanotubes on Human Pulmonary Cells. *J. Toxicol. Environ. Health A* **2009**, *72*, 60. [CrossRef]
67. Periasamy, V.S.; Athinarayanan, J.; Alfawaz, M.A.; Alshatwi, A.A. Carbon Nanoparticle Induced Cytotoxicity in Human Mesenchymal Stem Cells through Upregulation of TNF3, NFKBIA and BCL2L1 Genes. *Chemosphere* **2016**, *144*, 275–284. [CrossRef] [PubMed]
68. Gupta, S.S.; Singh, K.P.; Gupta, S.; Dusinska, M.; Rahman, Q. Do Carbon Nanotubes and Asbestos Fibers Exhibit Common Toxicity Mechanisms? *Nanomaterials* **2022**, *12*, 1708. [CrossRef] [PubMed]
69. Kharlamova, M.V.; Kramberger, C. Cytotoxicity of Carbon Nanotubes, Graphene, Fullerenes, and Dots. *Nanomaterials* **2023**, *13*, 1458. [CrossRef]
70. Coccini, T.; Manzo, L.; Roda, E. Safety Evaluation of Engineered Nanomaterials for Health Risk Assessment: An Experimental Tiered Testing Approach Using Pristine and Functionalized Carbon Nanotubes. *ISRN Toxicol.* **2013**, *2013*, 825427. [CrossRef]
71. Chowdhry, A.; Kaur, J.; Khatri, M.; Puri, V.; Tuli, R.; Puri, S. Characterization of Functionalized Multiwalled Carbon Nanotubes and Comparison of Their Cellular Toxicity between HEK 293 Cells and Zebra Fish in Vivo. *Heliyon* **2019**, *5*, e02605. [CrossRef]
72. Chen, B.; Liu, Y.; Song, W.M.; Hayashi, Y.; Ding, X.C.; Li, W.H. In Vitro Evaluation of Cytotoxicity and Oxidative Stress Induced by Multiwalled Carbon Nanotubes in Murine RAW 264.7 Macrophages and Human A549 Lung Cells. *Biomed. Environ. Sci.* **2011**, *24*, 593–601. [CrossRef]
73. Fox, J.R.; Cox, D.P.; Drury, B.E.; Gould, T.R.; Kavanagh, T.J.; Paulsen, M.H.; Sheppard, L.; Simpson, C.D.; Stewart, J.A.; Larson, T.V.; et al. Chemical Characterization and in Vitro Toxicity of Diesel Exhaust Particulate Matter Generated under Varying Conditions. *Air Qual. Atmos. Health* **2015**, *8*, 507–519. [CrossRef]
74. Demir, C.; Tasdemir, D.; Ilhan, S.; Isik, A.F.; Bayram, H. Toxicological and Inflammatory Effects of Engineered Nanoparticles and Diesel Exhaust Particles on Human Mesothelial Cells. *Eur. Respir. J.* **2017**, *50*, PA3918. [CrossRef]
75. Hakkarainen, H.; Järvinen, A.; Lepistö, T.; Salo, L.; Kuittinen, N.; Laakkonen, E.; Yang, M.; Martikainen, M.V.; Saarikoski, S.; Aurela, M.; et al. Toxicity of Exhaust Emissions from High Aromatic and Non-Aromatic Diesel Fuels Using in Vitro ALI Exposure System. *Sci. Total Environ.* **2023**, *890*, 164215. [CrossRef] [PubMed]
76. Jäntti, H.; Jonk, S.; Gómez Budia, M.; Ohtonen, S.; Fagerlund, I.; Fazaludeen, M.F.; Aakko-Saksa, P.; Pebay, A.; Lehtonen, Š.; Koistinaho, J.; et al. Particulate Matter from Car Exhaust Alters Function of Human IPSC-Derived Microglia. *Part. Fibre Toxicol.* **2024**, *21*, 6. [CrossRef] [PubMed]
77. Si, M.; Cheng, Q.; Song, J.; Liu, Y.; Tao, M.; Lou, C. Study on Inversion of Morphological Parameters of Soot Aggregates in Hydrocarbon Flames. *Combust. Flame* **2017**, *183*, 261–270. [CrossRef]
78. Mühlbauer, W.; Zöllner, C.; Lehmann, S.; Lorenz, S.; Brüggemann, D. Correlations between Physicochemical Properties of Emitted Diesel Particulate Matter and Its Reactivity. *Combust. Flame* **2016**, *167*, 39–51. [CrossRef]
79. Nováková, Z.; Novák, J.; Kitanovski, Z.; Kukučka, P.; Smutná, M.; Wietzoreck, M.; Lammel, G.; Hilscherová, K. Toxic Potentials of Particulate and Gaseous Air Pollutant Mixtures and the Role of PAHs and Their Derivatives. *Environ. Int.* **2020**, *139*, 105634. [CrossRef]

80. Lara, S.; Villanueva, F.; Cabañas, B.; Sagrario, S.; Aranda, I.; Soriano, J.A.; Martin, P. Determination of Policyclic Aromatic Compounds, (PAH, Nitro-PAH and Oxy-PAH) in Soot Collected from a Diesel Engine Operating with Different Fuels. *Sci. Total Environ.* **2023**, *900*, 165755. [CrossRef]
81. Majeed, W.; Bourdo, S.; Petibone, D.M.; Saini, V.; Vang, K.B.; Nima, Z.A.; Alghazali, K.M.; Darrigues, E.; Ghosh, A.; Watanabe, F.; et al. The Role of Surface Chemistry in the Cytotoxicity Profile of Graphene. *J. Appl. Toxicol.* **2017**, *37*, 462–470. [CrossRef]
82. Zerboni, A.; Bengalli, R.; Baeri, G.; Fiandra, L.; Catelani, T.; Mantecca, P. Mixture Effects of Diesel Exhaust and Metal Oxide Nanoparticles in Human Lung A549 Cells. *Nanomaterials* **2019**, *9*, 1302. [CrossRef] [PubMed]
83. Zerboni, A.; Bengalli, R.; Fiandra, L.; Catelani, T.; Mantecca, P. Cellular Mechanisms Involved in the Combined Toxic Effects of Diesel Exhaust and Metal Oxide Nanoparticles. *Nanomaterials* **2021**, *11*, 1437. [CrossRef]
84. Zhang, Z.H.; Balasubramanian, R. Effects of Cerium Oxide and Ferrocene Nanoparticles Addition As Fuel-Borne Catalysts on Diesel Engine Particulate Emissions: Environmental and Health Implications. *Environ. Sci. Technol.* **2017**, *51*, 4248–4258. [CrossRef]
85. Ursini, C.L.; Cavallo, D.; Fresegna, A.M.; Ciervo, A.; Maiello, R.; Buresti, G.; Casciardi, S.; Tombolini, F.; Bellucci, S.; Iavicoli, S. Comparative Cyto-Genotoxicity Assessment of Functionalized and Pristine Multiwalled Carbon Nanotubes on Human Lung Epithelial Cells. *Toxicol. In Vitro* **2012**, *26*, 831–840. [CrossRef] [PubMed]
86. Patlolla, A.; Patlolla, B.; Tchounwou, P. Evaluation of Cell Viability, DNA Damage, and Cell Death in Normal Human Dermal Fibroblast Cells Induced by Functionalized Multiwalled Carbon Nanotube. *Mol. Cell. Biochem.* **2010**, *338*, 225–232. [CrossRef]
87. Song, G.; Guo, X.; Zong, X.; Du, L.; Zhao, J.; Lai, C.; Jin, X. Toxicity of Functionalized Multi-Walled Carbon Nanotubes on Bone Mesenchymal Stem Cell in Rats. *Dent. Mater. J.* **2019**, *38*, 127–135. [CrossRef]
88. Cattani-Cavalieri, I.; Baarsma, H.A.; Oun, A.; Van Der Veen, M.; Dolga, A.; Valença, S.D.S.; Schmidt, M. Diesel Exhaust Particles Alter Mitochondrial Bioenergetics in Human Bronchial Epithelial Cells. *ERJ Open Res.* **2020**, *6*, 23. [CrossRef]
89. Lawal, A.O. Diesel Exhaust Particles and the Induction of Macrophage Activation and Dysfunction. *Inflammation* **2018**, *41*, 356–363. [CrossRef]
90. Taylor-Just, A.J.; Ihrie, M.D.; Duke, K.S.; Lee, H.Y.; You, D.J.; Hussain, S.; Kodali, V.K.; Ziemann, C.; Creutzenberg, O.; Vulpoi, A.; et al. The Pulmonary Toxicity of Carboxylated or Aminated Multi-Walled Carbon Nanotubes in Mice Is Determined by the Prior Purification Method. *Part. Fibre Toxicol.* **2020**, *17*, 60. [CrossRef]
91. Lanone, S.; Andujar, P.; Kermanizadeh, A.; Boczkowski, J. Determinants of Carbon Nanotube Toxicity. *Adv. Drug Deliv. Rev.* **2013**, *65*, 2063–2069. [CrossRef]
92. Tokonami, R.; Aoki, K.; Goto, T.; Takahashi, T. Surface Modification of Carbon Fiber for Enhancing the Mechanical Strength of Composites. *Polymers* **2022**, *14*, 3999. [CrossRef] [PubMed]
93. Checa, J.; Aran, J.M. Reactive Oxygen Species: Drivers of Physiological and Pathological Processes. *J. Inflamm. Res.* **2020**, *13*, 1057. [CrossRef]
94. Jomova, K.; Raptova, R.; Alomar, S.Y.; Alwasel, S.H.; Nepovimova, E.; Kuca, K.; Valko, M. Reactive Oxygen Species, Toxicity, Oxidative Stress, and Antioxidants: Chronic Diseases and Aging. *Arch. Toxicol.* **2023**, *97*, 2499–2574. [CrossRef] [PubMed]
95. Nel, A.; Xia, T.; Mädler, L.; Li, N. Toxic Potential of Materials at the Nanolevel. *Science* **2006**, *311*, 622–627. [CrossRef] [PubMed]
96. Chetyrkina, M.R.; Fedorov, F.S.; Nasibulin, A.G. In Vitro Toxicity of Carbon Nanotubes: A Systematic Review. *RSC Adv.* **2022**, *12*, 16235–16256. [CrossRef] [PubMed]
97. Guo, Y.Y.; Zhang, J.; Zheng, Y.F.; Yang, J.; Zhu, X.Q. Cytotoxic and Genotoxic Effects of Multi-Wall Carbon Nanotubes on Human Umbilical Vein Endothelial Cells in Vitro. *Mutat. Res.* **2011**, *721*, 184–191. [CrossRef] [PubMed]
98. Lee, K.; Lee, J.; Kwak, M.; Cho, Y.L.; Hwang, B.; Cho, M.J.; Lee, N.G.; Park, J.; Lee, S.H.; Park, J.G.; et al. Two Distinct Cellular Pathways Leading to Endothelial Cell Cytotoxicity by Silica Nanoparticle Size. *J. Nanobiotechnol.* **2019**, *17*, 24. [CrossRef] [PubMed]
99. Yu, Z.; Li, Q.; Wang, J.; Yu, Y.; Wang, Y.; Zhou, Q.; Li, P. Reactive Oxygen Species-Related Nanoparticle Toxicity in the Biomedical Field. *Nanoscale Res. Lett.* **2020**, *15*, 115. [CrossRef] [PubMed]
100. Vijayalakshmi, V.; Sadanandan, B.; Anjanapura, R.V. In Vitro Comparative Cytotoxic Assessment of Pristine and Carboxylic Functionalized Multiwalled Carbon Nanotubes on LN18 Cells. *J. Biochem. Mol. Toxicol.* **2023**, *37*, e23283. [CrossRef]
101. Kostarelos, K.; Lacerda, L.; Pastorin, G.; Wu, W.; Wieckowski, S.; Luangsivilay, J.; Godefroy, S.; Pantarotto, D.; Briand, J.P.; Muller, S.; et al. Cellular Uptake of Functionalized Carbon Nanotubes Is Independent of Functional Group and Cell Type. *Nat. Nanotechnol.* **2007**, *2*, 108–113. [CrossRef]
102. Höfinger, S.; Melle-Franco, M.; Gallo, T.; Cantelli, A.; Calvaresi, M.; Gomes, J.A.N.F.; Zerbetto, F. A Computational Analysis of the Insertion of Carbon Nanotubes into Cellular Membranes. *Biomaterials* **2011**, *32*, 7079–7085. [CrossRef] [PubMed]
103. AbouAitah, K.; Abdelaziz, A.M.; Higazy, I.M.; Swiderska-Sroda, A.; Hassan, A.M.E.; Shaker, O.G.; Szałaj, U.; Stobinski, L.; Malolepszy, A.; Lojkowski, W. Functionalized Carbon Nanotubes for Delivery of Ferulic Acid and Diosgenin Anticancer Natural Agents. *ACS Appl. Bio Mater.* **2024**, *7*, 791–811. [CrossRef]
104. Karimi, M.; Ghasemi, A.; Mirkiani, S.; Basri, S.M.M.; Hamblin, M.R. Mechanism of Carbon Nanotube Uptake by Cells. *Carbon Nanotub. Drug Gene Deliv.* **2017**, *5*, 5–6. [CrossRef]
105. Ijaz, H.; Mahmood, A.; Abdel-Daim, M.M.; Sarfraz, R.M.; Zaman, M.; Zafar, N.; Alshehery, S.; Salem-Bekhit, M.M.; Ali, M.A.; Eltayeb, L.B.; et al. Review on Carbon Nanotubes (CNTs) and Their Chemical and Physical Characteristics, with Particular Emphasis on Potential Applications in Biomedicine. *Inorg. Chem. Commun.* **2023**, *155*, 111020. [CrossRef]

106. Danielsen, P.H.; Poulsen, S.S.; Knudsen, K.B.; Clausen, P.A.; Jensen, K.A.; Wallin, H.; Vogel, U. Physicochemical Properties of 26 Carbon Nanotubes as Predictors for Pulmonary Inflammation and Acute Phase Response in Mice Following Intratracheal Lung Exposure. *Environ. Toxicol. Pharmacol.* **2024**, *107*, 104413. [CrossRef]
107. Shi, J.; Tian, F.; Ren, J.; Li, R.; Yang, M.; Li, W. Diesel Exhaust Particulate Matter Induces GC-1 Spg Cells Oxidative Stress by KEAP1-NRF2 Pathway and Inhibition of ATP5α1 S-Sulfhydration. *Food Chem. Toxicol.* **2024**, *189*, 114746. [CrossRef] [PubMed]
108. Pizzino, G.; Irrera, N.; Cucinotta, M.; Pallio, G.; Mannino, F.; Arcoraci, V.; Squadrito, F.; Altavilla, D.; Bitto, A. Oxidative Stress: Harms and Benefits for Human Health. *Oxid. Med. Cell. Longev.* **2017**, *2017*, 8416763. [CrossRef]
109. Kwon, M.; Jung, J.; Park, H.S.; Kim, N.H.; Lee, J.; Park, J.; Kim, Y.; Shin, S.; Lee, B.S.; Cheong, Y.H.; et al. Diesel Exhaust Particle Exposure Accelerates Oxidative DNA Damage and Cytotoxicity in Normal Human Bronchial Epithelial Cells through PD-L1. *Environ. Pollut.* **2023**, *317*, 120705. [CrossRef] [PubMed]
110. Lawal, A.O.; Davids, L.M.; Marnewick, J.L. Diesel Exhaust Particles and Endothelial Cells Dysfunction: An Update. *Toxicol. In Vitro* **2016**, *32*, 92–104. [CrossRef]
111. Wei, J.; Wang, Y. Effects of Biodiesels on the Physicochemical Properties and Oxidative Reactivity of Diesel Particulates: A Review. *Sci. Total Environ.* **2021**, *788*, 147753. [CrossRef]
112. Wong, J.Y.Y.; Imani, P.; Grigoryan, H.; Bassig, B.A.; Dai, Y.; Hu, W.; Blechter, B.; Rahman, M.L.; Ji, B.T.; Duan, H.; et al. Exposure to Diesel Engine Exhaust and Alterations to the Cys34/Lys525 Adductome of Human Serum Albumin. *Environ. Toxicol. Pharmacol.* **2022**, *95*, 103966. [CrossRef]
113. Pecchillo Cimmino, T.; Ammendola, R.; Cattaneo, F.; Esposito, G. NOX Dependent ROS Generation and Cell Metabolism. *Int. J. Mol. Sci.* **2023**, *24*, 2086. [CrossRef] [PubMed]
114. Jos, A.; Pichardo, S.; Puerto, M.; Sánchez, E.; Grilo, A.; Cameán, A.M. Cytotoxicity of Carboxylic Acid Functionalized Single Wall Carbon Nanotubes on the Human Intestinal Cell Line Caco-2. *Toxicol. In Vitro* **2009**, *23*, 1491–1496. [CrossRef] [PubMed]
115. Chatterjee, N.; Yang, J.S.; Choi, J. Differential Genotoxic and Epigenotoxic Effects of Graphene Family Nanomaterials (GFNs) in Human Bronchial Epithelial Cells. *Mutat. Res. Genet. Toxicol. Environ. Mutagen.* **2016**, *798–799*, 1–10. [CrossRef] [PubMed]
116. Jackson, P.; Kling, K.; Jensen, K.A.; Clausen, P.A.; Madsen, A.M.; Wallin, H.; Vogel, U. Characterization of Genotoxic Response to 15 Multiwalled Carbon Nanotubes with Variable Physicochemical Properties Including Surface Functionalizations in the FE1-Muta(TM) Mouse Lung Epithelial Cell Line. *Environ. Mol. Mutagen.* **2015**, *56*, 183–203. [CrossRef] [PubMed]
117. Alarifi, S.; Ali, D. Mechanisms of Multi-Walled Carbon Nanotubes-Induced Oxidative Stress and Genotoxicity in Mouse Fibroblast Cells. *Int. J. Toxicol.* **2015**, *34*, 258–265. [CrossRef] [PubMed]
118. Møller, P.; Jacobsen, N.R. Weight of Evidence Analysis for Assessing the Genotoxic Potential of Carbon Nanotubes. *Crit. Rev. Toxicol.* **2017**, *47*, 867–884. [CrossRef] [PubMed]
119. Ema, M.; Masumori, S.; Kobayashi, N.; Naya, M.; Endoh, S.; Maru, J.; Hosoi, M.; Uno, F.; Nakajima, M.; Hayashi, M.; et al. In Vivo Comet Assay of Multi-Walled Carbon Nanotubes Using Lung Cells of Rats Intratracheally Instilled. *J. Appl. Toxicol.* **2013**, *33*, 1053–1060. [CrossRef]
120. Lindberg, H.K.; Falck, G.C.M.; Singh, R.; Suhonen, S.; Järventaus, H.; Vanhala, E.; Catalán, J.; Farmer, P.B.; Savolainen, K.M.; Norppa, H. Genotoxicity of Short Single-Wall and Multi-Wall Carbon Nanotubes in Human Bronchial Epithelial and Mesothelial Cells in Vitro. *Toxicology* **2013**, *313*, 24–37. [CrossRef] [PubMed]
121. Migliore, L.; Saracino, D.; Bonelli, A.; Colognato, R.; D'Errico, M.R.; Magrini, A.; Bergamaschi, A.; Bergamaschi, E. Carbon Nanotubes Induce Oxidative DNA Damage in RAW 264.7 Cells. *Environ. Mol. Mutagen.* **2010**, *51*, 294–303. [CrossRef]
122. Jantzen, K.; Roursgaard, M.; Desler, C.; Loft, S.; Rasmussen, L.J.; Møller, P. Oxidative Damage to DNA by Diesel Exhaust Particle Exposure in Co-Cultures of Human Lung Epithelial Cells and Macrophages. *Mutagenesis* **2012**, *27*, 693–701. [CrossRef]
123. Uski, O.J.; Rankin, G.D.; Wingfors, H.; Magnusson, R.; Boman, C.; Muala, A.; Blomberg, A.; Bosson, J.; Sandström, T. In Vitro Toxicity Evaluation in A549 Cells of Diesel Particulate Matter from Two Different Particle Sampling Systems and Several Resuspension Media. *J. Appl. Toxicol.* 2024; *Epub ahead of print*. [CrossRef] [PubMed]
124. Risom, L.; Møller, P.; Loft, S. Oxidative Stress-Induced DNA Damage by Particulate Air Pollution. *Mutat. Res.* **2005**, *592*, 119–137. [CrossRef] [PubMed]
125. Kowalska, M.; Wegierek-Ciuk, A.; Brzoska, K.; Wojewodzka, M.; Meczynska-Wielgosz, S.; Gromadzka-Ostrowska, J.; Mruk, R.; Øvrevik, J.; Kruszewski, M.; Lankoff, A. Genotoxic Potential of Diesel Exhaust Particles from the Combustion of First- and Second-Generation Biodiesel Fuels—The FuelHealth Project. *Environ. Sci. Pollut. Res.* **2017**, *24*, 24223–24234. [CrossRef]
126. Tudek, B.; Winczura, A.; Janik, J.; Siomek, A.; Foksinski, M.; Oliński, R. Involvement of Oxidatively Damaged DNA and Repair in Cancer Development and Aging. *Am. J. Transl. Res.* **2010**, *2*, 254. [PubMed]
127. Li, N.; Nel, A.E. The Cellular Impacts of Diesel Exhaust Particles: Beyond Inflammation and Death. *Eur. Respir. J.* **2006**, *27*, 667–668. [CrossRef] [PubMed]
128. Danielsen, P.H.; Møller, P.; Jensen, K.A.; Sharma, A.K.; Wallin, H.; Bossi, R.; Autrup, H.; Mølhave, L.; Ravanat, J.L.; Briedé, J.J.; et al. Oxidative Stress, DNA Damage, and Inflammation Induced by Ambient Air and Wood Smoke Particulate Matter in Human A549 and THP-1 Cell Lines. *Chem. Res. Toxicol.* **2011**, *24*, 168–184. [CrossRef] [PubMed]
129. Yang, Z.; Liu, Q.; Liu, Y.; Qi, X.; Wang, X. Cell Cycle Arrest of Human Bronchial Epithelial Cells Modulated by Differences in Chemical Components of Particulate Matter. *RSC Adv.* **2021**, *11*, 10582–10591. [CrossRef] [PubMed]

130. Rönkkö, T.J.; Hirvonen, M.R.; Happo, M.S.; Ihantola, T.; Hakkarainen, H.; Martikainen, M.V.; Gu, C.; Wang, Q.; Jokiniemi, J.; Komppula, M.; et al. Inflammatory Responses of Urban Air PM Modulated by Chemical Composition and Different Air Quality Situations in Nanjing, China. *Environ. Res.* **2021**, *192*, 110382. [CrossRef] [PubMed]
131. Liu, F.; Joo, T.; Ditto, J.C.; Saavedra, M.G.; Takeuchi, M.; Boris, A.J.; Yang, Y.; Weber, R.J.; Dillner, A.M.; Gentner, D.R.; et al. Oxidized and Unsaturated: Key Organic Aerosol Traits Associated with Cellular Reactive Oxygen Species Production in the Southeastern United States. *Environ. Sci. Technol.* **2023**, *57*, 14150–14161. [CrossRef] [PubMed]
132. Yang, Y.; Battaglia, M.A.; Mohan, M.K.; Robinson, E.S.; DeCarlo, P.F.; Edwards, K.C.; Fang, T.; Kapur, S.; Shiraiwa, M.; Cesler-Maloney, M.; et al. Assessing the Oxidative Potential of Outdoor PM2.5 in Wintertime Fairbanks, Alaska. *ACS ES&T Air* **2024**, *1*, 175–187. [CrossRef]
133. Lobato, S.; Castillo-Granada, A.L.; Bucio-Pacheco, M.; Salomón-Soto, V.M.; Álvarez-Valenzuela, R.; Meza-Inostroza, P.M.; Villegas-Vizcaíno, R. PM2.5, Component Cause of Severe Metabolically Abnormal Obesity: An in Silico, Observational and Analytical Study. *Heliyon* **2024**, *10*, e28936. [CrossRef]
134. Billet, S.; Landkocz, Y.; Martin, P.J.; Verdin, A.; Ledoux, F.; Lepers, C.; André, V.; Cazier, F.; Sichel, F.; Shirali, P.; et al. Chemical Characterization of Fine and Ultrafine PM, Direct and Indirect Genotoxicity of PM and Their Organic Extracts on Pulmonary Cells. *J. Environ. Sci.* **2018**, *71*, 168–178. [CrossRef]
135. Schraufnagel, D.E. The Health Effects of Ultrafine Particles. *Exp. Mol. Med.* **2020**, *52*, 311–317. [CrossRef]
136. Kwon, H.S.; Ryu, M.H.; Carlsten, C. Ultrafine Particles: Unique Physicochemical Properties Relevant to Health and Disease. *Exp. Mol. Med.* **2020**, *52*, 318–328. [CrossRef]
137. Moreno-Ríos, A.L.; Tejeda-Benítez, L.P.; Bustillo-Lecompte, C.F. Sources, Characteristics, Toxicity, and Control of Ultrafine Particles: An Overview. *Geosci. Front.* **2022**, *13*, 101147. [CrossRef]
138. Meloni, E.; Rossomando, B.; De Falco, G.; Sirignano, M.; Arsie, I.; Palma, V. Effect of a Cu-Ferrite Catalyzed DPF on the Ultrafine Particle Emissions from a Light-Duty Diesel Engine. *Energies* **2023**, *16*, 4071. [CrossRef]
139. Hassan, T.; Rahman, M.M.; Rabbi, M.S.; Rahman, M.A.; Meraz, R.M. Recent Advancement in the Application of Metal Based Nanoadditive in Diesel/Biodiesel Fueled Compression Ignition Engine: A Comprehensive Review on Nanofluid Preparation and Stability, Fuel Property, Combustion, Performance, and Emission Characteristics. *Environ. Prog. Sustain. Energy* **2023**, *42*, e13976. [CrossRef]
140. Zhang, Y.; Niu, Q.; Xu, Y.; Chen, J.; Zhai, C.; Azhar, U.; Lu, Y.; Li, H.; Zong, C. Synthesis of Long Straight-Chain Alkane Substituted Ferrocene with Ultra-Robust Oil Solubility as Multifunctional Organometallic Additive. *Appl. Organomet. Chem.* **2024**, *38*, e7482. [CrossRef]
141. Costa, S.D.; Bukkarapu, K.R.; Fernandes, R.; Krishnasamy, A.; Morajkar, P.P. Advancements in In-Cylinder and After-Treatment Strategies for Mitigating Pollutants from Diesel Engines: A Critical Review. *Energy Fuels* **2024**, *38*, 9218–9261. [CrossRef]

nanomaterials

Article

Evaluation of TiO_2 Nanoparticle-Enhanced Palm and Soybean Biodiesel Blends for Emission Mitigation and Improved Combustion Efficiency

Ramozon Khujamberdiev and Haeng Muk Cho *

Department of Mechanical Engineering, Kongju National University, Cheonan 31080, Republic of Korea; khujamberdievramozon@gmail.com
* Correspondence: hmcho@kongju.ac.kr; Tel.: +82-(10)-87113252

Abstract: The use of biodiesel as an alternative to conventional diesel fuels has gained significant attention due to its potential for reducing greenhouse gas emissions and improving energy sustainability. This study explores the impact of TiO_2 nanoparticles on the emission characteristics and combustion efficiency of biodiesel blends in compression ignition (CI) engines. The fuels analyzed include diesel, SB20 (soybean biodiesel), SB20 + 50 TiO_2 ppm, SB20 + 75 TiO_2 ppm, PB20 (palm biodiesel), PB20 + 50 TiO_2 ppm, and PB20 + 75 TiO_2 ppm. Experiments were conducted under a consistent load of 50% across engine speeds ranging from 1000 to 1800 RPM. While TiO_2 nanoparticles have been widely recognized for their ability to enhance biodiesel properties, limited research exists on their specific effects on soybean and palm biofuels. This study addresses these gaps by providing a comprehensive analysis of emissions, including NO_X, CO, CO_2, and HC, as well as exhaust gas temperature (EGT), across various engine speeds and nanoparticle concentrations. The results demonstrate that TiO_2 nanoparticles lead to a reduction in CO emissions by up to 30% and a reduction in HC emissions by 21.5% at higher concentrations and engine speeds. However, this improvement in combustion efficiency is accompanied by a 15% increase in CO_2 emissions, indicating more complete fuel oxidation. Additionally, NO_X emissions, which typically increase with engine speed, were mitigated by 20% with the addition of TiO_2 nanoparticles. Exhaust gas temperatures (EGTs) were also lowered, indicating enhanced combustion stability. These findings highlight the potential of TiO_2 nanoparticles to optimize biodiesel blends for improved environmental performance in CI engines.

Keywords: blended biodiesel; nanoparticle; titanium oxide; performance; emission; biofuel

Citation: Khujamberdiev, R.; Cho, H.M. Evaluation of TiO_2 Nanoparticle-Enhanced Palm and Soybean Biodiesel Blends for Emission Mitigation and Improved Combustion Efficiency. *Nanomaterials* **2024**, *14*, 1570. https://doi.org/10.3390/nano14191570

Academic Editors: Robert A. Varin and Geng Zhong

Received: 2 September 2024
Revised: 22 September 2024
Accepted: 26 September 2024
Published: 28 September 2024

1. Introduction

Biofuels, derived from organic materials such as plants and animal fats, present a renewable and sustainable alternative to fossil fuels. They have garnered attention due to their potential to reduce greenhouse gas emissions, enhance energy security, and promote environmental sustainability. Biodiesel, a major category of biofuels, can be produced from various feedstocks including palm oil and soybean oil. These biofuels can be directly used in existing diesel engines with minimal modifications, offering a significant reduction in carbon footprint compared to conventional fossil fuels [1–3]. Despite these advantages, biofuels face several challenges such as higher nitrogen oxide (NO_X) emissions and lower oxidation stability compared to fossil fuels. These limitations require biofuel properties to be further optimized to meet emission standards and improve engine performance [4,5].

To address these challenges, the addition of nanoparticles to biofuels has shown promising results in enhancing fuel properties and engine performance. Nanoparticles, due to their small size and large surface area, act as effective catalysts and lubricants, leading to better combustion efficiency and reduced emissions. Nanoparticles, such as titanium dioxide (TiO_2), cerium oxide, and aluminum oxide, have been extensively studied for their ability to improve fuel atomization, air–fuel mixing, and ignition delay reduction. These improvements contribute to higher engine efficiency, increased power output, and reduced harmful emissions such as NO_X and particulate matter [1,6].

Titanium dioxide (TiO_2) nanoparticles have emerged as a particularly effective additive for enhancing biofuel properties. TiO_2 nanoparticles improve the thermal and oxidative stability of biodiesel, leading to more efficient combustion and reduced engine wear. The catalytic properties of TiO_2 nanoparticles facilitate oxidation reactions during combustion, promoting more complete fuel combustion and reducing the formation of unburned hydrocarbons (HC) and carbon deposits. Studies have shown that TiO_2 nanoparticles can significantly enhance the combustion characteristics of biodiesel, resulting in improved fuel economy and lower emissions of pollutants such as carbon monoxide (CO), hydrocarbons (HC), and nitrogen oxides (NO_X) [6,7].

The mechanism behind the improvement is largely attributed to the high surface area of TiO_2 nanoparticles, which provides more active sites for catalytic reactions. This leads to better oxidation of fuel molecules and higher combustion efficiency. Additionally, TiO_2 nanoparticles stabilize combustion temperatures, reducing NO_X formation, which is typically associated with high-temperature combustion [8,9]. Research indicates that the inclusion of TiO_2 nanoparticles in biodiesel blends results in significant improvements in engine performance and emission characteristics. For instance, a study on Mahua biodiesel blended with TiO_2 nanoparticles demonstrated enhanced combustion efficiency and reduced emissions at varying injection pressures [10]. Another study on tamarind oil biodiesel with TiO_2 and alcoholic fuel additives showed similar benefits, with notable reductions in NO_X and particulate emissions [9].

Moreover, TiO_2 nanoparticles have been used as catalysts in the synthesis of biodiesel, enhancing the conversion efficiency and quality of the biofuel. For instance, TiO_2-catalyzed biodiesel production using *Carthamus tinctorius* L. oil showed significant improvements in yield and purity [11].

Comparative studies have highlighted the superior performance of TiO_2 nanoparticle-enhanced biodiesel over traditional biodiesel and diesel blends. For example, the use of TiO_2 nanoparticles in biodiesel from *Eichhornia crassipes* led to improved engine performance and lower emissions across various operating conditions [12]. Similarly, an experimental evaluation of *Semecarpus anacardium* biodiesel with TiO_2 nanoparticles demonstrated enhanced stability and physicochemical properties, contributing to better engine performance [13].

The potential of TiO_2 nanoparticles in biodiesel applications is further supported by their role in reducing soot and NO_X emissions in advanced engine technologies. The incorporation of TiO_2 nanoparticles in biodiesel blends has shown to be effective in engines equipped with exhaust gas recirculation (EGR) systems, significantly lowering emissions without compromising engine performance [14].

The Objectives of This Study

This study aims to investigate the emission and performance characteristics of TiO_2 nanoparticles in palm and soybean biofuel in a compression ignition (CI) engine. The specific objectives are as follows:

1. To evaluate the influence of TiO_2 nanoparticles on the reduction in harmful emissions (CO, HC, and NO_X) and the enhancement in combustion efficiency in soybean and palm biodiesel blends.
2. To explore the potential of TiO_2 nanoparticles in mitigating NO_X emissions and lowering exhaust gas temperatures (EGTs), particularly at higher engine speeds.

2. Literature Review

Titanium dioxide (TiO_2) nanoparticles have been extensively studied for their potential to enhance the properties of biofuels. The integration of TiO_2 nanoparticles into biodiesel has shown significant improvements in combustion efficiency, emission reduction, and engine performance, largely due to their catalytic properties, which enhance oxidation stability. Several studies highlight the role of TiO_2 nanoparticles in acting as catalysts that improve the oxidation stability of biodiesel, which is crucial for efficient combustion and reduced emissions.

Several studies have highlighted the specific benefits of TiO_2 nanoparticles in biodiesel blends. Rajak et al. (2020) discussed the benefits of using TiO_2 nanoparticles in first-generation biodiesel blends, including palm and soybean oils, highlighting enhanced combustion characteristics and lower emissions [15]. Building on this, Lai et al. (2023) demonstrated that green fuel blends with TiO_2 nanoparticles significantly optimized diesel engine performance and reduced pollutants [6]. These findings suggest that TiO_2 nanoparticles not only improve the combustion process but also address the persistent issue of high NO_X emissions typically associated with biodiesel combustion, offering a more sustainable alternative to traditional diesel.

In comparative studies, TiO_2 nanoparticles have demonstrated superior performance when used in combination with other additives. Pullagura et al. (2022) conducted a comparative study on the use of TiO_2 nanoparticles and alcoholic fuel additives in biodiesel-diesel blends, showing that TiO_2 nanoparticles improved the combustion efficiency and reduced NO_X emissions [9]. Bayindirli et al. (2023) also reported that TiO_2 nanoparticles enhanced the thermophysical properties and combustion performance of biodiesel [2]. These studies underline the versatility of TiO_2 nanoparticles in improving various aspects of biodiesel performance, from thermal stability to emission control.

Previous studies have established the baseline performance of palm and soybean biodiesel in compression ignition (CI) engines, laying the groundwork for exploring further enhancements with TiO_2 nanoparticles. For instance, Bari and Zhang (2020) evaluated the performance of CI engines using palm oil biodiesel, reporting improved combustion efficiency and reduced emissions compared to traditional diesel fuels [16]. Similarly, Kumar et al. (2022) assessed the performance, emissions, and combustion attributes of CI engines using palm biodiesel blends and found significant improvements in engine efficiency and a reduction in harmful emissions such as NO_X and HC [17]. However, when TiO_2 nanoparticles are introduced into these biodiesel blends, the improvements become even more pronounced as the catalytic activity of the nanoparticles further enhances combustion stability and reduces emissions, particularly under high engine loads.

In the case of soybean oil-based biodiesel, extensive research has also confirmed its effectiveness in reducing emissions and improving engine performance. Seraç et al. (2020) conducted a comprehensive evaluation of soybean biodiesel blends in CI engines and reported enhanced performance and lower emissions compared to conventional diesel [18]. Vellaiyan (2020) investigated the combustion, performance, and emissions of a diesel engine fueled with soybean biodiesel and its water blends, finding significant reductions in NO_X and particulate matter emissions [19].

A study by Hussain et al. (2020) demonstrated that the addition of 50 ppm of TiO_2 nanoparticles to soybean biodiesel blends increased the brake thermal efficiency by 20.66% while reducing CO and HC emissions by 30% and 21.5%, respectively. However, their study also noted a slight increase in NO_X emissions due to the higher combustion temperatures associated with more efficient fuel oxidation [20].

Further research conducted by Lin and Lin in 2023 explored the performance of soybean biodiesel emulsions with TiO_2 nanoparticles, revealing a significant reduction in NO_X and particulate matter emissions. Their experiments indicated that TiO_2-enhanced soybean biodiesel reduced CO emissions by 25% and NO_X emissions by 3.75% compared to conventional diesel fuel [21]. The influence of TiO_2 nanoparticles, combined with biodiesel's inherent oxygen content, contributes to the reduction in unburned hydrocarbons, thus minimizing the environmental impact of biodiesel use in CI engines.

The addition of TiO_2 nanoparticles to palm biodiesel blends has been widely studied for its potential to reduce harmful emissions in diesel engines. For example, Venu et al. (2019) demonstrated that palm biodiesel blended with TiO_2 nanoparticles and exhaust gas recirculation (EGR) significantly reduced NO_X emissions, a common challenge in biodiesel combustion. The catalytic action of TiO_2 nanoparticles improves oxidation reactions, reducing unburned hydrocarbons (HCs) and carbon monoxide (CO) emissions in palm biodiesel blends [22]. In addition, Mujtaba et al., 2020 found that the inclusion of TiO_2 nanoparticles in palm–sesame biodiesel blends resulted in a 32.09% reduction in CO emissions and a 25.4% decrease in HC emissions compared to regular biodiesel [23]. These reductions in emissions are crucial for making biodiesel blends more viable for use in commercial diesel engines, where regulatory limits on pollutants are becoming increasingly stringent.

Furthermore, the impact of TiO_2 nanoparticles on particulate matter (PM) emissions has been explored. Fayad et al. (2023) showed that the addition of TiO_2 nanoparticles to biodiesel blends reduced the total concentration of particulate matter by 26.74% compared to standard diesel fuel [11]. The synergistic effect of TiO_2 nanoparticles and EGR systems has been shown to lower NO_X emissions by up to 21.83% and significantly reduce PM concentrations [14]. Research indicates that TiO_2 nanoparticles play a crucial role in improving both gaseous and particulate emissions, making palm biodiesel blends more environmentally sustainable when used in compression ignition engines.

Identification of Research Gaps

Despite extensive research indicating the benefits of TiO_2 nanoparticles in enhancing biodiesel properties, there remains a significant gap in understanding their specific impact on palm and soybean biofuels. Current studies often generalize the effects of TiO_2 without delving into how these nanoparticles influence the combustion efficiency of these biofuels. Furthermore, while reductions in emissions such as NO_X and particulate matter have been noted, a comprehensive analysis of all emission profiles, including CO, HC, and soot particles, is lacking. This study aims to fill these gaps by providing detailed evaluations of the combustion processes and emission characteristics of CI engines fueled with TiO_2-enhanced palm and soybean biodiesels.

In addition, while there are indications of improved engine performance metrics like power output and fuel consumption with TiO_2-enhanced biodiesels, specific data for palm and soybean biodiesel blends are insufficient. This study seeks to systematically assess these performance metrics and explore the long-term sustainability and economic feasibility of using TiO_2 nanoparticles in these biofuels. By addressing the stability, cost implications, and lifecycle impact of TiO_2 integration, this research will offer a comprehensive understanding of the potential benefits and challenges, contributing valuable insights into the optimization of biofuels for sustainable engine performance.

3. Methodology

3.1. Biofuel Production

In Figure 1, the production of biodiesel from palm oil utilizing a laboratory-scale transesterification process is illustrated. The experimental setup comprised a 1000 mL flask equipped with a thermometer and a magnetic stirrer (Topscien, Ningbo, China, Model MS300) to ensure uniform mixing. Initially, 1000 mL of palm oil was heated to 65 °C in the flask. Subsequently, a solution of potassium hydroxide (12.75 g from Sigma-Aldrich, Burlington, MA, USA) dissolved in 255 mL of methanol (Sigma-Aldrich, USA, ≥99.9% purity, suitable for high-performance liquid chromatography) was added to the preheated oil. The reaction mixture was maintained under continuous stirring for 2 h. Following the reaction, the mixture was transferred to a separating funnel to separate the glycerol layer. The resulting esters were washed twice: first with warm water containing 5% acetic acid and then with plain water to purify the methyl esters. The final step involved drying the ester at 100 °C to remove any remaining alcohol and water, culminating in the production of biodiesel from palm oil.

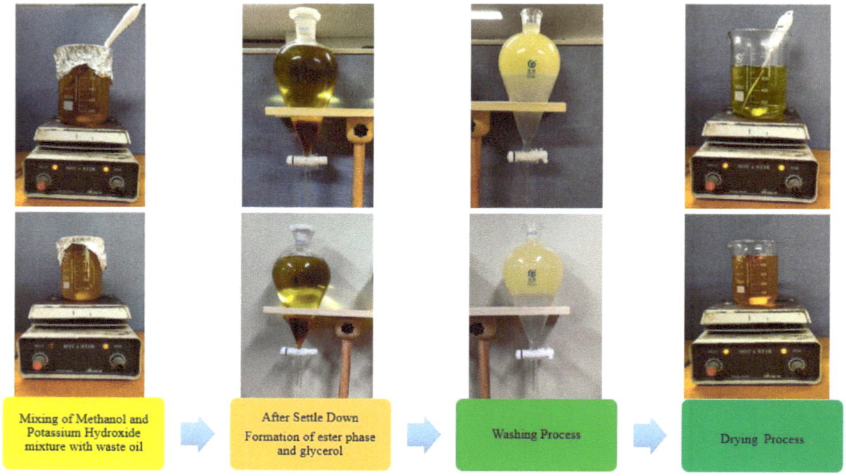

Figure 1. Biodiesel production: palm oil and soybean oil.

Similarly, the synthesis of biodiesel from soybean oil was conducted using a comparable one-stage alkaline transesterification method. The apparatus included a 500 mL flask, also equipped with a thermometer and magnetic stirrer. In this instance, 500 mL of soybean oil was combined with a solution of 135 mL methanol and 2.5 g potassium hydroxide, maintaining a 10:1 molar ratio. The mixture was heated to 55 °C and stirred at a constant rate of 700 RPM for 2 h. After the reaction, the mixture was subjected to the same separation process using a funnel to remove the glycerol layer. The esters were then purified through two washes: one with warm water mixed with 5% acetic acid and the other with plain water. The final drying step at 100 °C removed any residual alcohol and water, resulting in the production of biodiesel from soybean oil.

3.2. Incorporation of TiO_2 Nanoparticles

The incorporation of TiO_2 nanoparticles into soybean oil was a carefully controlled process designed to ensure a uniform dispersion of the nanoparticles within the biodiesel blend. The TiO_2 nanoparticles used were AEROXIDE TiO_2 P25 (Evonik Degussa, Essen, Germany). Initially, the TiO_2 nanoparticles were mixed with soybean oil using an ultrasonic mixer (Branson Ultrasonics, Model 2510E-MT, Danbury, CT, USA) for one hour. Ultrasonic mixing is a crucial step as it uses high-frequency sound waves to create cavitation bubbles in the liquid, which collapse and produce intense shear forces. These forces help to break apart any nanoparticle agglomerates, ensuring that the nanoparticles are evenly distributed throughout the soybean oil. This process was repeated to prevent any settling or clumping of the nanoparticles within the blend, as uniform dispersion is key to maintaining the desired properties of the fuel. Prior to testing, the nano blend was vigorously shaken and stirred for an additional two hours, further ensuring that the nanoparticles remained in suspension and did not settle. This method of incorporating TiO_2 nanoparticles aimed to improve the fuel's combustion characteristics and stability, potentially enhancing the overall performance of the biodiesel.

The process of incorporating TiO_2 nanoparticles into palm oil biodiesel followed a similar methodology to that used with soybean oil, with specific adjustments made to accommodate the properties of palm oil. TiO_2 nanoparticles, prepared through a chemical precipitation method, were first mixed into the palm oil using an ultrasonic mixer for one hour. The ultrasonic mixer facilitated the even distribution of nanoparticles by breaking down any potential clumps, thus preventing agglomeration. The mixing process was repeated to ensure the consistency and stability of the nano blend, as an uneven distribution could lead to performance inconsistencies. Additionally, to avoid settling of the nanoparticles, the mixture was subjected to vigorous shaking and stirring for two hours before any experimental testing. This step was critical in maintaining a stable dispersion as it ensured the nanoparticles were well suspended and did not settle at the bottom of the container. The incorporation of TiO_2 nanoparticles into palm oil was carried out with the aim of enhancing the biodiesel's properties, potentially improving the combustion efficiency, reducing emissions, and increasing fuel stability. Specific concentration levels of TiO_2 nanoparticles (50 ppm and 75 ppm) were selected to explore their impacts on the biodiesel's performance characteristics.

The quantities of TiO_2 nanoparticles selected for the study—50 ppm and 75 ppm—were carefully chosen to investigate the effects of varying concentrations on the biodiesel's performance. These specific concentrations were selected based on previous research and preliminary studies, which indicated that these levels are effective in enhancing fuel properties without causing adverse effects, such as the clogging of fuel injectors or increased viscosity:

- A concentration of 50 ppm: This moderate concentration was selected to evaluate the balance between catalytic enhancement and potential changes in fuel properties, such as viscosity and stability. It provides insights into optimal nanoparticle loading that can deliver noticeable benefits without compromising the fuel's quality.
- A concentration of 75 ppm: The highest concentration was chosen to explore the upper limits of TiO_2 incorporation. This level helps in understanding the maximum potential benefits and any possible drawbacks, such as increased costs or technical challenges, associated with higher nanoparticle loading.

The detailed characteristics of the TiO_2 nanoparticles and the soybean and palm oil blends are shown in Tables 1 and 2.

Table 1. Nanoparticle (TiO_2) properties.

Detail	Properties
Name (TiO_2)	Nanoparticles of titanium dioxide
Appearance	White
Purity of TiO_2	96%
Particle size	20–50 nm
Surface area	>42 m^2/g
Melting point	>233 °C

Table 2. Fuel characteristics.

Fuel Type	Density (kg/m^3)	Viscosity (cSt)	Flash Point (°C)	Cetane Number	Calorific Value (MJ/kg)	Ester Content (%)
Diesel	0.820	2.87	58	48.7	45.515	-
SB20	0.834	2.92	79	52	43.536	96.2
SB20 + 50 ppm TiO_2	0.841	3.19	86	53.5	43.607	96.5
SB20 + 75 ppm TiO_2	0.846	3.27	89	53.8	43.695	96.8
PB20	0.862	3.10	82	51.5	43.201	97.1
PB20 + 50 ppm TiO_2	0.868	3.36	88	52.9	43.272	97.4
PB20 + 75 ppm TiO_2	0.872	3.45	91	53.2	43.335	97.7

3.3. Experimental Setup

Figure 2 depicts the experimental setup used in this study, highlighting a detailed array of components essential for accurately measuring and analyzing the performance of a four-cylinder, four-stroke electronic engine. The core of the setup is the engine itself, which is connected to an eddy current dynamometer that applies varying loads to the engine. Torque measurements are precisely recorded using a load cell and a dedicated torque measurement device. Engine speed is tracked using an RPM sensor and displayed on an RPM display.

Exhaust gasses are routed through an exhaust pipe and analyzed for gas composition and smoke levels using a gas emission analyzer and a smoke analyzer, respectively.

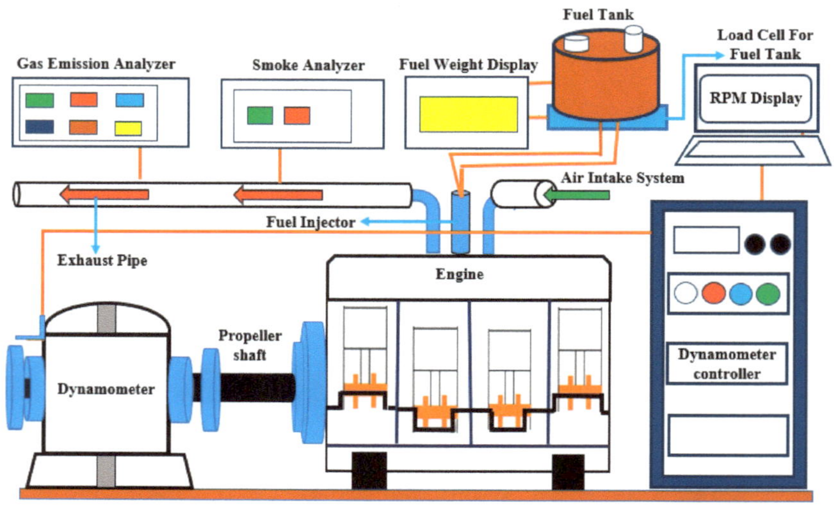

Figure 2. Experimental setup.

The engine receives fuel from a fuel tank, with fuel consumption being closely monitored by a load cell for fuel weight and shown on a fuel weight display. The air intake system and fuel injector work together to ensure the precise delivery of air and fuel to the engine. Additionally, a propeller shaft is included in the setup, likely serving the purpose of transmitting power from the engine. This comprehensive arrangement is meticulously designed to provide in-depth insights into the engine's performance across various operating conditions.

3.4. Engine Specifications

The engine specifications detailed in Table 3 describe a four-cylinder, four-stroke power unit with electronic control, a common setup in modern vehicles that effectively balances efficiency and performance. Manufactured by Hyundai Motor Co., Ltd. in Seoul, Republic of Korea, this engine utilizes a sequential fuel injection system and follows a 1-3-4-2 firing order. This specific firing sequence helps minimize engine vibrations and enhances operational smoothness. With a displacement of 1995 cc, the engine is within the mid-range capacity for contemporary automotive engines, offering a good compromise between power and fuel efficiency. The high compression ratio of 16:1 indicates that the engine is optimized for high efficiency and may be designed to run on premium fuels. The bore and stroke are measured at 84 mm and 90 mm, respectively, suggesting a slightly over-square design that can improve engine responsiveness and potentially allow for higher RPMs. Overall, the engine's features are tailored to provide a balanced mix of performance and efficiency, making it suitable for a wide range of driving scenarios.

Table 3. Engine specification and data points.

Parameters	Specifications
Engine Type	Electronic, four-cylinder, four-stroke type
Injection Sequence	1-3-4-2
Stroke Length (mm)	90
Bore (mm)	84
Compression Ratio	16:1
Displacement (cc)	1995

This research employed a standard internal combustion engine without any modifications to ensure that the findings accurately represent the performance and emissions characteristics of engines commonly found in real-world conditions. This approach allows for a precise evaluation of different biodiesel blends alongside regular diesel, reflecting how these fuels perform in engines as they are typically configured. By avoiding any alterations to the engine, this study maintains the original design specifications, making the results highly relevant and applicable to the majority of engines currently in use. This decision ensures that the conclusions drawn from this research are both reliable and widely transferable.

3.5. Error Analysis and Uncertainty

During the experimental trials, the lubricating oil temperature was carefully controlled and consistently maintained within a narrow range of 85 to 90 °C. The engine was operated continuously for 15 min, during which detailed observations were made and recordings systematically documented. To minimize potential discrepancies in the data, a rigorous evaluation of uncertainty was conducted. A critical aspect of this process was the calibration of all instruments involved, ensuring the accuracy and reliability of the experimental results.

To further enhance the reliability of the findings, measurements were taken multiple times, with a minimum of four repetitions for each experiment. These repeated readings were averaged to calculate the arithmetic mean, which formed the foundation for a subsequent data analysis. This approach helped mitigate the impact of any anomalies or outliers, providing a more accurate representation of the experimental outcomes.

The detailed analysis of error and uncertainty for the smoke meter and gas analyzer is presented in Table 4. This table highlights the precision of these instruments, accounting for possible deviations and inaccuracies inherent in the measurement processes. By quantifying the uncertainty ranges and potential error margins, this table offers a clear insight into the reliability and limitations of the collected data. This comprehensive assessment is crucial for interpreting the experimental results with the necessary scientific rigor, ensuring that the conclusions drawn are both accurate and valid within the specified uncertainty bounds.

Table 4. Measuring range and precision of smoke meter and gas analyzer.

Exhaust Emission	Range	Resolution	Accuracy and Uncertainties
CO	0.00–10.00	%	±0.001%
HC	0–10,000	ppm	±1 ppm
CO_2	0.0–20.0	%	±0.01%
O_2	0.00–25.00	%	±0.01%
NO_X	0–5000	ppm	±1 ppm
Smoke	0–100	%	±0.05%
Thermocouple (K-Type)	0–1200	°C	±0.1 °C

To assess the engine's emission parameters, a CGA-4500 gas analyzer from the Republic of Korea was utilized. This sophisticated device employs Non-Dispersive Infrared (NDIR) technology to measure carbon monoxide (CO) concentrations ranging from 0.00 to 10.00% and carbon dioxide (CO_2) levels between 0.0 and 20.0%. Additionally, it is capable of detecting hydrocarbons (HCs) within the range of 0 to 10,000 parts per million (ppm). The analyzer also features an electrochemical sensor for measuring oxygen (O_2) levels from 0.00 to 25.00% and another electrochemical sensor specifically for quantifying nitric oxide (NO_X) emissions, which can range from 0 to 5000 ppm.

For the acquisition of digital emission data, a probe was strategically positioned within the exhaust pipeline to ensure accurate sampling. Smoke emissions were measured using a dedicated smoke meter, providing data on the particulate content. To monitor the temperature of the exhaust gasses, a k-type thermocouple was employed. Together, these instruments allowed for a comprehensive analysis of the engine's emission characteristics across various operational conditions, enabling a thorough evaluation of both its performance and environmental impact.

4. Results and Discussion

4.1. Emission Analysis

4.1.1. Carbon Monoxide (CO) Emissions

Figure 3 indicates that CO emissions generally decrease as the RPM increases across all fuel types. Among the fuel types, the base diesel fuel (B0) consistently shows the highest CO emissions at each RPM level, while the biofuels, particularly those with added TiO_2 nanoparticles, exhibit lower CO emissions. For instance, at 1000 RPM, the amount of CO emissions for SB20 is 0.178%, but with the addition of 75 mg of TiO_2, the amount of emissions reduces significantly to 0.1%. A similar trend is observed with PB20, where CO emissions drop from 0.174% to 0.11% with the addition of 75 mg of TiO_2.

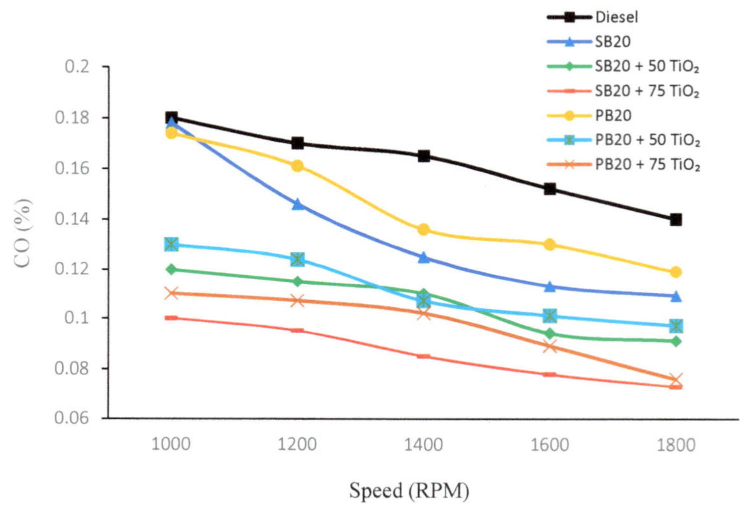

Figure 3. CO emission fluctuations with varying engine speed.

CO emissions reduced by 43.82% for SB20 with 75 mg of TiO_2 at 1000 RPM, while PB20 showed a 36.78% reduction under similar conditions.

These findings are consistent with previous studies, which have shown that TiO_2 nanoparticles enhance oxidative reactions, leading to a reduction in CO emissions. For instance, Jayabalaji and Shanmughasundaram in 2019 observed that incorporating TiO_2 into biodiesel blends markedly reduced CO emissions due to the catalytic properties of the nanoparticles, which improve the combustion process [24]. Likewise, the research conducted by Gunasekar et al. (2019) confirmed that the addition of TiO_2 nanoparticles in biodiesel blends can effectively lower CO emissions by promoting the oxidation of carbon monoxide to carbon dioxide during combustion [25].

4.1.2. Hydrocarbon (HC) Emissions

In Figure 4, it is evident that there is a consistent decrease in HC emissions as the RPM increases for all fuel types. Among the fuels, pure diesel (B0) exhibits the highest HC emissions at each RPM level, while the biofuels show lower HC emissions. The incorporation of TiO_2 nanoparticles into the biofuels further reduces HC emissions, with the highest concentration of TiO_2 (75 mg) yielding the lowest emissions. For instance, at 1000 RPM, SB20 emits 56 ppm of HC, while the SB20 + 75 TiO_2 blend reduces this value to 47 ppm. Similarly, PB20 shows 64 ppm of HC emissions at 1000 RPM, which drops to 44 ppm with the addition of 75 mg of TiO_2.

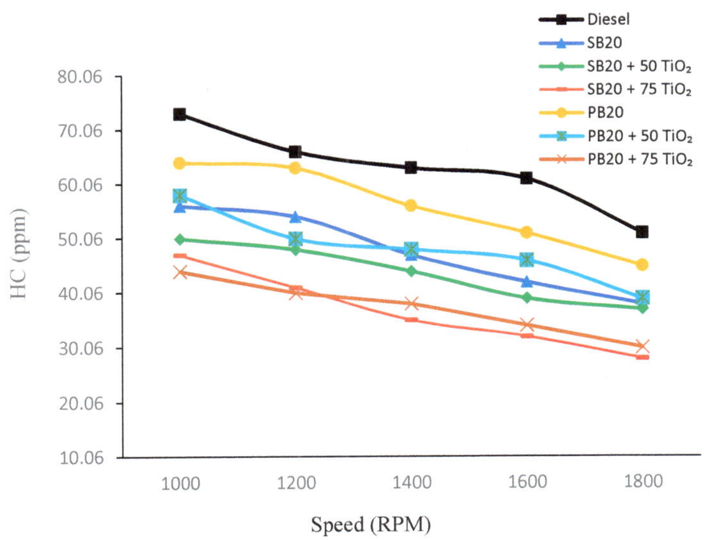

Figure 4. HC emission fluctuations with varying engine speed.

At the highest RPM of 1800, the trend remains clear, with SB20 + 75 TiO_2 and PB20 + 75 TiO_2 displaying the lowest HC emissions at 28 ppm and 30 ppm respectively. The addition of 75 mg TiO_2 to SB20 resulted in a 16.07% reduction in HC emissions at 1000 RPM, while PB20 with 75 mg of TiO_2 showed a 31.25% reduction. These results suggest that TiO_2 nanoparticles significantly improve the combustion efficiency, leading to lower hydrocarbon emissions across all tested biofuels.

The reduction in HC emissions with the addition of TiO_2 nanoparticles can be attributed to the improved oxidation and combustion efficiency provided by these nanoparticles. Nanoparticles help achieve more complete combustion, thereby reducing the amount of unburned hydrocarbons in the exhaust. The same trend was observed in previous research that demonstrated the effectiveness of TiO_2 nanoparticles in reducing HC emissions. For instance, the research conducted by Sarma et al. (2023) showed that the incorporation of TiO_2 nanoparticles into biodiesel blends led to a significant reduction in HC emissions due to enhanced combustion processes [26]. Similarly, Kurre et al. (2023) found that TiO_2 nanoparticles contributed to lower HC emissions by improving the oxidation of hydrocarbons during combustion [27].

4.1.3. Carbon Dioxide (CO_2) Emissions

In Figure 5, a clear trend is observed where CO_2 emissions increase as the RPM rises across all fuel types. Among the biofuels, the addition of TiO_2 nanoparticles results in higher CO_2 emissions compared to the base fuels, suggesting a more complete combustion process facilitated by the nanoparticles. For example, at 1000 RPM, SB20 emits 5.6% of CO_2, but this increases to 6.8% with the addition of 75 mg of TiO_2. Similarly, PB20 emits 5.2% of CO_2 at 1000 RPM, which increases to 6.7% when 75 mg of TiO_2 is added.

At the highest RPM of 1800, the trend is consistent, with SB20 + 75 TiO_2 emitting 8.8% of CO_2 compared to 4.4% for the base diesel (B0). Similarly, PB20 + 75 TiO_2 shows 8.7% of CO_2 emissions at 1800 RPM, highlighting the impact of TiO_2 in enhancing the combustion efficiency, thereby increasing CO_2 emissions due to more complete fuel oxidation.

CO_2 emissions increase by 21.43% for SB20 and 28.85% for PB20 with 75 mg of TiO_2 at 1000 RPM.

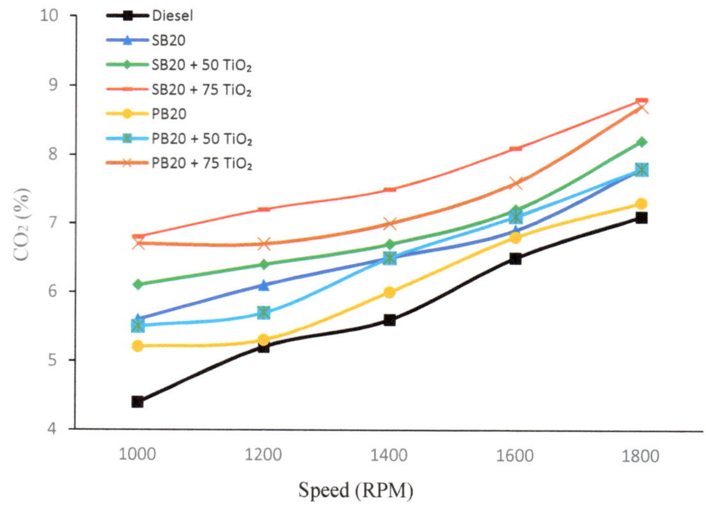

Figure 5. CO_2 emission fluctuations with varying engine speed.

These findings corroborate previous research showing that TiO_2 nanoparticles enhance combustion efficiency, leading to increased CO_2 emissions as a greater proportion of carbon is fully oxidized during the combustion process. This outcome was observed in studies by Fangsuwannarak et al. (2020) and Madhuri et al. (2023), where the addition of TiO_2 to biofuels resulted in higher CO_2 emissions due to the improved catalytic combustion [28,29].

4.1.4. Nitrogen Oxide (NO_X) Emissions

Figure 6 indicates that NO_X emissions increase with higher RPMs across all fuel types. Among the fuels, the base diesel (B0) consistently has the lowest NO_X emissions, while the pure SB20 biofuel exhibits the highest emissions, reaching 904 ppm at 1800 RPM. The introduction of TiO_2 nanoparticles generally leads to a reduction in NO_X emissions for both SB and PB biofuels. For example, at 1000 RPM, SB20 emits 722 ppm of NO_X, which decreases to 685 ppm when 75 mg of TiO_2 is added.

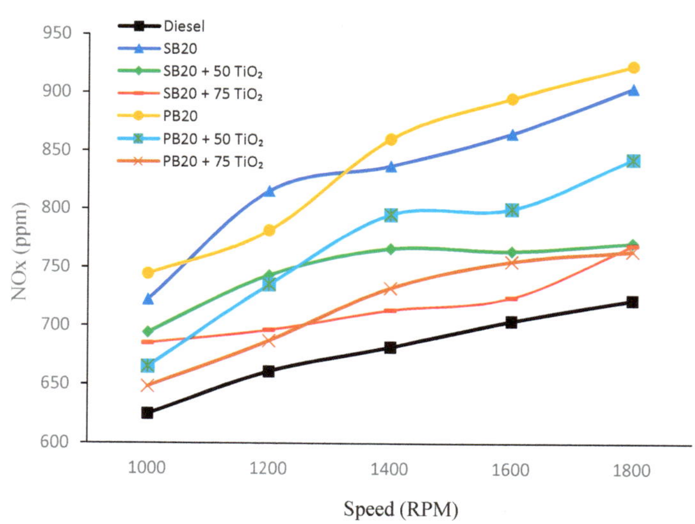

Figure 6. NO_X emission fluctuations with varying engine speed.

Similarly, PB20 follows this trend, with NO_X emissions decreasing from 781 ppm to 687 ppm at 1200 RPM with the addition of 75 mg of TiO_2. At higher RPMs, the reduction in NO_X emissions becomes more pronounced, particularly in the SB20 + 75 TiO_2 and PB20 + 75 TiO_2 blends, which consistently show lower emissions compared to their pure biofuel versions. The addition of TiO_2 resulted in a 5.12% reduction in NO_X emissions for SB20 and 12.03% for PB20 at 1000 RPM.

These data are compared with findings from other studies in biofuel research, where the addition of TiO_2 nanoparticles has been shown to improve combustion efficiency and reduce NO_X emissions due to the catalytic properties of the nanoparticles. For example, Prasetya et al. (2023) demonstrated that incorporating TiO_2 nanoparticles in palm biodiesel blends resulted in a reduction in NO_X emissions due to enhanced catalytic activity during combustion [30]. Similarly, Razzaq et al. (2023) support the idea that TiO_2 can effectively reduce NO_X emissions when used in biodiesel blends, particularly under high engine loads [31]. However, the data also align with the observations made by Mehregan and Moghiman in 2020, who noted that the impact of TiO_2 nanoparticles varies significantly depending on the specific conditions, such as the type of biodiesel and engine operating parameters [32]. Thus, while TiO_2 nanoparticles can be effective in reducing NO_X emissions, their efficacy is influenced by factors such as the biofuel type and engine speed.

4.1.5. Exhaust Gas Temperature (EGT)

Figure 7 reveals that the EGT increases as the RPM rises for all fuel types. Pure SB20 exhibits a consistently higher EGT compared to base diesel (B0), reaching up to 283 °C at 1800 RPM. However, when TiO_2 nanoparticles are added, the EGT decreases for both SB and PB biofuels. For example, at 1000 RPM, the EGT for SB20 is 112 °C, but it drops to 89 °C with the addition of 75 mg of TiO_2. Similarly, PB20 shows an EGT of 103 °C at 1000 RPM, which decreases to 82 °C when 75 mg of TiO_2 is added.

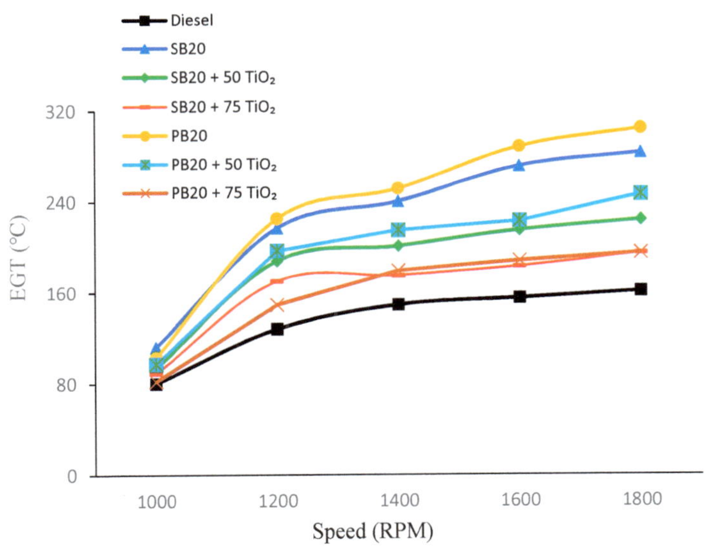

Figure 7. EGT fluctuations with varying engine speed.

As the engine speed increases, this trend of a reduced EGT with TiO_2 nanoparticle addition continues. At 1800 RPM, SB20 + 75 of TiO_2 results in an EGT of 195 °C compared to 283 °C for pure SB20. Likewise, PB20 + 75 of TiO_2 has an EGT of 195 °C, which is significantly lower than the 304 °C temperature observed for pure PB20. The addition of TiO_2 to SB20 at 1000 RPM resulted in a 20.54% reduction in the EGT, while PB20 showed a 25.24% reduction.

The reduction in EGT with the addition of TiO_2 nanoparticles can be attributed to the improved combustion efficiency provided by the nanoparticles, which promotes more complete combustion and less heat being carried away by exhaust gasses. This aligns with previous research findings, where TiO_2 nanoparticles were shown to enhance combustion characteristics, leading to lower EGTs. Studies such as those by Sarma et al. (2023) have demonstrated that the inclusion of TiO_2 nanoparticles in biodiesel blends effectively lowers EGTs by improving combustion efficiency [26].

5. Conclusions

The comprehensive analysis conducted in this study underscores the significant impact of TiO_2 nanoparticles on the emission characteristics and combustion efficiency of soybean (SB) and palm (PB) biodiesel blends in compression ignition (CI) engines. Across all test conditions, the incorporation of TiO_2 nanoparticles consistently resulted in lower emissions of carbon monoxide (CO) and hydrocarbons (HC). This reduction is attributed to the enhanced catalytic oxidation process facilitated by TiO_2 nanoparticles, which promotes more complete combustion. These findings are particularly noteworthy as they suggest that even at higher engine speeds, where emissions typically increase, the presence of TiO_2 can help maintain lower pollutant levels.

In addition to reducing CO and HC emissions, this study observed an increase in carbon dioxide (CO_2) emissions with the addition of TiO_2 nanoparticles. While higher CO_2 emissions might seem counterintuitive to the goals of emission reduction, they actually indicate more efficient fuel combustion as more carbon is being fully oxidized. This trade-off highlights the dual role of TiO_2 nanoparticles in both reducing incomplete combustion by-products and ensuring more complete fuel utilization.

A critical finding from this study is the reduction in nitrogen oxide (NO_X) emissions, which are typically challenging to control in biodiesel combustion. NO_X emissions generally rise with an increasing engine speed due to higher combustion temperatures. However, the addition of TiO_2 nanoparticles demonstrated a mitigating effect, particularly at higher concentrations and RPMs, suggesting that these nanoparticles can help manage peak combustion temperatures and reduce NO_X formation. This effect is crucial for meeting stringent emission regulations without compromising engine performance.

By improving combustion efficiency, reducing harmful emissions, and managing exhaust temperatures, these nanoparticles offer a promising pathway toward more sustainable and efficient biodiesel usage in diesel engines. Future research could explore the long-term effects of TiO_2 nanoparticle use on engine wear and the potential for further optimizing nanoparticle concentrations to balance emissions and performance. The insights gained from this study can inform the development of next-generation biodiesel formulations that align with global efforts to reduce carbon footprints and promote cleaner energy solutions in the transportation sector.

6. Future Scope

While this study offers important insights into the effects of TiO_2 nanoparticles on biofuel emissions and performance, several areas merit further research:

- The Fate of TiO_2 Nanoparticles: Future studies should investigate whether TiO_2 nanoparticles remain in the soot, are filtered, or are emitted into the atmosphere post-combustion.
- Fuel Consumption: Future research should evaluate the impact of TiO_2 on fuel economy, which was not assessed in this study.
- Smoke and Particulate Emissions: Measuring smoke opacity and particulate matter emissions will provide a more comprehensive understanding of TiO_2's environmental impact.
- Stability, Costs, and Life Cycle Impact: Studies should assess the long-term stability, cost-effectiveness, and environmental life cycle of TiO_2-enhanced biofuels.

- Nanoparticle Dispersion: While ultrasonic mixing was used to disperse TiO_2, future work should employ techniques like SEM or TEM to verify nanoparticle distribution and long-term stability in biofuels.
- Sample Testing: Testing each fuel sample three times would improve statistical reliability. Future studies should aim to carry out triplicate testing to enhance data robustness.
- The Uniformity of TiO_2 Dispersion: Variations in viscosity and calorific value may result from slight inconsistencies in nanoparticle dispersion. Future work will focus on advanced testing to ensure better uniformity in the samples.

These areas provide significant opportunities to further optimize the integration of TiO_2 nanoparticles into biofuels for improved engine performance and environmental sustainability.

Author Contributions: Conceptualization, R.K.; methodology, R.K.; software, R.K.; validation, R.K. and H.M.C.; formal analysis, R.K.; investigation, R.K.; resources, H.M.C.; data curation, R.K.; writing—original draft preparation, R.K.; writing—review and editing, R.K. and H.M.C.; visualization, R.K.; supervision, H.M.C.; project administration, H.M.C.; funding acquisition, H.M.C. All authors have read and agreed to the published version of the manuscript.

Funding: This work was supported by the National Research Foundation of Korea (NRF) grant funded by the Korean government (NRF-2022H1A7A2A02000033).

Data Availability Statement: Data are contained within the article.

Conflicts of Interest: The authors declare no conflicts of interest.

Abbreviations

kg/m^3	Kilograms per cubic meter (density)
cSt	Centistokes (viscosity)
°C	Degrees Celsius (flash point)
ppm	Parts per million (nanoparticle concentration)
MJ/kg	Megajoules per kilogram (calorific value)
%	Percentage
NO_X	Nitrogen oxide
CO	Carbon monoxide
HC	Hydrocarbon
CO_2	Carbon dioxide
EGT	Exhaust gas temperature
CI Engine	Compression ignition engine
TiO_2	Titanium dioxide
SB20	20% soybean biodiesel and 80% diesel blend
SB20 + 50 ppm TiO_2	SB20 blend with 50 ppm TiO_2 nanoparticles
SB20 + 75 ppm TiO_2	SB20 blend with 75 ppm TiO_2 nanoparticles
PB20	20% palm biodiesel and 80% diesel blend
PB20 + 50 ppm TiO_2	PB20 blend with 50 ppm TiO_2 nanoparticles
PB20 + 75 ppm TiO_2	PB20 blend with 75 ppm TiO_2 nanoparticles

References

1. El-Adawy, M. Effects of diesel-biodiesel fuel blends doped with zinc oxide nanoparticles on performance and combustion attributes of a diesel engine. *Alex. Eng. J.* **2023**, *80*, 269–281. [CrossRef]
2. Bayindirli, C.; Celik, M.; Zan, R. Optimizing the thermophysical properties and combustion performance of biodiesel by graphite and reduced graphene oxide nanoparticle fuel additive. *Eng. Sci. Technol. Int. J.* **2023**, *37*, 101295. [CrossRef]
3. Zheng, F.; Cho, H.M. The Effect of Different Mixing Proportions and Different Operating Conditions of Biodiesel Blended Fuel on Emissions and Performance of Compression Ignition Engines. *Energies* **2024**, *17*, 344. [CrossRef]
4. Lee, K.; Cho, H. Comparative Analysis of Performance and Emission Characteristics of Biodiesels from Animal Fats and Vegetable Oils as Fuel for Common Rail Engines. *Energies* **2024**, *17*, 1711. [CrossRef]
5. Khujamberdiev, R.; Cho, H. Impact of Biodiesel Blending on Emission Characteristics of One-Cylinder Engine Using Waste Swine Oil. *Energies* **2023**, *16*, 5489. [CrossRef]
6. Lai, C.-M.; Loo, D.L.; Teoh, Y.H.; How, H.G.; Le, T.D.; Nguyen, H.T.; Ghfar, A.A.; Sher, F. Optimization and performance characteristics of diesel engine using green fuel blends with nanoparticles additives. *Fuel* **2023**, *347*, 128462. [CrossRef]

7. Sandhi, R.S.; Chebattina, K.R.R.; Sambana, N.R.; Vadapalli, S.; Pullagura, G.; Pathem, U.C. Evaluation of TiO_2 nanoparticles as an additive in diesel-n-butanol-Bombax ceiba biodiesel blends for enhance performance and emissions control of a CI engine. *Int. J. Heat Technol.* **2021**, *39*, 1930–1936. [CrossRef]
8. Zheng, F.; Cho, H.M. The Comprehensive Effects of Nano Additives on Biodiesel Engines—A Review. *Energies* **2024**, *17*, 4126. [CrossRef]
9. Pullagura, G.; Bikkavolu, J.; Vadapalli, S.; Chebattina, K.R.R.; Kuchipudi, V. Comparative study of TiO_2 nanoparticles and alcoholic fuel additives-biodiesel-diesel blend for combustion, performance, and emission improvements. *Int. J. Heat Technol.* **2022**, *40*, 1249–1257. [CrossRef]
10. Simhadri, K.; Rao, P.S.; Paswan, M. Improving the combustion and emission performance of a diesel engine with TiO_2 nanoparticle blended Mahua biodiesel at different injection pressures. *Int. J. Thermofluids* **2024**, *21*, 100563. [CrossRef]
11. Jan, H.A.; Saqib, N.U.; Khusro, A.; Sahibzada, M.U.K.; Rauf, M.; Alghamdi, S.; Almehmadi, M.; Khandaker, M.U.; Bin Emran, T.; Mohafez, H. Synthesis of biodiesel from Carthamus tinctorius L. oil using TiO_2 nanoparticles as a catalyst. *J. King Saud Univ.-Sci.* **2022**, *34*, 102317. [CrossRef]
12. Jain, A.; Bora, B.J.; Kumar, R.; Sharma, P.; Medhi, B.J.; Farooque, A.A.; Tirth, V.; Senthilkumar, N.; Peyyala, P.K. Impact of titanium dioxide (TiO_2) nanoparticles addition in Eichhornia Crassipes biodiesel used to fuel compression ignition engine at variable injection pressures. *Case Stud. Therm. Eng.* **2023**, *49*, 103295. [CrossRef]
13. Illipilla, M.; Lankapalli, S.V.P.; Sagari, J. Influence of dispersant-mixed TiO_2 nanoparticles on stability and physicochemical properties of Semecarpus anacardium biodiesel blend. *Int. Nano Lett.* **2022**, *13*, 53–62. [CrossRef]
14. Fayad, M.A.; Sobhi, M.; Chaichan, M.T.; Badawy, T.; Abdul-Lateef, W.E.; Dhahad, H.A.; Yusaf, T.; Isahak, W.N.R.W.; Takriff, M.S.; Al-Amiery, A.A. Reducing soot nanoparticles and NOx emissions in CRDI diesel engine by incorporating TiO_2 nano-additives into biodiesel blends and using high rate of EGR. *Energies* **2023**, *16*, 3921. [CrossRef]
15. Rajak, U.; Verma, T.N. Influence of combustion and emission characteristics on a compression ignition engine from a different generation of biodiesel. *Eng. Sci. Technol. Int. J.* **2019**, *23*, 10–20. [CrossRef]
16. Bari, S.; Zhang, C. Analysis of Performance and Emission of Diesel Engines Operating on Palm Oil Biodiesel. *SAE Int. J. Adv. Curr. Pract. Mobil.* **2020**, *2*, 1862–1869. [CrossRef]
17. Kumar, P.V.; Ashok, B.; Vignesh, R.; Bhasker, J.P.; Kumar, A.N. Evaluation of performance, emissions and combustion attributes of CI engine using palmyra biodiesel blend with distinct compression ratios, EGR rates and nano additives. *Fuel* **2022**, *321*, 124092. [CrossRef]
18. Seraç, M.R.; Aydın, S.; Şevik, S. Evaluation of comparative combustion, performance, and emission of soybean-based alternative biodiesel fuel blends in a CI engine. *Renew. Energy* **2019**, *148*, 1065–1073. [CrossRef]
19. Vellaiyan, S. Combustion, performance and emission evaluation of a diesel engine fueled with soybean biodiesel and its water blends. *Energy* **2020**, *201*, 117633. [CrossRef]
20. Hussain, F.; Soudagar, M.E.; Afzal, A.; Mujtaba, M.A.; Fattah, I.; Naik, B.; Mulla, M.H.; Badruddin, I.A.; Khan, T.M.; Raju, V.D.; et al. Enhancement in Combustion, Performance, and Emission Characteristics of a Diesel Engine Fueled with Ce-ZnO Nanoparticle Additive Added to Soybean Biodiesel Blends. *Energies* **2020**, *13*, 4578. [CrossRef]
21. Lin, C.; Lin, K. Comparison of the Engine Performance of Soybean Oil Biodiesel Emulsions Prepared by Phase Inversion Temperature and Mechanical Homogenization Methods. *Processes* **2023**, *11*, 907. [CrossRef]
22. Venu, H.; Subramani, L.; Raju, V.D. Emission reduction in a DI diesel engine using exhaust gas recirculation (EGR) of palm biodiesel blended with TiO_2 nano additives. *Renew. Energy* **2019**, *140*, 245–263. [CrossRef]
23. Mujtaba, M.; Masjuki, H.H.; Kalam, M.A.; Noor, F.; Farooq, M.; Ong, H.C.; Gul, M.; Soudagar, M.E.M.; Bashir, S.; Fattah, I.M.R.; et al. Effect of Additivized Biodiesel Blends on Diesel Engine Performance, Emission, Tribological Characteristics, and Lubricant Tribology. *Energies* **2020**, *13*, 3375. [CrossRef]
24. Jayabalaji, G.; Shanmughasundaram, P. Effect of Titanium Dioxide (TiO_2) Nano-Fluid on Performance and Emission Features of a Diesel Engine Operated on Aphanizomenon Flos Biodiesel-Diesel Blend. *Mater. Sci. Forum* **2019**, *969*, 421–426. [CrossRef]
25. Gunasekar, P.; Manigandan, S.; Ilangovan, N.; Nithya, S.; Devipriya, J.; Saravanan, W.S. Effect of TiO_2 and nozzle geometry on diesel emissions fuelled with biodiesel blends. *Int. J. Ambient. Energy* **2019**, *40*, 477–481. [CrossRef]
26. Sarma, C.J.; Sharma, P.; Bora, B.J.; Bora, D.K.; Senthilkumar, N.; Balakrishnan, D.; Ayesh, A.I. Improving the combustion and emission performance of a diesel engine powered with mahua biodiesel and TiO_2 nanoparticles additive. *Alex. Eng. J.* **2023**, *72*, 387–398. [CrossRef]
27. Kurre, S.K.; Yadav, J.K.; Khan, M.K.; Yadav, P.; Munaf Dhebar, M.; Rawat, B.S. Experimental evaluation of performance and emission of diesel engine fueled with nano-material titanium dioxide (TiO_2) nanoparticles supplemented diesel-biodiesel-ethanol blends. *Mater. Today Proc.* **2023**. [CrossRef]
28. Fangsuwannarak, K.; Fangsuwannarak, T.; Khotbut, Y. Effect of Nano-TiO_2 Additives Blended in Palm Biodiesel on Compression Ignition Engine Performance. *J. Clean Energy Technol.* **2020**, *8*, 20–23. [CrossRef]
29. Madhuri, R.J.; Naik, J.V.; Roa, V. Effects of Aloe Vera Biodiesel Blends with TiO_2 on the Emission and Performance Characteristics of a DI Diesel Engine. *Biotechnol. J. Int.* **2023**, *27*, 1–7. [CrossRef]
30. Prasetya, H.; Apriyanto, H.; Agustina, L. Determination of optimum technologies implementation scenarios toward net zero emissions in palm oil biodiesel production. *IOP Conf. Ser. Earth Environ. Sci.* **2023**, *1187*, 012029. [CrossRef]

31. Razzaq, L.; Abbas, M.M.; Waseem, A.; Jauhar, T.A.; Fayaz, H.; Kalam, M.A.; Soudagar, M.E.; Silitonga, A.S.; Husnain, S.U.; Ishtiaq, U. Influence of varying concentrations of TiO_2 nanoparticles and engine speed on the performance and emissions of diesel engine operated on waste cooking oil biodiesel blends using response surface methodology. *Heliyon* **2023**, *9*, e17758. [CrossRef] [PubMed]
32. Mehregan, M.; Moghiman, M. Experimental investigation of the distinct effects of nanoparticles addition and urea-SCR after-treatment system on NOx emissions in a blended-biodiesel fueled internal combustion engine. *Fuel* **2020**, *262*, 116609. [CrossRef]

nanomaterials

MDPI

Article

Enhancing the Performance and Stability of Li-CO_2 Batteries Through LAGTP Solid Electrolyte and MWCNT/Ru Cathode Integration

Dan Na [1], Dohyeon Yu [1], Hwan Kim [1], Baeksang Yoon [1], David D. Lee [2] and Inseok Seo [1,*]

[1] Department of Electronic and Information Materials Engineering, Division of Advanced Materials Engineering, Research Center of Advanced Materials Development, Jeonbuk National University, Jeonju 54896, Republic of Korea; ld3310@jbnu.ac.kr (D.N.); yudohyeon@jbnu.ac.kr (D.Y.); rlaghks0202@jbnu.ac.kr (H.K.); qortkd2134@jbnu.ac.kr (B.Y.)

[2] Aerospace Engineering, Iowa State University, Ames, IA 50011, USA; daylee@iastate.edu

* Correspondence: isseo@jbnu.ac.kr; Fax: +82-63-270-2305

Abstract: Li-CO_2 batteries (LCBs) have emerged as promising solutions for energy storage, with the added benefit of contributing to carbon neutrality by capturing and utilizing CO_2 during operation. In this study, a high-performance LCB was developed using a Ge-doped LiAlGeTi $(PO_4)_3$ (LAGTP) solid electrolyte, which was synthesized via a solution-based method by doping Ge into NASICON-type LATP. The ionic conductivity of the LAGTP pellets was measured as 1.04×10^{-3} S/cm at 25 °C. The LCB utilizing LAGTP and an MWCNT/Ru cathode maintained a stable cycling performance over 200 cycles at a current density of 100 mA/g, with a cut-off capacity of 500 mAh/g. Post-cycle analysis confirmed the reversible electrochemical reactions at the cathode. The integration of LAGTP as a solid electrolyte effectively enhanced the ionic conductivity and improved the cycle life and performance of the LCB. This study highlights the potential of Ge-doped NASICON-type solid electrolytes for advanced energy-storage technologies and offers a pathway for developing sustainable and high-performance LCBs.

Keywords: Li-CO_2 battery; solid electrolyte; NASICON; Ru catalyst

Citation: Na, D.; Yu, D.; Kim, H.; Yoon, B.; Lee, D.D.; Seo, I. Enhancing the Performance and Stability of Li-CO_2 Batteries Through LAGTP Solid Electrolyte and MWCNT/Ru Cathode Integration. *Nanomaterials* **2024**, *14*, 1894. https://doi.org/10.3390/nano14231894

Academic Editor: Francis Leonard Deepak

Received: 22 October 2024
Revised: 13 November 2024
Accepted: 19 November 2024
Published: 26 November 2024

1. Introduction

With growing concerns regarding global warming and environmental deterioration, the reduction of CO_2 emissions has become an urgent global priority. As a greenhouse gas, CO_2 contributes significantly to climate change, highlighting the need for innovative strategies for its reduction and utilization [1,2]. Li-CO_2 batteries (LCBs) have emerged as a promising solution, offering the dual advantages of high energy density and CO_2 sequestration [3,4]. LCBs operate through reversible electrochemical reactions that enable simultaneous energy storage and carbon capture [5,6].

$$4Li^+ + 4e^- + 3CO_2 \rightarrow 2Li_2CO_3 + C \quad (1)$$

This reaction provides a high theoretical energy density of up to 1876 Wh/kg, making LCBs attractive for applications in CO_2-rich environments and contributing to efforts toward achieving carbon neutrality [7]. However, conventional LCBs typically employ organic liquid electrolytes, which present several challenges such as flammability, volatility, leakage risks, and inadequate suppression of lithium dendrite formation [8,9]. These issues compromise the safety and hinder the practical application of LCBs. Solid-state electrolytes have been introduced to address these limitations, offering enhanced safety and electrochemical performance in terms of safety and stability [10,11].

Lithium aluminum titanium phosphate (LiAlTi $(PO_4)_3$, LATP), which has a sodium superionic conductor (NASICON) structure, is a well-known solid electrolyte characterized by high ionic conductivity and excellent chemical and thermal stabilities [12]. The performance of LATP can be further enhanced by doping it with various metals. Doping expands the Li-ion migration pathways and optimizes the lattice structure, leading to increased ionic conductivity and an expanded electrochemical-stability window, which facilitates stable operation at higher voltages [13,14]. Building on these advantages, Ge doping has been explored to improve the properties of LATP further. By substituting Ti^{4+} with Ge^{4+} ions, lithium-aluminum-germanium-titanium phosphate (LiAlGeTi $(PO_4)_3$, LAGTP) can be formed from LATP. Ge doping induces lattice distortions that adjust lattice constants and improve ionic conductivity by reducing the migration energy barrier. This process facilitates lithium-ion migration by creating interconnected diffusion pathways [15].

In addition to optimizing the electrolyte, developing efficient cathode materials is essential for enhancing LCB performance. Multi-walled carbon nanotubes (MWCNTs) are widely used as cathodes because of their high electrical conductivity and large surface area; however, they have limited catalytic activity for the decomposition of Li_2CO_3 [16]. The catalytic activity of MWCNTs can be significantly enhanced by introducing an Ru catalyst, effectively reducing overpotentials and improving battery efficiency and cycle life. The Ru catalyst lowers the activation energy required for the decomposition of Li_2CO_3, reducing the polarization and further improving battery performance [17].

In this paper, we present a high-performance LCB that integrates LAGTP solid electrolytes with Ru-catalyzed MWCNT cathodes. Ge doping enhances the electrolyte's ionic conductivity and structural stability, and the Ru catalyst improves electrochemical reactions involving CO_2. We evaluated the electrochemical properties of the fabricated LCB, and the results demonstrated significant performance enhancements due to the synergistic effects of the optimized solid electrolyte and the MWCNT/Ru cathode material. This approach offers a promising pathway for developing safe, efficient, and high-energy-density LCBs for advanced energy storage applications.

2. Materials and Methods

2.1. Preparation of LAGTP Pellet

LAGTP powder was synthesized using a solution-based method. Initially, LiCl (Samchun, Seoul, Republic of Korea, 98.2%), $NH_4H_2PO_4$ (Samchun, 98.0%), and Al $(NO_3)_3{\cdot}9H_2O$ (Samchun, 98.0%) were dissolved stoichiometrically in deionized water with magnetic stirring. Subsequently, GeO_2 (99.9%, Sigma-Aldrich, St. Louis, MI, USA) was added to the mixture and pulverized in a high-energy mill (Taemyong Scientific, Uiwang-si, Republic of Korea). Titanium butoxide (Samchun, 97.0%) was added to the solution, followed by continuous magnetic stirring. The resulting precipitates were homogeneously mixed using a planetary mill (Pulverisette 5, Fritsch, Idar-Oberstein, Germany). After drying in an oven at 80 °C for 24 h, the mixture was calcined at 800 °C for 10 h to eliminate volatile impurities. Polyvinyl alcohol (3wt%) was then added to the mixture and pressed into pellets using a tungsten carbide die at 60 MPa. The pellets were sintered at 950 °C for 12 h and polished to achieve the desired thickness and surface smoothness for cell applications.

2.2. Preparation of the Cathode (MWCNT and MWCNT/Ru Powder)

$RuCl_3{\cdot}xH_2O$ (Sigma Aldrich, 50 mg) was added to ethylene glycol (Daejung, Goryeong-gun, Republic of Korea, 99.0%, 150 mL) and dissolved through magnetic stirring. Subsequently, MWCNT (Sigma Aldrich, 60 mg) was introduced into the solution and sonicated for 1 h. The mixture was then refluxed at 180 °C for 3 h. After cooling to room temperature, the supernatant was decanted, and the mixture was centrifuged to separate the MWCNT/Ru from solution, which were then rinsed with ethanol and deionized water. The collected solids were dried in a vacuum oven at 70 °C for 12 h. To prepare the cathode material, MWCNT and MWCNT/Ru powders (80 wt%) were separately blended with polyvinylidene fluoride (PVDF, 20 wt%) and dispersed in N-methyl-2-pyrrolidone (NMP) using a non-bubbling kneader. The resultant paste was uniformly applied onto a carbon cloth substrate with 14 mm diameter and dried in a vacuum oven at 70 °C.

2.3. Li-CO_2 Cell Assembly Procedure

For the electrochemical evaluation of the proposed LCB, we used a CR-2032 coin cell that featured a hole on its side for CO_2 gas exposure. The LCB was assembled in an Ar-filled glove box (with H_2O and O_2 concentration below 1.0 ppm), using MWCNT/Ru cathode, along with 30 μL of a 1.2 M solution of lithium bis (trifluoromethanesulfonyl)imide (LiTFSI) in tetraethylene glycol dimethyl ether (TEGDME), LAGTP solid-state electrolyte, and a Li metal anode (0.5 mm thickness, 11 mm diameter, 99.9% purity). Following the assembly, the coin cell was placed in a testing zig designed to facilitate CO_2 gas circulation and assess its electrochemical properties. CO_2 gas (99.9% purity) was used for electrochemical testing. The testing zig was flushed with CO_2 gas and linked to a battery-testing apparatus overnight to ensure LCB stabilization.

2.4. Characterization of Electrochemical Performance

Electrochemical performance tests were conducted using a battery test system (WBCS 3000S, WonATech, Seoul, Republic of Korea) under ambient conditions via galvanostatic discharge/charge tests. Furthermore, a cycle test was performed at a constant current density of 100 mA g^{-1} with a defined cut-off capacity of 500 mAh g^{-1}. The ionic conductivity of the LATP electrolyte was determined using an EIS tester (Zive SP1, WonATech) with 150 nm thick Au electrodes affixed to both sides of the electrolyte pellets as blocking electrodes. EIS measurements were conducted at an amplitude of 50 mV across a frequency range of 0.1 Hz–1 MHz, maintaining the test environment at room temperature.

2.5. Material Characterizations

Raman spectroscopy (InVia Qontor, RENISHAW Ltd., Wotton-under-Edge, United Kingdom) and X-ray diffraction (XRD, D8 ADVANCE, Bruker, Billerica, MA, USA) were used for the structural characterization of LAGTP and MWCNT-Ru. Morphological and nanostructural assessments were performed using scanning transmission electron microscopy (STEM, Tecnai G2 F20, FEI, Waltham, MA, USA) and field-emission scanning electron microscopy (FE-SEM, SU-70, HITACHI, Tokyo, Japan). Energy-dispersive X-ray spectroscopy (EDS, Tecnai G2 F20, FEI) was employed to characterize the elemental composition and distribution of the MWCNT/Ru composite and LAGTP. The surface area of the MWCNTs was determined from nitrogen adsorption–desorption isotherms using the Brunauer–Emmett–Teller (BET) method (BELSORP mini X, Microtracbel, Osaka, Japan). X-ray photoelectron spectroscopy (XPS, Nexsa XPS System, Thermo Scientific, Waltham, MA, USA), FE-SEM, EDS, and STEM were used to analyze the cathodes.

3. Results and Discussion

Figure 1 shows a comprehensive characterization of MWCNT-Ru in terms of its morphological, structural, chemical, and gas-adsorption properties. The STEM image in Figure 1a reveals that nanosized particles are uniformly distributed on MWCNT. EDS mapping (Figure 1b) confirmed that these nanoparticles were composed of Ru. The EDS spectrum (Figure S1) further shows the elemental composition of MWCNT/Ru, with carbon (C) at 98.68 at. % and ruthenium (Ru) at 1.32 at.%. Figure S2 shows the FE-SEM image of an MWCNT with Ru, which shows a dense MWCNT network. The crystallographic details of the Ru nanoparticles were further analyzed using STEM (Figure 1c) and fast Fourier transform (FFT) electron diffraction (Figure 1d). Lattice fringes corresponding to the (101) and (113) crystal planes of crystalline Ru, with spacings of 2.056 and 2.343 Å, respectively, are observed. From the FFT images, lattice points corresponding to the (100), (002), and (101) planes were identified [18]. Figure 1e shows the XRD patterns of MWCNT and MWCNT/Ru. The XRD pattern of the pristine MWCNTs exhibits a prominent peak at approximately 26°, corresponding to the (002) plane of hexagonal carbon. This peak is indicative of well-ordered graphitic layers typical of MWCNTs. Additionally, a less intense peak is observed at approximately 43°, attributed to the (100) plane of the hexagonal carbon structure. In the XRD pattern of MWCNT/Ru, in addition to the characteristic carbon peaks of MWCNTs, several new peaks emerge at 2θ values of approximately 38.2°, 42.2°, 43.8°, 58.3°, and 69.2°. These peaks correspond to the (100), (002), (101), (102), and (110) crystal planes of metallic Ru, respectively, in agreement with the reference pattern of hexagonal Ru (PDF 00-006-0663). The presence of these peaks confirmed the successful introduction of Ru nanoparticles on the MWCNTs. The additional Ru peaks were consistent with the hexagonal crystalline structure of Ru, corroborating the results obtained from the TEM and FFT analyses. Figure 1f shows the Raman spectra of MWCNT and MWCNT/Ru, which were used to analyze the structural properties of these materials. Both spectra exhibit two prominent peaks corresponding to the D and G bands at approximately 1350 cm^{-1} and 1580 cm^{-1}, respectively. The D band, indicative of defects and disorder in the carbon structure, arises from the breathing modes of sp^3-hybridized carbon atoms. By contrast, the G band is associated with the E_{2g} phonon mode of sp^2-hybridized carbon atoms, representing the graphitic structural characteristics of the carbon nanotubes [19]. Additionally, both spectra exhibit a minor peak corresponding to the 2D band at around 2700 cm^{-1}. The I_D/I_G ratios for the MWCNT and MWCNT/Ru samples were 1.09 and 1.03, respectively, indicating no significant differences, which suggests that the carbon network of the MWCNTs remained intact and well-preserved during the synthesis. Figure 1g shows the XPS survey spectrum of MWCNT/Ru, in which peaks corresponding to C, O, and Ru can be identified. Figure S3 presents the high-resolution C 1s spectrum of pristine MWCNT. Peaks are observed at 284.4, 285.5, 286.9, 289.2, and 291.5 eV, corresponding to sp^2-C, sp^3-C, C-O, -COO-, and π-π* bonds, respectively [20]. Figure 1h shows the high-resolution C 1s spectrum of MWCNT/Ru, where, in addition to the peaks exhibited by pristine MWCNT, the peaks corresponding to Ru $3d_{3/2}$ (284.5 eV) and Ru $3d_{5/2}$ (280.3 eV) are present. The spin-orbit splitting value of 4.2 eV between Ru $3d_{3/2}$ and Ru $3d_{5/2}$ and the binding energy of Ru $3d_{5/2}$ indicated no oxidation or other chemical-state changes, signifying the formation of pure metallic Ru [21]. The surface area, adsorption–desorption characteristics, and pore size distribution of the MWCNT and MWCNT/Ru samples were characterized via BET analysis, as shown in the nitrogen adsorption–desorption isotherm graph and pore size distribution. Figure 1i presents the nitrogen adsorption–desorption isotherms for MWCNT and MWCNT/Ru. Compared to MWCNT/Ru, MWCNT exhibits higher adsorption in the relative pressure range of 0.1 to 0.8. This can be attributed to the larger surface area and porous structure of MWCNT, which facilitates more active initial adsorption. In contrast, MWCNT/Ru shows relatively lower adsorption in this range, likely due to the presence of Ru particles. The BET surface areas of MWCNT and MWCNT/Ru are 261.81 m^2/g and 82.2 m^2/g, respectively. The reduction in BET surface area upon Ru addition is likely because Ru particles have a relatively low surface area and high density, partially blocking the pores of MWCNT. Despite the decrease in total BET surface area, both samples exhibit hysteresis in the isotherms, a characteristic of mesoporous materials, indicative of a typical Type IV isotherm. [22,23]. This suggests that the main adsorption–desorption characteristics are still governed by the mesoporous MWCNT structure, even with the addition of Ru. The pore size distribution is shown in Figure S4. The overall distribution pattern of MWCNT and MWCNT/Ru is similar, with pore sizes mainly distributed between 30 and 80 nm, but the pore volume of MWCNT/Ru is relatively reduced. The decrease in pore volume and size in the MWCNT/Ru sample can be attributed to Ru particles dispersing on the MWCNT surface, blocking some pore structures and thus reducing the overall pore volume and size.

Multiscale characterization confirmed the formation of nanosized crystalline Ru on the MWCNTs, which did not significantly affect the original carbon structure or gas adsorption properties of the MWCNTs. This suggests that the inherent characteristics of the MWCNTs are preserved, and the catalytic effects of Ru can be effectively utilized.

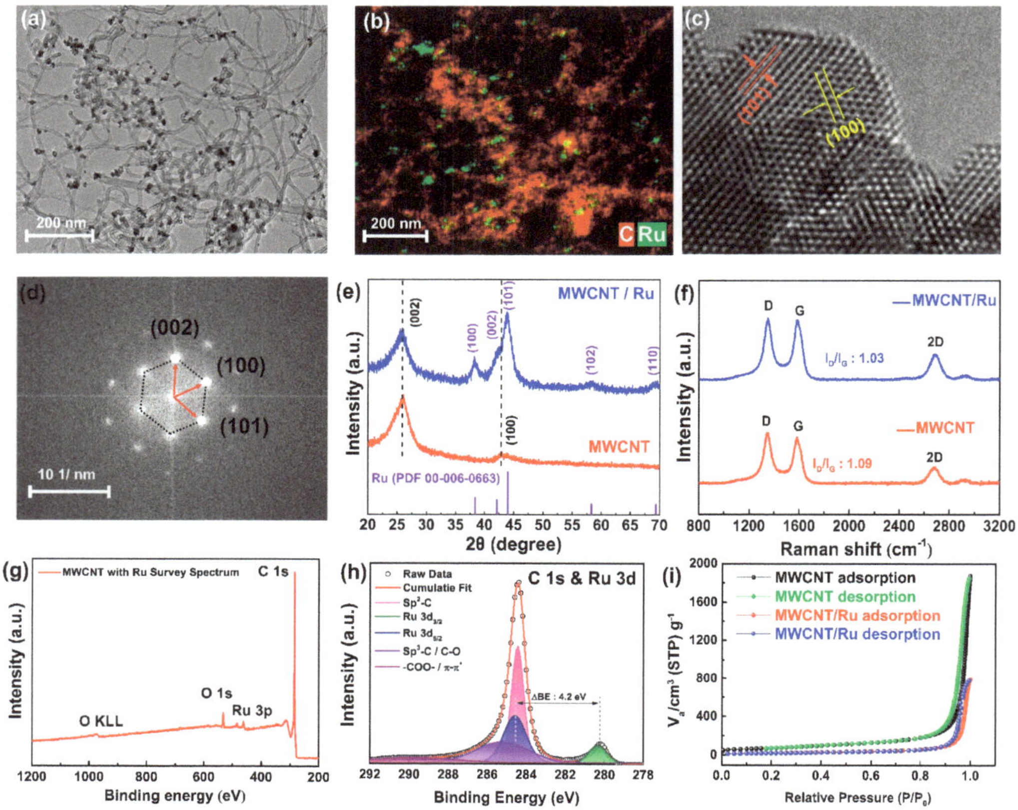

Figure 1. (**a**) TEM and (**b**) EDS mapping images of MWCNT/Ru. (**c**) Magnified lattice fringe image of Ru nanoparticles. (**d**) FFT pattern showing crystallographic planes of Ru. (**e**) XRD pattern of MWCNT and MWCNT/Ru. (**f**) Raman spectra of MWCNT and MWCNT/Ru. (**g**) XPS survey spectrum of MWCNT/Ru. (**h**) High-resolution C 1s and Ru 3d XPS spectra. (**i**) BET isotherm for MWCNT/Ru.

Figure 2a shows a STEM image of LAGTP powder, which reveals the general morphology and size of the particles. The particles exhibit clear nanoscale dimensions, providing an overview of their structural arrangement. Figure S5 shows the EDS map and spectrum of LAGTP, illustrating the spatial distribution and chemical composition of its constituent elements, confirming that each element was incorporated in the intended stoichiometric amounts. The individual elemental maps for Al, Ge, Ti, P, and O show that these elements are uniformly distributed throughout the material, confirming their homogeneous integration into the LAGTP structure. Figure 2b displays a magnified image of the lattice fringes of LAGTP, with the measured d-spacing of 0.6 nm, corresponding to the (012) crystallographic plane. The distinct and sharp lattice fringes confirm the material's high crystallinity, which indicates its well-ordered atomic structure. Figure S6 shows the FFT analysis results corresponding to Figure 2b, showing diffraction spots indexed to the (012) and (024) crystallographic planes of LAGTP. The calculated d-spacing values from the FFT analysis were 0.6 nm for the (012) plane and 0.3 nm for the (024) plane, demonstrating the well-ordered atomic arrangement of the material. Additionally, Figure 2c shows the selected area electron diffraction (SAED) pattern of LAGTP powder, where diffraction spots corresponding to the (012), (024), and (−123) planes can be observed.

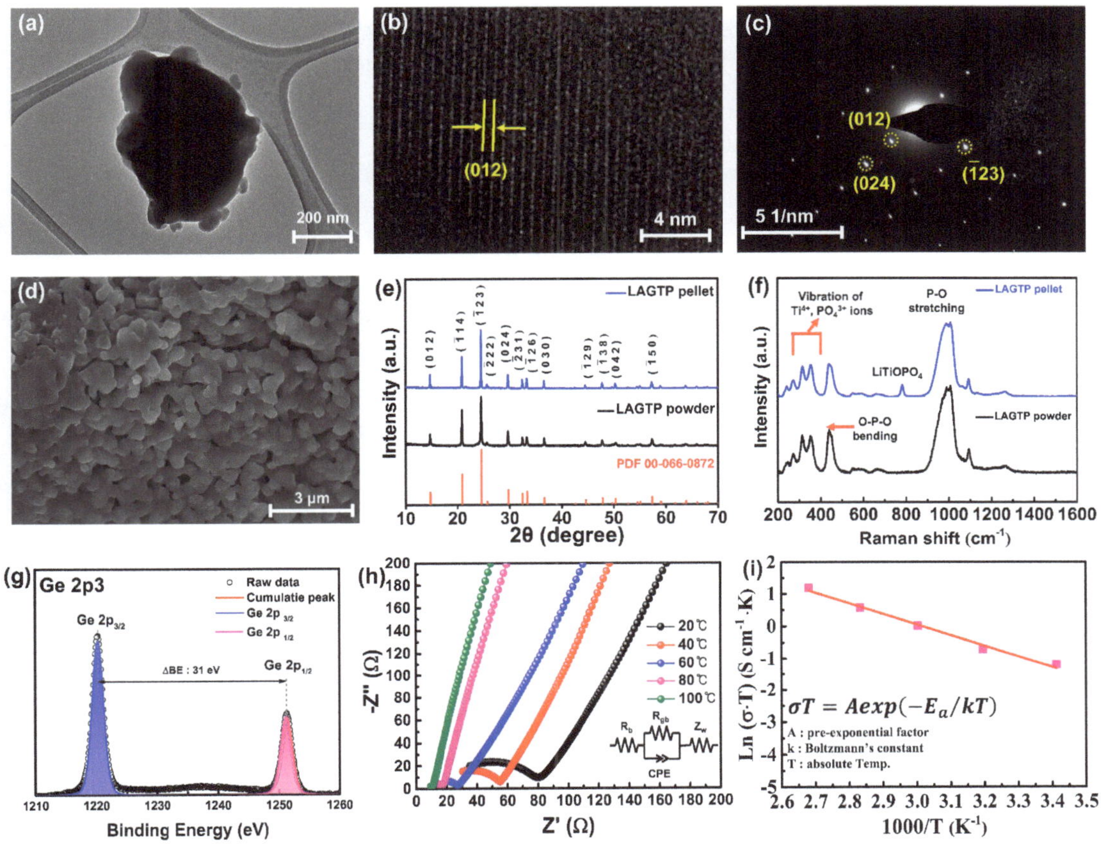

Figure 2. (**a**) TEM image, (**b**) magnified lattice fringe image, and (**c**) the SEAD pattern of LAGTP powder. (**d**) Magnified cross−sectional FE−SEM image of LAGTP pellets. (**e**) XRD patterns and (**f**) Raman spectra of LAGTP powder and pellets. (**g**) High−resolution XPS Ge 2p spectrum of LAGTP powder. (**h**) Nyquist plot of the LAGTP electrolyte at various temperatures, and (**i**) Arrhenius plot of the ionic conductivity of the LAGTP electrolyte.

The morphology of the LAGTP solid-electrolyte pellets was examined using FE-SEM. Figure S7 shows an overall image and a magnified view of the cross-section of the LAGTP pellets. The analysis showed that the pellets were dense at the microscale, with no visible pores or cracks. Figure 2d presents the cross-section of the LAGTP pellet after sintering. A comparison of the cross-sectional images of the LAGTP pellet before and after sintering is provided in Figure S8. In Figure S8a, the cross-section before sintering shows the powder compacted by mechanical pressure. In contrast, Figure S8b illustrates the grain growth and formation of an interconnected structure after sintering, providing continuous pathways for Li ions.

XRD analysis was conducted to examine the structural properties of the LAGTP powder synthesized using a solution-based method. Figure 2e shows the diffraction patterns of the LAGTP powder and pellets in the 2θ range of 10–70°, compared with LATP reference data (PDF 00-066-0872). The prominent peaks are observed at 20.9°, 24.5°, and 29.7°, corresponding to the (104), (−123), and (024) planes of LATP powder and pellets, respectively. The XRD patterns of the LAGTP powder and pellets are well consistent with the NASICON structure standard of LATP, which is also consistent with the TEM and SEAD results, confirming that LAGTP adopted a NASICON-type crystal structure. For a more detailed analysis of the structural characteristics of the LAGTP powder and pellets, the investigation focused on the 20–30° range, where the dominant peaks were located. Figure S9 shows this range's XRD patterns of LAGTP powders and pellets. Interestingly, several weak peaks that were absent in the XRD pattern of LAGTP powder were observed at 25.9°, 27.0°, 27.4°, and 27.9°. These peaks were identified as the secondary phases of GeO_2 and $LiTiOPO_4$, which were formed during the high-temperature sintering of the LAGTP pellet [24,25]. Despite the presence of these secondary phases, their minimal intensity in the XRD pattern suggests that their formation does not significantly affect the overall performance of the pellets.

Figure 2f illustrates the Raman spectra of LAGTP powder and the sintered pellet. When comparing the spectra of the LAGTP powder and the pellet, it is evident that apart from the peak at 782 cm^{-1}, which corresponds to $LiTiOPO_4$, the two samples exhibit similar patterns. The emergence of the $LiTiOPO_4$ peak after high-temperature sintering aligns with the previously discussed XRD results. Peaks observed at 241, 271, 315, and 355 cm^{-1} are associated with external vibrational modes, specifically the translational and librational motions of Ti^{4+} and $PO_4{}^{3-}$ ions. The peak at 439 cm^{-1} corresponds to the O-P-O bending vibration, while the internal vibrational modes observed at 969, 991, 1008, and 1093 cm^{-1} indicate the stretching vibrations within the PO_4 tetrahedral structure. These vibrational modes are primarily observed in LATP materials, and the Raman and XRD results confirm that, apart from the subtle formation of secondary phases, there are no significant structural changes after sintering.

XPS was used to investigate the elemental composition and chemical states of LAGTP. Figure S10 shows the XPS survey spectrum of LAGTP, where peaks corresponding to Li, Al, P, Ti, O, and Ge are observed at 55.4, 74.5, 133.2, 459.6, 531, and 1220.3 eV, respectively, confirming the presence of all the constituent elements of LAGTP. Figure 2g presents the high-resolution Ge 2p spectrum of LAGTP powder, which shows distinct orbital splitting at 1220.3 eV for Ge 2p3/2 and 1251.3 eV for Ge 2p1/2. The symmetric binding energy values of Ge 2p3/2 and Ge 2p1/2 and the energy difference of 31 eV (ΔBE) between these two peaks correspond to the expected spin-orbit coupling for Ge, confirming that Ge is in the Ge4+ oxidation state [26,27]. This indicates that Ge was not reduced to a lower-valence state and was adequately integrated into the LATP crystal structure.

The electrochemical properties of the LAGTP solid electrolyte were evaluated across a temperature range of 20–100 °C using EIS. Figure 2h shows the Nyquist plots of the LAGTP electrolyte at different temperatures and the corresponding equivalent circuits. Typically, Nyquist plots of ionically conductive solid electrolytes are modeled using an equivalent circuit consisting of the grain bulk resistance (R_b) and grain boundary resistance (R_{gb}) in series and parallel combinations, along with a constant phase element (CPE) and Warburg impedance (Zw). The ionic conductivity of the LAGTP electrolyte was calculated using Equation (2).

$$\sigma = \frac{1}{R} \times \frac{t}{A} \quad (2)$$

where σ represents the ionic conductivity of the electrolyte, R (R_b + R_{gb}) is the total resistance, t is the thickness of the pellets, and A refers to the area of the blocking electrode. The total resistance and ionic conductivity of the LAGTP electrolyte at various temperatures are listed in Table 1.

Table 1. Total resistance and total ionic conductivity of LAGTP at various test temperatures.

Temperature (°C)	Total Resistance (Ω)	Total Ionic Conductivity (mS cm^{-1})
20	79.9	1.04
40	53.2	1.56
60	27.3	3.03
80	16.5	5.02
100	9.4	8.80

Additionally, the activation energy was derived from temperature-dependent EIS measurements using the Arrhenius Equation (3):

$$\sigma T = Aexp\langle -\frac{E_a}{kT} \rangle \quad (3)$$

where A, E_a, k, and T denote the pre-exponential factor, activation energy, Boltzmann constant, and absolute temperature, respectively. From the slope of the Arrhenius plot in the Figure 2i, the activation energy for ion conduction in LAGTP was calculated to be approximately 0.25 eV, which is notably lower than the activation energy typically reported for LATP [28–30]. This reduction in activation energy suggests that Ge doping improves the ionic mobility in LAGTP. The ionic radius of Ge^{4+} (0.53 Å) is smaller than that of Ti^{4+} (0.605 Å). When Ge partially replaces Ti^{4+} in the LATP structure, the smaller Ge^{4+} ions induce a structural distortion. This structural adjustment helps create interconnected, low-energy pathways, facilitating Li-ion transport via a percolation network [31]. Introducing local distortions generates a distribution of site energies, enabling overlapping energy levels between neighboring sites. This overlap creates a percolation pathway, enabling Li ions to migrate more freely through the electrolyte and enhancing diffusion without requiring a significant increase in cell size.

A schematic illustration of the Li-CO_2 battery system is presented in Figure S11, highlighting the LAGTP solid electrolyte and the MWCNT/Ru cathode, along with the reaction mechanisms of CO_2 reduction during discharge and Li_2CO_3 decomposition during charge. This configuration was used in the electrochemical tests to assess the battery's performance. Cycling tests were conducted to evaluate the electrochemical performance of the LCB based on the LAGTP solid electrolyte and Ru-enhanced MWCNT cathode. The current density and specific capacity were calculated using the mass of the MWCNT/Ru cathode. The cycling tests were performed at a current density of 100 mA/g and a capacity limit of 500 mAh/g. Figure 3a shows the overall cycling profile of the LCB over 200 cycles, demonstrating its successful operation over 2000 h of testing. The battery components underwent minimal degradation during the cycling test. Figure 3b shows the galvanostatic charge–discharge profiles over various cycles (1st, 50th, 100th, 150th, and 200th cycles). As the cycles progressed, changes in the voltage plateaus for both discharging and charging were observed. Figure 3c illustrates the change in terminal voltage during cycling. Initially, the discharge voltage was approximately 2.6 V. The charge voltage was approximately 4.0 V, resulting in a polarization of approximately 1.4 V. After 200 cycles, the discharge voltage increased to 2.7 V, and the charge voltage increased to 4.6 V. Additionally, the polarization increased to 1.9 V, marking a 0.5 V increase from that at the initial cycle. While the discharge terminal voltage remains relatively stable, the charge terminal voltage steadily increased from 4.0 V to approximately 4.6 V after 200 cycles. The initial terminal voltage (4.0 V) was significantly lower than the typical charge-terminal voltage (4.5 V) observed in conventional MWCNT-based LCBs, which can be attributed to the catalytic activity of Ru [32,33]. NASICON-type solid electrolytes tend to degrade faster when high voltages are sustained [34]; therefore, the formation of a lower charge–voltage plateau due to the introduction of Ru contributes to the stability of the solid electrolyte in the LCB. Table 2 presents the performances of LCBs reported to date that use solid electrolytes. The LCB utilizing LAGTP and MWCNT/Ru exhibited excellent cycling performance. This improvement is attributed to the enhanced stability of the solid electrolyte, which was achieved through Ge doping [35], and the catalytic effects of Ru, enabling operation within a lower voltage range, leading to the development of a high-performance LCB.

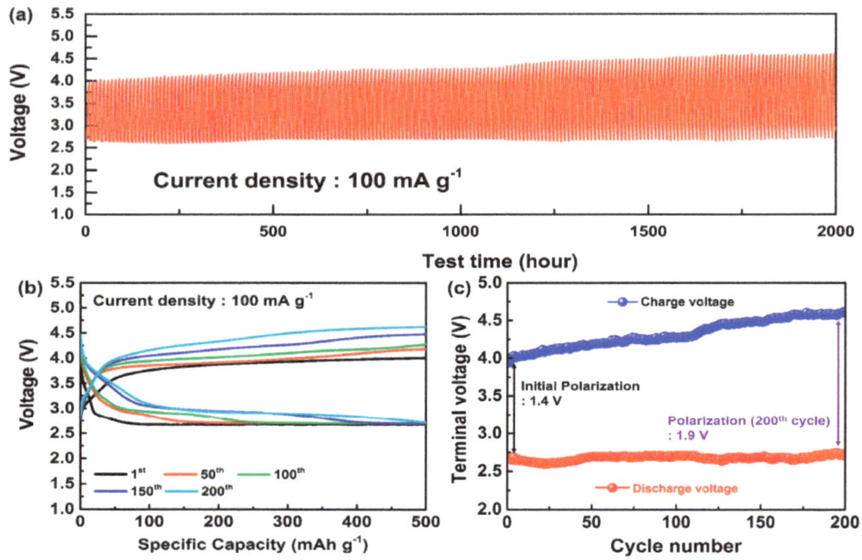

Figure 3. (**a**) Voltage−time, (**b**) voltage−specific capacity, and (**c**) terminal voltage-cycle number graph of LAGTP applied LCBs at a current density of 100 mA/g.

Table 2. Cycle performance of Li-CO_2 batteries with inorganic solid electrolytes at room temperature.

S.No	Cathode Material	Inorganic Electrolyte	Cycle Life	Ref.
1	MWCNT	LATP	50	[36]
2	SWCNT/RuO_2	LAGP	30	[37]
3	Ru/CNT	LAGP	45	[33]
4	MWCNT	LAGTP	60	[38]
5	MWCNT /Ru	LATP	50	[32]
6	Co_3O_4-derived MOF	LATP	100	[39]
7	Fe_3C/N-doped CNT	Zn-doped LATP	180	[40]
8	MWCNT/Ru	LAGTP	200	This work

A post-cycle analysis was conducted to validate the performance of the LCB. Figure 4a,b show FE-SEM images of the cathode after discharge. In Figure 4a, discharge products can be observed to form on the MWCNT matrix. Figure 4b provides a higher-magnification view, where granular discharge products covering the surface are visible. Figure 4e,f show the FE-SEM images of the cathode after charging. According to Equation (1), the discharge products decompose after charging, and the surface returns to that of pristine MWCNT/Ru, as shown in Figure S2. Figure 4c,g present the EDS mapping images of the cathode after discharging and charging, respectively. The mapping results illustrate the spatial distribution of C, O, and Ru across the cathode surface. C and Ru from MWCNT/Ru were observed in the discharged and charged states. In Figure 4c, which represents the discharged state, O from the discharge product Li_2CO_3 is dominant. Conversely, as shown in Figure 4g, the O concentration decreases significantly after charging, indicating the decomposition of the discharge products. However, the localized areas show the presence of irreversibly formed Li_2CO_3. As the cycling progressed, the accumulation of Li_2CO_3 contributed to increased polarization. XPS analysis was performed to verify the chemical states of the cathode and the discharge products after charge and discharge tests. Figure S12 presents high-resolution XPS spectra of Li 1s and Ru 3p in the cathode after charge and discharge. Figure S12a shows the Li 1s high-resolution XPS spectrum for the charged and discharged states of the cathode. In the discharged state, peaks at Li_2CO_3 at 54.5 eV corresponding to Li_2CO_3 and a trace peaks at 55.6 eV for LiTFSI are observed.

The presence of LiTFSI is attributed to residual lithium salt remaining after the cleaning process. After charging, the Li 1s spectrum reveals only a minor peak for LiTFSI, indicating that the discharge product, Li_2CO_3, has decomposed [41]. Figure S12b displays the Ru 3p high-resolution XPS spectrum for both charged and discharged states, with identical peak positions observed in both cases. The peaks are located at Ru^0 $3p_{3/2}$ (462.0 eV), Ru^{4+} $3p_{3/2}$ (465.0 eV), Ru^0 $3p_{1/2}$ (484.5 eV), and Ru^{4+} $3p_{1/2}$ (487.6 eV) [42]. Figure 4d,h show the high-resolution C 1s XPS spectra of the discharged and charged cathodes, respectively. In the discharged cathode, the C 1s spectrum shows peaks corresponding to sp^2–C (284.4 eV), sp^3–C (285.4 eV), and C–O (286.8 eV) bonds and Li_2CO_3 (289.3 eV), indicating the formation of lithium carbonate during discharge [43,44]. After charging, the XPS spectrum revealed a change in the chemical environment, with peaks corresponding to MWCNT and F-C-F from the PVDF binder [45], whereas the Li_2CO_3 peak diminished but remained. This indicates that, although Li_2CO_3 decomposed during the charging process, some irreversible Li_2CO_3 persisted. These characterization results confirm the occurrence of reversible reactions, as shown in Equation (1), while also indicating the presence of irreversible Li_2CO_3, which contributes to the increase in polarization during cycling.

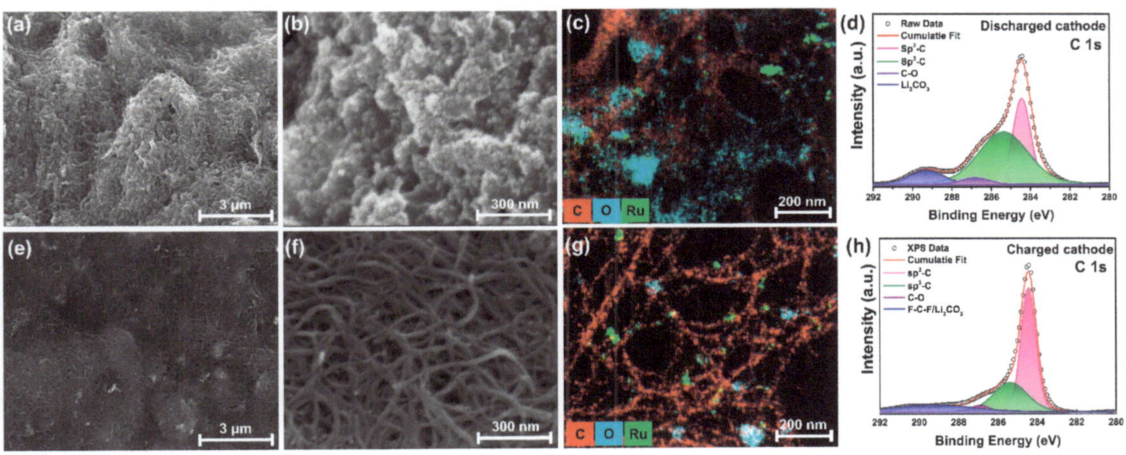

Figure 4. FE-SEM images of (**a**,**b**) the discharged cathode and (**e**,**f**) the charged cathode, showing the surface morphology. EDS mapping of (**c**) the discharged and (**g**) charged cathodes. XPS spectra of the C 1s peak for (**d**) the discharged and (**h**) charged cathodes.

4. Conclusions

In this study, we demonstrated the successful integration of Ge-doped LAGTP solid electrolytes and MWCNT/Ru cathodes in LCBs. Ge doping in LAGTP resulted in significantly higher ionic conductivity, achieving a value of 1.04×10^{-3} S/cm at 25 °C by lowering the activation energy and expanding the Li-ion migration channels while maintaining the NASICON-type structure. Ru nanoparticles anchored on the MWCNT enhanced the catalytic activity for Li_2CO_3 decomposition, reducing the overpotential and extending the cycling performance of the LCB. Electrochemical tests revealed that the LCB assembled with LAGTP and MWCNT/Ru maintained stable operation over 200 cycles, highlighting the combined effects of enhanced electrolyte stability and improved cathode performance. Furthermore, post-cycle analysis confirmed the reversible electrochemical reaction of CO_2 with Li, although a minor accumulation of irreversible Li_2CO_3 contributed to a gradual increase in polarization. This study demonstrates the potential of using Ge-doped NASICON-type electrolytes and Ru catalysts to enhance the electrochemical stability and performance of LCBs, offering a promising strategy for future energy storage technologies contributing to CO_2 sequestration.

Supplementary Materials: The following supporting information can be downloaded from: https://www.mdpi.com/article/10.3390/nano14231894/s1, Figure S1. EDS spectrum and elemental composition of MWCNT/Ru; Figure S2. FESEM images of the MWCNT with Ru; Figure S3. High-resolution C 1s spectrum of pristine MWCNT; Figure S4. Pore size distribution of MWCNT and MWCNT/Ru; Figure S5. Energy-dispersive X-ray spectroscopy (EDS) elemental mapping of LAGTP powders, indicating the distribution of (a) all detected elements (Al, Ge, P, Ti, and O), (b) Al, (c) Ge, (d) O, (e) P, and (f) Ti. (g) EDS spectrum of LAGTP and chemical composition; Figure S6. FFT image of the LASTP powder corresponding to Figure 2b; Figure S7. (a) Cross-sectional FESEM image of the LAGTP pellet attached to the carbon tape (a) and (b) magnified FESEM image of the pellet; Figure S8. Cross-sectional FESEM image of an LAGTP pellet (a) before and (b) after sintering; Figure S9. Magnified XRD spectra of LAGTP powder and pellets in the 20–30°; Figure S10. Survey spectrum of the LAGTP powder. Figure S11. Schematic illustration of the Li-CO2 battery and its electrochemical reactions; Figure S12. High-resolution XPS spectrum of (a) Li 1s and (b) Ru 3p in the cathode after charge and discharge.

Author Contributions: Conceptualization, methodology, investigation, formal analysis, writing—original draft, visualization, D.N.; Validation, investigation, visualization, D.Y.; Validation, investigation, H.K.; Resources, B.Y.; Conceptualization, supervision, project administration, D.D.L.; Supervision, writing—review and editing, resources, project administration, funding acquisition, I.S. All authors have read and agreed to the published version of the manuscript.

Funding: This work was supported by the Technology Innovation Program (RS-2024-00419781) funded by the Ministry of Trade Industry & Energy (MOTIE, Korea) and the BK21 FOUR Program through a Jeonbuk National University Research Grant.

Data Availability Statement: The original contributions presented in the study are included in the article/Supplementary Materials; further inquiries can be directed to the corresponding authors.

Conflicts of Interest: The authors declare no conflicts of interest.

References

1. Qiu, F.; Ren, S.; Mu, X.; Liu, Y.; Zhang, X.; He, P.; Zhou, H. Towards a Stable Li-CO_2 Battery: The Effects of CO_2 to the Li Metal Anode. *Energy Storage Mater.* **2020**, *26*, 443–447. [CrossRef]
2. Bhatti, U.A.; Bhatti, M.A.; Tang, H.; Syam, M.S.; Awwad, E.M.; Sharaf, M.; Ghadi, Y.Y. Global Production Patterns: Understanding the Relationship between Greenhouse Gas Emissions, Agriculture Greening and Climate Variability. *Environ. Res.* **2024**, *245*, 118049. [CrossRef] [PubMed]
3. Chen, H.; Li, X.; Xue, H.; Jia, L.; Xu, Y.; Tao, Y.; Yan, Y.; Fan, X.; He, J.; Wang, T. Recent Advances in the Mechanism and Catalyst Design in the Research of Aprotic, Photo-Assisted, and Solid-State Li-CO_2 Batteries. *Inorg. Chem. Front.* **2024**, *11*, 5833–5857. [CrossRef]
4. Cheng, Y.; Wang, Y.; Chen, B.; Han, X.; He, F.; He, C.; Hu, W.; Zhou, G.; Zhao, N. Routes to Bidirectional Cathodes for Reversible Aprotic Alkali Metal–CO_2 Batteries. *Adv. Mater.* **2024**, *36*, 2410704. [CrossRef] [PubMed]
5. Wang, Y.; Li, S.; Li, L.; Dan, B.; Wang, D.; Liu, X. Minireview on Achieving Low Charge Overpotential in Li-CO_2 Batteries with Advancing Cathode Materials and Electrolytes. *Energy Fuels* **2024**, *38*, 2743–2758. [CrossRef]
6. Pan, Q.; Ma, X.; Wang, H.; Shu, Y.; Liu, H.; Yang, L.; Li, W.; Liu, J.; Wu, Y.; Mao, Y.; et al. Approaching Splendid Catalysts for Li-CO_2 Battery from the Theory to Practical Designing: A Review. *Adv. Mater.* **2024**, *36*, 2406905. [CrossRef]
7. Yang, J.; Liu, Y.; Liu, Y.; Huang, Y.; Wang, K.; Liu, M.; Li, W. Enhanced Co_3O_4 Nanoflakes Reactivity via Integrated Al-Doping and Metal Vacancy Engineering for Large Capacity Li-CO_2 Batteries. *Nano Energy* **2024**, *129*, 109979. [CrossRef]
8. Liang, J.; Luo, J.; Sun, Q.; Yang, X.; Li, R.; Sun, X. Recent Progress on Solid-State Hybrid Electrolytes for Solid-State Lithium Batteries. *Energy Storage Mater.* **2019**, *21*, 308–334. [CrossRef]
9. Tang, X.; Zhou, D.; Li, P.; Guo, X.; Wang, C.; Kang, F.; Li, B.; Wang, G. High-Performance Quasi-Solid-State MXene-Based Li–I Batteries. *ACS Cent. Sci.* **2019**, *5*, 365–373. [CrossRef]
10. Ahmad, N.; Fan, C.; Faheem, M.; Liang, X.; Xiao, Y.; Cao, X.; Zeng, C.; Dong, Q.; Yang, W. Key Challenges and Advancements toward Fast-Charging All-Solid-State Lithium Batteries. *Green Chem.* **2024**, *26*, 9529–9553. [CrossRef]
11. Yang, S.-J.; Hu, J.-K.; Jiang, F.-N.; Yuan, H.; Park, H.S.; Huang, J.-Q. Safer Solid-State Lithium Metal Batteries: Mechanisms and Strategies. *InfoMat* **2024**, *6*, e12512. [CrossRef]
12. Xiao, W.; Wang, J.; Fan, L.; Zhang, J.; Li, X. Recent Advances in $Li_{1+x}Al_xTi_{2-x}(PO_4)_3$ Solid-State Electrolyte for Safe Lithium Batteries. *Energy Storage Mater.* **2019**, *19*, 379–400. [CrossRef]
13. Wu, P.; Zhou, W.; Su, X.; Li, J.; Su, M.; Zhou, X.; Sheldon, B.W.; Lu, W. Recent Advances in Conduction Mechanisms, Synthesis Metho ds, and Improvement Strategies for $Li_{1+x}Al_xTi_{2-x}(PO_4)_3$ Solid Electrolyte for All-Solid-State Lithium Batteries. *Adv. Energy Mater.* **2023**, *13*, 2203440. [CrossRef]

14. Yin, J.-H.; Zhu, H.; Yu, S.-J.; Dong, Y.-B.; Wei, Q.-Y.; Xu, G.-Q.; Xiong, Y.; Qian, Y. Recent Advances of LATP and Their NASICON Structure as a Solid-State Electrolyte for Lithium-Ion Batteries. *Adv. Eng. Mater.* **2023**, *25*, 2300566. [CrossRef]
15. Baek, J.; Yoon, B.; Jeong, H.; Jeong, J.; Mamidi, S.; Seo, H.-K.; Lee, C.-R.; Seo, I. Dependences of Ionic Conductivity and Activation Energy on Germanium Content in Superionic $Li_{1.4}Al_{0.4}Ge_xTi_{(1.6-x)}(PO_4)_3$ Solid Electrolytes. *J. Electroanal. Chem.* **2022**, *920*, 116631. [CrossRef]
16. Liu, Q.; Hu, Z.; Li, L.; Li, W.; Zou, C.; Jin, H.; Wang, S.; Chou, S.-L. Facile Synthesis of Birnessite δ-MnO_2 and Carbon Nanotube Composites as Effective Catalysts for Li-CO_2 Batteries. *ACS Appl. Mater. Interfaces* **2021**, *13*, 16585–16593. [CrossRef]
17. Yang, S.; Qiao, Y.; He, P.; Liu, Y.; Cheng, Z.; Zhu, J.; Zhou, H. A Reversible Lithium–CO_2 Battery with Ru Nanoparticles as a Cathode Catalyst. *Energy Environ. Sci.* **2017**, *10*, 972–978. [CrossRef]
18. Zhou, Y.; Xie, Z.; Jiang, J.; Wang, J.; Song, X.; He, Q.; Ding, W.; Wei, Z. Lattice-Confined Ru Clusters with High CO Tolerance and Activity for the Hydrogen Oxidation Reaction. *Nat. Catal.* **2020**, *3*, 454–462. [CrossRef]
19. Nakajima, H.; Kobashi, K.; Zhou, Y.; Zhang, M.; Okazaki, T. Quantitative Analysis of the Correlation between Sp^3 Bonds and Functional Groups in Covalently Functionalized Single-Walled Carbon Nanotubes. *Carbon* **2024**, *216*, 118495. [CrossRef]
20. Li, J.; Qiao, Y.; Li, D.; Zhang, S.; Liu, P. Improving Interfacial and Mechanical Properties of Glass Fabric/Polyphenylene Sulfide Composites via Grafting Multi-Walled Carbon Nanotubes. *RSC Adv.* **2019**, *9*, 32634–32643. [CrossRef]
21. Mayor, L.C.; Ben Taylor, J.; Magnano, G.; Rienzo, A.; Satterley, C.J.; O'Shea, J.N.; Schnadt, J. Photoemission, Resonant Photoemission, and x-Ray Absorption of a Ru(II) Complex Adsorbed on Rutile $TiO_2(110)$ Prepared by in Situ Electrospray Deposition. *J. Chem. Phys.* **2008**, *129*, 114701. [CrossRef] [PubMed]
22. Wang, Z.; Jiang, X.; Pan, M.; Shi, Y. Nano-Scale Pore Structure and Its Multi-Fractal Characteristics of Tight Sandstone by N_2 Adsorption/Desorption Analyses: A Case Study of Shihezi Formation from the Sulige Gas Filed, Ordos Basin, China. *Minerals* **2020**, *10*, 377. [CrossRef]
23. Tóth, A.; Voitko, K.V.; Bakalinska, O.; Prykhod'ko, G.P.; Bertóti, I.; Martínez-Alonso, A.; Tascón, J.M.D.; Gun'ko, V.M.; László, K. Morphology and Adsorption Properties of Chemically Modified MWCNT Probed by Nitrogen, *n*-Propane and Water Vapor. *Carbon* **2012**, *50*, 577–585. [CrossRef]
24. Ma, B.; Li, D.; Wang, X.; Lin, K. Fast and Safe Synthesis of Micron Germanium in an Ammonia Atmosphere Using Mo_2N as Catalyst. *RSC Adv.* **2018**, *8*, 35753–35758. [CrossRef] [PubMed]
25. Hupfer, T.; Bucharsky, E.C.; Schell, K.G.; Hoffmann, M.J. Influence of the Secondary Phase $LiTiOPO_4$ on the Properties of $Li_{1+x}Al_xTi_{2-x}(PO_4)_3$ (x = 0; 0.3). *Solid State Ion.* **2017**, *302*, 49–53. [CrossRef]
26. Tkalcevic, M.; Boršćak, D.; Periša, I.; Radovic, I.; Saric, I.; Bernstorff, S.; Micetic, M. Multiple Exciton Generation in 3D-Ordered Networks of Ge Quantum Wires in Alumina Matrix. *Materials* **2022**, *15*, 5353. [CrossRef]
27. Gao, A.; Jiang, P.; Duan, M.; Lai, H.; Zhou, Y.; Zhang, X.; Yang, M.; Gong, L.; Chen, J.; Liu, S.; et al. Interphase Design Enabling Stable Cycling of All-Solid-State Lithium Metal Batteries by in-Situ X-Ray Photoelectron Spectroscopy Lithium Metal Sputtering. *J. Power Sources* **2024**, *602*, 234299. [CrossRef]
28. Jackman, S.D.; Cutler, R.A. Effect of Microcracking on Ionic Conductivity in LATP. *J. Power Sources* **2012**, *218*, 65–72. [CrossRef]
29. Vinod Chandran, C.; Pristat, S.; Witt, E.; Tietz, F.; Heitjans, P. Solid-State NMR Investigations on the Structure and Dynamics of the Ionic Conductor $Li_{1+x}Al_xTi_{2-x}(PO_4)_3$ ($0.0 \leq x \leq 1.0$). *J. Phys. Chem. C* **2016**, *120*, 8436–8442. [CrossRef]
30. Kang, J.; Gu, R.; Guo, X.; Li, J.; Sun, H.; Zhang, L.; Jing, R.; Jin, L.; Wei, X. Effect of SnO–P_2O_5–MgO Glass Addition on the Ionic Conductivity of Li1.3Al0.3Ti1.7(PO4)3 Solid Electrolyte. *Ceram. Int.* **2022**, *48*, 157–163. [CrossRef]
31. Zeng, Y.; Ouyang, B.; Liu, J.; Byeon, Y.-W.; Cai, Z.; Miara, L.J.; Wang, Y.; Ceder, G. High-Entropy Mechanism to Boost Ionic Conductivity. *Science* **2022**, *378*, 1320–1324. [CrossRef] [PubMed]
32. Na, D.; Kampara, R.K.; Yu, D.; Yoon, B.; Martin, S.W.; Seo, I. $Li_{1.4}Al_{0.4}Ti_{1.6}(PO_4)_3$ Inorganic Solid Electrolyte for All-Solid-State Li-CO_2 Batteries with MWCNT and Ru Nanoparticle Catalysts. *Mater. Today Energy* **2023**, 101418. [CrossRef]
33. Savunthari, K.V.; Chen, C.-H.; Chen, Y.-R.; Tong, Z.; Iputera, K.; Wang, F.-M.; Hsu, C.-C.; Wei, D.-H.; Hu, S.-F.; Liu, R.-S. Effective Ru/CNT Cathode for Rechargeable Solid-State Li-CO_2 Batteries. *ACS Appl. Mater. Interfaces* **2021**, *13*, 44266–44273. [CrossRef] [PubMed]
34. Benabed, Y.; Rioux, M.; Rousselot, S.; Hautier, G.; Dollé, M. Assessing the Electrochemical Stability Window of NASICON-Type Solid Electrolytes. *Front. Energy Res.* **2021**, *9*, 682008. [CrossRef]
35. Wang, J.; He, T.; Yang, X.; Cai, Z.; Wang, Y.; Lacivita, V.; Kim, H.; Ouyang, B.; Ceder, G. Design Principles for NASICON Super-Ionic Conductors. *Nat. Commun.* **2023**, *14*, 5210. [CrossRef]
36. Na, D.; Jeong, H.; Baek, J.; Yu, H.; Lee, S.-M.; Lee, C.-R.; Seo, H.-K.; Kim, J.-K.; Seo, I. Highly Safe and Stable Li-CO_2 Batteries Using Conducting Ceramic Solid Electrolyte and MWCNT Composite Cathode. *Electrochim. Acta* **2022**, *419*, 140408. [CrossRef]
37. Du, Y.; Liu, Y.; Yang, S.; Li, C.; Cheng, Z.; Qiu, F.; He, P.; Zhou, H. A Rechargeable All-Solid-State Li-CO_2 Battery Using a $Li_{1.5}Al_{0.5}Ge_{1.5}(PO_4)_3$ Ceramic Electrolyte and Nanoscale RuO_2 Catalyst. *J. Mater. Chem. A* **2021**, *9*, 9581–9585. [CrossRef]
38. Yoon, B.; Baek, J.; Na, D.; Yu, D.; Kampara, R.K.; Seo, H.-K.; Lee, D.Y.; Seo, I. $Li_{1.4}Al_{0.4}Ge_{0.1}Ti_{1.5}(PO_4)_3$: A Stable Solid Electrolyte for Li-CO_2 Batteries. *Mater. Chem. Phys.* **2024**, *322*, 129583. [CrossRef]
39. Mamidi, S.; Na, D.; Yoon, B.; Sharma, H.; Pathak, A.D.; Sahu, K.K.; Lee, D.Y.; Lee, C.-R.; Seo, I. Safe and Stable Li-CO_2 Battery with Metal-Organic Framework Derived Cathode Composite and Solid Electrolyte. *J. Power Sources* **2024**, *591*, 233867. [CrossRef]
40. Zhu, Q.-C.; Ma, J.; Huang, J.-H.; Mao, D.-Y.; Wang, K.-X. Realizing Long-Cycling Solid-State Li-CO_2 Batteries Using Zn-Doped LATP Ceramic Electrolytes. *Chem. Eng. J.* **2024**, *482*, 148977. [CrossRef]

41. Pipes, R.; Bhargav, A.; Manthiram, A. Phenyl Disulfide Additive for Solution-Mediated Carbon Dioxide Utilization in Li-CO_2 Batteries. *Adv. Energy Mater.* **2019**, *9*, 1900453. [CrossRef]
42. Zhang, J.; Li, Y.; Yang, L.; Zhang, F.; Li, R.; Dong, H. Ruthenium Nanosheets Decorated Cobalt Foam for Controllable Hydrogen Production from Sodium Borohydride Hydrolysis. *Catal. Lett.* **2022**, *152*, 1386–1391. [CrossRef]
43. Hornsveld, N.; Put, B.; Kessels, W.M.M.; Vereecken, P.M.; Creatore, M. Plasma-Assisted and Thermal Atomic Layer Deposition of Electrochemically Active Li_2CO_3. *RSC Adv.* **2017**, *7*, 41359–41368. [CrossRef]
44. Na, D.; Kampara, R.K.; Yu, D.; Yoon, B.; Lee, D.Y.; Seo, I. Exploring Li-CO_2 Batteries with Electrospun PAN-Derived Carbon Nanofibers and $Li_{1.4}Al_{0.4}Ti_{1.6}(PO_4)_3$ Solid-State Electrolyte. *J. Alloys Compd.* **2024**, *970*, 172559. [CrossRef]
45. Viswanath, P.; Yoshimura, M. Light-Induced Reversible Phase Transition in Polyvinylidene Fluoride-Based Nanocomposites. *SN Appl. Sci.* **2019**, *1*, 1519. [CrossRef]

nanomaterials

MDPI

Article

Modified Fe_3O_4 Nanoparticles for Foam Stabilization: Mechanisms and Applications for Enhanced Oil Recovery

Dandan Yin [1,2], Judong Qiu [1,2], Dongfeng Zhao [1,2,*], Yongzheng Wang [1,2,*], Tao Huang [1,2], Yunqian Long [1,2] and Xiaohe Huang [1,2]

1 College of Petrochemical Engineering & Environment, Zhejiang Ocean University, Zhoushan 316021, China; aayindan@163.com (D.Y.); qiujudong@163.com (J.Q.); huangtaozjou@126.com (T.H.); longyunqian@zjou.edu.cn (Y.L.); hxh_0258@163.com (X.H.)

2 Zhejiang Key Laboratory of Pollution Control for Port-Petrochemical Industry, Zhoushan 316022, China

* Correspondence: dongfeng117@126.com (D.Z.); sdwangyongzheng@163.com (Y.W.)

Abstract: Nanoparticles (NPs) have shown great potential in stabilizing foam for enhanced oil recovery (EOR). However, conventional NPs are difficult to recover and may contaminate produced oil, increasing operational costs. In contrast, superparamagnetic Fe_3O_4 NPs can be efficiently recovered using external magnetic fields, offering a sustainable solution for foam stabilization. In this study, Fe_3O_4 NPs were coated with SiO_2 using tetraethyl orthosilicate (TEOS) and further modified with dodecyltrimethoxysilane to enhance their hydrophobicity. The modification effects were characterized, and the optimal foam-stabilizing $Fe_3O_4@SiO_2$ NPs were found to have a contact angle of 77.01°. The foam system formed with α-olefin sulfonate (0.2 wt%) as the foaming agent and the optimal modified NPs exhibited a drainage half-life of 452 s. After foam-stabilization experiments, the NPs were recovered and reused, with the results indicating that three recovery cycles were optimal. Finally, visual microscopic displacement experiments demonstrated that the foam stabilized by modified NPs effectively mobilized clustered, membranous, and dead-end residual oil, increasing the recovery rate by 17.01% compared with unmodified NPs. This study identifies key areas for future investigation into the application of magnetic nanoparticles for enhanced oil recovery.

Keywords: Fe_3O_4 nanoparticles; foam stabilization; enhanced oil recovery; magnetic nanoparticles; hydrophobic modification

Academic Editors: Robert A. Varin and Geng Zhong

Received: 5 February 2025
Revised: 25 February 2025
Accepted: 26 February 2025
Published: 4 March 2025

Citation: Yin, D.; Qiu, J.; Zhao, D.; Wang, Y.; Huang, T.; Long, Y.; Huang, X. Modified Fe_3O_4 Nanoparticles for Foam Stabilization: Mechanisms and Applications for Enhanced Oil Recovery. *Nanomaterials* **2025**, *15*, 395. https://doi.org/10.3390/nano15050395

1. Introduction

In late-stage reservoir development, reservoir heterogeneity intensifies and the distribution of remaining oil becomes increasingly scattered, leading to a significant decline in oil recovery [1]. Foam flooding has been proposed as an effective method to enhance oil recovery by blocking high-permeability zones and increasing the viscosity of the displacing fluid, thereby improving sweep efficiency [2,3]. However, the thermodynamic instability of foam severely limits its industrial application [4,5].

To enhance foam stability, surfactants are commonly used to reduce surface tension, while polymers can enhance the mechanical strength of the liquid film [6–9]. Although these methods can temporarily stabilize foam, the degradation and loss of surfactants and polymers at high temperatures in the reservoir make it difficult to maintain the foam's stability for a long time [10]. Nanoparticles (NPs) can enhance foam stability by adsorbing at the bubble surface, preventing coalescence and coarsening, and by reinforcing the liquid film, delaying foam drainage [11–14]. Moreover, NPs maintain stability in high-temperature and high-salinity reservoirs [15–19]. However, conventional NPs cannot be recovered after use, increasing operational costs and complicating the treatment of produced oil.

Fe_3O_4 nanoparticles, with their superparamagnetic properties, can rapidly respond to external magnetic fields, enabling efficient recovery and reuse [20,21], which significantly reduces costs and environmental impact. Nevertheless, unmodified Fe_3O_4 NPs tend to aggregate and exhibit poor oxidation stability, limiting their effectiveness in foam stabilization. To address these limitations, surface modification has been researched. Research on the modification of Fe_3O_4 NPs primarily focuses on improving their dispersibility, stability, interfacial adsorption properties [22], and adaptability to complex reservoir conditions [23]. Common modification methods include surface coating (e.g., SiO_2, polymers) [24], chemical modification (e.g., silane coupling agents), and functionalization (e.g., hydrophobic modification) [25,26]. SiO_2 coating significantly enhances the oxidation resistance and dispersibility of Fe_3O_4 NPs while improving their stability under high-temperature and high-salinity conditions [18,27]. Hydrophobic modification using silane coupling agents (e.g., WD-10) can regulate the wettability of Fe_3O_4 NPs [28,29], enabling them to adsorb at the gas–liquid interface and form a physical barrier that prevents bubble coalescence and coarsening [30,31], while also strengthening the liquid film and delaying foam drainage, thereby enhancing foam stability. Additionally, polymer modification (e.g., polyethylene glycol) can improve the biocompatibility and environmental friendliness of Fe_3O_4 NPs, reducing their environmental impact in oilfield applications [32,33].

Despite the progress in foam-stabilization technology based on Fe_3O_4 NPs, challenges such as particle aggregation and insufficient surface modification remain [34]. Further optimization through shell design and functional modification is required. Moreover, most studies have focused on macroscopic foam stability and oil displacement efficiency, failing to reveal the processes of foam generation, migration, and residual oil mobilization using modified Fe_3O_4 NPs [35]. To address these issues, this study focuses on the modification of Fe_3O_4 NPs through silica coating and hydrophobic functionalization, aiming to enhance their foam-stabilization performance and recyclability. The modified NPs are characterized, and their foam-stabilization mechanisms are investigated through static foam stability tests and microscopic visualization experiments. Additionally, the recyclability of the modified NPs is evaluated, and their performance in mobilizing clustered, film, and dead-end residual oil is systematically studied using a microfluidic model that mimics real reservoir conditions.

The novelty of this work lies in its comprehensive investigation of the foam-stabilization and enhanced oil recovery mechanisms of modified Fe_3O_4 NPs, particularly their ability to mobilize different types of residual oil in complex pore networks. By combining macroscopic foam stability tests with microscopic visualization experiments, this study provides new insights into the pore-scale mechanisms of NP-stabilized foam flooding, offering valuable theoretical and technical support for the application of magnetic NPs in enhanced oil recovery.

2. Materials and Methods

2.1. Materials

This experimental research used Fe_3O_4 nanoparticles (NPs) with an average particle size of 20 nm, purchased from Macklin Biochemical Co., Ltd. (Shanghai, China). Additionally, hydrochloric acid (36.0–38.0%), absolute ethanol (99.7%), and ammonia solution (25.0–28.0%) were obtained from Sinopharm Chemical Reagent Co., Ltd. (Shanghai, China). Other chemicals included tetraethyl orthosilicate (TEOS, 99%), n-hexane (97%), KH-304 ($C_{15}H_{34}O_3Si$, 97%), and α-olefin sulfonate (AOS, 92%), all sourced from Macklin Biochemical Co., Ltd. All reagents were of analytical grade and used without further purification.

2.2. Methods

2.2.1. Modification of Fe_3O_4 NPs

To mitigate the mutual attraction between Fe_3O_4 NPs, a silica layer was deposited onto the Fe_3O_4 NPs using tetraethyl orthosilicate (TEOS) as the precursor. The synthesis of $Fe_3O_4@SiO_2$ NPs was carried out via a modified Stöber method based on the sol–gel process, as illustrated in Figure 1. Specifically, 1 g of Fe_3O_4 NPs was dispersed in 50 mL of 0.1 mol/L HCl solution under ultrasonic (pulse mode: 5 s on/2 s off) treatment for 10 min to activate the hydroxyl groups on the surface of NPs, thereby facilitating the subsequent SiO_2 coating. The Fe_3O_4 NPs were then washed five times with deionized water and ethanol to remove impurities, followed by dispersion in 80 mL of an ethanol solution (with a volume fraction of 80%). After ultrasonic dispersion for 10 min, the mixture was transferred to a stirring apparatus and stirred at 720 r/min for 20 min. Subsequently, 12 mL of $NH_3{\cdot}H_2O$ (with a concentration of 25~28%) was added, and the stirring was continued for an additional 30 min. Following this, varying volumes of tetraethyl orthosilicate (TEOS) (800 μL, 1000 μL, and 1200 μL) were introduced into C_2H_5OH/H_2O solution (with a volume fraction of 80%) and stirred at room temperature for 5 h. The resulting solution was centrifuged, washed repeatedly with ethanol and deionized water (3–4 times), and dried at 70 °C for 12 h to obtain the $Fe_3O_4@SiO_2$ NPs.

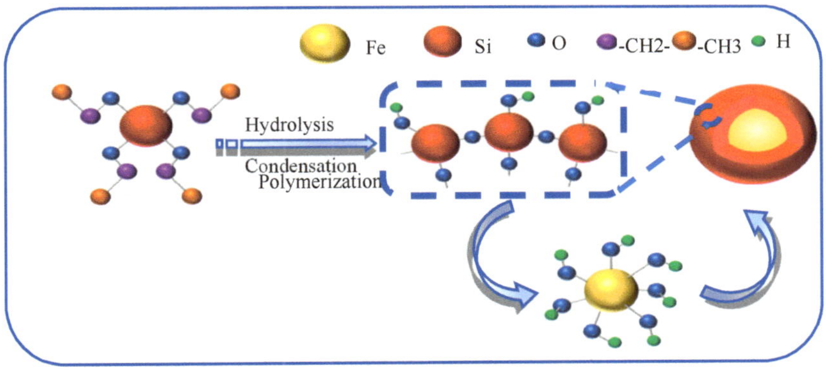

Figure 1. Schematic representation of the synthesis mechanism of $Fe_3O_4@SiO_2$ NPs.

The synthesized $Fe_3O_4@SiO_2$ nanoparticles (NPs) were further functionalized using dodecyltrimethoxysilane (KH-304), as depicted in Figure 2. Specifically, a measured quantity of KH-304 was introduced into 50 mL of 90% ethanol solution and subjected to hydrolysis under continuous stirring at 75 °C for 20 min, resulting in the formation of solution D. Concurrently, 0.2 g of $Fe_3O_4@SiO_2$ NPs was dispersed in 50 mL of 90% ethanol solution via ultrasonication for 10 min to obtain solution E. Solutions D and E were then combined and stirred at 75 °C with a stirring speed of 400–450 rpm for 5 h. Following the reaction, the mixture was centrifuged, washed 3–4 times with ethanol and deionized water, and dried at 70 °C for 6 h to yield the surface-modified $Fe_3O_4@SiO_2$ NPs.

2.2.2. Characterization Techniques

The morphological characteristics of the NPs were examined using scanning electron microscopy (SEM; ZEISS SUPRA-55, Oberkochen, Germany), and the particle size distribution was quantitatively assessed using Image-J software (V1.8.0.112). Surface functional groups of the NPs were analyzed by Fourier-transform infrared spectroscopy (FTIR; Nicolet 6700, Thermo Scientific, Waltham, MA, USA) employing the potassium bromide (KBr) pellet method. Additionally, the hydrophobicity of the modified NPs was evaluated by measuring the contact angle using a contact angle goniometer (JY-82A).

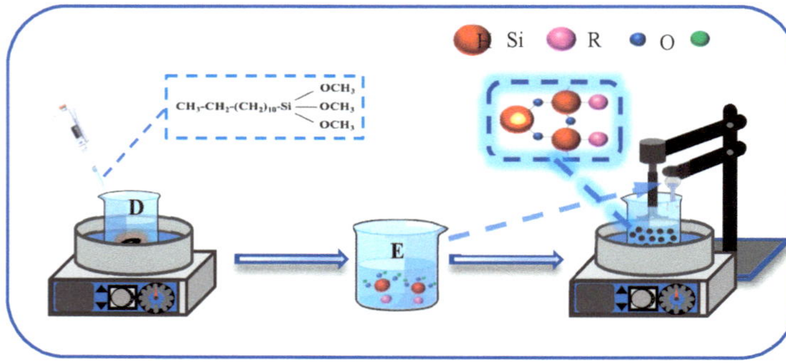

Figure 2. Synthesis process of modified Fe_3O_4@SiO_2 NPs.

2.2.3. Evaluation of Foam Stabilization and NPs Recovery

The foam stability was assessed using the Waring blender method [36]. Specifically, Fe_3O_4@SiO_2 NPs with different wettability modifications (0.2 wt%) were dispersed in 200 mL of a 0.2 wt% AOS solution using an ultrasonic disperser to prepare a homogeneous Fe_3O_4@SiO_2 (0.2 wt%)-AOS (0.2 wt%) dispersion. The dispersion was then foamed using a high-speed blender (8011ES, Waring, Stamford, CT, USA) at 5000 rpm for 3 min, and the resulting foam was transferred to a graduated cylinder. Foamability was characterized by measuring the initial foam volume, while foam stability was evaluated based on the drainage half-life, defined as the time required for half of the liquid to drain. The NPs exhibiting the optimal contact angle were selected for further foam-stabilization experiments. After foam collapse, the mixed solution was subjected to an external magnetic field to facilitate NP sedimentation. The recovered NPs were washed with distilled water and reused for subsequent foam-stabilization tests. The changes in foam half-life were recorded to evaluate the reusability and performance of NPs.

2.2.4. Adsorption of NPs on Liquid Films

The distribution of NPs within the foam system was investigated using a fluorescence microscope (Trim Scope, Wilnsdorf, Germany) to elucidate the stabilization mechanism of Fe_3O_4@SiO_2 NPs in foam. For this purpose, the fluorescent probe rhodamine B, which carries a negative charge and exhibits a maximum excitation wavelength of 543 nm, was employed to label the NPs in the dispersion. The stained NPs were subsequently centrifuged and washed repeatedly with distilled water until the supernatant became clear. The labeled NPs were then utilized to generate foam, and fluorescence images of the foam were captured using the microscope to analyze the spatial distribution and behavior of NPs within the foam structure.

2.2.5. Microscopic Oil Displacement Experiment

Microscopic oil displacement experiments were conducted using a micro-etched porous media model designed to replicate the natural pore structure of cores from the Daqing oilfield. The pore network was etched onto a glass plate using photolithography, and the etched plate was bonded to a smooth glass plate to create the microscopic model. The fabricated glass micromodel exhibited a porosity of 25%, with an average pore depth of approximately 40 μm and a pore width of approximately 100 μm. The experimental setup, as illustrated in Figure 3, comprised a foam generation device, a microscopic pore model, and an observation data acquisition system.

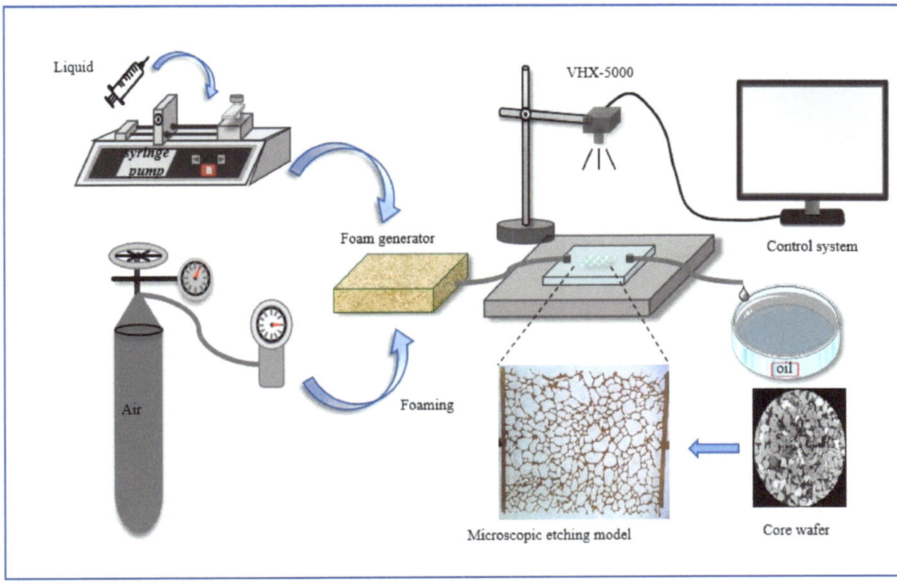

Figure 3. Flow chart of microscopic oil displacement experiment.

Prior to the experiment, the micromodel was evacuated and saturated with crude oil (viscosity = 7.69 mPa·s at 25 °C). Water flooding was initiated by injecting 5 pore volumes (PVs) of water until water without oil was produced. Foam was generated by co-injecting a foam solution and air at a 1:1 volume ratio into the foam generator. Subsequently, 2.5 PVs of either AOS foam or Fe_3O_4@SiO_2-AOS-stabilized foam were injected, followed by 2.5 PVs of water flooding at a constant injection rate of 1 μL/min. The entire displacement process was monitored and recorded using a microscope (VHX-5000, Keyence, Osaka, Japan) equipped with a high-resolution CCD camera and a parallel light source. The experiments focused on investigating the mobilization of various types of residual oil by Fe_3O_4@SiO_2-stabilized foam under high water-saturation conditions at 25 °C.

After the experiment, the micromodel was cleaned by injecting petroleum ether and ethanol to remove residual oil and ensure all pores and throats were free of contaminants. The micromodel was then dried in a constant temperature oven at 45 °C for subsequent use.

3. Results

3.1. Characterization of NPs

Following the modification process, a portion of the hydroxyl (-OH) groups on the surface of the NPs were substituted with carbon chains, resulting in a reduction in saturation magnetization. The successful formation of the modified NPs was confirmed through comprehensive characterization using scanning electron microscopy (SEM), transmission electron microscopy (TEM), and Fourier-transform infrared spectroscopy (FTIR). These techniques were employed to evaluate the morphological and structural changes, as well as to analyze the surface functional groups of the modified NPs.

3.1.1. Morphological Analysis

The SEM images of Fe_3O_4@SiO_2 NPs synthesized with varying amounts of TEOS (800 μL, 1000 μL, and 1200 μL) are presented in Figure 4, illustrating the morphological and size characteristics of the NPs. Following the SiO_2 coating process, the particle size increased significantly from 20 nm to approximately 300 nm. The SEM analysis also revealed the enhanced dispersibility of NPs after coating. As depicted in Figure 4a, bare Fe_3O_4 NPs displayed irregular shapes and pronounced agglomeration. In contrast, Figure 4b shows Fe_3O_4@SiO_2-800 NPs, which exhibited a flocculent structure with reduced agglomeration, suggesting that the SiO_2 coating mitigated the interparticle interactions of Fe_3O_4 NPs. However, due to the insufficient amount of TEOS, a well-defined spherical core–shell structure was not achieved. Figure 4c,d display Fe_3O_4@SiO_2-1000 NPs and Fe_3O_4@SiO_2-1200 NPs, respectively. These NPs demonstrated spherical morphologies, significantly larger particle sizes, reduced agglomeration, and more uniform structures, confirming the successful formation of an SiO_2 coating on the Fe_3O_4 NPs' surface.

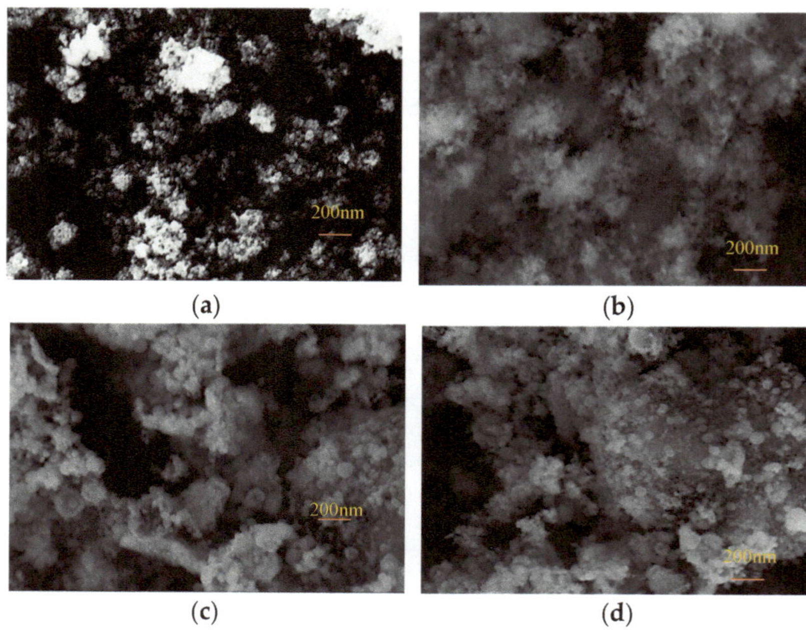

Figure 4. SEM images of NPs: (**a**) Fe_3O_4; (**b**) Fe_3O_4@SiO_2-800; (**c**) Fe_3O_4@SiO_2-1000; (**d**) Fe_3O_4@SiO_2-1200.

To evaluate the size distribution of Fe_3O_4@SiO_2-1000 NPs and Fe_3O_4@SiO_2-1200 NPs, particle size analysis was conducted using Image-J software by randomly selecting 80 points from each SEM image. The results are presented in Figure 5. As depicted in Figure 5a, the particle size distribution of bare Fe_3O_4 NPs ranges from 20 to 200 nm, with an average diameter of 57.663 nm. Figure 5b illustrates that the particle size distribution of Fe_3O_4@SiO_2-1000 NPs spans from 60 to 240 nm, with an average diameter of 147.252 nm. Similarly, Figure 5c demonstrates that the particle size distribution of Fe_3O_4@SiO_2-1200 NPs extends from 60 to 320 nm, with an average diameter of 158.021 nm. These findings indicate a significant increase in the diameter of the NPs following the core–shell coating process, with the particle size positively correlated with the amount of TEOS used. Furthermore, compared with Fe_3O_4@SiO_2-1200 NPs, Fe_3O_4@SiO_2-1000 NPs exhibit a narrower particle size distribution, a smaller average particle size, and enhanced uniformity. Consequently, 1000 μL of TEOS was determined to be the optimal amount for coating the Fe_3O_4 NPs.

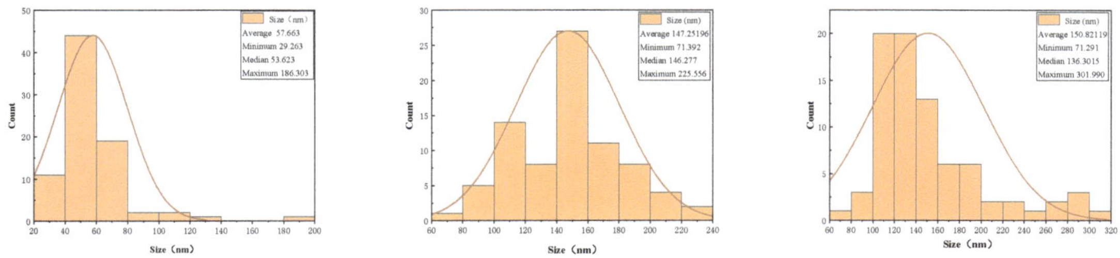

(a) (b) (c)

Figure 5. Size distribution of NPs: (**a**) Naked Fe_3O_4; (**b**) $Fe_3O_4@SiO_2$-1000; (**c**) $Fe_3O_4@SiO_2$-1200.

To visually confirm the formation of the core–shell structure, the NPs were characterized using TEM and Energy-Dispersive X-ray Spectroscopy (EDS), as illustrated in Figure 6. In Figure 6a, the bare Fe_3O_4 NPs exhibit a uniform granular morphology, consistent with the observations from SEM imaging. Figure 6b reveals the core–shell structure of $Fe_3O_4@SiO_2$-1000 NPs, where the core (Fe_3O_4) appears darker, smaller, and denser, while the shell (SiO_2) appears lighter, larger, and forms a well-defined spherical coating. This confirms the successful deposition of a silica shell on the surface of the Fe_3O_4 NPs.

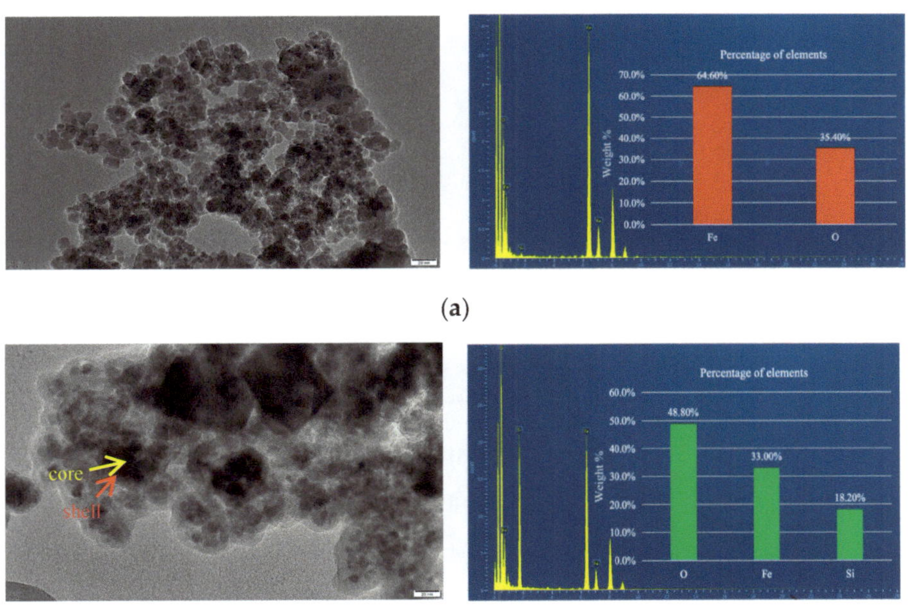

(b)

Figure 6. TEM images and EDS of nanoparticles (**a**) Fe_3O_4; (**b**) $Fe_3O_4@SiO_2$-1000.

To further validate the chemical composition of the materials, EDS analysis was performed. The results indicate that the bare Fe_3O_4 NPs consist of 64.6% Fe and 35.4% O, whereas the SiO_2-coated Fe_3O_4 NPs comprise 48.8% O, 33.0% Fe, and 18.2% Si. The presence of a weak Si peak alongside a strong Fe peak provides additional evidence for the successful formation of a silica shell on the surface of the Fe_3O_4 NPs. These findings collectively demonstrate the effective synthesis of the core–shell structure.

3.1.2. FTIR Analysis

FTIR analysis was employed to confirm the presence of grafted functional groups and the successful formation of modified NPs, as depicted in Figure 7. In all four spectra, the characteristic Fe-O bond vibration peak was observed at 586 cm^{-1} and 630 cm^{-1}, confirming the presence of Fe_3O_4. Additionally, peaks corresponding to -OH stretching and bending vibrations were identified at 3310 cm^{-1} and 1627 cm^{-1}, respectively [37]. A strong absorption peak at 1082 cm^{-1} was attributed to the asymmetric stretching vibration of the Si-O-Si bond, while the peak at 462 cm^{-1} represented the bending vibration of the Si-O-Si bond, both of which are indicative of the SiO_2 coating. The peak at 1123 cm^{-1}, corresponding to the Fe-O-Si stretching vibration, further confirmed the successful formation of the SiO_2 coating and the core–shell structure. In the FTIR spectrum of Fe_3O_4@SiO_2+KH-304, the peaks observed at 2924 cm^{-1} and 2831 cm^{-1} were assigned to the asymmetric stretching vibrations of the -CH_3 and -CH_2 groups, respectively, indicating the presence of carbon chains on the NP surface [38]. Since these carbon chains originate from the hydrophobic modifier KH-304, these results provide clear evidence of the successful hydrophobic modification of the NPs.

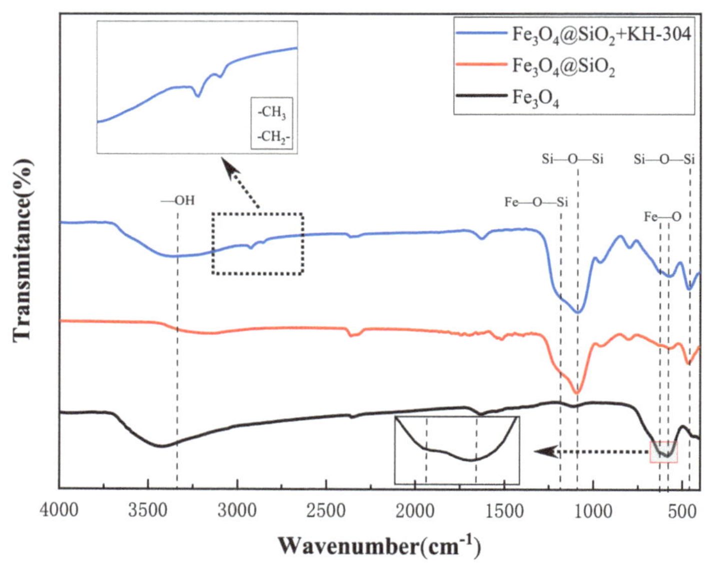

Figure 7. Infrared spectrum of NPs.

3.2. Analysis of Surface Hydrophobicity and Foam-Stabilization Ability of Modified NPs

3.2.1. Analysis of Surface Hydrophobicity of Modified NPs

NPs must exhibit moderate hydrophobicity to ensure strong adsorption at the gas–liquid interface. However, excessively hydrophobic NPs are ineffective in stabilizing foam, as they fail to adsorb onto the lamellae and prevent liquid drainage [39]. Therefore, it is crucial to select NPs with an optimal level of hydrophobicity.

The contact angles of Fe_3O_4@SiO_2-1000-KH-304 NPs, modified with varying amounts of KH-304, were measured, as illustrated in Figure 8. As the concentration of KH-304 increased, the contact angle between the NPs and water progressively rose from 57.18° to 121.34°, after which it plateaued, indicating that further increases in the modifier did not alter the contact angle. To investigate the influence of NPs with different wettabilities on foam stability, NPs with contact angles of 57.18°, 77.01°, 105.51°, and 121.34° were selected for subsequent foam stability experiments.

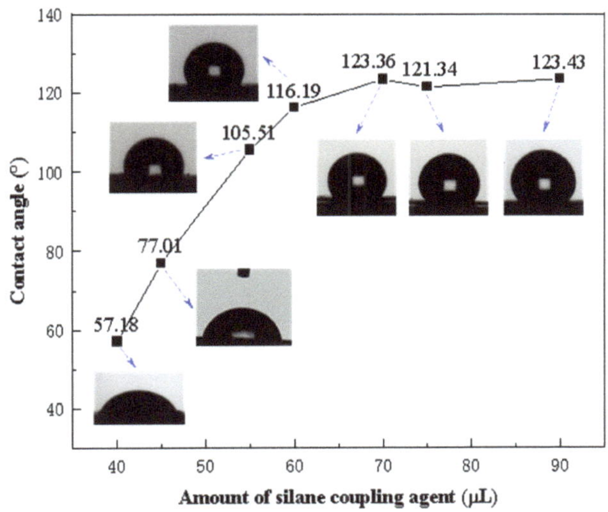

Figure 8. The contact angles of NPs with water as a function of amounts of silane coupling agent.

3.2.2. Analysis of Foam-Stabilization Ability of Modified NPs

Foam was generated by high-speed stirring of a foaming solution containing Fe_3O_4@SiO_2-1000-KH-304 NPs with varying hydrophobicities and alpha-olefin sulfonate (AOS) at a concentration of 0.2 wt%. The drainage half-life and initial foam volume were measured and are presented in Figure 9. The results demonstrated that the foam-stabilization effect was optimal at an NP concentration of 1.0 wt% and a contact angle of 77.01°, achieving a drainage half-life of 452 s and an initial foam volume of 664 mL. In contrast, at the same NP concentration of 1.0 wt%, the foam-stabilization capability was weakest for NPs with a contact angle of 123.3°, exhibiting a drainage half-life of only 374 s, which was lower than that of unmodified NPs.

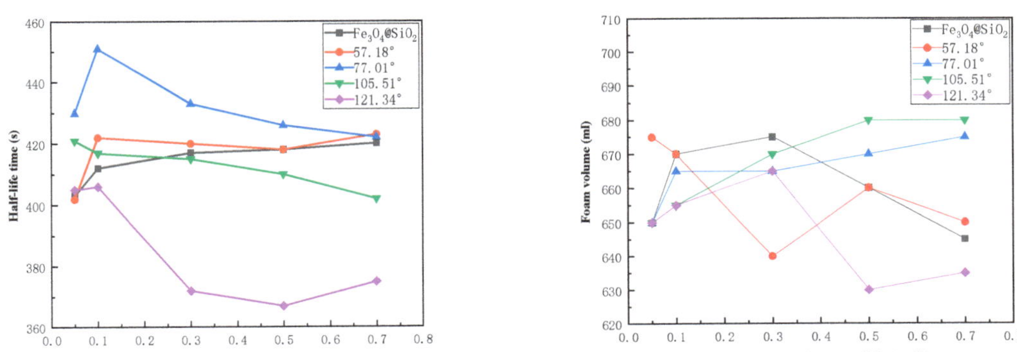

(a) (b)

Figure 9. The different wettability nanoparticles on stabilizing foam: (**a**) drainage half-life time; (**b**) foam volume.

3.2.3. Adsorption of NPs on Foam Surfaces

To elucidate the mechanism by which Fe_3O_4@SiO_2 NPs stabilize foam, the NPs were labeled with a fluorescent dye, and their distribution within the foam system was observed using fluorescence microscopy, as illustrated in Figure 10. Figure 10a reveals that the NPs were predominantly distributed in the liquid phase between bubbles. To more clearly visualize the particle distribution, the foam was drained to form dry foam. In the $Fe_3O_4@SiO_2$ NPs-AOS (0.2 wt%) system, significant NP loss was observed, with few NPs adsorbed on the liquid film, as shown in Figure 10b. In contrast, the majority of $Fe_3O_4@SiO_2$ NPs with a contact angle of 77.01° were adsorbed onto the liquid film of the bubbles and remained in place even after liquid drainage, as depicted in Figure 10c. Conversely, $Fe_3O_4@SiO_2$ NPs with a contact angle of 121.34° tended to migrate out of the liquid phase and into the gas phase, with minimal adsorption on the liquid film, as shown in Figure 10d.

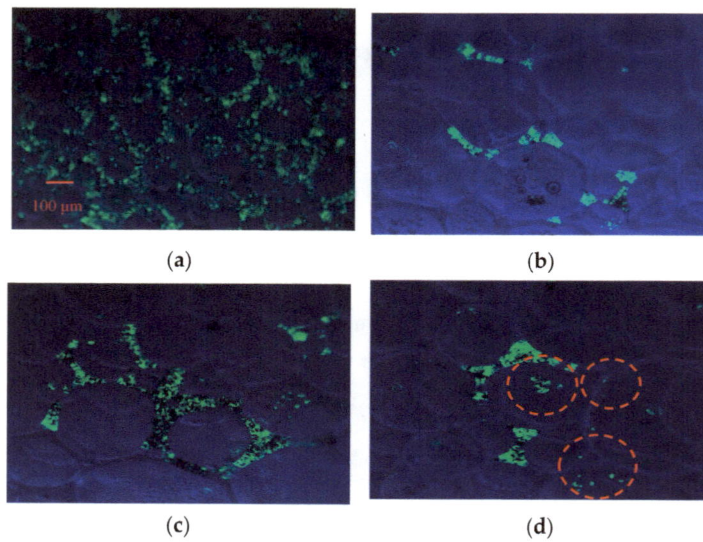

Figure 10. Confocal fluorescence image for the foams ($Fe_3O_4@SiO_2$ NPs-AOS (0.2wt%)). (**a**) Wet foam; (**b**) dry foam; (**c**) dry foam stabilized by NPs with a contact angle of 77.01; (**d**) dry foam stabilized by NPs with a contact angle of 121.34°. The red circles are bubbles.

These observations indicate that NPs located in the surrounding continuous liquid phase are carried away during liquid drainage, whereas NPs adsorbed at the gas–liquid interface remain within the dry foam skeleton after drainage. Specifically, NPs with a contact angle of 77.01° adsorb onto the bubble surface, forming a protective particle armor. Unmodified core–shell NPs, being more hydrophilic, predominantly reside in the liquid phase, while NPs with a contact angle of 121.34°, due to their stronger hydrophobicity, preferentially remain in the gas phase. These findings highlight the critical role of NPs' hydrophobicity in determining their distribution and foam-stabilizing behavior.

3.3. *Recyclability of NPs*

Fe_3O_4 NPs exhibit a high responsiveness to external magnetic fields. Consequently, after assessing foam stability, the NPs were recovered using an external magnetic field and reused for foam stabilization. The foam solution employed for recyclability evaluation consisted of 1 wt% $Fe_3O_4@SiO_2$ NPs (77.01°) and 0.2 wt% AOS, representing the optimal system identified earlier. Following foam generation, the foam half-life was recorded. The NPs were then recovered by applying a magnetic field, washed multiple times with ethanol and deionized water, and reused for foam stabilization until a significant reduction in drainage half-life was observed.

As shown in Figure 11a, the mixed solution exhibited a noticeable lightening in color after 5 min, with a mound of NPs settling at the bottom after 10 min. The solution became clear after 20 min, and the NPs were completely aggregated at the bottom of the container under the influence of the external magnetic field. This demonstrates that the modified NPs retain a high sensitivity to external magnetic fields and can be efficiently recovered. Figure 11b illustrates that foam stability was minimally affected during the first three recovery cycles, with a reduction in drainage half-life of only 2–7%. However, after the fourth recovery cycle, the drainage half-life decreased significantly, negatively impacting foam stability, with a 19% reduction observed by the fifth cycle. Therefore, to maintain optimal foam stability, the NPs should be recovered and reused no more than three times.

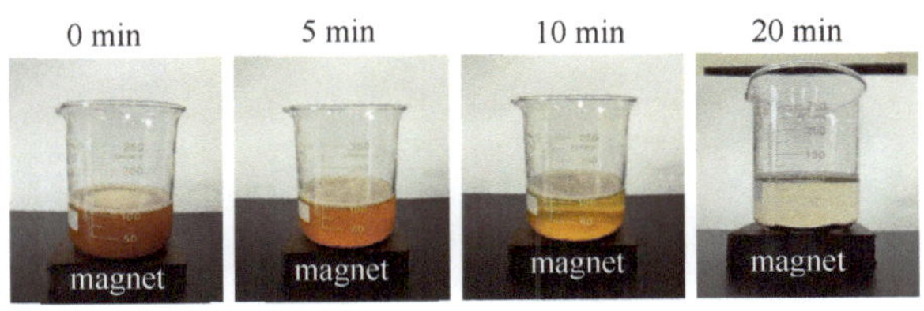

(a)

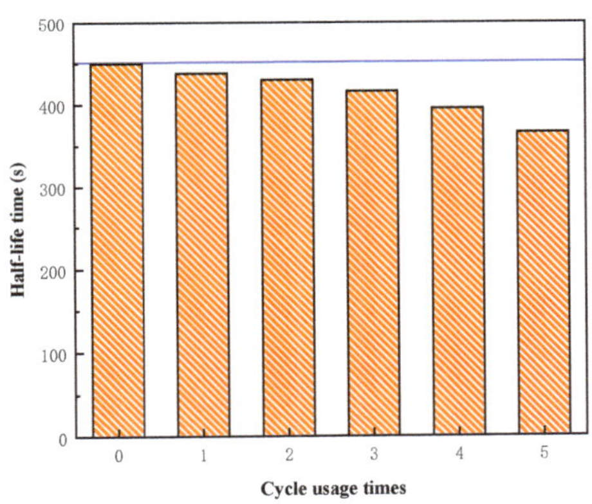

(b)

Figure 11. Evaluation of recyclability of NPs: (**a**) the responsiveness of NPs to a magnet; (**b**) the drainage half-life time changes with the recycle number of NPs.

3.4. Microscopic Oil Displacement Mechanism of Modified NPs

Water, AOS foam (0.2 wt%), and Fe_3O_4@SiO_2 (0.1 wt%)-AOS (0.2 wt%) foam with a contact angle of 77.01° were employed for oil displacement experiments. The displacement process of specific types of residual oil within a fixed area was monitored in real-time using microscopy. Figure 12 illustrates the effects of the three displacement agents on clustered, membranous, and dead-end residual oil. Due to the high water–oil mobility ratio, water flooding tends to bypass oil, causing water channeling through larger pores and leaving oil trapped in smaller pores, resulting in the formation of clustered residual oil, as depicted in Figure 12a. Membranous residual oil is predominantly distributed in water-flooded regions, where oil adheres to pore surfaces and is not fully displaced by water or surfactant solutions, as shown in Figure 12b. Dead-end residual oil is commonly found in rock formations, characterized by pores connected to the pore-throat network at only one end. Owing to the low viscosity of water, it cannot penetrate dead-end pores, leading to the retention of dead-end residual oil, as illustrated in Figure 12c.

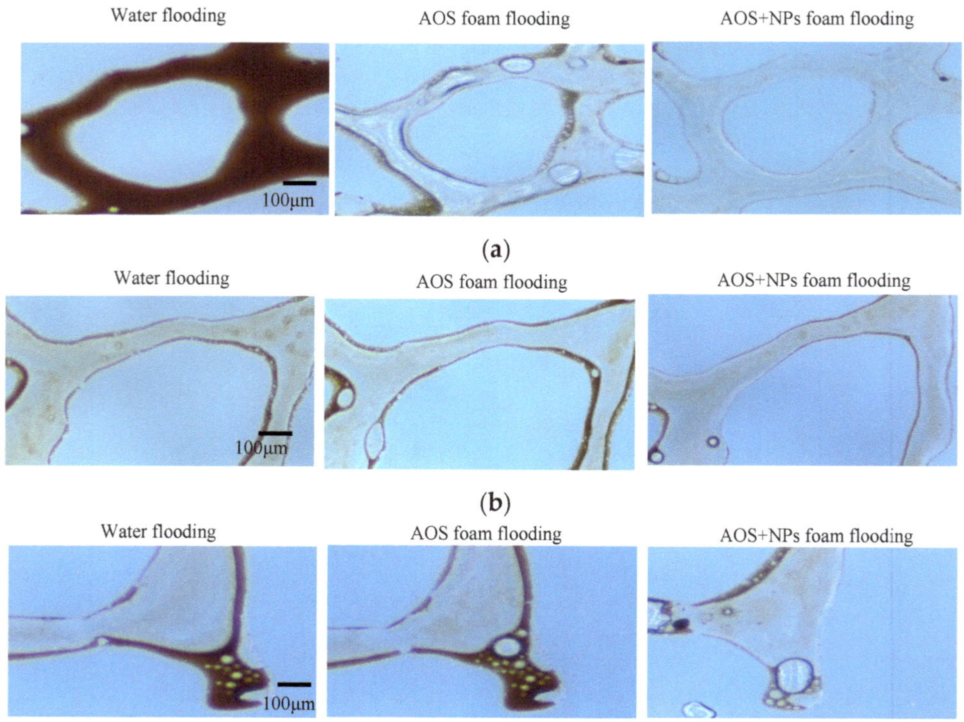

(c)

Figure 12. Microscopic residual oil after flooding: (**a**) cluster residual oil; (**b**) membranous residual oil; (**c**) dead-end residual oil.

Compared with water flooding, AOS foam demonstrates a stronger capability to mobilize these three types of residual oil. However, Fe_3O_4@SiO_2 (0.1 wt%)-AOS (0.2 wt%) foam exhibits an even more effective displacement performance for these residual oils. As illustrated in the injection pressure curve (Figure 13), the pressure generated during the injection of AOS foam (0.2 wt%) is significantly higher than that of water injection. Furthermore, the injection pressure of Fe_3O_4@SiO_2 (0.1 wt%)-AOS (0.2 wt%) foam is markedly greater than that of AOS foam alone. This indicates that Fe_3O_4@SiO_2 (0.1 wt%)-AOS (0.2 wt%) foam can more effectively perform plugging and profile control functions, thereby enhancing the displacement efficiency of the injected fluid. The enhanced performance is attributed to the synergistic effects of the NPs and foam, which improve the stability and viscoelasticity of the displacing fluid, thereby enhancing its ability to access and displace oil from complex pore structures.

To investigate the mechanism by which Fe_3O_4@SiO_2-(0.1 wt%)-AOS (0.2 wt%) foam with a contact angle of 77.01° mobilizes the three types of residual oil, typical images of the displacement process are shown in Figure 14. As shown in Figure 14a, the foam stabilized by modified Fe_3O_4@SiO_2 NPs blocked the water channeling paths around the clustered residual oil, allowing bubbles to enter low-permeability pores, emulsifying and dispersing the clustered residual oil into membranous oil or small oil droplets, which were then displaced. As shown in Figure 14b, the mobilization of membranous residual oil by modified Fe_3O_4@SiO_2 NP-stabilized foam occurred in two stages: first, the foam emulsified and peeled off the oil attached to the pore surfaces, reducing the difficulty of displacement; second, due to the adsorption of NPs, foam stability was improved, reducing gas diffusion between bubbles and the surface area of the dispersed residual oil, allowing the foam to continuously regenerate and propagate in narrow pores, pushing the oil droplets out. As shown in Figure 14c, the foam stabilized by Fe_3O_4@SiO_2-AOS had good stability, maintaining relatively small bubbles for a longer time, reducing the flow resistance when entering dead-end pores, and allowing deeper penetration into oil-containing dead-end pores. In the dead-end pores, the bubbles deformed under the driving force due to the higher interfacial viscoelasticity of Fe_3O_4@SiO_2-AOS bubbles. The deformed bubbles tended to restore their original shape, gradually replacing the residual oil with films or droplets, and effectively displacing oil from the dead-end pores.

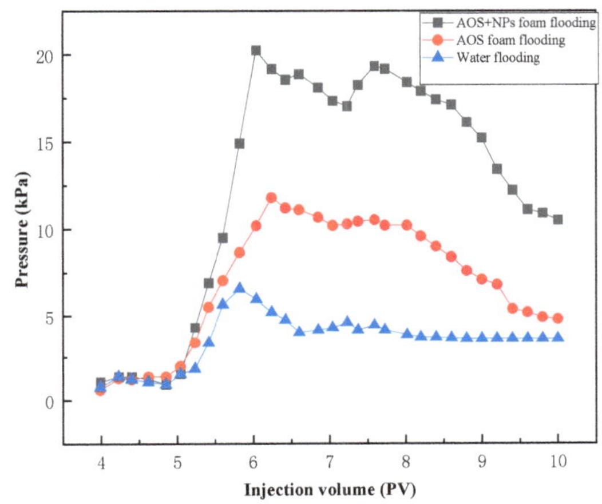

Figure 13. The pressure of flooding experiments.

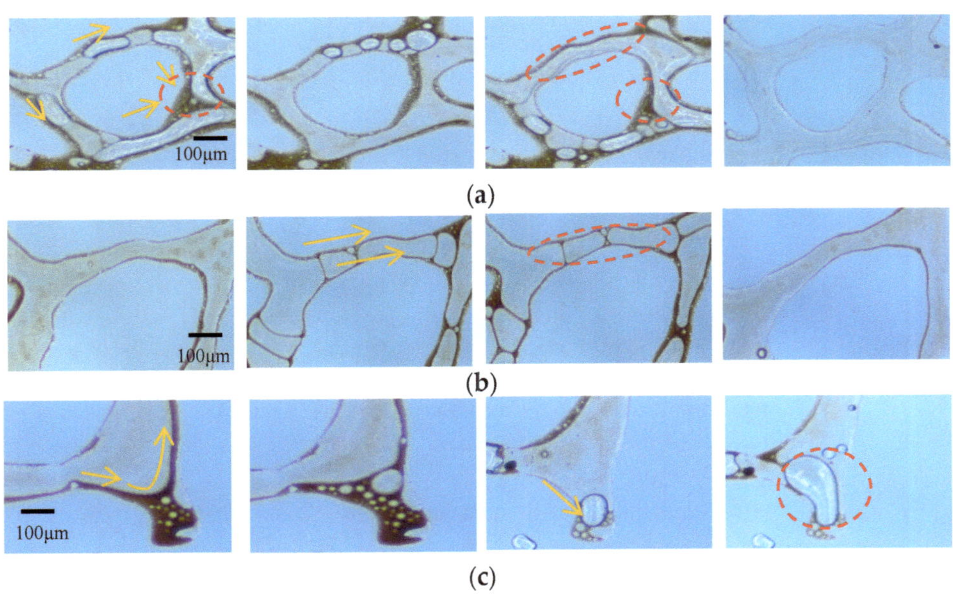

Figure 14. Microscopic residual oil after Fe_3O_4@SiO_2-AOS foam flooding: (**a**) cluster residual oil; (**b**) membranous residual oil; (**c**) dead-end residual oil.

The residual oil in the micromodel was recorded using a microscope, and the images were processed using Image-J software, as shown in Figure 15, to quantitatively calculate the oil recovery rate. Figure 15a,e show the initial oil-saturated state of the micromodel, while Figure 15b–d,f–h show the residual oil after water flooding, foam flooding, and modified NP-stabilized foam flooding, respectively. Compared with the water flooding recovery rate of 42.07%, foam flooding effectively improved the sweep efficiency, compensating for the shortcomings of water flooding and increasing the recovery rate by approximately 16%. However, significant residual oil remained after foam flooding. In contrast, the modified NP-stabilized foam flooding had a larger sweep efficiency and higher displacement efficiency, achieving a recovery rate of 75.40%, an increase of 33.33% compared with water flooding.

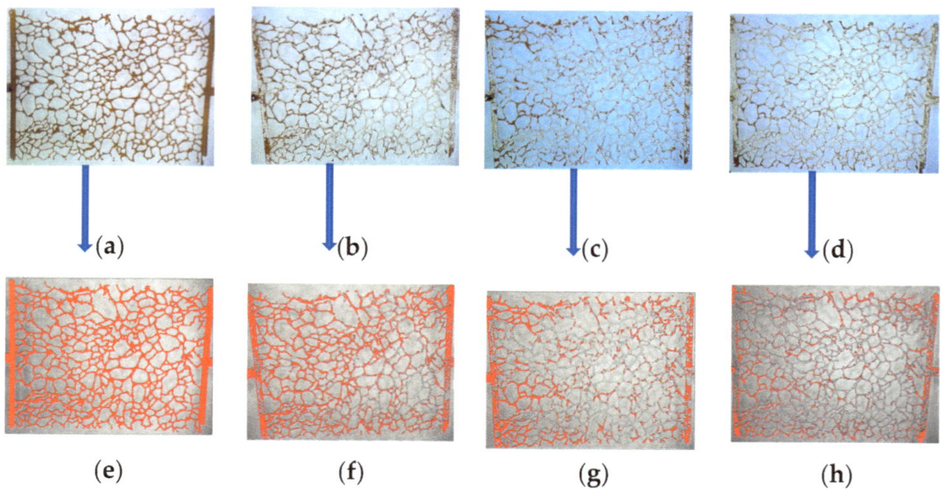

Figure 15. Microscopic images of residual oil distributions in 2D micromodel. (**a**) The raw images of initial oil distribution, (**b**) residual oil distributions after water flooding, (**c**) AOS foam flooding and extended water flooding, and (**d**) Fe_3O_4@SiO_2-AOS foam flooding and extended water flooding, respectively, while (**e–h**) were processed versions of images (**a–d**) using Image J software to estimate trapped oil. The dark brown color is oil and the milky color is displacement fluid in (**a–d**) whereas red is oil and the ash color is displacement fluid in (**e–h**). The flow direction is from left to right.

4. Conclusions

Through shell coating and hydrophobic modification experiments on Fe_3O_4 NPs, Fe_3O_4@SiO_2 NPs with different wettabilities were prepared, and their foam-stabilization performance was evaluated. The effects of modified NPs on foam stability were assessed through static foam stability experiments, including drainage half-life and initial foam volume. The foam-stabilization mechanism of Fe_3O_4@SiO_2 NPs was investigated using confocal laser scanning microscopy. Microscopic oil displacement experiments were conducted to evaluate the enhanced oil recovery mechanism of modified NP-stabilized foam.

The main findings are as follows:

(1) The characterization of the prepared NPs and foam stability evaluation showed that Fe_3O_4@SiO_2-1.0 NPs with a contact angle of 77.01° had the best foam-stabilization performance, significantly improving foam stability. The optimal foam system consisted of 1 wt% NPs (77.01°) + 0.2 wt% SDS, with a drainage half-life of 452 s and an initial foam volume of 510 mL.

(2) Confocal laser scanning microscopy experiments showed that the modified Fe_3O_4@SiO_2 NPs were adsorbed on the bubble surface, forming a three-dimensional network structure between armored bubbles, thereby enhancing foam stability. A static foam

stability evaluation indicated that the optimal number of NP recovery cycles was three, and $Fe_3O_4@SiO_2$ NPs responded quickly to magnetic fields.

(3) Microscopic visual oil displacement experiments demonstrated that, compared with AOS foam alone, $Fe_3O_4@SiO_2$-1000 NP-stabilized foam had a higher ability to mobilize residual oil. The foam's strong stability blocked large pores, allowing subsequent fluids to enter small pores, emulsifying and mobilizing clustered residual oil. The dense adsorption of modified Fe_3O_4 NPs at the liquid film interface significantly enhances film strength, enabling bubbles to undergo elastic deformation rather than rupture when passing through pore throats, emulsifying and peeling off membranous residual oil and pushing it out. Bubbles entered dead-end pores through high viscoelastic deformation, carrying out residual oil. Compared with the water flooding recovery rate of 42.07%, the modified NP-stabilized foam achieved a recovery rate of 75.40%, an increase of 33.33%, effectively mobilizing residual oil.

Author Contributions: Conceptualization, D.Y. and D.Z.; methodology, Y.W.; formal analysis, Y.W.; investigation, J.Q.; resources, D.Z.; data curation, J.Q.; writing—original draft preparation, D.Z.; writing—review and editing, D.Y.; visualization, T.H.; supervision, X.H.; project administration, Y.L.; funding acquisition, D.Y. All authors have read and agreed to the published version of the manuscript.

Funding: This research was funded by the National Natural Science Foundation of China (Nos. 52004247 and 51604243) and Zhoushan Science and Technology Bureau (No. 2021C21024), China Scholarship Council.

Data Availability Statement: The authors confirm that the data supporting the findings of this study are available within the article.

Conflicts of Interest: The authors declare no conflicts of interest. The funders had no role in the design of the study; in the collection, analyses, or interpretation of data; in the writing of the manuscript; or in the decision to publish the results.

References

1. Lake, L.W.; Johns, R.; Rossen, B.; Pope, G.A. *Fundamentals of Enhanced Oil Recovery*; Society of Petroleum Engineers: Richardson, TX, USA, 2014; Volume 1.
2. Kovscek, A.R.; Radke, C.J. *Fundamentals of Foam Transport in Porous Media*; Lawrence Berkeley National Lab. (LBNL): Berkeley, CA, USA, 1993.
3. Rossen, W.R. Foams in enhanced oil recovery. In *Foams*; Routledge: London, UK, 2017; pp. 413–464.
4. Farajzadeh, R.; Andrianov, A.; Krastev, R.; Rossen, W.; Hirasaki, G. Foam-oil interaction in porous media-Implications for foam-assisted enhanced oil recovery (SPE 154197). In Proceedings of the 74th EAGE Conference and Exhibition Incorporating EUROPEC, Copenhagen, Denmark, 4–7 June 2012; p. cp-293-00225.
5. Simjoo, M.; Rezaei, T.; Andrianov, A.; Zitha, P. Foam stability in the presence of oil: Effect of surfactant concentration and oil type. *Colloids Surf. A Physicochem. Eng. Asp.* **2013**, *438*, 148–158. [CrossRef]
6. Schramm, L.L. *Foams: Fundamentals and Applications in the Petroleum Industry*; ACS Publications: Washington, DC, USA, 1994.
7. Nowrouzi, I.; Mohammadi, A.H.; Khaksar Manshad, A. A non-ionic green surfactant extracted from the Anabasis setifera plant for improving bulk properties of CO_2-foam in the process of enhanced oil recovery from carbonate reservoirs. *Can. J. Chem. Eng.* **2024**, *103*, 590–605. [CrossRef]
8. Peng, B.; Zhang, L.; Luo, J.; Wang, P.; Ding, B.; Zeng, M.; Cheng, Z. A review of nanomaterials for nanofluid enhanced oil recovery. *RSC Adv.* **2017**, *7*, 32246–32254. [CrossRef]
9. Wang, Z.; Xu, Y.; Khan, N.; Zhu, C.; Gao, Y. Effects of the surfactant, polymer, and crude oil properties on the formation and stabilization of oil-based foam liquid films: Insights from the microscale. *J. Mol. Liq.* **2023**, *373*, 121194. [CrossRef]
10. Zhang, Z.; Feng, X.; Zeng, G.; Liu, H.; Shen, L.; Yang, Y.; Yuan, H.; Yan, X.; Mi, Y. Hyperbranched poly (amido amine) demulsifiers using diaminonaphthalene as the central core and their demulsification performance in oil-in-water and water-in-oil emulsions. *Energy Fuels* **2021**, *35*, 3095–3103. [CrossRef]
11. Binks, B.P.; Horozov, T.S. *Colloidal Particles at Liquid Interfaces*; Cambridge University Press: Cambridge, UK, 2006.

12. Fameau, A.-L.; Salonen, A. Effect of particles and aggregated structures on the foam stability and aging. *Comptes Rendus. Phys.* **2014**, *15*, 748–760. [CrossRef]
13. Hunter, T.N.; Pugh, R.J.; Franks, G.V.; Jameson, G.J. The role of particles in stabilising foams and emulsions. *Adv. Colloid Interface Sci.* **2008**, *137*, 57–81. [CrossRef]
14. Chanzab, F.F.; Ahmadi, M.; Sharifi, M. Investigation of the interfacial phenomena in the presence of nonionic surfactants and a silica nanoparticle at the n-decane-water interface: Insights from molecular dynamics simulation. *J. Mol. Liq.* **2024**, *394*, 123789. [CrossRef]
15. Nowrouzi, I.; Manshad, A.K.; Mohammadi, A.H. Effects of concentration and size of TiO_2 nano-particles on the performance of smart water in wettability alteration and oil production under spontaneous imbibition. *J. Pet. Sci. Eng.* **2019**, *183*, 106357. [CrossRef]
16. Pang, J.; Mohanty, K. Surfactant–Nanoparticle Foam to Increase CO_2 Storage in High-Salinity Carbonate Reservoirs. *Energy Fuels* **2024**, *38*, 9967–9979. [CrossRef]
17. Yekeen, N.; Padmanabhan, E.; Idris, A.K. Synergistic effects of nanoparticles and surfactants on n-decane-water interfacial tension and bulk foam stability at high temperature. *J. Pet. Sci. Eng.* **2019**, *179*, 814–830. [CrossRef]
18. Khoramian, R.; Issakhov, M.; Pourafshary, P.; Gabdullin, M.; Sharipova, A. Surface modification of nanoparticles for enhanced applicability of nanofluids in harsh reservoir conditions: A comprehensive review for improved oil recovery. *Adv. Colloid Interface Sci.* **2024**, 103296. [CrossRef] [PubMed]
19. Ilkhani, M.; Bayat, A.E.; Harati, S. Applicability of methane foam stabilized via Nanoparticles for enhanced oil recovery from carbonate porous media at various temperatures. *J. Mol. Liq.* **2022**, *367*, 120576. [CrossRef]
20. Setiawan, A.; Sari, E.K.; Mahardhika, L.J.; Jayanti, P.D.; Rini, N.P.; Istiqomah, N.I.; Aliah, H.; Asri, N.S.; Angel, J.; Suharyadi, E. Magnetically separable and reusable Fe_3O_4/rGO photocatalyst synthesized through green approach for heavy metal ion reduction application. *Diam. Relat. Mater.* **2025**, *151*, 111779. [CrossRef]
21. Gogoi, B.; Das, U. Magnetization dynamics of iron oxide super paramagnetic nanoparticles above blocking temperature. *Mater. Today Proc.* **2022**, *65*, 2636–2644. [CrossRef]
22. Kulkarni, S.A.; Sawadh, P.; Palei, P.K. Synthesis and characterization of superparamagnetic Fe_3O_4@ SiO_2 nanoparticles. *J. Korean Chem. Soc.* **2014**, *58*, 100–104. [CrossRef]
23. Salah, T. Effect of Fe_3O_4 Nanoparticles on Performance of Shape Memory Polymers Foam Using Solid State Foaming Process. 2019. Available online: https://scholarworks.uaeu.ac.ae/mechan_theses/7/ (accessed on 25 February 2025).
24. Chaudhry, A.U.; Muneer, R.; Lashari, Z.A.; Hashmet, M.R.; Osei-Bonsu, K.; Abdala, A.; Rabbani, H.S. Recent advancements in novel nanoparticles as foam stabilizer: Prospects in EOR and CO_2 sequestration. *J. Mol. Liq.* **2024**, *407*, 125209. [CrossRef]
25. Zandahvifard, M.J.; Elhambakhsh, A.; Ghasemi, M.N.; Esmaeilzadeh, F.; Parsaei, R.; Keshavarz, P.; Wang, X. Effect of Modified Fe_3O_4 Magnetic NPs on the Absorption Capacity of CO_2 in Water, Wettability Alteration of Carbonate Rock Surface, and Water–Oil Interfacial Tension for Oilfield Applications. *Ind. Eng. Chem. Res.* **2021**, *60*, 3421–3434. [CrossRef]
26. Yu, F.; Ma, H.; Sun, C.; Xia, S. Recyclable Fe_3O_4 Nanoparticles Responsive to Temperature for Heavy Oil Viscosity Reduction: Synthesis, Characterization, and Performance. *Energy Fuels* **2023**, *37*, 14752–14763. [CrossRef]
27. Yi, M.; Lin, R.; Wang, Q.; Wang, Y.; Wang, X. Numerical simulation study of Fe_3O_4-nanofluid-assisted electromagnetic heating for heavy oil reservoirs. *Geoenergy Sci. Eng.* **2025**, *247*, 213671. [CrossRef]
28. Liu, Q.; Zhang, Y.; Zhao, X.; Ye, H.; Luo, D. Enhanced oil recovery by foam flooding using foam stabilized with modified Fe_3O_4 nanoparticles. *J. Pet. Sci. Eng.* **2022**, *209*, 109850. [CrossRef]
29. Liu, Q.; Qu, H.; Liu, S.; Zhang, Y.; Zhang, S.; Liu, J.; Peng, B.; Luo, D. Modified Fe_3O_4 nanoparticle used for stabilizing foam flooding for enhanced oil recovery. *Colloids Surf. A Physicochem. Eng. Asp.* **2020**, *605*, 125383. [CrossRef]
30. Zhou, J.; Wang, L.; Qiao, X.; Binks, B.P.; Sun, K. Pickering emulsions stabilized by surface-modified Fe_3O_4 nanoparticles. *J. Colloid Interface Sci.* **2012**, *367*, 213–224. [CrossRef] [PubMed]
31. Wang, L.; Guo, J.-X.; Chen, X.-W.; Li, C.; Kiyingi, W.; Xiong, R.-Y.; Zhang, X.-J.; Gao, C.-H. Fe_3O_4/AM-PAA/Ni nanomagnetic spheres: A breakthrough in in-situ catalytic reduction of heavy oil viscosity. *J. Anal. Appl. Pyrolysis* **2024**, *181*, 106664. [CrossRef]
32. Patel, N.N.; Mulla, N.R.; Khot, V.M.; Patil, R.S. Anticancer activity of surface functionalized magnetite (Fe_3O_4) nanoparticles—Effect of polymer coating. *Emergent Mater.* **2024**, *7*, 1071–1080. [CrossRef]
33. Liu, Y.; Zhang, H.; Yu, Y.; Yu, M.; Long, S.; Yang, W.; Li, W.; Hu, Y. Study on the stability and magnetically induced demulsification performance of Pickering emulsions based on arginine-modified lignin/Fe_3O_4 nanoparticles. *Int. J. Biol. Macromol.* **2025**, *285*, 138315. [CrossRef]
34. Worthen, A.J.; Bagaria, H.G.; Chen, Y.; Bryant, S.L.; Huh, C.; Johnston, K.P. Nanoparticle-stabilized carbon dioxide-in-water foams with fine texture. *J. Colloid Interface Sci.* **2013**, *391*, 142–151. [CrossRef]
35. Yu, J.; Khalil, M.; Liu, N.; Lee, R. Effect of particle hydrophobicity on CO_2 foam generation and foam flow behavior in porous media. *Fuel* **2014**, *126*, 104–108. [CrossRef]

36. Duan, M.; Hu, X.; Ren, D.; Guo, H. Studies on foam stability by the actions of hydrophobically modified polyacrylamides. *Colloid Polym. Sci.* **2004**, *282*, 1292–1296. [CrossRef]
37. Saranya, T.; Parasuraman, K.; Anbarasu, M.; Balamurugan, K. XRD, FT-IR and SEM study of magnetite (Fe_3O_4) nanoparticles prepared by hydrothermal method. *Nano Vis.* **2015**, *5*, 149–154.
38. Liang, J.; Wang, L.; Liu, L.; Tian, L.; Zhang, L.; Wang, W. The Modification of Fe_3O_4@ SiO_2 by Silane Coupling Agent and Performance. *J. Liaoning Univ. Pet. Chem. Technol.* **2019**, *39*, 21.
39. Li, S.; Li, Z.; Wang, P. Experimental study of the stabilization of CO_2 foam by sodium dodecyl sulfate and hydrophobic nanoparticles. *Ind. Eng. Chem. Res.* **2016**, *55*, 1243–1253. [CrossRef]

MDPI AG
Grosspeteranlage 5
4052 Basel
Switzerland
Tel.: +41 61 683 77 34

MDPI Books Editorial Office
E-mail: books@mdpi.com
www.mdpi.com/books

www.ingramcontent.com/pod-product-compliance
Lightning Source LLC
Chambersburg PA
CBHW041421210926
51850CB00013B/3322
* 9 7 8 3 7 2 5 8 3 8 8 5 1 *